Animal Evolution

Animal Evolution
Interrelationships of the Living Phyla

CLAUS NIELSEN

Natural History Museum of Denmark,
University of Copenhagen

THIRD EDITION

OXFORD

UNIVERSITY PRESS

9/13

OXFORD
UNIVERSITY PRESS

Great Clarendon Street, Oxford OX2 6DP

Oxford University Press is a department of the University of Oxford.
It furthers the University's objective of excellence in research, scholarship,
and education by publishing worldwide in

Oxford New York

Auckland Cape Town Dar es Salaam Hong Kong Karachi
Kuala Lumpur Madrid Melbourne Mexico City Nairobi
New Delhi Shanghai Taipei Toronto

With offices in

Argentina Austria Brazil Chile Czech Republic France Greece
Guatemala Hungary Italy Japan South Korea Poland Portugal
Singapore Switzerland Thailand Turkey Ukraine Vietnam

Oxford is a registered trade mark of Oxford University Press
in the UK and in certain other countries

Published in the United States
by Oxford University Press Inc., New York

First edition published 1995
Second edition published 2001
This edition published 2012

British Library Cataloguing in Publication Data
Data available

Library of Congress Cataloging in Publication Data
Library of Congress Control Number: 2011941928

Typeset by SPI Publisher Services, Pondicherry, India
Printed and bound by
CPI Group (UK) Ltd, Croydon, CR0 4YY

ISBN 978–0–19–960602–3 (Hbk)
978–0–19–960603–0 (Pbk)

3 5 7 9 10 8 6 4 2

Preface to third edition

Studies of molecular phylogeny have dominated the biological literature during the last decade, but important new information from morphology and embryology has been obtained, especially with the use of new immunostaining methods. For a period, the molecular results of higher animal phylogeny looked rather chaotic, with many conflicting trees based on studies of mitochondrial and ribosomal genes. However, the trees obtained through the new studies of large numbers of various sequences, of expressed sequence tags, and even on whole genomes, now seem to converge, and a congruence with morphology-based trees seems possible. This has inspired me to make this new edition.

Revisions of larger texts are always in danger of growing, because new material is added and old material not discarded. I have tried to avoid this, for example by using the excellent series *Microscopic Anatomy of Invertebrates*, edited by Frederick W. Harrison (Wiley-Liss, New York, 1991–1999) as a general reference for the anatomy of the various groups; references that can be found in these volumes have generally been excluded. My two papers on the development of trochophora larvae (Nielsen, C. 2004, 2005. *J. Exp. Zool. (Mol. Dev. Evol.)* **302**B: 35–68 and **304**B: 401–447) have been used in the same way.

Many colleagues have helped in various ways; some have sent me illustrations for use in the book and others have read one or more chapters and given good suggestions: S. Amano (Kanazawa University), M. Blaxter (University of Edinburgh), M.J. Dayel (University of California, Berkeley), G. Edgecombe (The Natural History Museum, London), D.E.K. Ferrier (University of St Andrews), D. Gordon (NIWA, Wellington), S. Harzsch (University of Greifswald), L.Z. Holland (Scripps Research Institute), T.W. Holstein (University of Heidelberg), G. Jékely (University of Tübingen), S. Karpov (University of Southampton), N. King (University of California, Berkeley), R.M. Kristensen (University of Copenhagen), B.S.C. Leadbeater (University of Southampton), S.P. Leys (University of Alberta), C.J. Lowe (Hopkins Marine Station), C. Lüter (Museum für Naturkunde, Berlin), J. Olesen (University of Copenhagen), G. Purschke (University of Osnabrück), H. Ruhberg (University of Hamburg), E. Schierenberg (University of Köln), A. Schmidt-Rhaesa (University of Hamburg), G. Shinn (Truman State University), A.B. Smith (The Natural History Museum, London), T. Stach (Freie University, Berlin), B.J. Swalla (University of Washington), J.M. Turbeville (Virginia Commonwealth University), R.M. Woollacott (Harvard University). A special thank you to my 'neighbour' in the Zoological Museum, Dr Martin Vinther Sørensen, for many good discussions. All these colleagues (and others that I may have forgotten in the list) have been of great help, but don't blame them for my mistakes.

Mrs Birgitte Rubæk is thanked for fine help with the new illustrations.

Copenhagen
March 2011

PREFACE TO SECOND EDITION

During the years since the text for the first edition of this book was concluded (in 1992), a wealth of new morphological information has become available, including both histological/ultrastructural and embryological data, and new areas, such as numerical cladistic analyses, DNA sequencing, and developmental biology, have become prominent in phylogenetic studies. I have tried to update the information about morphology, but the other fields have been more difficult to deal with; numerical cladistic analyses and molecular phylogeny are discussed in separate chapters, but following my conclusions in these two chapters, I have in general refrained from discussing results obtained through these methods.

I am fully aware that my coverage of molecular studies, including the extremely promising evolutionary developmental biology, is very incomplete. I have tried to select information from studies that appear to describe consistent phylogenetic signals, but my choice is biased by my background as a morphologist. The interested reader is strongly advised to consult a recent textbook or review articles on the subject.

Once again it is my pleasure to thank the many generous colleagues who have helped me in various ways, especially those who have read drafts of various chapters and given many good comments: André Adoutte (Paris), Wim J.A.G. Dictus (Utrecht), Andriaan Dorresteijn (Mainz), Danny-Eibye-Jacobsen (Copenhagen), Peter W.H. Holland (Reading), Reinhardt Møbjerg Kristensen (Copenhagen), Thurston C. Lacalli (Saskatoon), George O. Mackie (Victoria), Mark Q. Martindale (Hawaii), Rudolf Meier (Copenhagen), Edward E. Ruppert (Clemson), George L. Shinn (Kirksville), Nikolaj Scharff (Copenhagen), Gerhard Scholtz (Berlin), Ralf Sommer (Tübingen), Martin Winther Sørensen (Copenhagen), Gregory A. Wray (Durham), Russel L. Zimmer (Los Angeles). None of them should be held responsible for the ideas expressed here.

My special thanks are due to Mr Gert Brovad (Zoological Museum, Copenhagen) who has provided the new photos and to Mrs Birgitte Rubæk (Zoological Museum, Copenhagen) who has spared no effort in preparing the many new drawings and diagrams.

Preface to first edition

No naturalist can avoid being fascinated by the diversity of the animal kingdom, and by the sometimes quite bizarre specializations that have made it possible for the innumerable species to inhabit almost all conceivable ecological niches.

However, comparative anatomy, embryology, and especially molecular biology, demonstrate a striking unity among organisms, and show that the sometimes quite bewildering diversity is the result of variations over a series of basic themes, some of which are even common to all living beings.

To me, this unity of the animal kingdom is just as fascinating as the diversity, and in this book I will try to demonstrate the unity by tracing the evolution of all of the 31 living phyla from their unicellular ancestor.

All modern books on systematic zoology emphasize phylogeny, but space limitations usually preclude thorough discussions of the characteristics used to construct the various phylogenetic trees. I will try to document and discuss all the characters that have been considered in constructing the phylogeny—both those that corroborate my ideas and those that appear to detract from their probability.

In the study of many phyla, I have come across several important areas in which the available information is incomplete or uncertain, and yet other areas that have not been studied at all; on the basis of this I have, for each phylum, given a list of some interesting subjects for future research, and I hope that these lists will serve as incentives to further investigations.

It should be stressed that this book is not meant as an alternative to the several recent textbooks of systematic zoology, but as a supplement, one that I hope will inspire not only discussions between colleagues but also seminars on phylogeny—of the whole animal kingdom or of selected groups—as an integrated part of the teaching of systematic zoology.

The ideas put forward in this book have developed over a number of years, and during that period I have benefitted greatly from interactions with many colleagues. Some have been good listeners when I have felt the need to talk about my latest discovery; some have discussed new or alternative ideas, names or concepts with me; some have provided eagerly sought pieces of literature or given me access to their unpublished results; and some have sent me photos for publication. To all these friends I extend my warmest thanks; no names are mentioned, because such a list will inevitably be incomplete. A number of colleagues have read one to several chapters (the late Robert D. Barnes (Gettysburg) and Andrew Campbell (London) have read them all) and given very valuable and constructive comments that I have often but not always followed; I want to mention them all, not to make them in any way responsible, but to thank them for the help and support that is necessary during an undertaking such as this: Quentin Bone (Plymouth), Kristian Fauchald (Washington, DC), Gary Freemann (Austin), Jens T. Høeg (Copenhagen), Åse Jespersen (Copenhagen), Niels Peder Kristensen (Copenhagen), Margit Jensen (Copenhagen), Reinhardt Møbjerg Kristensen (Copenhagen), Barry S.C. Leadbeater (Birmingham), Jørgen Lützen (Copenhagen), George O. Mackie (Victoria), Mary E. Petersen (Copenhagen), Mary E. Rice (Fort Pierce), Edward E. Ruppert (Clemson), Amelie H. Scheltema (Woods Hole), George L. Shinn (Kirksville), Volker Storch (Heidelberg), Ole S. Tendal (Copenhagen), and Russell L. Zimmer (Los Angeles).

The Danish Natural Science Research Council and the Carlsberg Foundation are thanked heartily for their continued support covering travel expenses, instrumentation and laboratory assistance; the Carlsberg Foundation has given a special grant to cover the expenses of the illustrations for this book.

Financial support from 'Højesteretssagfører C.L. Davids Legat for Slægt og Venner' is gratefully acknowledged.

Mrs Birgitte Rubæk and Mrs Beth Beyerholm are thanked for their excellent collaboration on the artwork.

My warmest thanks go to Kai and Hanne (Olsen & Olsen, Fredensborg) for a congenial undertaking of the typesetting of the book and for fine work with the layout and lettering of the illustrations.

Dr Mary E. Petersen (Copenhagen) is thanked for her meticulous reading of the first set of proofs.

Finally, my thanks go to Oxford University Press, and especially Dr Cathy Kennedy, for a positive and constructive collaboration.

CONTENTS

Introduction

Modern understanding of biological diversity goes back to Darwin (1859), who created a revolution in biological thought by regarding the origins of species as the result of 'descent with modification'. As a consequence of this idea he also stated that the 'natural system' (i.e. the classification) of the organisms must be strictly genealogical ('like a pedigree'), and that the 'propinquity of descent' is the cause for the degree of similarity between organisms. The term 'homology' had already been in use for some time, and Owen (1848) had used it in a practical attempt to create a common anatomical nomenclature for the vertebrates, but it was Darwin's ideas about evolution that gave the word its present meaning and importance: structures are homologous in two or more species when they are derived from one structure in the species' most recent common ancestor. This phylogenetic or historical, morphology-based definition of homology (Hall 1994) is used throughout this book. It should be emphasized that homology can be proved only in very special cases, such as in vertebrate skeletons, where the evolution can be followed directly in the fossil record. In all other cases homology can be inferred from morphological or molecular characteristics that point to the evolution from a common ancestor.

Haeckel (1866) drew the first phylogenetic tree ('Stammbaum'; Fig. 1.1) based on Darwin's ideas, and coined the words 'phylogeny' and 'ontogeny'. His tree was labelled 'monophyletic', and his definition of a phylum as consisting of an ancestor and all its living and extinct descendants agrees completely with the cladistic use of the word 'monophyletic'. Haeckel leaned toward the opinion that the 19 phyla in his tree had evolved separately from unorganized organic substances, but a common ancestry was also considered a possibility (and was proposed soon after; see Haeckel 1870); this should not detract from the general validity of his definition of the term 'monophyletic', which is now used at all systematic levels.

The conceptual base for phylogenetic work is thus more than a century old, and it could perhaps be expected that such studies had reached a level where only details of genealogy remained to be cleared up, but this is far from being the case. There are several reasons for this.

Darwin's comprehensive theory of evolution was actually five interwoven theories (Mayr 1982), and his theories about speciation and selection mechanisms were soon attacked from several sides. So, although the idea of evolution and speciation became accepted rather easily, the attacks on his explanation of speciation focussed on one side of the theory that has turned out to need a good deal of modification. Some of Darwin's followers carried their arguments to extremes, which undoubtedly detracted from the credibility of the whole field. In addition, the growing interest in experimental biology turned the spotlight away from phylogeny.

However, since the 1950s, a revival of the phylogenetic interests has taken place and the field is again producing a strong flow of interesting results. New sources of information have been added, the most

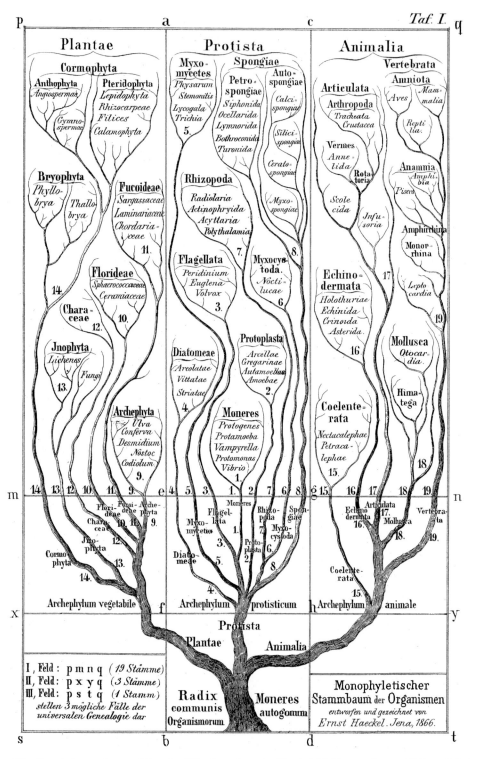

Figure 1.1. The first phylogenetic tree of the animal kingdom, drawn by Haeckel (1866); note the word 'Monophyletischer' in the legend.

important being ultrastructure, immunocytochemistry, and gene sequencing. Also, there has been important progress in methods of analysing data, and the phylogenetic reasoning has been sharpened by the methods proposed by Hennig (1950), called 'cladistics' or 'phylogenetic systematics'. All phylogenetic analyses should now be based on 'tree thinking' (Baum *et al.* 2005), i.e. argumentations based on characters of common ancestors and their modification in the descendents, and not on 'horizontal' comparisons of living species. Ideally, every phylogeny should be checked for functional continuity between the hypothetical ancestors and their descendants (Budd 2006). With the wording of Frazzetta (1975, p. 20), the analysis should ascertain if the proposed evolution has proceeded like 'the gradual improvement of a machine *while it is running*' [my italics]. This principle has been applied to the basal part of the tree, from the unicellular to the eubilaterian level, and to the early evolution of the protostomes (see Chapters 4 and 22). However, some areas of the tree are unresolved, which makes it impossible to make this test on all levels.

One of the important reasons for the disagreements between earlier morphology-based trees has been that only narrow sets of characters were used, often with strong emphasis on either adult or larval characters instead of considering all characters of whole life cycles (as already pointed out by Darwin). It should be evident that all characters are of importance: the only question is at which level they contain phylogenetic information. Another weakness has been that 'advanced' characters, such as coelom and metanephridia, have been used to characterize higher taxa without discussing whether these characters have evolved more than once. For example, it is now clear that coeloms have evolved several times (Chapter 21), and dealing with this character in only two states, absent and present, is bound to lead to unreliable results. Furthermore, it has turned out that character losses are more important than has been expected by most morphologists (Jenner 2004), and some of the proposed losses, which at first have appeared improbable, have turned out to find support from new methods. An example is segmentation in chordates, where the loss of this character in the urochordates is indicated by the molecular phylogenies (for example,

Hejnol *et al.* 2009), and where the demonstration of a massive loss of genes gives strong support to this, at first sight, improbable hypothesis (Holland *et al.* 2008).

Morphological characters have been analysed with various computer programs, and this has given fine results on the lower systematic levels, but on the level discussed here the data matrices have been very subjective; only states of homologous characters can be coded, the homology is in many cases problematic, and the character state 'absent', may cover both primary absence and loss (see above).

The analyses of DNA/RNA sequences have produced an astonishing number of new phylogenies for all the living beings as well as for narrower categories, such as phyla, classes, and orders. Many areas of the metazoan tree have been shuffled around, and a number of bizarre phylogenies have been proposed. This has been discouraging for a traditional morphologist. However, after analyses of a large dataset, Abouheif *et al.* (1998, p. 404) concluded that '…the 18S rRNA molecule alone is an unsuitable candidate for reconstructing a phylogeny of the Metazoa…'. This molecule, in combination with other molecules, such as 28S rRNA, cytochrome-c oxidase I, and elongation factor-1α, has provided very good analyses of lower groups, such as bivalves (Giribet and Wheeler 2002), where morphology and molecular data can be brought into good accordance. The monophyly of a number of morphologically well-established clades, e.g. Metazoa, Eumetazoa, Bilateria, Protostomia, and Deuterostomia, has found very consistent support in these studies, whereas the radiations of other more comprehensive clades, such as Protostomia and Lophotrochozoa, have turned out to be highly problematic. Studies of whole mitochondrial genomes (for example, Chen *et al.* 2009; Jang and Hwang 2009) and mitochondrial gene order (for example, Webster *et al.* 2007) have shown very confusing results. A discussion of all these analyses would require a special publication, and they will only be mentioned in a few places in the following discussions. However, new analyses using, for example, 'household genes' (Peterson *et al.* 2008), expressed sequence tags (EST) (Dunn *et al.* 2008), 'phylogenomics' (analyses of selected parts of genomes)

(Hejnol *et al.* 2009), and analyses based on whole genomes, (Srivastava *et al.* 2008, 2010), now seem to bring much more stability to the more basal parts of the phylogenetic tree. Both Hox genes (see Fig. 21.3) and microRNAs appear to be of special value in the studies of evolution of the major metazoan groups (de Rosa *et al.* 1999; Peterson *et al.* 2009).

It should be emphasized that the classification of Bilaterians in Protostomia and Deuterostomia dates back to Grobben (1908), so it is not 'the new phylogeny' as proposed, especially in a number of papers on molecular phylogenetics, but a return to the classical system after a period with the popular classification of Bilateria = Acoela + Pseudocoela + Coelomata, which is a misunderstanding of the presentation of Hyman (1940) (see Nielsen 2010).

It should also be remembered that the molecular phylogenies are 'naked' trees without morphological characters, and, as stated by Raff *et al.* (1989, p. 258), 'The use of rRNA sequences to infer distant phylogenetic relationships will not displace morphology and embryology from the study of the evolutionary history of animal life: *after all, it is the history of morphological change that we wish to explain*' [my italics].

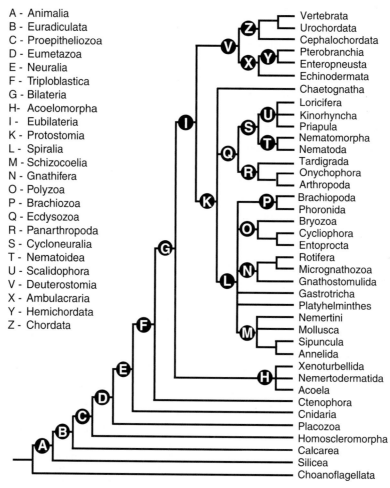

A - Animalia
B - Euradiculata
C - Proepitheliozoa
D - Eumetazoa
E - Neuralia
F - Triploblastica
G - Bilateria
H- Acoelomorpha
I - Eubilateria
K - Protostomia
L - Spiralia
M - Schizocoelia
N - Gnathifera
O - Polyzoa
P - Brachiozoa
Q - Ecdysozoa
R - Panarthropoda
S - Cycloneuralia
T - Nematoidea
U - Scalidophora
V - Deuterostomia
X - Ambulacraria
Y - Hemichordata
Z - Chordata

Figure 1.2. The animal phylogeny proposed in this book; the method is described in the text. Named supraphyletic clades are indicated.

In this book phylogenetic/cladistic principles have been used to infer the topology of the basal metazoan radiation, i.e. the interrelationships between the ancestors of the 38 monophyletic groups, here called phyla (Fig. 1.2). The levels of all categories above the species are of course highly subjective, and some zoologists have abandoned all suprageneric categories (for example Ax 1996, 2001, 2003). However, it seems practical to use the term phylum for the units in the present discussions. The important question is whether the 'phyla' are monophyletic groups, and this has been discussed for each of them. In constructing the tree, as many characters as possible have been taken into account, and the phyla identified by morphological methods have been compared with the results from molecular studies. The method has been iterative, but a number of steps can be outlined. The first step has been traditional phylogenetic analyses seeking identification of monophyletic phyla; if a phylum has been found to be polyphyletic it has been broken up and the new phyla defined (for example, Porifera divided into Silicea, Calcarea, and Homoscleromorpha). If a phylum has turned out to be an ingroup of another phylum, it has been included in that phylum (for example, Acanthocephala included in Rotifera). The next step has been the identification of ancestral characters of the phylum. The phyla have then been compared and sister groups identified, and increasingly more comprehensive, monophyletic groups defined. The new results from molecular phylogeny and Hox genes have been taken into consideration. In many areas of the tree, congruence between morphology and molecules is now being reached, but some groups are still quite problematic. For example, the interrelationships of the lophotrochozoan phyla are still largely unresolved. However, it is very gratifying that the newest phylogeny of the metazoans, mainly based on EST data (Edgecombe *et al.* 2011), is remarkably similar to that obtained in the present study (Fig. 1.2).

Fossils have played an important part in the analyses of phyla such as molluscs, arthropods, echinoderms, and vertebrates, which have a rich fossil record and sometimes an astonishing preservation of fine details. However, a number of the earliest fossils cannot be assigned to any of the living phyla, and some authors refer most of them to a separate kingdom, the Vendobionta (see, for example, Seilacher 2007); they have in general been excluded from the discussions.

The focus of this book is on the origin of the living phyla, which are now separated from each other by clear morphological gaps. Some of these gaps are bridged by fossils—as seen, for example, in the panarthropods, where several 'lobopode' fossils give the impression of a more coherent Cambrian group from which the three living phyla have evolved (Hou *et al.* 2004). A 'Cambrian systematist' without knowledge of the later radiations and extinctions would probably have classified the panarthropod stem groups as one phylum. Some fossils show mosaics of characters that may confuse our concepts of some of the living groups (Shubin 1998). So I have only discussed information from fossils that appear to throw light over the origin of the living phyla, but in general have abstained from trying to place many fossils in the phylogeny.

References

Abouheif, E., Zardoya, R. and Meyer, A. 1998. Limitations of metazoan 18S rRNA sequence data: implications for reconstructing a phylogeny of the animal kingdom and inferring the reality of the Cambrian explosion. *J. Mol. Evol.* **47**: 394–405.

Ax, P. 1996. *Multicellular Animals: Vol. 1. A New Approach to the Phylogenetic Order in Nature*. Springer, Berlin.

Ax, P. 2001. *Multicellular Animals: Vol. 2. The Phylogenetic System of the Metazoa*. Springer, Berlin.

Ax, P. 2003. *Multicellular Animals: Vol. 3. Order in Nature— System Made by Man*. Springer, Berlin.

Baum, D.A., Smith, S.D. and Donovan, S.S. 2005. The tree-thinking challenge. *Science* **310**: 979–980.

Budd, G. 2006. On the origin and evolution of major morphological characters. *Biol. Rev.* **81**: 609–628.

Chen, H.-X., Sundberg, P., Norenburg, J.L. and Sun, S.-C. 2009. The complete mitochondrial genome of *Cephalotbrix simula* (Iwata) (Nemertea: Palaeonemertini). *Gene* **442**: 8–17.

Darwin, C. 1859. *On the Origin of Species by Means of Natural Selection*. John Murray, London.

de Rosa, R., Grenier, J.K., Andreeva, T., *et al.* 1999. Hox genes in brachiopods and priapulids and protostome evolution. *Nature* **399**: 772–776.

Dunn, C.W., Hejnol, A., Matus, D.Q., *et al.* 2008. Broad phylogenomic sampling improves resolution of the animal tree of life. *Nature* **452**: 745–749.

Edgecombe, G.D., Giribet, G., Dunn, C.W., *et al.* 2011. Higher-level metazoan relationships: recent progress and remaining questions. *Org. Divers. Evol.* **11**: 151–172.

Frazzetta, T.H. 1975. *Complex Adaptations in Evolving Populations.* Sinauer Associates, Sunderland, MA.

Giribet, G. and Wheeler, W. 2002. On bivalve phylogeny: a high-level analysis of the Bivalvia (Mollusca) based on combined morphology and DNA sequence data. *Invert. Biol.* **121**: 271–324.

Grobben, K. 1908. Die systematische Einteilung des Tierreichs. *Verh. Zool.-Bot. Ges. Wien* **58**: 491–511.

Haeckel, E. 1866. *Generelle Morphologie der Organismen. 2 vols.* Georg Reimer, Berlin.

Haeckel, E. 1870. *Natürliche Schöpfungsgeschichte, 2nd ed.* Georg Reimer, Berlin.

Hall, B.K. 1994. Introduction. In B.K. Hall (ed.): *Homology. The Hiererchical Basis of Comparative Biology*, pp. 1–19. Academic Press, San Diego.

Hejnol, A., Obst, M., Stamatakis, A., *et al.* 2009. Assessing the root of bilaterian animals with scalable phylogenomic methods. *Proc. R. Soc. Lond. B* **276**: 4261–4270.

Hennig, W. 1950. *Grundzüge einer Theorie der phylogenetischen Systematik.* Deutsche Zentralverlag, Berlin.

Holland, L.Z., Albalat, R., Azumi, K., *et al.* 2008. The amphioxus genome illuminates vertebrate origins and cephalochordate biology. *Genome Res.* **18**: 1100–1111.

Hou, X.-G., Aldridge, R.J., Bergström, J., *et al.* 2004. *The Cambrian Fossils of Chengjiang, China.* Blackwell, Malden, MA.

Hyman, L.H. 1940. *The Invertebrates, vol. 1. Protozoa through Ctenophora.* McGraw-Hill, New York.

Jang, K.H. and Hwang, U.W. 2009. Complete mitochondrial genome of *Bugula neritina* (Bryozoa, Gymnolaemata, Cheilostomata): phylogenetic position of Bryozoa and phylogeny of lophophorates within the Lophotrochozoa. *BMC Genomics* **10**: 167.

Jenner, R.A. 2004. When molecules and morphology clash: reconciling phylogenies of the Metazoa by considering secondary character loss. *Evol. Dev.* **6**: 372–376.

Mayr, E. 1982. *The Growth of Biological Thought. Diversity, Evolution, and Inheritance.* Harvard Univ. Press, Cambridge, MA.

Nielsen, C. 2010. The 'new phylogeny'. What is new about it? *Palaeodiversity* **3 (Suppl.)**: 149–150.

Owen, R. 1848. *On the Archetype and Homologies of the Vertebrate Skeleton.* Richard & John E. Taylor, London.

Peterson, K.J., Cotton, J.A., Gehling, J.G. and Pisani, D. 2008. The Ediacaran emergence of bilaterians: congruence between the genetic and the geological fossil records. *Phil. Trans. R. Soc. Lond. B* **363**: 1435–1443.

Peterson, K.J., Dietrich, M.R. and McPeek, M.A. 2009. MicroRNAs and metazoan macroevolution: insight into canalization, complexity, and the Cambrian explosion. *BioEssays* **31**: 736–747.

Raff, R.A., Field, K.G., Olsen, G.J., *et al.* 1989. Metazoan phylogeny based on analysis of 18S ribosomal RNA. In B. Fernholm, K. Bremer and H. Jörnvall (eds): *The Hierarchy of Life. Molecules and Morphology in Phylogenetic Analysis*, pp. 247–260. Excerpta Medica/Elsevier, Amsterdam.

Seilacher, A. 2007. The nature of vendobionts. In P. Vickers-Rich and P. Komarower (eds): *The Rise and Fall of the Ediacaran Biota (Geological Society, Special Publication 286)*, pp. 387–397. Geological Society, London.

Shubin, N. 1998. Evolutionary cut and paste. *Nature* **394**: 12–13.

Srivastava, M., Begovic, E., Chapman, J., *et al.* 2008. The *Trichoplax* genome and the nature of placozoans. *Nature* **454**: 955–960.

Srivastava, M., Simakov, O., Chapman, J., *et al.* 2010. The *Amphimedon queenslandica* genome and the volution of animal complexity. *Nature* **466**: 720–727.

Webster, B.L., Mackenzie-Dodds, J.A., Telford, M.J. and Littlewood, D.T.J. 2007. The mitochondrial genome of *Priapulus caudatus* Lamarck (Priapulida: Priapulidae). *Gene* **389**: 96–105.

2

ANIMALIA (METAZOA)

In the first edition of *Systema Naturæ*, Linnaeus (1735) defined the Kingdom Animalia as natural objects that grow, live, and sense, in contrast to plants, which grow and live but do not sense, and minerals, which grow but neither live nor sense. This definition of the animal kingdom, which goes back to the antiquity, was retained almost unchanged in the 10th edition of *Systema Naturæ* (Linnaeus 1758), which forms the baseline for zoological nomenclature. His arrangement of the species in classes, families and genera reflects the similarity of the organisms, but of course without a causal explanation. His division of the organisms into animals and plants was almost unchallenged for more than a century.

The first classification of living beings based on Darwin's (1859) evolutionary thoughts was presented by Haeckel (1866) (Fig. 1.1). He gave a remarkably modern definition of the Kingdom Animalia, which was separated from the new Kingdom Protista by the possession of tissues and organs. His definition excluded the sponges from the animal kingdom, but he later included the sponges (Haeckel 1874).

The word 'animal' is still used in the wide, Linnaean sense, but in the scientific literature the Kingdom Animalia is now restricted to multicellular animals, i.e. the Metazoa. Both the monophyly of the Animalia and its sister-group relationship with the Choanoflagellata are now generally well documented. A few of the metazoan apomorphies are discussed below.

The most conspicuous synapomorphy of the metazoans is their multicellularity, as opposed to the colo-

niality shown by many choanoflagellates. In colonies the cells may have different shapes and functions, but each cell feeds individually because there is no transport of nutrients between cells. In multicellular organisms the cells are in contact with each other through junction molecules (Adell *et al.* 2004), some of which make transport of nutrients between cells possible. This enables a division of labour, because some cells do not have to feed, and these cells can therefore be specialized to serve specific functions, such as digestion, sensation, contraction, or secretion.

The sexual reproduction and life cycle with differentiation—embryology—of the multicellular organism from the zygote is another important apomorphy (Buss 1987).

All metazoan cells, except eggs and sperm, are diploid, and it appears beyond doubt that the metazoan ancestor was diploid with meiosis directly preceding the differentiation of eggs and sperm. Sexual reproduction has never been observed in choanoflagellates, but as the genes involved in meiosis have been identified in one species (Carr *et al.* 2010), the reproductive stage has probably just been overlooked.

Metazoan spermatozoa show many specializations in the various phyla, but the supposedly ancestral type consists of an ovoid head, a mid piece with mitochondrial spheres surrounding the basal part of a long cilium with a perpendicular accessory centriole, and a tail which is the long undulatory cilium (Franzén 1987). This type is found in many free spawners. Sperm with an elongate, fusiform-to-filiform head is seen in many groups with free-swimming sperm and internal fertilization, and some authors argue that this is the primitive type of fertilization (Buckland-Nicks and Scheltema 1995). Non-motile sperm are found in many species with copulation.

Metazoan eggs develop from one of the four cells of meiosis, whereas the other three cells become polar bodies and degenerate.

After fertilization, the zygote divides and forms an embryo. During embryogenesis the cells become organized in functional units, usually forming layers that give rise to tissues and organs. The blastomeres form cleavage patterns and, later on, embryos/larvae of types characteristic of larger systematic groups. Representatives of many phyla have a blastula stage, and this has been considered one of the important apomorphies of the Animalia (Margulis and Chapman 2009). However, the classical blastula consisting of non-feeding cells (Haeckel 1875) cannot represent an ancestor, and it has been proposed that the ancestor consisted of choanocytes (Buss 1987; a 'choanoblastaea', see Chapter 4). The embryos hatch, either as larvae (indirect development), which may be planktotrophic or lecithotrophic, or as juveniles (direct development). Planktotrophic larvae usually have feeding structures that are completely different from those of the adults.

The organization of various cells in an organism depends on a number of molecules, which fulfil various functions, such as cell recognition, cell adhesion, and cell signalling (Richards and Degnan 2009), all coordinated by a network of regulatory genes (Davidson 2006) and transcription factors (Degnan *et al.* 2009). Interestingly, it has turned out that a number of the molecules involved in these functions have been found in choanoflagellates, where they must serve other purposes (King *et al.* 2008).

The whole genome study, including the demosponge *Amphimedon* (Srivastava *et al.* 2010), has revealed that a very high number of genes involved in organizing the metazoan body are common to all metazoans.

A few morphological characters that have been used in phylogenetic analyses should be discussed here.

Cilia The organelles called cilia and flagella are widespread among the eukaryotes, and the terms are currently used for essentially identical structures (characterized by the presence of an axoneme) and probably of common origin (Jékely and Arendt 2006; Satir *et al.* 2008). The term 'flagella' is used also for the much simpler structures (without microtubuli) found in bacteria. Botanists have preferred the word 'flagella' for the structures found in algae (Moestrup 1982), whereas zoologists have generally used the term 'flagella' when only one or a few appendages are found per cell, and the term 'cilia' when many occur. I have chosen to follow the general trend among molecular biologists and use the word 'cilia' about all the structures that contain one axoneme (Nielsen 1987; Satir *et al.* 2008).

A single cilium usually develops at the end of a cell division, when one of the centrioles templates the formation of the axoneme. Motile cilia usually have an axoneme with a $(9 \times 2) + 2$ pattern of microtubules with radial spokes and dynein arms on the outer tubules, and perform stereotypic movements. Primary cilia, found in most vertebrate tissues but probably present in many invertebrates too, lack both the central tubules, radial spokes, and dynein arms, and are non-motile (Praetorius and Spring 2005; Singla and Reiter 2006). They host many sensory functions, and most sensory structures (for example, many eyes) have the sensitive molecular structures located in the cell membrane of modified cilia.

Two main types of locomotory cilia—undulatory and effective-stroke cilia—can be recognized in metazoans (Fig. 2.1). The undulatory cilia of choanoflagellates, choanocytes, protonephridia, and spermatozoa transport water away from the cell body, or propel the cell through the water with the cell body in front (characteristic of the clade Opisthokonta), whereas most protists swim in the opposite direction. The undulatory cilia usually have an accessory centriole but no long root. The simple structure with only an axoneme, and no hairs or other extracellular specializations (except the vane observed in choanoflagellates and some sponges, see Chapters 3, 5, and 7), is characteristic of choanoflagellates and metazoans, whereas most of the unicellular algae and the swarmers of multicellular algae have extra rods or other structures

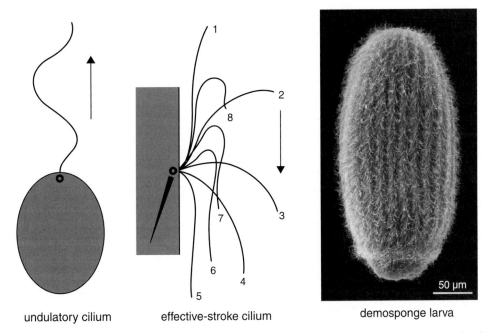

undulatory cilium   effective-stroke cilium   demosponge larva

Figure 2.1. The two types of cilia: undulatory cilia, which propel the water away from the cell, and effective-stroke cilia, which propel the water parallel to the apical cell surface. The arrows indicate the direction of the water currents. The effective-stroke cilia have a long ciliary root. Cells with effective-stroke cilia are arranged in ciliated epithelia, where the ciliary beat forms meta-chronal waves, as shown in the larva of the demosponge *Haliclona*.

along the axoneme, or intricate extracellular orna-mentations (Moestrup 1982).

The basal structures of cilia of some choanoflag-ellates, metazoan spermatozoa and almost all mono-ciliate metazoan cells show specific similarities, with an accessory centriole situated perpendicular to the basal body of the cilium (Fig. 6.1). None of the other unicellular organisms show a similar structure. Star-shaped arrangements of microtubuli, possibly with an 'anchoring and strengthening' function, surround the basal bodies in choanoflagellates (Hibberd 1975), whereas similar patterns in spermatozoa of many metazoans are formed by microfilaments (Franzén 1987). Undulatory cilia of choanoflagellates and cho-anocytes lack a well-defined root, whereas the effec-tive-stroke cilia have roots and rootlets. Larvae of calcareous and homoscleromorph sponges, and adult placozoans and bilaterians, have a cross-striated root, but these structures have only occasionally been observed in adult sponges (see Chapter 6). The various types of ciliated epithelia show well-defined

patterns of ciliary movement with various types of metachronal waves (Nielsen 1987) (Fig. 2.1). Such patterns are rare in the protists, with the ciliates as a conspicuous exception.

Choanocytes (Fig. 4.1) are cells with a ring of long microvilli surrounding an undulatory cilium and engaged in particle capture. However, cells with a ring of shorter or longer microvilli surrounding one or more cilia are known from most metazoan phyla. Certain types, often with long microvilli of various spe-cializations, function in excretion (cyrtocytes of the protonephridia; see, for example, Ruppert and Smith 1988), while the types with shorter microvilli have vari-ous functions (Cantell *et al.* 1982). One type of such cells line coelomic cavities in several phyla, where they generate circulation in the coelomic fluid, probably aiding gas exchange. Mucus cells with a ring of short microvilli are known from many phyla, and the nema-tocytes containing the nematocysts have a modified cilium surrounded by microvilli. Many of these cells are known to be sensory: for example, the hair-bundle mechanoreceptors of cnidarian nematocysts and the mammalian inner ear, which are both sensitive to vibrations of specific frequencies (Watson and Mire 1999); and some ganglion cells have a small complex with a cilium surrounded by inverted microvilli (Westfall and Hessinger 1988). It is important to note that all these cells have functions different from those of the choanocytes and consequently different structures; all the just-mentioned choanocyte-like cells have microvilli that are not retractile and, at least in most cases, have various supporting intracellular structures. Contractile microvilli containing actin, like those of the collared units, occur on intestinal cells of many animals (Revenu *et al.* 2004), but they form a thick 'brush border' and are not engaged in particle capture. Many cell types can thus form various types of microvillar structures, and it is highly questionable to propose a homology between all the cell types that have the microvilli arranged in a circle. I prefer to restrict the term 'choanocyte' to the structures found in choanoflagellates and sponges. Also Cantell *et al.* (1982) suggested a similar restriction of the term, and they proposed to use the term 'collar cells' for all cells with a ring of microvilli around one or more cilia; this change in terminology would remove inaccuracy of many comparative discussions and is highly recommended.

Cell junctions and epithelia Cells of multicellular organisms are connected by various types of cell junctions. The terminology is somewhat confusing, but three functional types can be recognized (Leys *et al.* 2009): (1) adhesion junctions, which can be between cells (zonula adherens or macula adherens), or between a cell and the basement membrane (hemidesmosomes); (2) sealing junctions or zonula occludens, such as septate or tight junctions; and (3) communication junctions (gap junctions), which form small pores between cells (discussed further in Chapter 12).

Adhesion junctions are, of course, found in all metazoans, but several adhesion genes, such as cadherins (Abedin and King 2008), are found in the choanoflagellate *Monosiga*, where they must have functions other than cell adhesion.

Layers of cells with similar orientation and connected by various types of junctions are called epithelia.

It has been customary to state that 'sponges' do not have epithelia, because they are 'leaky' with non-sealed cell layers. However, this view has been challenged because a number of the epithelial genes that are characteristic of true epithelia are found in the sponge *Amphimedon* (Leys *et al.* 2009; Fahey and Degnan 2010; Srivastava *et al.* 2010). However, the stepwise addition of the genes along the evolutionary line, as shown in Fig. 4.2, demonstrates the evolution of one important group of genes at the level of the Silicea (the other groups of sponges have not been sequenced), and another important group of genes at the level of the eumetazoans (with true epithelia). So although genes that encode sealing cell-junction proteins have been found in the genome of a silicean, some of the genes coding for septate junctions are missing, and this may be used to distinguish 'sealed epithelia' of sponges from 'true epithelia' of the eumetazoans that have extracellular digestion. This, as well as several other examples, indicates that absence of a gene is sometimes more phylogenetically informative than presence. The definition of 'epithelium' may of course be extended to cover the consolidated cell layers of the sponges, but I will stick to a more restrictive, classical definition that excludes the outer layers, i.e. pinacoderm, of the sponges.

Many families of molecules are now being used in phylogenetic analyses. An example is the collagens, which are a large family of proteins that are secreted from the cells. Certain types have been present in the common ancestor of choanoflagellates and metazoans, but the 'canonical metazoan adhesion-protein architectures' apparently evolved in the metazoan line (King *et al.* 2008). An extracellular matrix with fibronectin and collagen A is found in the siliceans and the eumetazoans (Müller 1997; Exposito *et al.* 2008). Collagen IV is of a characteristic 'chicken-wire' shape and forms the felt-like basement membrane in all major eumetazoan groups and in the homosclero-morphs, collectively called Proepitheliozoa (see Chapter 8). Other collagens are fibrillar, and it appears that the genes specifying the various types found, for example, in sponges and vertebrates have evolved from one ancestral gene (Exposito *et al.* 1993).

Several genes have been used to infer phylogeny. A special group is the Hox genes (and further the whole Antp family), which are discussed in Chapter 21. Presence or absence of genes is of course an important phylogenetic characteristic, but one caveat should be mentioned. Organs or structures that show expression of a specific gene are not necessarily homologous (homocracy; see Nielsen and Martinez 2003), because most genes are involved in several processes (pleiotrophy; see e.g. Carroll *et al.* 2005), and it is a recurrent theme that the genes involved in the organization of an organ evolve long before the organ, and are then co-opted into organizing the organ. Good examples are the many 'neurogenic' genes present in the 'sponges' that lack nerves (Galliot *et al.* 2009), and the cell-recognition and adhesion genes found in the unicellular choanoflagellate *Monosiga* (King *et al.* 2008).

References

Abedin, M. and King, N. 2008. The premetazoan ancestry of cadherins. *Science* **319**: 946–948.

Adell, T., Gamulin, V., Perovic-Ottstadt, S., *et al.* 2004. Evolution of metazoan cell junction proteins: the scaffold protein MAGI and the transmembrane receptor tetraspanin in the demosponge *Suberites domuncula*. *J. Mol. Evol.* **59**: 41–50.

Buckland-Nicks, J. and Scheltema, A. 1995. Was external fertilization an innovation of early Bilateria? Evidence from sperm ultrastructure of a mollusc. *Proc. R. Soc. Lond. B* **261**: 11–18.

Buss, L.W. 1987. *The Evolution of Individuality*. Princeton University Press, Princeton.

Cantell, C.E., Franzén, Å. and Sensenbaugh, T. 1982. Ultrastructure of multiciliated collar cells in the pilidium larva of *Lineus bilineatus* (Nemertini). *Zoomorphology* **101**: 1–15.

Carr, M., Leadbeater, B.S.C. and Baldauf, S.L. 2010. Conserved meiotic genes point to sex in the choanoflagellates. *J. Eukaryot. Microbiol.* **57**: 56–62.

Carroll, S.B., Grenier, J.K. and Weatherbee, S.D. 2005. *From DNA to Diversity. Molecular Genetics and the Evolution of Animal Design*, 2nd ed. Blackwell Publishing, Malden, MA.

Darwin, C. 1859. *On the Origin of Species by Means of Natural Selection*. John Murray, London.

Davidson, E.H. 2006. *The Regulatory Genome. Gene Regulatory Networks in Development and Evolution*. Academic Press, Amsterdam.

Degnan, B.M., Vervoort, M., Larroux, C. and Richards, G.S. 2009. Early evolution of metazoan transcription factors. *Curr. Opin. Genet. Dev.* **19**: 591–599.

Exposito, J.-Y., Larroux, C., Cluzel, C., *et al.* 2008. Demosponge and sea anemone fibrillar collagen diversity reveals the early emergence of A/C clades and the maintenance of the modular structure of Type V/XI collagens from sponge to human. *J. Biol. Cherm.* **283**: 28226–28235.

Exposito, J.Y., van der Rest, M. and Garrone, R. 1993. The complete intron/exon structure of *Ephydatia mülleri* fibrillar collagen gene suggests a mechanism for the evolution of an ancestral gene molecule. *J. Mol. Evol.* **37**: 254–259.

Fahey, B. and Degnan, B.M. 2010. Origin of animal epithelia: insights from the sponge genome. *Evol. Dev.* **12**: 601–617.

Franzén, Å. 1987. Spermatogenesis. In A.C. Giese, J.S. Pearse and V.B. Pearse (eds): *Reproduction of Marine Invertebrates*, Vol. 9, pp. 1–47. Blackwell/Boxwood, Pacific Grove, CA.

Galliot, B., Quiquand, M., Ghila, L., *et al.* 2009. Origins of neurogenesis, a cnidarian view. *Dev. Biol.* **332**: 2–24.

Haeckel, E. 1866. *Generelle Morphologie der Organismen*. 2 vols. Georg Reimer, Berlin.

Haeckel, E. 1874. Die Gastraea-Theorie, die phylogenetische Classification des Thierreichs und die Homologie der Keimblätter. *Jena. Z. Naturw.* **8**: 1–55, 51 pl.

Haeckel, E. 1875. Die Gastrula und die Eifurchung der Thiere. *Jena. Z. Naturw.* **9**: 402–508.

Hibberd, D.J. 1975. Observations on the ultrastructure of the choanoflagellate *Codosiga botrytis* (Ehr.) Saville-Kent with special reference to the flagellar apparatus. *J. Cell Sci.* **17**: 191–219.

Jékely, G. and Arendt, D. 2006. Evolution of intraflagellar transport from coated vesicles and autogenous origin of the eukaryotic cilium. *BioEssays* **28**: 191–198.

King, N., Westbrook, M.J., Young, S.L., *et al.* 2008. The genome of the choanoflagellate *Monosiga brevicollis* and the origin of metazoans. *Nature* **451**: 783–788.

Leys, S.P., Nichols, S.A. and Adams, E.D.M. 2009. Epithelia and integration in sponges. *Integr. Comp. Biol.* **49**: 167–177.

Linnaeus, C. 1735. *Systema Naturæ sive Regna Tria Naturæ systematice proposita per Classæ, Ordines, Genera, & Species*. Theod. Haack, Lugdunum Batavorum.

Linnaeus, C. 1758. *Systema Naturæ*, 10th edition. 10 vols. Laurentius Salvius, Stockholm.

Margulis, L. and Chapman, M.J. 2009. *Kingdoms & Domains. An Illustrated Guide to the Phyla of Life on Earth*. Academic Press, Amsterdam.

Moestrup, Ø. 1982. Flagellar structure in algae: a review, with new observations particularly on the Chrysophyceae, Phaeophyceae (Fucophyceae), Euglenophyceae, and *Reckertia*. *Phycologia* **21**: 427–528.

Müller, W.E.G. 1997. Origin of metazoan adhesion molecules and adhesion receptors as deduced from cDNA analyses in the marine sponge *Geodia cydonium*: a review. *Cell Tissue Res.* **289**: 383–395.

Nielsen, C. 1987. Structure and function of metazoan ciliary bands and their phylogenetic significance. *Acta Zool. (Stockh.)* **68**: 205–262.

Nielsen, C. and Martinez, P. 2003. Patterns of gene expression: homology or homocracy? *Dev. Genes Evol.* **213**: 149–154.

Praetorius, H.A. and Spring, K.R. 2005. A physiological view of the primary cilium. *Annu. Rev. Physiol.* **67**: 519–529.

Revenu, C., Athman, R., Robine, S. and Louvard, D. 2004. The co-workers of actin filaments: from cell structure to signals. *Nat. Rev. Mol. Cell Biol.* **5**: 1–12.

Richards, G.S. and Degnan, B.M. 2009. The dawn of developmental signaling in the Metazoa. *Cold Spring Harbor Symp. Quant. Biol.* **74**: 1–10.

Ruppert, E.E. and Smith, P.R. 1988. The functional organization of filtration nephridia. *Biol. Rev.* **63**: 231–258.

Satir, P., Mitchell, D.R. and Jékely, G. 2008. How did the cilium evolve? *Curr. Topics Dev. Biol.* **85**: 63–82.

Singla, V. and Reiter, J.F. 2006. The primary cilium as the cell's antenna: signaling at a sensory organelle. *Science* **313**: 629–633.

Srivastava, M., Simakov, O., Chapman, J., *et al.* 2010. The *Amphimedon queenslandica* genome and the evolution of animal complexity. *Nature* **466**: 720–727.

Watson, G.M. and Mire, P. 1999. A comparison of hair bundle mechanoreceptors in sea anemones and vertebrate systems. *Curr. Topics Dev. Biol.* **43**: 51–84.

Westfall, J.A. and Hessinger, D.A. 1988. Presumed neurone-matocyte synapses and possible pathways controlling discharge of a battery of nematocysts in *Hydra*. In D.A. Hessinger and H.M. Lenhoff (eds): *The Biology of Nematocysts*, pp. 41–51. Academic Press, San Diego, CA.

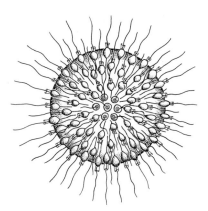

3

Prelude:
Phylum **Choanoflagellata**

The Choanoflagellata is a small phylum, containing only about 200 species of unicellular, solitary, or colony-forming 'flagellates' without chloroplasts. The cells of some of the colonial species are united by cytoplasmic bridges, which have not been studied in detail, but which appear to be the remains of incomplete cell divisions so that the colonies are actually plasmodia; cell junctions have not been reported. The colonies are plate-like or spherical; the more well-known spherical types, such as *Sphaeroeca* (see the chapter vignette), have the collar complexes on the outer side of the sphere, but *Diaphanoeca* has the collar complexes facing an internal cavity, so that the colonies resemble free-swimming collar chambers of a sponge. The monophyly of the group seems unquestioned (Carr *et al.* 2008).

Choanoflagellates occur in most aquatic habitats and are either pelagic or sessile. The apparently most primitive species are naked or sheathed in a gelatinous envelope, but a large group of marine and brackish-water species have an elaborate lorica consisting of siliceous costae united into an elegant bell-shaped meshwork. The costae are secreted in small vacuoles and transported to the surface of the cell, where they become arranged in species-specific patterns (Leadbeater 1994).

The ovoid cell body has a circle of 15–50 microvilli or tentacles, forming a funnel surrounding a long undulating cilium. The microvilli are retractile and contain actin, and the whole structure is held in shape by a mucous or fibrillar meshwork (Karpov and Leadbeater 1998; King 2005). The single cilium (usually called a flagellum) is usually much longer than the collar and has a dainty extracellular fibrillar vane (Leadbeater 2006). The cilium forms sinusoidal waves travelling towards the tip of the cilium, pumping water between the microvilli; bacteria and other food items are retained on the outside of the collar and become engulfed by pseudopodia formed from the base of the collar (Leadbeater 1977, 1983a). In pelagic species, the ciliary activity is also locomotory. The basal body of the ciliary axoneme is surrounded by an intricate system of radiating microtubules, and there is an accessory centriole that is oriented at right angles to the basal body in some species (Karpov and Leadbeater 1997).

Reproduction is by binary fission. Sexual reproduction has not been observed, but the presence of conserved meiotic genes indicates that the process has just gone unnoticed (Carr *et al.* 2010). Also the presence of a complicated life cycle, with different cell and colony types in one species, (Leadbeater 1983b) indicates the

---

Chapter vignette: *Sphaeroeca volvox*. (Redrawn from Leadbeater 1983b.)

presence of sexual reproduction. The ploidy level is unknown.

The sister-group relationship of choanoflagellates and metazoans is supported by numerous morphological and molecular analyses, including analyses of whole genomes (King *et al.* 2008). Many parts of the molecular machinery involved in multicellularity are present, but with unknown functions and have supposedly become co-opted into their cell–cell interactions in the metazoans (King 2004).

Interesting subjects for future research

1. Sexual reproduction

References

Carr, M., Leadbeater, B.S.C. and Baldauf, S.L. 2010. Conserved meiotic genes point to sex in the choanoflagellates. *J. Eukaryot. Microbiol.* **57**: 56–62.

Carr, M., Leadbeater, B.S.C., Hassan, R., Nelson, M. and Baldauf, S.L. 2008. Molecular phylogeny of choanoflagellates, the sister group to Metazoa. *Proc. Natl. Acad. Sci. USA* **105**: 16641–16646.

Karpov, S.A. and Leadbeater, B.S.C. 1997. Cell and nuclear division in a freshwater choanoflagellate, *Monosiga ovata* Kent. *Eur. J. Protistol.* **33**: 323–334.

Karpov, S.A. and Leadbeater, B.S.C. 1998. Cytoskeleton structure and composition in choanoflagellates. *J. Eukaryot. Microbiol.* **45**: 361–367.

King, N. 2004. The unicellular ancestry of animal development. *Dev. Cell* **7**: 313–325.

King, N. 2005. Choanoflagellates. *Curr. Biol.* **15**: R113–R114.

King, N., Westbrook, M.J., Young, S.L., *et al.* 2008. The genome of the choanoflagellate *Monosiga brevicollis* and the origin of metazoans. *Nature* **451**: 783–788.

Leadbeater, B.S.C. 1977. Observations on the life-history and ultrastructure of the marine choanoflagellate *Choanoeca perplexa* Ellis. *J. Mar. Biol. Assoc. U.K.* **57**: 285–301.

Leadbeater, B.S.C. 1983a. Distribution and chemistry of microfilaments in choanoflagellates, with special reference to the collar and other tentacle systems. *Protistologia* **19**: 157–166.

Leadbeater, B.S.C. 1983b. Life-history and ultrastructure of a new marine species of *Proterospongia* (Choanoflagellida). *J. Mar. Biol. Assoc. U.K.* **63**: 135–160.

Leadbeater, B.S.C. 1994. Developmental studies on the loricate choanoflagellate *Stephanoeca diplocostata* Ellis. VIII. Nuclear division and cytokinesis. *Eur. J. Protistol.* **30**: 171–183.

Leadbeater, B.S.C. 2006. The 'mystery' of the flagellar vane in choanoflagellates. *Nova Hedwigia* Beiheft **130**: 213–223.

4

Early animal radiation

There have been many ideas about metazoan origin. Haeckel's ideas developed in steps during a series of papers that used slightly varying names for the earliest phylogenetic stages, and most of the theories discussed subsequently were actually proposed in his early papers (see, for example, Haeckel 1868, 1870, 1873, 1875). Two main types of theories have been prominent: the colonial theory and the cellularization theory.

The colonial theory proposes that the first multicellular organism evolved from a colony of cells, derived from a zygote, which developed cell contacts enabling exchange of nutrients between the cells and, subsequently, specialization of cells. Haeckel (1868) originally thought that the earliest metazoan ancestor, *Synamoebium*, was formed by amoeboid cells that later became ciliated. The idea of a ciliated hollow blastaea was proposed later (Haeckel 1875). He mentioned the colonial flagellates *Volvox* and *Synura* as living examples of organisms showing the same type of organization (Haeckel 1889). Different authors have mentioned various colonial protists as ancestors, but Metschnikoff (1886) seems to have been the first to discuss a choanoflagellate origin of the metazoans (in the light of the newly described *Proterospongia*). The question moved out of focus for more than half a century, but Remane (1963) argued explicitly for a spherical choanoflagellate colony as an ancestor of the monophyletic Metazoa.

A sister-group relationship between choanoflagellates is supported both by the detailed similarity between choanoflagellates and the choanocytes of the 'sponges' (Fig. 4.1), and with several newer molecular phylogenies (Chapter 3). These results have been combined and presented in a previous paper (Nielsen 2008), and will be the backbone of the phylogeny explained below and in the following chapters.

The alternative 'cellularization' theories, which derive a turbellariform-metazoan ancestor through compartmentalization of a ciliate, or a ciliate-like organism (see e.g. Hadzi 1953; Steinböck 1963), is now only of historical interest.

Recent molecular studies (for example, Sperling *et al.* 2007, 2009; Peterson *et al.* 2008), and a few morphological studies (Cavalier-Smith *et al.* 1996; Borchiellini *et al.* 2001) have cast doubt about the monophyly of the Porifera, and indicated a phylogeny like that shown in Fig. 4.2 (Nielsen 2008). This evolutionary scenario is based on a combination of molecular data and studies on embryology and ultrastructure, combined with considerations of the possible adaptive value of each of the proposed evolutionary steps. The scenario is in full accordance with the phylogeny inferred by Degnan *et al.* (2009), based on transcription factors of the choanoflagellate *Monosiga*, the siliceous sponge *Amphimedon*, *Trichoplax*, cnidarians (*Nematostella*, *Hydra*), and bilaterians, and with the whole-genome studies of Srivastava *et al.* (2008, 2010), based on *Neurospora*, *Monosiga*, *Amphimedon*, *Trichoplax*, *Nematostella*, *Hydra*, and a number of bilaterians. However, it is not in agreement with the numerous studies based on ribosomal genes or various collections of other smaller genes—but the very incongruent

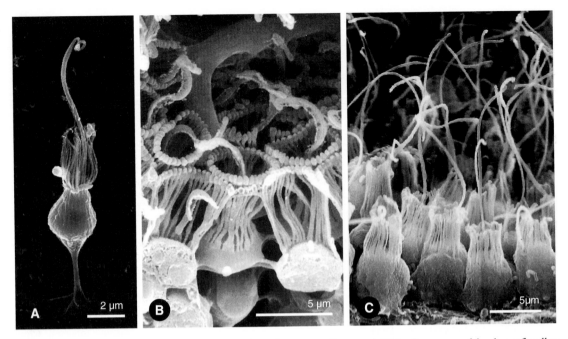

Figure 4.1. SEM of a choanoflagellate and collar chambers of two types of 'sponges'. (A) A solitary stage of the choanoflagellate *Salpingoeca* sp. (courtesy of Drs M.J. Dayel and N. King, University of California, Berkeley, CA, USA.) (B) The demosponge *Callyspongia diffusa* (courtesy of Dr I.S. Johnston, Bethel College, MI, USA; see Johnston and Hildemann 1982). (C) The calcarean *Sycon* sp. (Friday Harbor Laboratories, WA, USA, July 1988.)

topologies presented by these papers indicate that these genes are not sufficiently informative for resolving the deep-branching pattern of the metazoans. A further reason for this is that the taxon sampling has only recently begun to include 'sponges' other than demosponges.

The ancestral metazoan was probably a choanoblastaea (Fig. 4.3; see also Chapter 2 and Nielsen 2008), which had developed so close cell contacts that nutrients could be exchanged between cells; this made specialization of cells possible, because not all cells had to feed. It probably had a life cycle with sexual reproduction and haploid eggs and sperm, characteristic of the living metazoans. Sexual reproduction originated at an early step in the evolution of eukaryotes (Dacks and Roger 1999), but it could of course have been lost in the choanoflagellates.

The new understanding of the early metazoan radiation with a paraphyletic Porifera implies that the life cycle found in the sponge grade, with a ciliated free-swimming larva and a sessile adult form with a water-canal system and choanocyte chambers, was ancestral. The evolution of this type of organization is visualized in Fig. 4.3.

A choanoblastaea lineage developed a number of internalized cells (the 'advanced choanoblastaea'), and established a polarity with special cells without collars at the anterior pole. These cells were used in settling, and the choanocytes of the benthic stage became arranged in a longitudinal groove, which established a common water current along the groove, enhancing the current and preventing recirculation of already processed water. This 'ancestral sponge' retained the pelagic stage as a dispersal larval stage, and a pelago-benthic life cycle had become established. The groove with the vulnerable choanocytes became protected by overarching from the lateral cells so that a tubular 'choanocyte chamber' was formed, further enhancing the water currents ('the primitive sponge'). This is an organization resembling the living calcareans of the

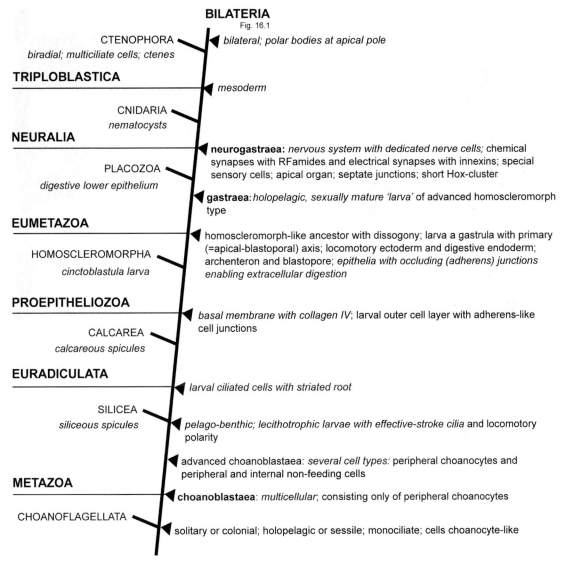

Figure 4.2. Origin and early evolution of the Metazoa. (Modified from Nielsen 2008.)

Ascon-type (see Chapter 7), except that it had no skeleton. This form could grow to a larger size by organizing the layer of choanocytes into numerous separate choanocyte chambers, and the larger size made it possible to invest more yolk in the eggs, so that development could become lecithotrophic. These non-feeding larvae abandoned the external choanocytes with the undulatory cilium, but developed a new

type of locomotory cilium, the effective-stroke cilium. This type of cilia has a long, more or less organized root (Woollacott and Pinto 1995; Fig. 6.1), and shows coordinated movements between neighbouring cells, creating characteristic metachronal waves (Fig. 2.1). This is the life cycle seen in most living 'sponges'.

Two lineages can be recognized from the ancestral sponge: one lineage leading to the Silicea, which

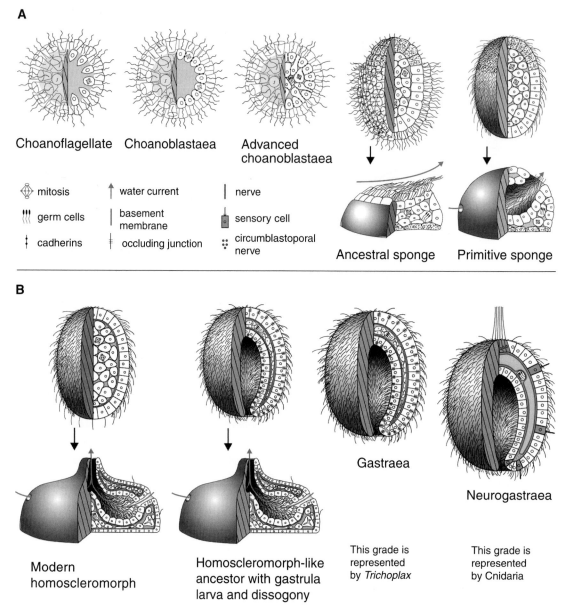

A

Choanoflagellate Choanoblastaea Advanced choanoblastaea

◇ mitosis   ↑ water current | nerve

‖‖‖ germ cells | basement membrane   ▢ sensory cell

| cadherins   ‡ occluding junction   ∷ circumblastoporal nerve

Ancestral sponge   Primitive sponge

B

Gastraea

Neurogastraea

Modern homoscleromorph

Homoscleromorph-like ancestor with gastrula larva and dissogony

This grade is represented by *Trichoplax*

This grade is represented by Cnidaria

Figure 4.3 A,B. The evolutionary scenario (see text). The water currents are indicated with blue arrows, the basement membrane is red, and the nervous cells green. The muscle cells of the gastraea have been omitted for clarity. See Colour Plate 1.

developed a siliceous skeleton (Chapter 5); and another lineage leading to the Euradiculata (Calcarea + Proepitheliozoa; Chapter 6) that developed cross-striated ciliary roots in the epithelium-like outer cell layer of the larvae (Fig. 6.1). The Calcareans developed calcareous spicules (Chapter 7).

The Proepitheliozoa (Homoscleromorpha + Eumetazoa) developed collagen IV in the basal membrane, and the first stages of occluding junctions became established. One lineage led to the Homoscleromorpha (Chapter 9), which remained at this stage. The other lineage led to the Eumetazoa, which will be discussed in Chapter 10.

References

Borchiellini, C., Manuel, M., Alivon, E., *et al.* 2001. Sponge paraphyly and the origin of the Metazoa. *J. Evol. Biol.* **14**: 171–179.

Cavalier-Smith, T., Allsopp, M.T.E.P., Chao, E.E., Boury-Esnault, N. and Vacelet, J. 1996. Sponge phylogeny, animal monophyly, and the origin of the nervous system: 18S rRNA evidence. *Can. J. Zool.* **74**: 2031–2045.

Dacks, J. and Roger, A.J. 1999. The first sexual lineage and the relevance of facultative sex. *J. Mol. Evol.* **48**: 779–783.

Degnan, B.M., Vervoort, M., Larroux, C. and Richards, G.S. 2009. Early evolution of metazoan transcription factors. *Curr. Opin. Genet. Dev.* **19**: 591–599.

Hadzi, J. 1953. An attempt to reconstruct the system of animal classification. *Syst. Zool.* **2**: 145–154.

Haeckel, E. 1868. *Natürliche Schöpfungsgeschichte*. Georg Reimer, Berlin.

Haeckel, E. 1870. *Natürliche Schöpfungsgeschichte*, 2nd ed. Georg Reimer, Berlin.

Haeckel, E. 1873. *Natürliche Schöpfungsgeschichte*, 4th ed. Georg Reimer, Berlin.

Haeckel, E. 1875. *Natürliche Schöpfungsgeschichte*, 6th ed. Georg Reimer, Berlin.

Haeckel, E. 1889. *Natürliche Schöpfungsgeschichte*, 8th ed. Georg Reimer, Berlin.

Johnston, I.S. and Hildemann, W.H. 1982. Cellular organization in the marine demosponge *Callyspongia diffusa. Mar. Biol.* **67**: 1–7.

Metschnikoff, E. 1886. *Embryologische Studien an Medusen*. A. Hölder, Wien.

Nielsen, C. 2008. Six major steps in animal evolution—are we derived sponge larvae? *Evol. Dev.* **10**: 241–257.

Peterson, K.J., Cotton, J.A., Gehling, J.G. and Pisani, D. 2008. The Ediacaran emergence of bilaterians: congruence between the genetic and the geological fossil records. *Phil. Trans. R. Soc. Lond. B* **363**: 1435–1443.

Remane, A. 1963. The evolution of the Metazoa from colonial flagellates *vs.* plasmodial ciliates. In C.E. Dougherty (ed.): *The Lower Metazoa. Comparative Biology and Phylogeny*, pp. 23–32. Univ. California Press, Berkeley.

Sperling, E.A., Peterson, K.J. and Pisani, D. 2009. Phylogenetic-signal dissection of nuclear housekeeping genes supports the paraphyly of sponges and the monophyly of Eumetazoa *Mol. Biol. Evol.* **26**: 2261–2274.

Sperling, E.A., Pisani, D. and Peterson, K.J. 2007. Poriferan paraphyly and its implications for Precambrian palaeobiology. *Geol. Soc. Lond. Spec. Publ.* **286**: 355–368.

Srivastava, M., Begovic, E., Chapman, J., *et al.* 2008. The *Trichoplax* genome and the nature of placozoans. *Nature* **454**: 955–960.

Srivastava, M., Simakov, O., Chapman, J., *et al.* 2010. The *Amphimedon queenslandica* genome and the evolution of animal complexity. *Nature* **466**: 720–727.

Steinböck, O. 1963. Origin and affinities of the lower Metazoa. The 'acoeloid' ancestry of the Eumetazoa. In E.C. Dougherty (ed.): *The Lower Metazoa*, pp. 40–54. University of California Press, Berkeley.

Woollacott, R.M. and Pinto, R.L. 1995. Flagellar basal apparatus and its utility in phylogenetic analyses of the Porifera. *J. Morphol.* **226**: 247–265.

Phylum Silicea

The Silicea (Demospongiae + Hexactinellida) is a phylum of sessile, aquatic metazoans; about 4800 living species are recognized. Sterols characteristic of demosponges have been found in Late Precambrian (Cryogenian) deposits (Love *et al.* 2009), but fossil spicules from both groups first appear in the Lower Cambrian (Brasier *et al.* 1997; Sperling *et al.* 2010). The 'sponge grade' of organization, seen in Silicea, Calcarea (Chapter 7), and Homoscleromorpha (Chapter 9), with a water-canal system and collar chambers with particle-collecting collared cells transporting water through the body, but with neither mouth nor anus, has traditionally been interpreted as an apomorphy of a monophyletic Porifera. Recent molecular studies combined with studies on ultrastructure and biochemistry indicate that the three groups are separate clades, with the eumetazoans as an ingroup of a paraphyletic Porifera (Nielsen 2008; Peterson *et al.* 2008). However, some molecular studies still regard a monophyletic Porifera as a probable model (Philippe *et al.* 2009). The microscopic anatomy is reviewed by Harrison and de Vos (1991), with the three phyla treated together. The concept of a 'diploblast' clade of sponges, *Trichoplax*, ctenophores, and cnidarians (Schierwater *et al.* 2009) is refuted by the whole-genome study of *Amphimedon* (Srivastava *et al.* 2010), which shows a phylogeny like that in Fig. 4.2 (although only one sponge and no ctenophore is included).

Demospongiae and Hexactinellida are structurally rather different, but their sister-group relationship is strongly supported by molecular phylogeny (Sperling *et al.* 2009). They will be described separately below.

The cells of adult demosponges belong to several types, such as choanocytes, myocytes, sclerocytes, porocytes, archaeocytes, and spongocytes, but contrary to the non-reversible differentiation of almost all cell types in the eumetazoans, some cell types in demosponges are able to de-differentiate and subsequently re-differentiate into other types; this is well established, for example, from choanocytes that can de-differentiate and become oocytes or spermatocytes. Other cell types become amoeboid if the sponge is dissociated, and the cells move around and rearrange into a new sponge; however, such cells may re-differentiate into their original type (Simpson 1984). The ciliated outer cells of some larvae are able to differentiate to all cell types in a whole adult (Ereskovsky *et al.* 2007).

The body of the sponge is covered by the pinacoderm, an epithelium-like layer of cells (see discussion in Chapter 2) that is not connected by structurally well-defined, occluding cell junctions, although some eumetazoan cell-junction proteins have been found (Adell *et al.* 2004). Cell junctions of various other types have been reported, and some junctions are characterized as tight (Leys *et al.* 2009). It appears that the

---

Chapter vignette: *Euplectella aspergillum*. (Redrawn from Schulze 1887.)

pinacoderm exerts some control of the extracellular fluids, and the freshwater sponge *Spongilla* has very close contacts between the cells, perhaps with cell-membrane fusion, so that the pinacoderm can control the transport of ions (Adams *et al.* 2010). It was further observed that an electric potential was maintained across the pinacoderm, but also that this potential was not calcium dependent, as that of the eumetazoans, which indicates a different mechanism. It remains to be seen whether this is a mechanism common to all siliceans, or even to all 'sponges'. A basal membrane has not been reported. Some freshwater sponges show scattered cilia on the pinacoderm. Their function and structure are unknown (Leys *et al.* 2009). Contractile cells—myocytes—contain actin filaments, and their contraction can change the shape of areas of the sponge (Elliott and Leys 2007); myosin has been isolated from *Halichondria* (Kanzawa *et al.* 1995).

The middle layer of cells, mesohyl, consists of several cell types embedded in a matrix secreted by the spongocytes. Sclerocytes secrete siliceous spicules already in the larvae, each spicule being secreted in a small vacuole (Leys 2003). The spicules are embedded in an organic skeleton of the demosponge-specific spongin (Aouacheria *et al.* 2006). A few types, for example *Halisarca*, are completely devoid of a skeleton (Bergquist and Cook 2002). The calcareous spherolites of the 'stromatoporid' *Astrosclera* are formed by use of enzymes from digested bacteria and are not endogenous as in the calcareans (Jackson *et al.* 2010).

The choanocytes show a collar complex consisting of a narrow tubular collar of microvilli surrounding an undulatory cilium (Fig. 4.1B) (Gonobobleva and Maldonado 2009). The choanocyte collars consist of about 20–40 long microvilli, the whole structure being stabilized by a fine glycocalyx meshwork; the microvilli contain a core of actin (in *Ephydatia*; personal communication, Dr S. P. Leys, University of Alberta). The collar surrounds a long cilium, which in some species has an accessory centriole at the base, but which always lacks a long striated root; *Halisarca* has small rootlets spreading over the apical part of the nucleus (Gonobobleva and Maldonado 2009). An extracellular structure in the shape of a pair of lateral wings forming a fibrillar vane on the basal portion of the cilium has been reported in some species (Weissenfels 1992). The undulatory movements of the cilium propel water away from the cell body, thus creating a current between the microvilli into the collar and away from the cell. The general orientation of the collared units in the collared chambers and their water currents create a flow of water through the sponge. Particles are captured almost exclusively by the collars. Captured particles become ingested by pseudopodia formed from an area around the funnel (Langenbruch 1985). Captured particles may be passed to archaeocytes, which transport the nutrients through the body (Leys and Reiswig 1998). The Cladorhizidae lack choanocytes and trap small crustaceans from the passing water currents, but their digestion is not extracellular, with secreted digestive enzymes like that of the eumetazoans (Vacelet and Duport 2004).

Nerve cells conducting electrical impulses and gap junctions have not been observed, but demosponges are clearly able to react to stimuli (Leys and Meech 2006; Tompkins-MacDonald *et al.* 2010). The conduction of the stimuli is presumed to be through small molecules. Acetylcholine and cholinesterase have been reported, with the cholinesterase restricted to the myocytes (Bergquist 1978; Horiuchi *et al.* 2003), but there is no evidence that they are involved in cell communication, and similar molecules are also found in plants (Mackie 1990). It appears that the myocytes form a network of contractile cells that can conduct stimuli and thereby coordinate, for example, rhythmic activity (Reiswig 1971).

The sexual reproduction of demosponges has been studied in a number of species (review in Ereskovsky 2010). Sperm and eggs develop from archaeocytes or choanocytes, which lose their cilium and collar. In some species, single choanocytes move away from the choanocyte chamber and differentiate into spermatocysts with spermatogonia, whereas in other species whole choanocyte chambers differentiate into spermatocysts. The spermatozoa have the elements of a typical metazoan spermatozoon but without an organized acrosome; some species have 'proacrosomal vesicles' (Maldonado and Riesgo 2009). The sperm is shed into the exhalant channels and expelled from the sponge. Most species have brood protection, but some species shed the eggs freely, and fertilization takes place in the water (Maldonado and Riesgo 2009). Cleavage is total and the 2- and 4-cell stages show the polar bodies situated at the periphery (Maldonado and Riesgo 2009). The early development shows considerable variation between species. The developing embryo is surrounded by a thin follicle of flattened cells. After various types of cell rearrangement, the larvae have an outer cell layer, which consists of monociliate cells, except at the anterior and posterior poles, where the cells are naked. In some species, the posterior zone of ciliated cells forms a ring of compound cilia. In *Haliclona* and *Amphimedon*, these cilia are light sensitive and change their position when stimulated, thereby changing the direction of swimming (Leys and Degnan 2001). Some species show a gastrulation-like invagination, but this process does not lead to the formation of an endodermal gut, so it would be best to avoid the term 'gastrulation' (Ereskovsky and Dondua 2006). The larvae swim for a short period and their cilia show the effective-stroke-beating pattern characteristic of all planktonic larvae, with conspicuous metachronal waves (Fig. 2.1). The cilia lack the accessory centriole in some species, whereas others have it; long root structures are present, but they show considerable variation and are not striated (Woollacott and Pinto 1995; Gonobobleva and Maldonado 2009). Gonobobleva (2007) reported a short striated root in an early stage of a *Halisarca* larva, but the later stages appeared unstriated. The larva of the carnivorous *Asbestopluma* shows multiciliated cells with weakly striated ciliary roots (Riesgo *et al.* 2007), but with a much finer periodicity than that found in calcarean and eumetazoan ciliary roots. The larvae have a clear polarity, but the anterior pole is devoid of a ciliated sensory organ. Some of the inner cells secrete spicules, and small choanocyte chambers develop precociously in freshwater sponges. The larvae settle with the anterior pole, the ciliated cells resorb the cilia and migrate to the inner of the body or are discarded. The choanocytes differentiate from these cells in some species, but from archaeocytes in other species.

Asexual reproduction through budding, fission, and the special resting stages called gemmulae are important in the life cycles of several species.

The structure of hexactinellids is thoroughly reviewed in Leys *et al.* (2007). It consists of a syncytial trabecular tissue, which covers the outer surface of the sponge, forming the main bulk of the inner tissue, secreting the spicules, and carrying the choanosyncytium, a term that may be misleading because new investigations indicate that it actually consists of highly branched mononucleate cells. The spicules consist of silica and may be loose and form a rather elastic structure (*Euplectella*), or more or less fused, forming a very compact and hard skeleton (*Aphrocallistes*). Collared units rise from a fine reticular choanosyncytium and protrude through openings in an overlying, more robust meshwork formed by the trabecular tissue (the meshwork is absent in *Dactylocalyx*). The cilia have a well-developed fibrillar vane; accessory centriole and roots are absent. The microvilli of the collar are connected by fine transverse fibrils, possibly of glycoproteins. The small cytoplasmic masses that bear a collared unit (collar bodies) lack nuclei and are isolated from the common syncytium by peculiar intracellular plugs (perforate septal partitions) (absent in *Dactylocalyx*). It appears that the collared units together with their basal cytoplasmic masses degenerate periodically and become replaced through budding from the choanosyncytium. Septate junctions occur between the collar bodies and the adjacent trabecular tissue. Nerve cells have not been observed, but electrical impulses travel along the syncytia, and these impulses arrest the activity of the cilia.

Reproduction and larval development have only been studied in detail in the cave-dwelling *Oopsacus* (Boury-Esnault *et al.* 1999; Leys *et al.* 2006). Sperm develops from groups of archaeocytes, but free sperm has not been described (Leys *et al.* 2007). The first five cleavages are total and almost equal, so that a coeloblastula is formed. The following cleavages result in an outer layer of cells of various sizes, and an inner layer of large yolk-filled cells. Some cells fuse to form the outer layer, and the inner cells fuse to form the trabecular tissue. The larvae have unciliated anterior and posterior poles, and a median zone with multiciliate cells covered by a thin layer of outer cells perforated by the cilia. The interior cell mass comprises a special larval type of siliceous spicules and choanocyte chambers resembling those of the adults, but lacking the trabecular tissue. The metamorphosis has not been studied.

There are many detailed similarities between the ciliary apparatus of choanoflagellates and choanocytes (Fig. 4.1) (Maldonado 2004), and the morphology of the silicean sponge with a canal system with choanocytes with intracellular digestion is characteristic of the 'sponge grade' of organization. The lack of striated ciliary roots and recent molecular studies indicate a sister-group relationship with the remaining metazoans, i.e. the Euradiculata (Chapter 6).

Interesting subjects for future research

1. Metamorphosis of hexactinellids

References

Adams, E.D.M., Goss, G.G. and Leys, S.P. 2010. Freshwater sponges have functional, sealing epithelia with high transepithelial resistance and negative transepithelial potential. *PLoS ONE* **5(11)**: e15040.

Adell, T., Gamulin, V., Perovic-Ottstadt, S., *et al.* 2004. Evolution of metazoan cell junction proteins: the scaffold protein MAGI and the transmembrane receptor tetraspanin in the demosponge *Suberites domuncula. J. Mol. Evol.* **59**: 41–50.

Aouacheria, A., Geourjon, C., Aghajari, N., *et al.* 2006. Insights into early extracellular matrix evolution: spongin short chain collagen-related proteins are homologous to basement membrane type IV collagens and form a novel family widely distributed in invertebrates. *Mol. Biol. Evol.* **23**: 2288–2302.

Bergquist, P.R. 1978. *Sponges.* Hutchinson, London.

Bergquist, P.R. and Cook, S.D.C. 2002. Family Halisarcidae Schmidt, 1862. In J.N.A. Hooper, R.W.M. van Soest and P. Willenz (eds): *Systema Porifera: a Guide to the Classification of Sponges,* Vol. 1, p. 1078. Kluwer/Plenum, New York.

Boury-Esnault, N., Efremova, S., Bézac, C. and Vacelet, J. 1999. Reproduction of a hexactinellid sponge: first description of gastrulation by cellular delamination in the Porifera. *Invertebr. Reprod. Dev.* **35**: 187–201.

Brasier, M., Green, O. and Shields, G. 1997. Ediacaran sponge spicule clusters from southwestern Mongolia and the origins of the Cambrian fauna. *Geology* **25**: 303–306.

Elliott, G.R.D. and Leys, S.P. 2007. Coordinated contractions effectively expel water from the aquiferous system of a freshwater sponge. *J. Exp. Biol.* **210**: 3736–3748.

Ereskovsky, A.V. 2010. *The Comparative Embryology of Sponges.* Springer, Dordrecht.

Ereskovsky, A.V. and Dondua, A.K. 2006. The problem of germ layers in sponges (Porifera) and some issues concerning early metazoan evolution. *Zool. Anz.* **245**: 65–76.

Ereskovsky, A.V., Konjukov, P. and Willenz, P. 2007. Experimental metamorphosis of *Halisarca dujardini* larvae (Demospongiae, Halisarcida): evidence of flagellated cell totipotentiality. *J. Morphol.* **268**: 529–536.

Gonoboblev, E. 2007. Basal apparatus formation in external flagellated cells of *Halisarca dujardini* larvae (Demospongiae: Halisarcida) in the course of embryonic development. In M.R. Custódio, G. Lôbo-Hajdu, E. Hajdu and G. Muricy (eds): *Porifera Research: Biodiversity, Innovation and Sustainability (Série Livros 28),* pp. 345–351. Museu Nacional Rio de Janeiro, Rio de Janeiro.

Gonobobleva, E. and Maldonado, M. 2009. Choanocyte ultrastructure in *Halisarca dujardini* (Demospongiae, Halisarcida). *J. Morphol.* **270**: 615–627.

Harrison, F.W. and de Vos, L. 1991. Porifera. In F.W. Harrison (ed.): *Microscopic Anatomy of Invertebrates,* Vol. 2, pp. 29–89. Wiley-Liss, New York.

Horiuchi, Y., Kimura, R., Kato, N., *et al.* 2003. Evolutional study on acetylcholine expression. *Life Sci.* **72**: 1745–1756.

Jackson, D.J., Thiel, V. and Wörheide, G. 2010. An evolutionary fast-track to biocalcification. *Geobiology* **8**: 191–196.

Kanzawa, N., Takano-Ohmuro, H. and Maruyama, K. 1995. Isolation and characterization of sea sponge myosin. *Zool. Sci. (Tokyo)* **12**: 765–769.

Langenbruch, P.F. 1985. Die Aufnahme partikulärer Nahrung bei *Reniera* sp. (Porifera). *Helgol. Wiss. Meeresunters.* **39**: 263–272.

Leys, S.P. 2003. Comparative study of spiculogenesis in demosponges and hexactinellid larvae. *Microsc. Res. Tech.* **62**: 300–311.

Leys, S.P. and Degnan, B.M. 2001. Cytological basis of photoresponsive behavior in a sponge larva. *Biol. Bull.* **201**: 323–338.

Leys, S.P. and Meech, R.W. 2006. Physiology of coordination in sponges. *Can. J. Zool.* **84**: 288–306.

Leys, S.P. and Reiswig, H.M. 1998. Transport pathways in the neotropical sponge *Aplysina. Biol. Bull.* **195**: 30–42.

Leys, S.P., Cheung, E. and Boury-Esnault, N. 2006. Embryogenesis in the glass sponge *Oopsacus minuta*: formation of syncytia by fusion of blastomeres. *Integr. Comp. Biol.* **46**: 104–117.

Leys, S.P., Mackie, G.O. and Reiswig, H.M. 2007. The biology of glass sponges. *Adv. Mar. Biol* **52**: 1–145.

Leys, S.P., Nichols, S.A. and Adams, E.D.M. 2009. Epithelia and integration in sponges. *Integr. Comp. Biol.* **49**: 167–177.

Love, G.D., Grosjean, E., Stalvies, C., *et al.* 2009. Fossil steroids record the appearance of Demospongiae during the Cryogenian period. *Nature* **457**: 718–721.

Mackie, G.O. 1990. The elementary nervous system revisited. *Am. Zool.* **30**: 907–920.

Maldonado, M. 2004. Choanoflagellates, choanocytes, and animal multicellularity. *Invert. Biol.* **123**: 1–22.

Maldonado, M. and Riesgo, A. 2009. Gametogenesis, embryogenesis, and larval features of the oviparous sponge *Petrosia ficiformis* (Haplosclerida, Demospongiae). *Mar. Biol.* **150**: 2181–2197.

Nielsen, C. 2008. Six major steps in animal evolution—are we derived sponge larvae? *Evol. Dev.* **10**: 241–257.

Peterson, K.J., Cotton, J.A., Gehling, J.G. and Pisani, D. 2008. The Ediacaran emergence of bilaterians: congruence between the genetic and the geological fossil records. *Phil. Trans. R. Soc. Lond. B* **363**: 1435–1443.

Philippe, H., Derelle, R., Lopez, P., *et al.* 2009. Phylogenomics revives traditional views on deep animal relationships. *Curr. Biol.* **19**: 706–712.

Reiswig, H.M. 1971. Particle feeding in natural populations of three marine demosponges. *Biol. Bull.* **141**: 568–591.

Riesgo, A., Taylor, C. and Leys, S.P. 2007. Reproduction in a carnivorous sponge: the significance of the absence of an aquiferous system to the sponge body plan. *Evol. Dev.* **9**: 618–631.

Schierwater, B., Eitel, M., Jakob, W., *et al.* 2009. Concatenated analysis sheds light on early metazoan evolution and fuels a modern 'urmetazoan' hypothesis. *PLoS Biol.* **7(1)**: e20.

Schulze, F.E. 1887. Report on the Hexactinellida. *Report on the Scientific Results of the Voyage of H.M.S. Challenger, Zoology* **21 (part 53)**: 1–514.

Simpson, T.L. 1984. *The Cell Biology of Sponges.* Springer-Verlag, New York.

Sperling, E.A., Peterson, K.J. and Pisani, D. 2009. Phylogenetic-signal dissection of nuclear housekeeping genes supports the paraphyly of sponges and the monophyly of Eumetazoa *Mol. Biol. Evol.* **26**: 2261–2274.

Sperling, E.A., Robinson, J.M., Pisani, D. and Peterson, K.J. 2010. Where's the glass? Biomarkers, molecular clocks and microRNAs suggest a 200 million year missing Precambrian fossil record of siliceous sponge spicules. *Geobiology* **8**: 24–36.

Srivastava, M., Simakov, O., Chapman, J., *et al.* 2010. The *Amphimedon queenslandica* genome and the evolution of animal complexity. *Nature* **466**: 720–727.

Tompkins-MacDonald, G.J., Gallin, W.G., Sakarya, O., *et al.* 2010. Expression of a poriferan potassium channel: insights into the evolution of ion channels in metazoans *J. Exp. Biol.* **212**: 761–767.

Vacelet, J. and Duport, E. 2004. Prey capture and digestion in the carnivorous sponge *Asbestopluma hypogea* (Porifera: Demospongiae). *Zoomorphology* **123**: 179–190.

Weissenfels, N. 1992. The filtration apparatus for food collection in freshwater sponges (Porifera, Spongillidae). *Zoomorphology* **112**: 51–55.

Woollacott, R.M. and Pinto, R.L. 1995. Flagellar basal apparatus and its utility in phylogenetic analyses of the Porifera. *J. Morphol.* **226**: 247–265.

6

EURADICULATA

This clade, which consists of Calcarea and Proepitheliozoa, is indicated in the phylogenetic diagram in Nielsen (2008, Fig. 3), but without a name and without an apomorphy. The same sister-group relationship has been indicated by some molecular studies (Sperling *et al.* 2007, 2009, 2010; Peterson *et al.* 2008). Before the phylogenetic importance of the Homoscleromorpha was realized, a sister-group relationship between Calcarea and Eumetazoa had been suggested by several molecular studies (for example, Cavalier-Smith *et al.* 1996; Zrzavý *et al.* 1998; Borchiellini *et al.* 2001; Manuel *et al.* 2003). The clade is not supported by many morphological characters, but the characteristic, cross-striated ciliary root seen in all the ciliated cells of the larvae of calcareans and homoscleromorphs, and in monociliated epithelial eumetazoan cells in general (Fig. 6.1), is used to name the group. The adaptive value of this character could be a more stable attachment and orientation of the effective-stroke cilia.

References

Amano, S. and Hori, I. 2001. Metamorphosis of coeloblastula performed by multipotential larval flagellated cells in the calcareous sponge *Leucosolenia laxa*. *Biol. Bull.* **200**: 20–32.

Borchiellini, C., Manuel, M., Alivon, E., *et al.* 2001. Sponge paraphyly and the origin of the Metazoa. *J. Evol. Biol.* **14**: 171–179.

Boury-Esnault, N., de Vos, L., Donadey, C. and Vacelet, J. 1984. Comparative study of the choanosome of Porifera. I. The Homosclerophora. *J. Morphol.* **180**: 3–17.

Boury-Esnault, N., Ereskovsky, A., Bézac, C. and Tokina, D. 2003. Larval development in the Homoscleromophora (Porifera, Demospongiae). *Invert. Biol.* **122**: 187–202.

Cavalier-Smith, T., Allsopp, M.T.E.P., Chao, E.E., Boury-Esnault, N. and Vacelet, J. 1996. Sponge phylogeny, animal monophyly, and the origin of the nervous system: 18S rRNA evidence. *Can. J. Zool.* **74**: 2031–2045.

Manuel, M., Borchiellini, C., Alivon, E., *et al.* 2003. Phylogeny and evolution of calcareous sponges: monophyly of Calcinea and Calcaronea, high level of morphological homoplasy, and the primitive nature of axial symmetry. *Syst. Biol.* **52**: 311–333.

Nielsen, C. 2008. Six major steps in animal evolution—are we derived sponge larvae? *Evol. Dev.* **10**: 241–257.

Peterson, K.J., Cotton, J.A., Gehling, J.G. and Pisani, D. 2008. The Ediacaran emergence of bilaterians: congruence between the genetic and the geological fossil records. *Phil. Trans. R. Soc. Lond. B* **363**: 1435–1443.

Ruthmann, A., Behrendt, G. and Wahl, R. 1986. The ventral epithelium of *Trichoplax adhaerens* (Placozoa): Cytoskeletal structures, cell contacts and endocytosis. *Zoomorphology* **106**: 115–122.

Sperling, E.A., Peterson, K.J. and Pisani, D. 2009. Phylogenetic-signal dissection of nuclear housekeeping genes supports the paraphyly of sponges and the monophyly of Eumetazoa. *Mol. Biol. Evol.* **26**: 2261–2274.

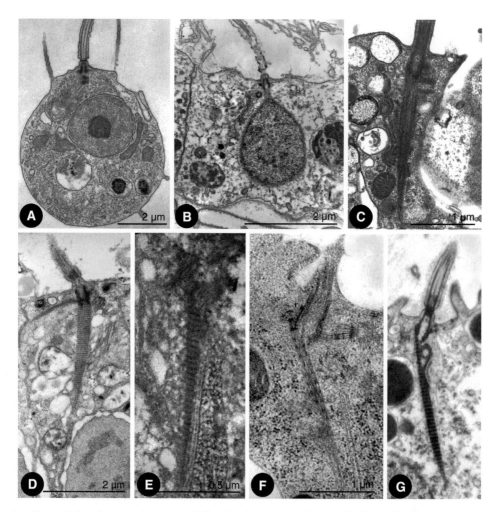

Figure 6.1. Transmission electron microscopy of ciliary basal structures. (A) and (B) Choanoflagellate and choanocyte. (A) Choanoflagellate *Monosiga*. (Courtesy of Drs S. Karpov and B.S.C. Leadbeater, University of Birmingham.) (B) Choanocyte of the homoscleromorph *Corticium*. (Courtesy of Dr N. Boury-Esnault, Museum National d'Histoire Naturelle, Paris; see also Boury-Esnault *et al.* 1984.) (C–G) Ciliated cells of larvae. (C) Larval Silicea: *Aplysilla*. (Courtesy of Dr R.M. Woollacott; see also Woollacott and Pinto 1995.) (D) Larval Calcarea: *Leucosolenia*. (Courtesy of Dr S. Amano, Kanazawa University; see also Amano and Hori 2001.) (E) Larval Homoscleromorpha: *Plakina*. (Courtesy of Dr A. Ereskovsky, University of St Petersburg; see also Boury-Esnault *et al.* 2003.) (F) Adult *Trichoplax*. (Modified from Ruthmann *et al.* 1986.) (G) Larval Cnidaria: *Nematostella*. (Courtesy of Drs Y. Kraus and U. Technau, University of Vienna.)

Sperling, E.A., Pisani, D. and Peterson, K.J. 2007. Poriferan paraphyly and its implications for Precambrian palaeobiology. *Geol. Soc. Lond. Spec. Publ.* **286**: 355–368.

Sperling, E.A., Robinson, J.M., Pisani, D. and Peterson, K.J. 2010. Where's the glass? Biomarkers, molecular clocks and microRNAs suggest a 200 million year missing Precambrian fossil record of siliceous sponge spicules. *Geobiology* **8**: 24–36.

Woollacott, R.M. and Pinto, R.L. 1995. Flagellar basal apparatus and its utility in phylogenetic analyses of the Porifera. *J. Morphol.* **226**: 247–265.

Zrzavý, J., Mihulka, S., Kepka, P., Bezděk, A. and Tietz, D. 1998. Phylogeny of the Metazoa based on morphological and 18S ribosomal DNA evidence. *Cladistics* **14**: 249–285.

Phylum Calcarea

The Calcarea is a small phylum of marine 'sponges', easily identified by their calcareous spicules. About 400 living species are recognized. The fossil record is problematic, but fossils presumed to belong to the Calcarea are known from the Lower Cambrian (Ediacaran) (Pickett 2002). The monophyly of the phylum seems unquestioned. Two classes are recognized: Calcinea and Calcaronea, which show differences both in adult morphology and in embryology (Ereskovsky and Willenz 2008).

The very uncomplicated structure of genera such as *Leucosolenia*, called the *Ascon* type, was used for decades as the starting point for descriptions of sponge evolution. It consists of branched tubes with an inner layer of choanocytes, an outer layer of pinacocytes, and a thin mesohyl with sclerocytes. The *Sycon* type has a central excurrent chamber with the choanocytes arranged in peripheral pockets. The *Leucon* type has the most complex morphology, with the choanocytes arranged in choanocyte chambers with branched incurrent and excurrent canals. However, this simple scheme does not reflect calcarean phylogeny (Dohrmann *et al.* 2006; Manuel 2006).

The pinacoderm covers the outside, and the incurrent and excurrent canals. The cells are in close apposition, and the structures are interpreted as sealing junctions (Eerkes-Medrano and Leys 2006). A loose collagenous mesohyl contains amoeboid cells and

sclerocytes. The spicules are single crystals (or groups of crystals) of calcium/magnesium carbonate formed in an extracellular space lined by sclerocytes connected by septate junctions (Ledger and Jones 1977; Sethmann and Wörheide 2008). The choanocytes (Fig. 4.1) have a long undulatory cilium; a fibrillar vane was reported by Afzelius (1961) but has not been seen in later studies (Eerkes-Medrano and Leys 2006). The basal body shows small feet but lacks an accessory centriole and roots. There are varying types of fine connections between the microvilli (Eerkes-Medrano and Leys 2006). Choanocytes are tightly arranged with some cell junctions of uncertain type (Eerkes-Medrano and Leys 2006). The choanocytes create a water current through the sponge, and particles are captured from this current, mostly by lamellipodia from the apical surface of the choanocytes, but also some by the microvilli of the collar (Leys and Eerkes-Medrano 2006).

Spermatogonia differentiate from choanocytes, but spermiogenesis is still poorly known (Lanna and Klautau 2010). An unciliated sperm cell was described by Anakina and Drozdov (2001) and Lanna and Klautau (2010), but this has to be reinvestigated. The sperm is shed freely and fertilization is internal.

Calcarean embryology was studied in detail before 1800, and the embryology of *Sycon* (Fig. 7.1), order Calcaronea, has been used as the textbook

---

Chapter vignette: *Sycon ciliatum*. (Drawn by Stine Elle.)

example of sponge embryology until recently (review in Ereskovsky 2010). More recent studies have generally confirmed the old observations. The oocytes differentiate from choanocytes, and early oocytes can be recognized at the base of the choanocyte layer at the excurrent opening of the choanocyte chamber (Lanna and Klautau 2010). The future main axis of the larva is perpendicular to the choanocyte layer, with the anterior pole in contact with the choanocyte layer; the polar bodies are given off at the equator. Sperm becomes trapped by a choanocyte, which sheds the collar and cilium and becomes a carrier cell, which transports the sperm head to an egg. Cleavage leads to the formation of a coeloblastula with non-ciliated cells at the pole facing the maternal choanocytes and mono-ciliated cells on the opposite half, with the cilia at the interior side of the sphere. An opening forms at the unciliated pole and the blastula turns inside out. The embryo is now an 'amphiblastula'. This developmental type seems characteristic of the group Calcaronea, whereas the Calcinea have a more usual embryology, with direct development of a coeloblastula, which may have modified maternal

choanocytes in the blastocoel. The fully developed larvae break through the choanocyte layer and escape through the excurrent canals. Their cilia beat with the normal effective stroke, and have an accessory centriole and a long striated ciliary root (Woollacott and Pinto 1996; Amano and Hori 2001) (Fig. 6.1). The larva of *Sycon* has a group of unciliated cells at the posterior pole, whereas that of *Leucosolenia* has unciliated cells scattered between the ciliated cells. The larvae settle after a shorter or longer free life by the ciliated anterior pole, or a more lateral area (in *Sycon*), or with a more unspecified area (in *Leucosolenia*). The gastrulation-like invagination of the ciliated cells reported by Haeckel (1872) is seen just before the settling (Leys and Eerkes-Medrano 2005). In *Sycon* and *Leucandra*, the cells reorganize so that the compact juvenile consists of a central mass of cells, with remains of ciliary roots surrounded by pinacocytes with an ultrastructure resembling that of the inner or granular cells of the larvae. The origin of the mesohyl with sclerocytes has not been demonstrated.

The calcareous sponges are obviously a monophyletic group at the 'sponge grade' of organization.

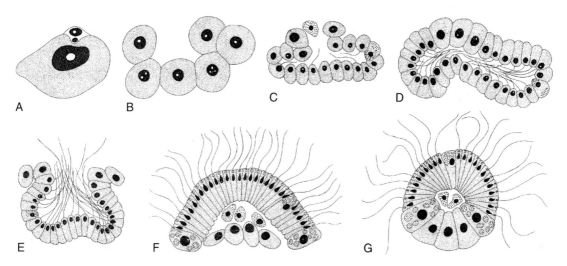

A B C D

E F G

Figure 7.1. Embryology of the calcarean sponge *Sycon ciliatum*; transverse sections; the upper sides are facing the collar chamber. (A) Fertilization with the sperm cell still inside the carrier cell (above). (B) The 8-cell stage. (C) Young blastula. (D) Blastula with cilia on the inside. (E) Inversion of the blastula that brings the cilia to the outer side of the larva. (F) Amphiblastula, the ciliated half of the larva now bulges into the collar chamber. (G) Free-swimming larva. (Modified from Franzen 1988.)

Both molecular phylogeny and morphological studies emphasizing the presence of striated roots of the larval cilia indicate that Calcarea is the sister group of the Eumetazoa (when Homoscleromorpha is not included in the study) (Cavalier-Smith *et al.* 1996; Borchiellini *et al.* 2001) or of the Proepitheliozoa (Homoscleromorpha + Eumetazoa) (Nielsen 2008; Peterson *et al.* 2008; Sperling *et al.* 2009).

Interesting subjects for future research

1. Spermiogenesis and sperm

References

Afzelius, B.A. 1961. Flimmer-flagellum of the sponge. *Nature* **191**: 1318–1319.

Amano, S. and Hori, I. 2001. Metamorphosis of coeloblastula performed by multipotential larval flagellated cells in the calcareous sponge *Leucosolenia laxa*. *Biol. Bull.* **200**: 20–32.

Anakina, R.P. and Drozdov, A.L. 2001. Gamete structure and fertilization in the Barents Sea sponge *Leucosolenia complicata*. *Russ. J. Mar. Biol.* **27**: 143–150.

Borchiellini, C., Manuel, M., Alivon, E., *et al.* 2001. Sponge paraphyly and the origin of the Metazoa. *J. Evol. Biol.* **14**: 171–179.

Cavalier-Smith, T., Allsopp, M.T.E.P., Chao, E.E., Boury-Esnault, N. and Vacelet, J. 1996. Sponge phylogeny, animal monophyly, and the origin of the nervous system: 18S rRNA evidence. *Can. J. Zool.* **74**: 2031–2045.

Dohrmann, M., Voigt, O., Erpenbeck, D. and Wörheide, G. 2006. Non-monophyly of most supraspecific taxa of calcareous sponges (Porifera, Calcarea) revealed by increased taxon sampling and partitioned Bayesian analysis of ribosomal DNA. *Mol. Phylogenet. Evol.* **40**: 830–843.

Eerkes-Medrano, D.I. and Leys, S.P. 2006. Ultrastructure and embryonic development of a syconid calcareous sponge. *Invert. Biol.* **125**: 177–194.

Ereskovsky, A.V. 2010. *The Comparative Embryology of Sponges.* Springer, Dordrecht.

Ereskovsky, A.V. and Willenz, P. 2008. Larval development in *Guancha arnesenae* (Porifera, Calcispongia, Calcinea). *Zoomorphology* **127**: 175–187.

Franzen, W. 1988. Oogenesis and larval development of *Scypha ciliata* (Porifera, Calcarea). *Zoomorphology* **107**: 349–357.

Haeckel, E. 1872. *Die Kalkschwämme. Eine Monographie.* 3 vols. Georg Reimer, Berlin.

Lanna, E. and Klautau, M. 2010. Oogenesis and spermatogenesis in *Paraleucilla magna* (Porifera, Calcarea). *Zoomorphology* **129**: 249–261.

Ledger, P.W. and Jones, W.C. 1977. Spicule formation in the calcareous sponge *Sycon ciliatum*. *Cell Tissue Res.* **181**: 553–567.

Leys, S.P. and Eerkes-Medrano, D.I. 2005. Gastrulation in calcareous sponges: in search of Haeckel's gastraea. *Integr. Comp. Biol.* **45**: 342–351.

Leys, S.P. and Eerkes-Medrano, D.I. 2006. Feeding in a calcareous sponge: particle uptake by pseudopodia. *Biol. Bull.* **211**: 157–171.

Manuel, M. 2006. Phylogeny and evolution of calcareous sponges. *Can. J. Zool.* **84**: 225–241.

Nielsen, C. 2008. Six major steps in animal evolution—are we derived sponge larvae? *Evol. Dev.* **10**: 241–257.

Peterson, K.J., Cotton, J.A., Gehling, J.G. and Pisani, D. 2008. The Ediacaran emergence of bilaterians: congruence between the genetic and the geological fossil records. *Phil. Trans. R. Soc. Lond. B* **363**: 1435–1443.

Pickett, J. 2002. Fossil Calcarea. An overview. In J.N.A. Hooper and R.W.M. Van Soest (eds): *Systema Porifera*, Vol. 2, pp. 1117–1119. Kluwer Academic/Plenum Publishers, New York.

Sethmann, I. and Wörheide, G. 2008. Structure and composition of calcareous sponge spicules: a review and comparison to structurally related biominerals. *Micron* **39**: 209–228.

Sperling, E.A., Peterson, K.J. and Pisani, D. 2009. Phylogenetic-signal dissection of nuclear housekeeping genes supports the paraphyly of sponges and the monophyly of Eumetazoa *Mol. Biol. Evol.* **26**: 2261–2274.

Woollacott, R.M. and Pinto, R.L. 1996. Flagellar basal apparatus and its utility in phylogenetic analyses of the Porifera. *J. Morphol.* **226**: 247–265.

8

PROEPITHELIOZOA

All 'sponges' have epithelia-like cell layers that have joined cells with parallel orientation (Leys *et al.* 2009), but the homoscleromorphs have a basal membrane with collagen IV, like that found in the eumetazoans, and the name Proepitheliozoa was introduced for this clade by Nielsen (2008). The same clade was called Epitheliozoa by Sperling *et al.* (2009). The silicean *Amphimedon* has most of the genes that are involved in organizing the epithelia of the eumetazoans (Leys *et al.* 2009; Srivastava *et al.* 2010), but the epithelium making extracellular digestion possible is nevertheless absent in the 'sponges'.

Homoscleromorphs have sperm with an acrosome (Baccetti *et al.* 1986). This could be a further apomorphy of the Proepitheliozoa, but the cnidarians have an acrosomal vesicle instead of an acrosome (Chapter 13).

Homoscleromorphs and eumetazoans share an astonishing number of cell-signalling and adhesion genes (Nichols *et al.* 2006), but a number of these genes occur in other 'sponges' too.

Collagen IV probably adds stability to the various cell layers.

The morphological support for this clade is weak, so more studies are much needed.

References

Baccetti, B., Gaino, E. and Sará, M. 1986. A sponge with acrosome: *Oscarella lobularis*. *J. Ultrastruct. Mol. Struct. Res.* **94**: 195–198.

Leys, S.P., Nichols, S.A. and Adams, E.D.M. 2009. Epithelia and integration in sponges. *Integr. Comp. Biol.* **49**: 167–177.

Nichols, S.A., Dirks, W., Pearse, J.S. and King, N. 2006. Early evolution of animal cell signaling and adhesion genes. *Proc. Natl. Acad. Sci. USA* **103**: 12451–12456.

Nielsen, C. 2008. Six major steps in animal evolution—are we derived sponge larvae? *Evol. Dev.* **10**: 241–257.

Sperling, E.A., Peterson, K.J. and Pisani, D. 2009. Phylogenetic-signal dissection of nuclear housekeeping genes supports the paraphyly of sponges and the monophyly of Eumetazoa. *Mol. Biol. Evol.* **26**: 2261–2274.

Srivastava, M., Simakov, O., Chapman, J., *et al.* 2010. The *Amphimedon queenslandica* genome and the evolution of animal complexity. Nature **466**: 720–727.

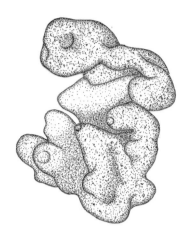

Phylum **Homoscleromorpha**

9

This small phylum, containing about 60 living species of marine 'sponges', has until recently been placed within the Demospongiae, but both morphological and molecular studies now indicate that it is a separate phylum, the sister group of the eumetazoans, although some studies indicate the traditional monophyletic Porifera (Philippe *et al.* 2009; Pick *et al.* 2010). The plakinids have siliceous spicules, the diagnostic type being the calthrop tetractine, whereas the oscarellids lack a skeleton (Muricy and Díaz 2002). *Oscarella* is now especially well studied (Ereskovsky *et al.* 2009). The fossil record is somewhat uncertain, but the earliest forms are probably carboniferous.

The general organization of the homoscleromorph body is of the 'sponge' type, and some of the species are difficult to distinguish from demosponges unless the spicules are studied. The pinacoderm consists of rather flat monociliate cells with specialized cell junctions (Boury-Esnault *et al.* 1984; Ereskovsky and Tokina 2007). The cilia have an accessory centriole, but a root system has not been reported. Both pinacoderm and choanoderm are lined by a basal membrane with collagen IV (Boute *et al.* 1996; Leys and Ereskovsky 2006). Sclerocytes in the pinacoderm and in the mesohyl secrete intracellular siliceous spicules with an axial filament (Maldonado and Riesgo 2007). The choanocytes have a collar of microvilli and a long undulatory cilium with an accessory centriole but no ciliary roots (Boury-Esnault *et al.* 1984) (Fig. 6.1).

The sexual reproduction of the homoscleromorphs has been reviewed by Ereskovsky (2010). Spermatogonia develop from choanocytes, with whole choanocyte chambers differentiating into spermiocysts. The spermatozoa have an acrosome (Maldonado and Riesgo 2009). Oocytes are found in the mesohyl, but they may have originated from choanocytes (Gaino *et al.* 1986). Fertilization is internal, and the embryo becomes surrounded by a follicle of flattened cells originating from choanocytes. Cleavage leads to a compact morula, but the cells later rearrange as a monolayered coeloblastula of monociliate cells with an accessory centriole and a long striated root (Fig. 6.1). Desmosome- or septate-like junctions develop between the apical parts of the ciliated cells. The basal membrane with collagen IV is present at this stage. The cells become increasingly tall and narrow and the larger embryos develop deep folds (wrinkles). When released, the wrinkles unfold and the cinctoblastula larva swims for a short period before settling on the anterior pole. Metamorphosis is a variable process, which internalizes parts of the ciliated outer layer into the interior of the body; other areas differentiate into the pinacoderm. Groups of internalized cells differentiate into choanocytes;

Chapter vignette: *Oscarella lobularis*. (Based on various sources.)

they retain their cilium, but their striated root disappears.

A large number of cell-signalling and adhesion genes are common to homoscleromorphs and eumetazoans (Nichols *et al.* 2006); some of these genes are absent in *Amphimedon* (Srivastava *et al.* 2010), but these studies are still in their infancy.

The presence of collagen IV in the basal membrane indicates that the homoscleromorphs are the sister group of the eumetazoans, and this has found support from recent molecular studies (Sperling *et al.* 2007, 2009, 2010; Peterson *et al.* 2008).

Interesting subjects for future research

1. The genome to search for cell adhesion and communication genes

References

Boury-Esnault, N., de Vos, L., Donadey, C. and Vacelet, J. 1984. Comparative study of the choanosome of Porifera. I. The Homosclerophora. *J. Morphol.* **180**: 3–17.

Boute, N., Exposito, J.Y., Boury-Esnault, N., *et al.* 1996. Type IV collagen in sponges, the missing link in basement membrane ubiquity. *Biol. Cell* **88**: 37–44.

Ereskovsky, A.V. 2010. *The Comparative Embryology of Sponges.* Springer, Dordrecht.

Ereskovsky, A.V. and Tokina, D.B. 2007. Asexual reproduction in homoscleromorph sponges (Porifera: Homoscleromorpha). *Mar. Biol.* **151**: 425–434.

Ereskovsky, A.V., Borchiellini, C., Gazave, E., *et al.* 2009. The homoscleromorph sponge *Oscarella lobularis*, a promising sponge model in evolutionary and developmental biology. *BioEssays* **31**: 89–97.

Gaino, E., Burlando, B. and Buffa, P. 1986. Contribution to the study of egg development and derivation in *Oscarella lobularis* (Porifera, Demospongiae). *Int. J. Invertebr. Reprod. Dev.* **9**: 59–69.

Leys, S.P. and Ereskovsky, A.V. 2006. Embryogenesis and larval differentiation in sponges. *Can. J. Zool.* **84**: 262–287.

Maldonado, M. and Riesgo, A. 2007. Intra-epithelial spicules in a homoscleromorphid sponge. *Cell Tissue Res.* **328**: 639–650.

Maldonado, M. and Riesgo, A. 2009. Gametogenesis, embryogenesis, and larval features of the oviparous sponge *Petrosia ficiformis* (Haplosclerida, Demospongiae). *Mar. Biol.* **150**: 2181–2197.

Muricy, G. and Díaz, M.C. 2002. Order Homosclerophorida Dendy, 1905, Family Plakinidae Schulze, 1880. In J.N.A. Hooper and R.W.M. Van Soest (eds): *Systema Porifera: a Guide to the Classification of Sponges*, pp. 71–82. Kluwer/Plenum, New York.

Nichols, S.A., Dirks, W., Pearse, J.S. and King, N. 2006. Early evolution of animal cell signaling and adhesion genes. *Proc. Natl. Acad. Sci. USA* **103**: 12451–12456.

Peterson, K.J., Cotton, J.A., Gehling, J.G. and Pisani, D. 2008. The Ediacaran emergence of bilaterians: congruence between the genetic and the geological fossil records. *Phil. Trans. R. Soc. Lond. B* **363**: 1435–1443.

Philippe, H., Derelle, R., Lopez, P., *et al.* 2009. Phylogenomics revives traditional views on deep animal relationships. *Curr. Biol.* **19**: 706–712.

Pick, K.S., Philippe, H., Schreiber, F., *et al.* 2010. Improved phylogenomic taxon sampling noticeably affects nonbilaterian relationships. *Mol. Biol. Evol.* **27**: 1983–1987.

Sperling, E.A., Peterson, K.J. and Pisani, D. 2009. Phylogenetic-signal dissection of nuclear housekeeping genes supports the paraphyly of sponges and the monophyly of Eumetazoa *Mol. Biol. Evol.* **26**: 2261–2274.

Sperling, E.A., Pisani, D. and Peterson, K.J. 2007. Poriferan paraphyly and its implications for Precambrian palaeobiology. *Geol. Soc. Lond. Spec. Publ.* **286**: 355–368.

Sperling, E.A., Robinson, J.M., Pisani, D. and Peterson, K.J. 2010. Where's the glass? Biomarkers, molecular clocks and microRNAs suggest a 200 million year missing Precambrian fossil record of siliceous sponge spicules. *Geobiology* **8**: 24–36.

Srivastava, M., Simakov, O., Chapman, J., *et al.* 2010. The *Amphimedon queenslandica* genome and the evolution of animal complexity. *Nature* **466**: 720–727.

10

EUMETAZOA (GASTRAEOZOA)

The eumetazoans constitute a very well-defined group that is now considered monophyletic by almost all morphologists and most molecular biologists. The only problematic groups are the Placozoa (*Trichoplax*) and the ctenophores. *Trichoplax* (Chapter 11) has been much discussed, but both morphological and newer molecular studies indicate that it is the sister group of the Neuralia (Cnidaria + Bilateria). It can be interpreted as a flattened gastrula with the endoderm functioning both in locomotion and digestion. Its ontogeny is unfortunately unknown, but its lack of a nervous system clearly sets it apart from the neuralians. The ctenophores (Chapter 15) have a mesoderm and a nervous system, but have nevertheless been placed in the basal metazoans, or in various other basal positions in most molecular studies. In the following, eumetazoan characters are discussed in general.

The most conspicuous apomorphy of the eumetazoans is that they consist of characteristic cell layers, i.e. epithelia, with polarized cells joined together by cell junctions, such as septate junctions, tight junctions, and belt-adherens junctions (Magie and Martindale 2008), and special molecules characteristic of sealing junctions (Leys *et al.* 2009). They rest on a basal membrane with various structural molecules, such as collagens (especially collagen IV), proteoglycans, and fibronectins (Tyler 2003). Only the acoels lack a basal membrane, except perhaps around the statocyst (Chapter 18). The high integration of the cells enables development of special tissues, the earliest probably being the outer ectoderm and the inner endoderm. The ancestral organism with this organization was

called gastraea by Haeckel (1874; 1875), who emphasized that this organization is found in all cnidarians and in early embryos of many other phyla.

As in the sponges, the ontogeny of the eumetazoans leads from the zygote through various embryological and larval stages before the adult form is reached. In species with free spawning, the cleavage often results in the formation of a sphere of mono- or multiciliate cells. The sphere may consist of only one layer of cells surrounding a cavity (a coeloblastula with the blastocoel), or be compact (a sterroblastula). Septate and gap junctions develop already after a few cleavages both in cnidarians and bilaterians (Kraus and Technau 2006; Burgess 2008).

In species with planktotrophic larvae, the subsequent ontogenetic stage is often a gastrula formed through an invagination of one side of a coeloblastula. This process leads to differentiation of two areas of the ciliated epithelium, viz. an outer epithelium (i.e. ectoderm), which in many species retains the locomotory cilia, and an inner epithelium (i.e. endoderm), which surrounds the archenteron and becomes the digestive epithelium. The two cell layers, often called the primary germ layers, are only connected around the edge of the blastopore. The blastocoel often becomes completely obliterated so that the basal membranes of ectoderm and endoderm become closely apposed.

The animals of the 'sponge grade' of organization have intracellular digestion and are, therefore, restricted to small food particles, but the evolution of sealed epithelia and the archenteron made the

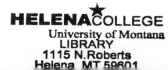

digestion of larger particles in a closed space possible. The ciliated epithelia with effective-stroke cilia in both larvae and adults ensures not only efficient swimming, but also other types of movement, such as creeping and burrowing, not seen in the sponges.

Free ciliated gastrulae swim with the apical pole (i.e. opposite the blastopore) in front. A concentration of sensory cells with long cilia, the apical organ (Fig. 10.1), can be recognized in almost all free ciliated larvae and is an important marker for the orientation of the ontogenetic stages. Cnidarian apical organs consist of tall ciliated cells with an FMRFamide-positive basal plexus (Chapter 13). They generally disappear at metamorphosis. Spiralian larvae generally have complicated apical organs, comprising an apical ganglion with cilia and paired cerebral ganglia (Chapter 23). The apical ganglion degenerates before or at metamorphosis, whereas the cerebral ganglia become the main part of the adult brain (see for example Fig. 27.3). The apical organs of ciliated deuterostome larvae (echinoderms and enteropneusts) comprise a bilaterally symmetric plexus of serotonergic neurons that becomes incorporated into the apical region of the ciliary band (neotroch); they are rather diffuse in echinoderms but well defined in enteropneusts

(Chapter 60). They are lost at metamorphosis, together with the ciliary bands.

Both the position of apical organs, with a tuft of cilia at the pole opposite the blastopore, and the sensory cells involved in metamorphosis indicate homology, and fibroblast-growth-factor signalling has been shown in apical organs of both cnidarians, annelids, sea urchins, and enteropneusts, where they may be important in connection with metamorphosis (Rentsch *et al.* 2008). Hadfield *et al.* (2000) showed that competent larvae of the gastropod *Phestilla* become unable to settle when certain cells in the apical ganglion are ablated. However, Dunn *et al.* (2007) showed that genes *NK2.1* and *HNF6* together are necessary for the development of apical tuft cilia in the sea urchin *Strongylocentrotus*, whereas this is not the case in the gastropod *Haliotis*. This has not been investigated in the cnidarians, so it may represent a deuterostome specialization. It appears that apical organs/ganglia are homologous, but some specializations have evolved in the three main groups.

Adult cnidarians are organized as gastrulae, whereas the Triploblastica (Ctenophora + Bilateria; Chapter 16) differentiate further by developing a third cell layer, i.e. the mesoderm, surrounded by the basal membranes of the two primary cell layers. Archenteron and

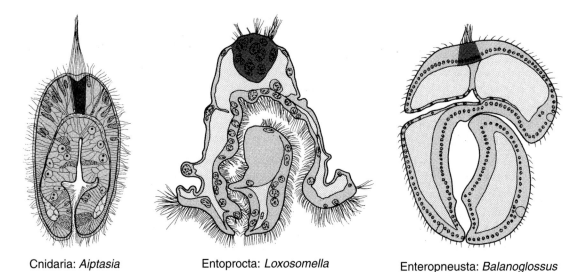

| Cnidaria: *Aiptasia* | Entoprocta: *Loxosomella* | Enteropneusta: *Balanoglossus* |

Figure 10.1. Median sections of larvae showing the apical organs (dark shadings). Planula larva (gastrula) of the cnidarian *Aiptasia mutabilis*. (Modified from Widersten 1968.) Trochophora larva of the entoproct *Loxosoma pectinaricola*. (After Nielsen 1971.) Tornaria larva of the enteropneust *Balanoglossus clavigerus*. (Redrawn from Stiasny 1914.)

blastopore directly become gut and mouth in cnidarians and ctenophores, where the mouth simultaneously functions as the anus. In bilaterians the archenteron develops into the midgut, and the fate of the blastopore varies between the major groups.

The early ontogeny just described is considered characteristic of all eumetazoans, but in fact there is enormous variation in both the processes leading from one stage to the next, and between comparable stages of different species. Like the blastulae, the gastrulae may be hollow or solid, and the processes leading to a two-layered stage with or without an archenteron show many variations. The inner cell layer, i.e. the endoderm, may be formed through invagination, delamination, or ingression, or through combinations of these. It is important to notice, however, that this variation can be observed within many phyla, and the variation can even be observed within the major cnidarian classes (Fig. 13.2). This shows that the developmental types cannot be used in phylogenetic analyses. The hollow blastula, and the invagination leading to a gastrula with an archenteron, are observed within many phyla and are presumed to be primitive (see below).

Larval and adult cnidarians and larval stages of many 'higher' metazoans are of a gastrula-like structure, and this led Haeckel (1874) to propose that all metazoans had evolved from a two-layered ancestor. The gastraea theory was seen as a causative explanation for the similarities in organization and embryology observed among the metazoans. He included the poriferans in his Metazoa or Gastraeozoa, interpreting the inner cells of the sponges as endoderm, but it is now generally accepted that the sponges do not have a gut, and therefore no endoderm (Ereskovsky and Dondua 2006). Haeckel (1875) proposed that the cells of a blastaea became differentiated into two types: anterior, locomotory, cells and posterior, digesting, cells, and that the posterior cells then became invaginated to form an archenteron. The advantage of this organization should be that food particles could remain there for a longer time and be 'assimilated' better. The gastraea theory has been accepted by very many authors, and is more or less clearly expressed in most of the modern textbooks (e.g. Brusca and Brusca 2003; Westheide and Rieger 2007; Hickman *et al.* 2008).

The alternative 'planula theories' actually grew out of Haeckel's early papers too (Haeckel 1870, 1873), the name planula being taken from a hydrozoan larva described by Dalyell (1847). Haeckel proposed that the earliest stage in animal phylogeny should have been synamoeba, a morula-like, spherical organism consisting of amoeba-like cells, followed by planaea, a compact, planula-like organism consisting of outer ciliated cells and inner, non-ciliated cells, and finally gastraea, the gastrula-like form described above. Haeckel soon abandoned the planula as a phylogenetic stage, but the idea was taken up, for example, by Lankester (1877) and many subsequent authors.

The feeding mode of the planula has usually not been considered. Metschnikoff (1886) proposed that particles captured by the outer cells (in some unexplained way) should be transported to the inner cells for digestion, but the origins of the archenteron and the blastopore were not explained. Willmer (1990, p. 169) adhered to the planula theory but admitted that 'It is not clear how the initial planula stage was supposed to feed, having no mouth or gut—existing planula larvae are transient non-feeding stages'. An archenteron without a blastopore appears to have no function, and the establishing of an archenteron with an 'occasional' mouth appears to be without any adaptational advantage. The planula larvae of living organisms are all lecithotrophic developmental stages, but the phylogenetic stage must have been a feeding adult, and its feeding mode remains unexplained (Nielsen 2009). The planuloid-acoeloid theory proposed a gut-less, ciliated organism (a planula) as the ancestor of the acoel turbellarians, which should then give rise to the 'higher' bilaterians. The theory was, of course, favoured by proponents of the cellularization theory, but the origin of the planula has not been discussed by most of the later authors. The theory was forcefully promoted by Hyman (1951), and several authors have followed her (for example, Salvini-Plawen 1978; Ivanova-Kazas 1987; Willmer 1990; Ax 1995); it is implicit in phylogenetic theories that regard lecithotrophy as ancestral in the Metazoa (for example, Rouse 1999; Sly *et al.* 2003). However, as the planula is apparently unable to feed, it is not a probable metazoan ancestor. Compact, ciliated, non-planktotrophic developmental stages are known in many bilaterian phyla, but many of these stages are enclosed by a fertilization membrane, so feeding is excluded, and they probably represent ontogenetic 'short cuts'. Direct development through compact, ciliated larval stages are known, for example, in echinoids, where all available information suggests that ancestral development was through an invagination gastrula to a planktotrophic larva (Nielsen 1998). Similar observations of multiple origins of non-planktonic larvae have been reported: for example, from the large gastropod genus *Conus* (Duda and Palumbi 1999). Platyhelminths superficially resemble planulae, but it should be stressed that they are in no way 'primitive', having, for example, spiral cleavage, multiciliate cells, and complicated reproductive organs (Chapter 29). Cladistic analyses of morphological characters that try to deduce the larval type of larger systematic groups appear problematic because they regard gain and loss of planktotrophy as equally probable, although there are numerous examples of loss of planktotrophy within clades with planktotrophic development, but very few, if any, well-documented examples of an evolution in the opposite direction (Nielsen 1998, 2009).

The new concept of early animal evolution, with the eumetazoans evolving from an organism with a sponge-grade life cycle (Figs. 4.2 and 4.3), more specifically a homoscleromorph-like ancestor, must change the idea about the eumetazoan ancestor. The evolution of a eumetazoan from a sponge-type, sessile adult that has collar chambers with undulatory cilia seems highly improbable, whereas the sponge larvae have an outer, epithelium-like layer of cells with effective-stroke cilia showing metachronal patterns (Fig. 2.1), just like many eumetazoan larvae. The homoscleromorphs have a basal membrane with collagen IV and cell junctions resembling those of the eubilaterians (Chapter 9), and it has been proposed that the eumetazoans evolved from a larva with these characteristics (Nielsen 2008). This implies that the eumetazoans evolved from a larva that became sexually mature and abandoned the sessile adult stage. An evolutionary process of this type has not been described in the literature and will indeed be difficult to demonstrate. However, one possible way of obtaining this type of evolution could be through a process indicated by the development of ctenophores (Chapter 15; Fig. 10.2). The newly hatched juvenile ctenophore is already sexually mature, and sperm and small eggs are shed and a normal development follows (the small life cycle). The gonads of the spent juvenile regress, but the full-grown stage develop gonads again and a new reproductive cycle begins (the large life cycle). This type of 'double' life cycle, called dissogony, enables the evolution of a simple life cycle based on the small life cycle by abandonment of the large life cycle of the ancestor, resulting in a sexually mature adult, which is 'homologous' to the larva of the ancestor. This is not a

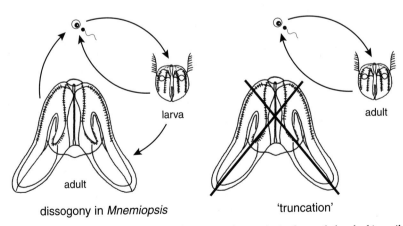

dissogony in *Mnemiopsis*   'truncation'

Figure 10.2. Truncation. The ctenophores show dissogony, i.e. sexual maturity in the newly hatched juveniles and again in the adult stage. If the adult stage is lost permanently, the juvenile becomes the adult, and this evolutionary type is named 'truncation'.

case of heterochrony, which denotes gradual changes in the relative timing of organ development (Klingenberg 1998). This hypothetical type of evolution is here named 'truncation'.

If the eumetazoan ancestor was a homoscleromorph-like larva, the question of the feeding of the ancestor becomes the question about the feeding or non-feeding of this 'larva'. The larvae of the living homoscleromorphs are non-feeding, so the earliest ancestor of the eumetazoans must have been a 'sponge' which developed a feeding larva, which could then become sexually mature—perhaps through truncation, as described above. The evolution of a feeding gastraea from the non-feeding homoscleromorph-like larva is difficult to explain, but some of the homoscleromorph embryos form temporary invaginations, and if one of these (probably at the posterior pole of the larva) became permanent, it could ingest cells from the maternal sponge. Similar ingestion of cells from the parent is found for example in some annelids and many gastropod molluscs, where nurse cells/eggs are engulfed by early embryological stages (Rasmussen 1973; Rivest 1983).

The eumetazoan ancestor was probably planktonic, but a benthic stage was subsequently added both in the cnidarians (the sessile polyp) and in the bilaterians, where the adults show an enormous variation of life styles. Thus, the addition of a benthic adult stage to an ancestral holopelagic life cycle happened both at the origin of the sponge grade, at the origin of the cnidarians (Chapter 13), and (once or twice) at the origin of the bilaterians (Chapters 16 and 21) (Fig. 10.3).

The organisms of the sponge grade are microphagous, with intracellular digestion of small particles

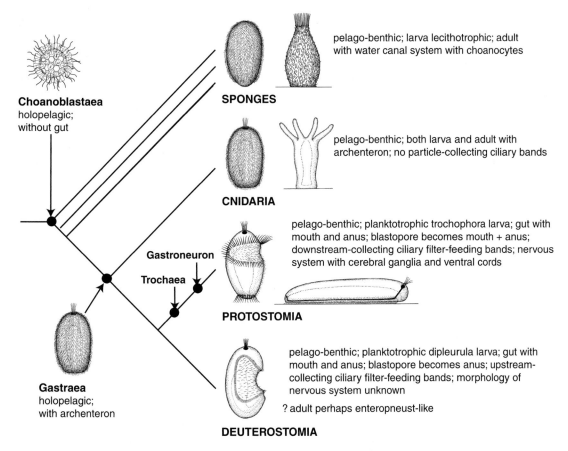

Choanoblastaea
holopelagic;
without gut

SPONGES
pelago-benthic; larva lecithotrophic; adult with water canal system with choanocytes

CNIDARIA
pelago-benthic; both larva and adult with archenteron; no particle-collecting ciliary bands

Gastroneuron

Trochaea

PROTOSTOMIA
pelago-benthic; planktotrophic trochophora larva; gut with mouth and anus; blastopore becomes mouth + anus; downstream-collecting ciliary filter-feeding bands; nervous system with cerebral ganglia and ventral cords

Gastraea
holopelagic;
with archenteron

DEUTEROSTOMIA
pelago-benthic; planktotrophic dipleurula larva; gut with mouth and anus; blastopore becomes anus; upstream-collecting ciliary filter-feeding bands; morphology of nervous system unknown

? adult perhaps enteropneust-like

Figure 10.3. Evolution of pelago-benthic life cycles in the metazoans. (From Nielsen 2009.)

caught by the choanocytes or other cells in the water canal system. The evolution of sealed epithelia, which enable extracellular digestion, made it possible for the early eumetazoans to capture and digest larger particles. This seems to be an obvious advantage. The evolution of permanent ciliated epithelia in the adults made it possible to engage in more active lifestyles, not only with more efficient swimming but also, for example, creeping and burrowing, so that new ecospaces could become inhabited (Xiao and Laflamme 2008).

References

Ax, P. 1995. *Das System der Metazoa I.* Gustav Fischer, Stuttgart.

Brusca, R.C. and Brusca, G.J. 2003. *Invertebrates*, 2nd ed. Sinauer, Sunderland, MA.

Burgess, D.R. 2008. Cytokinesis and the establishment of early embryonic cell polarity. *Biochem. Soc. Trans.* **36**: 384–386.

Dalyell, J.G. 1847. *Rare and Remarkable Animals of Scotland*, Vol. 1. John van Voorst, London.

Duda, T.F.J. and Palumbi, S.R. 1999. Developmental shifts and species selection in gastropods. *Proc. Natl. Acad. Sci. USA* **96**: 10272–10277.

Dunn, E.F., Moy, V.N., Angerer, L.M., Morris, R.L. and Peterson, K.J. 2007. Molecular paleoecology: using gene regulatory analysis to address the origins of complex life cycles in the late Precambrian. *Evol. Dev.* **9**: 10–24.

Ereskovsky, A.V. and Dondua, A.K. 2006. The problem of germ layers in sponges (Porifera) and some issues concerning early metazoan evolution. *Zool. Anz.* **245**: 65–76.

Hadfield, M.G., Meleshkevitch, E.A. and Boudko, D.Y. 2000. The apical sensory organ of a gastropod veliger is a receptor for settlement cues. *Biol. Bull.* **198**: 67–76.

Haeckel, E. 1870. *Natürliche Schöpfungsgeschichte*, 2nd ed. Georg Reimer, Berlin.

Haeckel, E. 1873. *Natürliche Schöpfungsgeschichte*, 4th ed. Georg Reimer, Berlin.

Haeckel, E. 1874. Die Gastraea-Theorie, die phylogenetische Classification des Thierreichs und die Homologie der Keimblätter. *Jena. Z. Naturw.* **8**: 1–55.

Haeckel, E. 1875. Die Gastrula und die Eifurchung der Thiere. *Jena. Z. Naturw.* **9**: 402–508.

Hickman, C.P., Roberts, L.S., Keen, S.L., *et al.* 2008. *Integrated Principles of Zoology*, 14th ed. McGraw-Hill, Boston.

Hyman, L.H. 1951. *The Invertebrates*, Vol. 2. *Platyhelminthes and Rhynchocoela. The Acoelomate Bilateria.* McGraw-Hill, New York.

Ivanova-Kazas, O.M. 1987. Origin, evolution and phylogenetic significance of ciliated larvae. *Zool. Zh.* **66**: 325–338.

Klingenberg, C.P. 1998. Heterochrony and allometry: the analysis of evolutionary change in ontogeny. *Biol. Rev.* **73**: 79–123.

Kraus, Y. and Technau, U. 2006. Gastrulation in the sea anemone *Nematostella vectensis* occurs by invagination and immigration: an ultrastructural study. *Dev. Genes Evol.* **216**: 119–132.

Lankester, E.R. 1877. Notes on the embryology and classification of the animal kingdom: comprising a review of speculations relative to the origin and significance of the germ-layers. *Q. J. Microsc. Sci., N. S.* **17**: 399–454.

Leys, S.P., Nichols, S.A. and Adams, E.D.M. 2009. Epithelia and integration in sponges. *Integr. Comp. Biol.* **49**: 167–177.

Magie, C.R. and Martindale, M.Q. 2008. Cell–cell adhesion in the Cnidaria: insights into the evolution of tissue morphogenesis. *Biol. Bull.* **214**: 218–232.

Metschnikoff, E. 1886. *Embryologische Studien an Medusen.* A. Hölder, Vienna.

Nielsen, C. 1971. Entoproct life-cycles and the entoproct/ectoproct relationship. *Ophelia* **9**: 209–341.

Nielsen, C. 1998. Origin and evolution of animal life cycles. *Biol. Rev.* **73**: 125–155.

Nielsen, C. 2008. Six major steps in animal evolution—are we derived sponge larvae? *Evol. Dev.* **10**: 241–257.

Nielsen, C. 2009. How did indirect development with planktotrophic larvae evolve? *Biol. Bull.* **216**: 203–215.

Rasmussen, E. 1973. Systematics and ecology of the Isefjord marine fauna. *Ophelia* **11**: 1–495.

Rentsch, F., Fritzenwanker, J.H., Scholz, C.B. and Technau, U. 2008. FGF signalling controls formation of the apical sensory organ in the cnidarian *Nematostella vectensis*. *Development* **135**: 1761–1769.

Rivest, B.R. 1983. Development and the influence of nurse egg allotment on hatching size in *Searlesia dira* (Reeve, 1846) (Prosobranchia: Buccinidae). *J. Exp. Mar. Biol. Ecol.* **69**: 217–241.

Rouse, G.W. 1999. Trochophore concepts: ciliary bands and the evolution of larvae in spiralian Metazoa. *Biol. J. Linn. Soc.* **66**: 411–464.

Salvini-Plawen, L.v. 1978. On the origin and evolution of the lower Metazoa. *Z. Zool. Syst. Evolutionsforsch.* **16**: 40–88.

Sly, B.J., Snoke, M.S. and Raff, R.A. 2003. Who came first—larvae or adults? Origins of bilaterian metazoan larvae. *Int. J. Dev. Biol.* **47**: 623–632.

Stiasny, G. 1914. Studien über die Entwicklung des *Balanoglossus clavigerus* Delle Chiaje. I. Die Entwicklung der Tornaria. *Z. Wiss. Zool.* **110**: 36–75.

Tyler, S. 2003. Epithelium—the primary building block for metazoan complexity. *Integr. Comp. Biol.* **43**: 55–63.

Westheide, W. and Rieger, R. 2007. *Spezielle Zoologie. Teil 1: Einzeller und Wirbellose Tiere*, 2nd ed. Elsevier, München.

Widersten, B. 1968. On the morphology and development in some cnidarian larvae. *Zool. Bidr. Upps.* **37**: 139–182.

Willmer, P. 1990. *Invertebrate Relationships. Patterns in Animal Evolution.* Cambridge University Press, Cambridge.

Xiao, S. and Laflamme, M. 2008. On the eve of animal radiation: phylogeny, ecology and evolution of the Ediacara biota. *Trends Ecol. Evol.* **24**: 31–40.

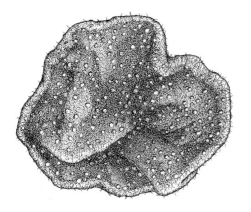

11

Phylum **Placozoa**

The phylum Placozoa comprises only one described species, *Trichoplax adhaerens*, but genetic studies have shown that many species that should represent several genera or families are present (Voigt *et al.* 2004; Buchsbaum-Pearse and Voigt 2007; Signorovitch *et al.* 2007). The usually 1–2 mm large, rounded, flat organisms creep on algae, and have been recorded from warm waters from many parts of the world. The Ediacaran fossil *Dickinsonia* has recently been interpreted as a placozoan, based on the fossil 'trails' that resemble tracks of material digested by *Trichoplax* (Sperling and Vinther 2010) (Fig. 11.1).

The microscopic anatomy of *Trichoplax* was reviewed by Grell and Ruthmann (1991). The flat body consists of one layer of epithelial cells surrounding a rather narrow, flat space containing a mesh-like, syncytial network of fibre 'cells' with actin filaments (Buchholz and Ruthmann 1995). The epithelial cells are diploid and the fibre syncytium is tetraploid; the cells can be dispersed after various treatments and reaggregate to apparently normal individuals with epithelial cells surrounding a fibre syncytium.

The epithelium of the lower side (facing the substratum) consists of rather tall ciliated cells and glandular cells. The ciliated cells have one cilium, which rises from a pit with a ring of supporting rods; the basal complex comprises a long, cross-striated root, short lateral rootlets, and a perpendicular accessory centriole (Fig. 6.1F). The glandular cells are filled with secretion droplets. The upper epithelium consists of flat monociliate cells and spectacular cells with refringent inclusions originating from degenerating cells; they may function as a protection against predators (Jackson and Buss 2009). The epithelial cells are connected by belt desmosomes and by what look like septate desmosomes. A basal membrane appears to be lacking (Schierwater *et al.* 2009), but Srivastava *et al.* (2008) found several genes coding for extracellular-matrix proteins—for example collagen IV—and it was suggested that the matrix is present but in a form that has not been seen with the methods used so far.

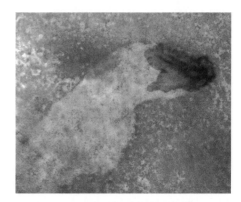

Figure 11.1. *Trichoplax* in front of its feeding trail (the light grey area). The animal is 2–3 mm long. Photo courtesy of Dr Jakob Vinther (Yale University).

Chapter vignette: *Trichoplax adhaerens*. (Based on Rassat and Ruthmann 1979.)

Some cells situated near the margin of the body contain RFamide (Schuchert 1993), which is characteristic of nerve cells in eumetazoans (Chapter 12), but the cells have not been identified by transmission electron microscopy. A *PaxB* gene is expressed in the same region; it appears to be basal to a number of Pax genes and indicates the position of *Trichoplax* as sister group to the neuralians (Hadrys *et al.* 2005). Cells with Hox/Parahox gene activity, regulating growth and fission, are located around the periphery (Jakob *et al.* 2004).

Extensions from the fibre syncytium are here and there connected by small disc-shaped, osmiophilic structures that may be temporary, but which in some cases appear to be intracellular plugs surrounded by the cell membrane.

Trichoplax feeds on the microbial mat of the substratum (Sperling and Vinther 2010), but larger organisms may be ingested too. The microbial mat is digested extracellularly (see Fig. 11.1) by the lower epithelium, where indications of endocytosis are seen in the ciliated cells. Other food items may be transported to the upper side by the cilia and through the epithelium to the fibre cells, where digestion occurs.

Asexual reproduction is by fission, or by formation of spherical swarmers from the upper side. The swarmers are surrounded by cells resembling the upper epithelium and have a central cavity lined by cells resembling the lower cells; fibre cells occur between the two layers of ciliated cells. The spheres open at one side and stretch out so that the normal upper and lower epithelia become established.

Knowledge of sexual reproduction is very incomplete. In cultures, one or a few large eggs have been observed to develop in each animal; they differentiated from a cell of the lower epithelium and became surrounded by a layer of fibre cells functioning as nurse cells. Meiosis and spermatozoa have not been observed. Some of the egg cells formed a fertilization membrane and started dividing, but the embryos soon degenerated (Grell 1972; Grell and Benwitz 1974).

Analyses of morphological characters and combined analyses have in several cases placed *Trichoplax* as the sister group of the neuralians (Nielsen *et al.* 1996; Zrzavý *et al.* 1998). Analyses of whole genomes show

Trichoplax between 'sponges' (the silicean *Amphimedon*) and Eumetazoa (Srivastava *et al.* 2008, 2010). Analyses of nuclear housekeeping genes of a large set of basal animal groups show it as a sister group of the eumetazoans, with the homoscleromorphs as the first outgroup (Sperling *et al.* 2009), in full agreement with the phylogeny proposed here (see Fig. 4.2 and 4.3). Other analyses, mostly with emphasis on bilaterians, show *Trichoplax* in various positions, for example as sister group to a monophyletic 'Diploblastica' (Porifera + Cnidaria) (Signorovitch *et al.* 2007), 'between' Demospongia and Homoscleromorpha (Hejnol *et al.* 2009), as sister group to the Bilateria (Pick *et al.* 2010) and as basal to all other metazoans (Dellaporta *et al.* 2006). None of these hypotheses have been discussed in relation to morphology. The *Trichoplax* genome shows several remarkable similarities with the human genome and must be regarded as highly conserved (Srivastava *et al.* 2008). The position of *Trichoplax* as the sister group of the Neuralia is shown in the whole-genome analyses (Srivastava *et al.* 2008, 2010).

The combination of morphological characters, especially the sealed epithelium enabling extracellular digestion and the lack of any nervous structures, and the molecular data strongly indicates the position of Placozoa as the sister group of the Neuralia (Fig. 4.2).

The body plan of *Trichoplax* has been interpreted in two ways: either as an unfolded gastrula, i.e. as a specialized gastraea, or as a two-layered organism, called plakula, which should have given rise to the other eumetazoans through invagination (Bütschli 1884). Both interpretations homologize the special cells of the periphery with the blastopore region of the gastraea, and are in accordance with the phylogenetic position favoured here, and because a gastrula stage is observed in so many eumetazoan/neuralian phyla, I prefer the first-mentioned theory. Information about the embryology of *Trichoplax* may throw further light on this question.

Interesting subjects for future research

1. Sexual reproduction (spermatozoa, cleavage, cell differentiation, larva)

References

Buchholz, K. and Ruthmann, A. 1995. The mesenchyme-like layer of the fiber cells of *Trichoplax adhaerens* (Placozoa), a syncytium. *Z. Naturforsch.* **C 50**: 282–285.

Buchsbaum-Pearse, V. and Voigt, O. 2007. Field biology of placozoans (*Trichoplax*): distribution, diversity, biotic interactions. *Integr. Comp. Biol.* **47**: 677–692.

Bütschli, O. 1884. Bemerkungen zur Gastraeatheorie. *Morph. Jb.* **9**: 415–427.

Dellaporta, S.L., Xu, A., Sagasser, S., *et al.* 2006. Mitochondrial genome of *Trichoplax adhaerens* supports Placozoa as the basal lower metazoan phylum. *Proc. Natl. Acad. Sci. USA* **103**: 8751–8756.

Grell, K.G. 1972. Eibildung und Furchung von *Trichoplax adhaerens* F.E. Schultze (Placozoa). *Z. Morph. Tiere* **73**: 297–314.

Grell, K.G. and Benwitz, G. 1974. Elektronenmikroskopische Untersuchungen über das Wachstum der Eizelle und die Bildung der 'Befruchtungsmembran' von *Trichoplax adhaerens* F.E. Schulze (Placozoa). *Z. Morph. Tiere* **79**: 295–310.

Grell, K.G. and Ruthmann, A. 1991. Placozoa. In F.W. Harrison (ed.): *Microscopic Anatomy of Invertebrates*, vol. 2, pp. 13–27. Wiley-Liss, New York.

Hadrys, T., DeSalle, R., Sagasser, S., Fischer, N. and Schierwater, B. 2005. The *Trichoplax PaxB* gene: a putative proto-*PaxA/B/C* gene predating the origin of nerve and sensory cells. *Mol. Biol. Evol.* **22**: 1569–1578.

Hejnol, A., Obst, M., Stamatakis, A., *et al.* 2009. Assessing the root of bilaterian animals with scalable phylogenomic methods. *Proc. R. Soc. Lond. B* **276**: 4261–4270.

Jackson, A.M. and Buss, L.W. 2009. Shiny spheres of placozoans (*Trichoplax*) function in anti-predator defense. *Invert. Biol.* **128**: 205–212.

Jakob, W., Sagasser, S., Dellaporta, S., *et al.* 2004. The *Trox-2* Hox/Parahox gene of *Trichoplax* (Placozoa) marks an epithelial boundary. *Dev. Genes Evol.* **214**: 170–175.

Nielsen, C., Scharff, N. and Eibye-Jacobsen, D. 1996. Cladistic analyses of the animal kingdom. *Biol. J. Linn. Soc.* **57**: 385–410.

Pick, K., Philippe, H., Schreiber, F., *et al.* 2010. Improved phylogenomic taxon sampling noticeably affects non-bilaterian relationships. *Mol. Biol. Evol.* **27**: 1983–1987.

Rassat, J. and Ruthmann, A. 1979. *Trichoplax adhaerens* F.E. Schulze (Placozoa) in the scanning microscope. *Zoomorphology* **93**: 59–72.

Schierwater, B., De Jong, D. and DeSalle, R. 2009. Placozoa and the evolution of Metazoa and intrasomatic cell differentiation. *Int. J. Biochem. Cell Biol.* **41**: 370–379.

Schuchert, P. 1993. *Trichoplax adhaerens* (Phylum Placozoa) has cells that react with antibodies against the neuropeptide RFamide. *Acta Zool. (Stockh.)* **74**: 115–117.

Signorovitch, A.Y., Buss, L.W. and Dellaporta, S.L. 2007. Comparative genomics of large mitochondria in placozoans. *PLoS Genet.* **3(1)**: e13.

Sperling, E.A. and Vinther, J. 2010. A placozoan affinity for *Dickinsonia* and the evolution of late Proterozoic metazoan feeding modes. *Evol. Dev.* **12**: 201–209.

Sperling, E.A., Peterson, K.J. and Pisani, D. 2009. Phylogenetic-signal dissection of nuclear housekeeping genes supports the paraphyly of sponges and the monophyly of Eumetazoa *Mol. Biol. Evol.* **26**: 2261–2274.

Srivastava, M., Begovic, E., Chapman, J., *et al.* 2008. The *Trichoplax* genome and the nature of placozoans. *Nature* **454**: 955–960.

Srivastava, M., Simakov, O., Chapman, J., *et al.* 2010. The *Amphimedon queenslandica* genome and the evolution of animal complexity. *Nature* **466**: 720–727.

Voigt, O., Collins, A.G., Pearse, V.B., *et al.* 2004. Placozoa—no longer a phylum of one. *Curr. Biol.* **14**: R944–R945.

Zrzavý, J., Mihulka, S., Kepka, P., Bezděk, A. and Tietz, D. 1998. Phylogeny of the Metazoa based on morphological and 18S ribosomal DNA evidence. *Cladistics* **14**: 249–285.

12

NEURALIA

The most conspicuous apomorphy of the Neuralia (Cnidaria + Ctenophora + Bilateria; see Nielsen 2008) is the presence of specialized communication cells, neurons with axons, signal propagation via action potentials using Na⁺/K⁺ channels, electrical synapses (gap junctions), and chemical synapses with various neurotransmitters (Marlow *et al.* 2009; Watanabe *et al.* 2009; Nickel 2010). Further apomorphies include the presence of special sensory cells, often organized in sensory organs, and special muscle cells, which are innervated and which make it possible for the animal to perform more complex movements. *Trichoplax* (Chapter 11) has contractile elements in the inner syncytial meshwork, but special nervous, sensory, or muscle cells are not found.

Unicellular organisms are of course sensory cells and effector cells at the same time, and this type of cell is found in some silicean larvae (Chapter 5) and in cubozoan larvae (Chapter 13). In a few cases, the two functions are separated in two cells. In the tentacle epidermis of the sea anemone *Aiptasia*, ciliated sensory cells communicate directly with a muscle cell (Westfall *et al.* 2002). In the eye-prototroch system of the larvae of the annelid *Platynereis*, an extension from the photoreceptor cell directly contacts two of the prototroch cells and can alter their ciliary beat (Jékely *et al.* 2008). However, the normal organization is a sensory cell, one or more nerve cells, and the effector organ.

The origin of special nerve cells has been discussed in several recent reviews (Galliot *et al.* 2009; Ryan and Grant 2009; Watanabe *et al.* 2009; Nickel 2010). Much

of the genetic circuitry expressed in neuronal cells has been observed in various sponges, and some even in choanoflagellates and other unicellular organisms (Galliot *et al.* 2009; Ryan and Grant 2009; Tompkins-MacDonald *et al.* 2010). Action potentials have long been known to occur in 'protozoans', such as ciliates (Schwab *et al.* 2008), so most of the building blocks for the construction of a nervous system were present before the actual structure became established.

Cells in non-neuralian organisms are able to communicate via a large number of cell-signalling molecules (Nichols *et al.* 2006), but neurons and synapses are present only in the neuralians. There are various ideas about the evolution of the neurons, but the 'classical' theory of Mackie (1970; see also Nickel 2010), still seems to be the most preferred one. It explains both the origin of neurons, and of myocytes, through specialization of sensory myoepithelial cells, like those seen in many living cnidarians (Lichtneckert and Reichert 2007). For the cell to become a muscle cell it sinks in from the epithelium and loses the sensory function, but retains connection with a neighbouring sensory cell. Further internalization can then result in the evolution of a primitive neuron, which is a cell specialized for transmitting information from a sensory cell to an effector cell.

Gap junctions are specialized contacts between cell membranes that permit transport of small molecules and direct conduction of action potentials without a synapse between neighbouring cells. Each gap junction consists of six protein molecules arranged in a ring in each cell membrane, together forming an intercellular channel that can be opened and closed (Hertzberg 1985). The proteins are 'tetra-span' transmembrane proteins of two types, innexins and connexins. Gap junctions and innexins have been found in the cnidarians *Hydra* and *Nematostella* (although the structure has not been observed microscopically in *Nematostella*), and in several protostomes, whereas connexins have been found only in urochordates and vertebrates (Garré and Bennett 2009; Scemes *et al.* 2009; Phelan *et al.* 1998). Surprisingly, the echinoderms appear to lack both gap junctions and junction proteins, but there is both electric and dye coupling in these animals, so another structure must be involved (Garré and Bennett 2009). Gap junctions are formed as early as at the 2-cell stage in some molluscs (van den Biggelaar *et al.* 1981), when transport of small molecules between the blastomeres can be detected. They are believed to be important in regulating developmental processes (Caveney 1985).

In chemical synapses, action potentials in the presynaptic cell membrane of axon terminals trigger the release of neuropeptides or neurotransmitters from synaptic vesicles into the synaptic cleft, where they excite the postsynaptic cell membrane. Neuropeptides, such as FMRFamides, are known from the synapses of all neuralians, and neurotransmitters, such as acetylcholine, dopamine, GABA and serotonin, are known to be involved in signalling in all neuralians, but their association with synapses has yet to be demonstrated in the cnidarians. A number of the genes are present in cnidarians (Anctil 2009), and the neurotransmitters are known to be involved, for example, in metamorphosis (Zega *et al.* 2007), but, for example, acetylcholine is found also in bacteria, plants, and fungi (Kawashima *et al.* 2007; Anctil 2009; Watanabe *et al.* 2009; Chapman *et al.* 2010). 'Post-synaptic scaffolds' are present both in choanoflagellates and sponges, which do not have nervous systems (Ryan and Grant 2009). The silicean *Amphimedon* has most of the genes needed for the organization of a nervous system with synapses (Srivastava *et al.* 2010), but it is only in the neural-

ians that these genes get co-opted into the gene network needed for building a synapse. This is an example of the general principle of 'preadaptation' (Marshall and Valentine 2010).

There are many types of dedicated sensory cells, and the phylogeny of photoreceptors has been much discussed. At an early point, it was suggested that the two main types of photoreceptor cells, ciliary and rhabdomeric, were characteristic of two lineages. However, it was soon realized that both types of cells are found in several species—for example, the annelid *Platynereis* (Arendt *et al.* 2004)—and in some cases even in the same eye—for example, in the mantle eyes of the bivalve *Pecten* (Nilsson 1994). Both cell types are present in cnidarians, so phylogenetic inference has to be based on details in each type. The *Pax6* gene has been dubbed the 'master control gene' for eye development in bilaterians (Gehring and Ikeo 1999) because it can induce ectopic eyes in various organisms, but the structurally related *PaxB*, which is able to induce ectopic eyes in *Drosophila*, is present in the cnidarian *Tripedalia* (Kozmik *et al.* 2003). A *PaxB*-like gene is found in the silicean *Chalinula*, but this gene is not able to induce ectopic eyes (Hill *et al.* 2010). It appears that *Pax6* is a gene involved in basal photoreception.

Almost all ciliated neuralian larvae have an apical ganglion/organ at the apical pole (Fig. 10.2). It usually carries a group of longer cilia, the apical tuft, and because they are at the front when the larva swims they have always been supposed to be sensory. It always degenerates before or at metamorphosis (see Chapter 10).

Most of the newest molecular phylogenies based on whole genomes interpret the Neuralia (Cnidaria + Bilateria) as a monophyletic group (Srivastava *et al.* 2008, 2010). The ctenophores are placed more basally in almost all phylogenies obtained by molecular analyses (see Chapter 15), but a whole ctenophore genome is not yet available.

Lichtneckert and Reichert (2007, p. 291) very precisely stated that: 'Complex, coordinated behaviour controlled by a primitive nervous system in early metazoan animals must have conferred strong selective advantages and thus contributed significantly to the evolutionary success of nervous systems within metazoan animals.'

References

Anctil, M. 2009. Chemical transmission in the sea anemone *Nematostella vectensis*: a genomic perspective. *Comp. Biochem. Physiol. D* **4**: 268–289.

Arendt, D., Tessmar-Raible, K., Snyman, H., Dorresteijn, A.W. and Wittbrodt, J. 2004. Ciliary photoreceptors with a vertebrate-type opsin in an invertebrate brain. *Science* **306**: 869–871.

Caveney, S. 1985. The role of gap junctions in development. *Annu. Rev. Physiol.* **47**: 319–335.

Chapman, J.A., Kirkness, E.F., Simakov, O., *et al.* 2010. The dynamic genome of *Hydra*. *Nature* **464**: 592–596.

Galliot, B., Quiquand, M., Ghila, L., *et al.* 2009. Origins of neurogenesis, a cnidarian view. *Dev. Biol.* **332**: 2–24.

Garré, J.M. and Bennett, M.V.L. 2009. Gap junctions as electrical synapses. In M. Hortsch and H. Umemori (eds): *The Sticky Synapse*, pp. 423–439. Springer, New York.

Gehring, W. and Ikeo, K. 1999. *Pax 6* mastering eye morphogenesis and eye evolution. *Trends Genet.* **15**: 371–377.

Hertzberg, E.L. 1985. Antibody probes in the study of gap junctional communication. *Annu. Rev. Physiol.* **47**: 305–318.

Hill, A., Boll, W., Ries, C., *et al.* 2010. Origin of Pax and Six gene families in sponges: single PaxB and Six1/2 orthologs in *Chalinula loosanoffi*. *Dev. Biol.* **343**: 106–123.

Jékely, G., Colombelli, J., Hausen, H., *et al.* 2008. Mechanism of phototaxis in marine zooplankton. *Nature* **456**: 395–399.

Kawashima, K., Misawa, H., Moriwaki, Y., *et al.* 2007. Ubiquitous expression of acetylcholine and its biological functions in life forms without nervous systems. *Life Sci.* **80**: 2206–2209.

Kozmik, Z., Daube, M., Frei, E., *et al.* 2003. Role of Pax genes in eye evolution: a cnidarian *PaxB* gene uniting Pax2 and Pax6 functions. *Dev. Cell* **5**: 773–785.

Lichtneckert, R. and Reichert, H. 2007. Origin and evolution of the first nervous system. In J.H. Kaas (ed.): *Evolution of Nervous Systems. A Comprehensive Reference*, Vol. 1, pp. 289–315. Academic Press, Amsterdam.

Mackie, G.O. 1970. Neuroid conduction and the evolution of conducting tissues. *Q. Rev. Biol.* **45**: 319–332.

Marlow, H.Q., Srivastava, M., Matus, D.Q., Rokhsar, D. and Martindale, M.Q. 2009. Anatomy and development of the nervous system of *Nematostella vectensis*, an anthozoan cnidarian. *Dev. Neurobiol.* **69**: 235–254.

Marshall, C.R. and Valentine, J.W. 2010. The importance of preadapted genomes on the origin of the animal bodyplans and the Cambrian explosion. *Evolution* **64**: 1189–1201.

Nichols, S.A., Dirks, W., Pearse, J.S. and King, N. 2006. Early evolution of animal cell signaling and adhesion genes. *Proc. Natl. Acad. Sci. USA* **103**: 12451–12456.

Nickel, M. 2010. Evolutionary emergence of synaptic nervous systems: what can we learn from the non-synaptic, nerveless Porifera? *Integr. Comp. Biol.* **129**: 1–16.

Nielsen, C. 2008. Six major steps in animal evolution – are we derived sponge larvae? *Evol. Dev.* **10**: 241–257.

Nilsson, D.E. 1994. Eyes as optical alarm systems in fan worms and ark clams. *Phil. Trans. R. Soc. Lond. B* **346**: 195–212.

Phelan, P., Stebbings, L.A., Baines, R.A., *et al.* 1998. *Drosophila* shaking-B protein forms gap junctions in paired *Xenopus* oocytes. *Nature* **391**: 181-184.

Ryan, T.J. and Grant, S.G.N. 2009. The origin and evolution of synapses. *Nature Rev. Neurosci.* **10**: 701–712.

Scemes, E., Spray, D.C. and Meda, P. 2009. Connexins, pannexins, innexins: novel roles of 'hemi-channels'. *Pflugers Arch. Eur. J. Physiol.* **457**: 1207–1226.

Schwab, A., Hanley, P., Fabian, A. and Stock, C. 2008. Potassium channels keep mobile cells on the go. *Physiology* **23**: 212–220.

Srivastava, M., Begovic, E., Chapman, J., *et al.* 2008. The *Trichoplax* genome and the nature of placozoans. *Nature* **454**: 955–960.

Srivastava, M., Simakov, O., Chapman, J., *et al.* 2010. The *Amphimedon queenslandica* genome and the evolution of animal complexity. *Nature* **466**: 720–727.

Tompkins-MacDonald, G.J., Gallin, W.G., Sakarya, O., *et al.* 2010. Expression of a poriferan potassium channel: insights into the evolution of ion channels in metazoans *J. Exp. Biol.* **212**: 761–767.

van den Biggelaar, J.A.M., Dorresteijn, A.W.C., de Laat, S.W. and Bluemink, J.G. 1981. The role of topographical factors in cell interaction and determination of cell lines in molluscan development. In G.H. Schweiger (ed.): *International Cell Biology 1980–1981*, pp. 526–538. Springer-Verlag, Berlin.

Watanabe, H., Fujisawa, T. and Holstein, T.W. 2009. Cnidarians and the evolutionary origin of the nervous system. *Dev. Growth Differ.* **51**: 167–183.

Westfall, J.A., Elliott, C.F. and Carlin, R.W. 2002. Ultrastructural evidence for two-cell and three-cell neuronal pathways in the tentacle epidermis of the sea anemone *Aiptasia pallida*. *J. Morphol.* **251**: 83–92.

Zega, G., Pennati, R., Fanzago, A. and De Bernardi, F. 2007. Serotonin involvement in the metamorphosis of the hydroid *Eudendrium racemosum*. *Int. J. Dev. Biol.* **51**: 307–313.

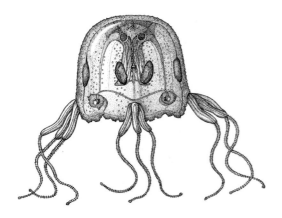

13

Phylum Cnidaria

Cnidarians are a well-defined phylum comprising about 10000 living, aquatic, mainly marine species, and their monophyly is supported in all newer molecular analyses (the approximate 1300 myxozoan species are discussed below). The freshwater polyp *Hydra* has been a favourite experimental organism ever since Trembley (1744) cut the polyp into small pieces and observed their complete regeneration; its genome has now been sequenced (Chapman *et al.* 2010). However *Hydra* is a highly specialized cnidarian and the marine starlet anemone *Nematostella* is now a popular new cnidarian model organism (Darling *et al.* 2005; Putnam *et al.* 2007). The radiation of the phylum is now well-established through both morphological and molecular studies (Collins *et al.* 2006; Pick *et al.* 2010). Anthozoans and medusozoans are sister groups, and the discussions about whether the polyp or the medusa represents the ancestral body plan seems settled by the morphology of the mitochondrial RNA (mtRNA) in the two groups, where the anthozoans have the usual (plesiomorphic) metazoan circular mtRNA, whereas the medusozoans have a linear mtDNA, which is clearly an apomorphy (Bridge *et al.* 1992). The derived character of the medusozoans is further supported by the several additional types of nematocysts and minicollagens (David *et al.* 2008).

Fossils of gelatinous organisms, such as medusae and naked polyps, are difficult to identify with cer-

tainty, and a number of Ediacaran fossils may belong to this phylum. The Uppermost Ediacaran *Olivooides* (Zhao and Bengtson 1999) is very similar to scyphozoan polyps with a strong perisarc. The earliest unquestionable anthozoan appears to be the sea anemone *Xianguangia* from Chengjiang (Hou *et al.* 2004). Putative fossils of scypho- and hydromedusae are known from Chengjiang (Sun and Hou 1987), and more certain forms from the Middle Cambrian (Cartwright *et al.* 2007).

The most conspicuous apomorphy of the phylum is the nematocysts (cnidae) (David *et al.* 2008), which are highly complicated structures formed inside special cells called cnidocytes (nematocytes) (Tardent 1995), and are differentiated from interstitial cells (Holstein 1981; Galliot *et al.* 2009). The nematocyst is formed as a small cup-shaped structure inside the cell (Fig. 13.1); it increases in size and becomes pear-shaped; a long hollow thread forms from the narrow end of the capsule, and this thread, which in most types has rows of spines at the base, finally invaginates and becomes coiled up inside the capsule. The nematocyst is now fully formed and when the cell has reached the position where it is to function, the final differentiation of the cell with the cnidocil (a modified cilium) takes place; when the nematocyst discharges, the tube everts again (Özbek *et al.* 2009). A few other animals contain nematocysts, but they have come from cnidarian prey that

Chapter vignette: The cubomedusa *Tripedalia cystophora*. (Redrawn from Werner 1973.)

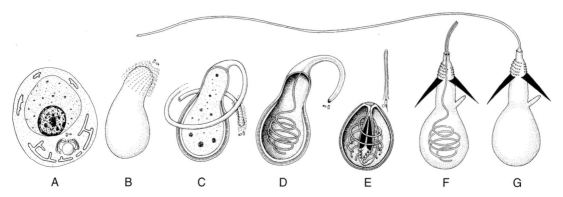

Figure 13.1. Formation and discharge of a nematocyst. (Redrawn from Holstein 1981 and Tardent and Holstein 1982.)

may be identified by studying the 'cleptocnidia' (stolen nematocysts). It has been discussed whether the nematocysts are symbiotic organelles originating, for example, from dinoflagellates (Hwang *et al.* 2008), but if this is true, the symbiosis must have been established already in the stem lineage of the living cnidarians. The colloblasts of the ctenophores are cells of a completely different structure (see Chapter 15).

The microscopic anatomy was reviewed by Fautin and Mariscal (1991: Anthozoa), Thomas and Edwards (1991: Hydrozoa) and Lesh-Laurie and Suchy (1991: Scyphozoa (incl. Staurozoa) and Cubozoa).

A characteristic that sets the cnidarians apart from all other neuralians is their primitive (plesiomorphic) body plan, which in principle is that of a gastrula, i.e. only an outer cell layer, i.e. the ectoderm (epidermis), and an inner cell layer, i.e. the endoderm (gastrodermis), the latter surrounding the digestive cavity (archenteron); the blastopore functions both as mouth and anus. Interstitial stem cells originate in the endoderm and can differentiate into neurons and nematocytes (Galliot *et al.* 2009).

The two cell layers are separated by a basal membrane that, especially in the medusa, is elaborated to form a thick gelatinous mesogloea (see below). Medusozoans are generally radial, but bilaterality is seen in certain hydroids (for example *Branchiocerianthus*) and the siphonophores. The anthozoans are all more or less bilateral, but a head with a brain is not developed. The larvae have an apical nervous concentration

and the polyps have a concentration of neurons around the blastopore (see below). The main axis of the cnidarians is the primary, apical-blastoporal axis. The quasiradial body plan with the sac-shaped gut and the lack of a brain apparently limits the life styles to sessile, pelagic or parasitic.

All anthozoans have a simple life cycle, with a larva and an adult polyp. Medusozoans have life cycles comprising both an asexual polyp and sexual medusa stages, although many groups especially of the hydrozoans have only polyps or medusae. Most species have asexual reproduction through budding that may occur both from polyps and medusae. The buds become released, or the budding leads to the formation of colonies of species-specific shapes. Some hydrozoans, for example the siphonophores, consist of several types of units that resemble polyps and medusae serving different functions.

Ectodermal and endodermal cells form sealed epithelia with septate junctions and belt desmosomes (Chapman *et al.* 2010). Thecate hydroids and some scyphozoan polyps have an exoskeleton called perisarc, consisting of chitin and proteins, and many anthozoans have a conspicuous calcareous exoskeleton, or an 'inner' skeleton, formed by fused calcareous spicules formed by sclerocytes embedded in a proteinaceous matrix.

Many of the cells carry one cilium with an accessory centriole and a striated root; some of these cilia are motile while others are sensory. Scattered examples of cells with more than one cilium are known: the endodermal cells of certain hydropolyps and anthozoans have two or several cilia per cell. The ectoderm of labial tentacles of some anthozoans comprises cells with several cilia each with an accessory centriole. The tentacles of the hydromedusa *Aglantha* have a pair of lateral ciliary bands that are formed from multiciliate cells (Mackie *et al.* 1989); these ciliary bands propel water past the tentacles, but do not collect particles like the ciliary bands of many bilaterian larvae (Chapter 21). Cilia generally beat as individual units, but compound cilia in the shape of wide, oblique membranelles occur in the anthozoan larva called zoanthina (Nielsen 1984).

The nervous system is intraepithelial, forming nerve nets both in ectoderm and endoderm, that may be concentrated in nerve rings in both polyps and medusa, and in ganglia in scyphomedusae and cubomedusae (Garm *et al.* 2006; Watanabe *et al.* 2009). Special sensory cells of various types occur at characteristic positions, for example in ocelli and statocysts, in both polyps and medusae. Interneuronal and neuromuscular chemical synapses of the unidirectional type (the usual type in the 'higher' metazoans) are known from all cnidarian classes, as are bidirectional (symmetrical) interneuronal synapses (Lichtneckert and Reichert 2007). Synapses are generally few and difficult to fix (Westfall 1996). Neurons may fuse to form neurosyncytia, which may be more widespread than first realized (Grimmelikhuijzen and Westfall 1995). Acetylcholine, the neurotransmitter found in ctenophores and bilaterians, has been observed in *Nematostella*, but its possible function in synapses has not been demonstrated; *Hydra* appears to lack typical acetylcholinesterases (Chapman *et al.* 2010). Gap junctions have not been observed in anthozoans, but both the structure and the gap-junction innexins (Phelan 2005) have been identified in interneuronal, neuromuscular and intermuscular junctions in *Hydra*.

Epithelio-muscular cells occur in both ectoderm and endoderm in all groups and is the only type of muscle cell in several groups. Special muscle cells (myocytes) without an epithelial portion occur in scyphozoan and cubozoan polyps. Some cubozoan genera (for example *Carybdea*) have only myocytes, while *Tripedalia* has both types of contractile cells with all types of intermediary stages (Chapman 1978). All muscle cells of the scyphistoma of *Aurelia* originate from the ectoderm (Chia *et al.* 1984). The muscle cells are generally of the smooth type, but striated muscle cells are found in the subumbrella of some hydromedusae. They develop from the so-called entocodon during budding, which has been interpreted as a type of mesodermal pockets (Seipel and Schmid 2005). However, the entocodon is just an ectodermal invagination, like those seen during budding of, for example, bryozoans, entoprocts, and ascidians. The striated muscle cells must be interpreted as a specialization of the subumbrellar epithelium (Burton 2007).

The epithelial cells are sometimes anchored both to the mesogloea by hemidesmosomes like those of the other eumetazoans (Chapman *et al.* 2010) and to the perisarc by tonofilaments. Ectoderm and endoderm are rather closely apposed in most scyphopolyps and hydropolyps, but are separated in all other forms by a more or less thick, gelatinous to almost cartilaginous, hyaline layer, i.e. the mesogloea. Both ecto- and endoderm participate in secretion of collagen, which forms the web-like organic matrix of the mesogloea in hydropolyps. The mesogloea is generally without cells in hydrozoans and cubozoans, but cells of various types enter the mesogloea from both ectoderm and endoderm in scyphozoans and especially in the anthozoans. The cells of the anthozoan mesogloea enter the gelatinous matrix in the form of tubes or solid cell strings and differentiate into isolated amoebocytes, star-shaped cells, scleroblasts, myocytes, and a number of other cell types, some of which secrete collagen fibres. It seems clear from all recent descriptions that the mesogloea, with its varying content of cells moving in from both ectoderm and endoderm, is different from the mesoderm of the bilaterians. The 'real' mesoderm forms epithelia and other tissues in which the cells are connected by cell junctions, and that are isolated from ectoderm and endoderm by basement membranes.

Eggs and sperm differentiate from endodermal cells in anthozoans, scyphozoans, and cubozoans, and

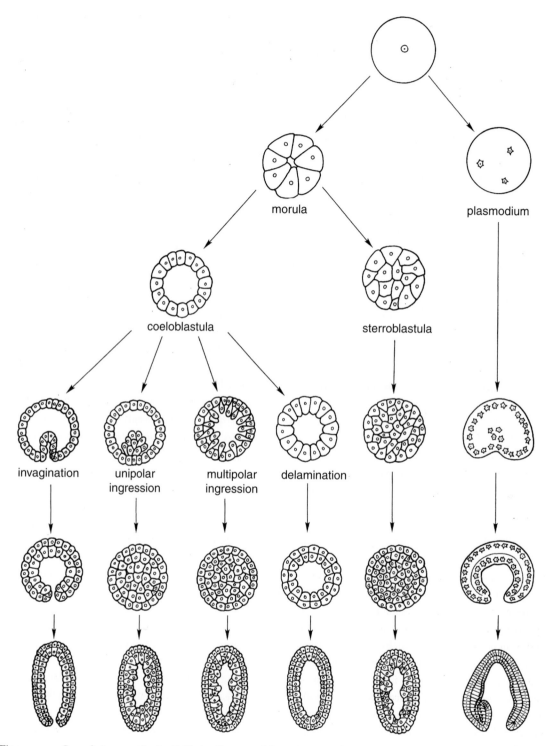

Figure 13.2. Gastrulation types in the Cnidaria; diagrams of the main types of gastrulation following the holoblastic cleavage are shown, together with one of the several types of plasmodial development. (Modified from Tardent 1978.)

from ectodermal cells in hydrozoans; they are usually shed freely in the water. The spermatozoa are of the primitive metazoan type, but have acrosomal vesicles instead of an acrosome (Ehlers 1993; Tiemann and Jarms 2010). In hydrozoans the main (apical-blastoporal) axis of the larva is determined already during oogenesis, with the presumptive apical pole situated at the side of the egg that is in contact with the endoderm or mesogloea (Freeman 1990). Fertilization and polar-body formation take place at the oral (blastoporal) pole where cleavage is also initiated (this end of the embryo is therefore called the animal pole). It appears that a similar orientation is found in anthozoans and scyphozoans (Tardent 1978), but the details are not so well documented. The position of the polar bodies at the blastoporal pole also appears to be a characteristic of cnidarians (and ctenophores, see Chapter 15); almost all bilaterians have the polar bodies situated at the apical pole (Chapter 16).

Cleavage is usually total, but the early developmental stages are plasmodial in some hydrozoans (Fig. 13.2). In species with total cleavage, the first cleavage cuts towards the apical pole, and a very unusual stage, with the two incipient blastomeres connected with a handle-like apical bridge, is reached just before the separation of the blastomeres (Tardent 1978). There is considerable variation in the subsequent development (Fig. 13.2). Several species have the presumably primitive coeloblastula, which invaginates to form a gastrula. The larvae are ciliated, often with an apical tuft of longer cilia (Fig. 10.1). Some of these larvae are planktotrophic, but the feeding structures never include ciliary bands engaged in filter feeding. Development through a feeding-gastrula larva occurs in several anthozoans (Fritzenwanker et al. 2007) and scyphozoans (Fadlallah 1983). In other species, the blastula develops into a ciliated, lecithotrophic planula larva that consists of ectoderm and endoderm; the endoderm may be compact or surround an archenteron; a mouth is usually lacking in the early stages but develops before the larva settles. In the anthozoan Nematostella (Marlow et al. 2009) that has a planktotrophic larva, the early gastrula stages develop a diffuse nerve net of both ectodermal and endodermal

components, with an apical organ with a tuft of cilia and oral and pharyngeal nerve rings. Tentacles develop after metamorphosis, and the nervous system becomes more concentrated around the mouth. In the scyphozoan Aurelia (Nakanishi et al. 2008; Yuan et al. 2008) that has a lecithotrophic larva, the development of the nervous system is slower, but essentially similar. The larva usually settles with the apical pole and becomes a polyp, but the larvae of a few hydrozoans and scyphozoans develop directly into medusae, and a polyp stage is lacking. The apical nervous concentration in the larva disappears at metamorphosis, but a nervous concentration is found at the base of the stalk in Hydra (Grimmelikhuijzen 1985), possibly related to its ability to detach from and reattach to the substratum. All primary polyps are capable of one or more types of asexual reproduction, for example lateral budding (often leading to the formation of colonies), frustule formation from a basal plate, or transverse fission giving rise to medusae developing through different processes in the three medusozoan groups:

In cubozoans, the polyp goes through a metamorphosis and becomes the medusa; the tentacles of the polyp lose the nematocysts and become sense organs, and new tentacles develop (Werner 1973, 1984).

In scyphozoans, the medusae are typically formed through transverse fission (strobilation) of the polyp, but, for example, Pelagia is holopelagic. Staurozoans lack the medusa stage; they are sometimes regarded as a separate group.

In hydrozoans, the polyps typically form lateral medusa buds, which detach as medusae, but there is a wide variation in the life cycles (Petersen 1990). Some groups are holopelagic without a polyp generation. Other groups show reduction of the medusae, which can be followed in many transformational series from the eumedusoid stage, where the medusa is reduced relatively little, through cryptomedusoid and styloid stages to the type represented by Hydra, which has testes and ovaries situated in the ectoderm of the polyp. Polypodium has a very unusual polyp stage, which is parasitic in the eggs of sturgeons, and that is everted so that the endoderm is on the outside and ectoderm with tentaculated buds on the inside; when the sturgeon sheds the eggs the polyps turn inside out and small medusae are budded off (Raikova 1994). It appears that *Polypodium* shows a link to the parasitic myxozoans (Lom 1990; Kent *et al.* 2001) that represent a further specialization to parasitic life, with extreme miniaturization, complete loss of nervous system, and highly modified sexual reproduction, but metazoan characteristics, such as septate junctions, have been retained. Their nematocysts resemble those of other cnidarians, and their systematic position as parasitic hydrozoans is supported by molecular analyses, which place the investigated species as sister group of *Polypodium* when this species is included (Siddall *et al.* 1995; Zrzavý *et al.* 1998; Siddall and Whiting 1999; Jiménez-Guri *et al.* 2007). Also the presence of cnidarians-specific minicollagens in the myxozoan *Tetracapsuloides* (Holland *et al.* 2011) also support their inclusion in the Cnidaria. Analyses that do not include *Polypodium* show myxozoans in a basal position within the metazoans (Schlegel *et al.* 1996; Evans *et al.* 2008), but this may be owing to both *Polypodium* and the myxozoans being 'long branch taxa'.

The cnidarians are at the gastraea stage with only ectoderm and endoderm, and have the nematocysts as the unquestionable apomorphy; they form the logical sister group of the Triploblastica, which all have mesoderm (Chapter 14). The lecithotrophic planula larvae can easily be interpreted as a specialization from the planktotrophic larval type; similar transitions from planktotrophic to lecithotrophic development with concurrent delay of development of the gut and various feeding structures are well known in almost all invertebrate groups (Nielsen 1998).

The ancestral cnidarian was probably a holopelagic, advanced gastrula, probably with nematocysts on small tentacles, and was thus one of the very first metazoan carnivores. If the adult of this gastrula type attached by the apical pole, we have an organism resembling Werner's (1984) hypothetical ancestor of the cnidarians.

Recent molecular studies almost unanimously show Cnidaria as a monophyletic group, but the relative position of the basal metazoan taxa is still not settled. Most studies now place the 'sponges' at the base of the metazoan tree, but the interrelationships of Cnidaria,

Trichoplax, Ctenophora, and Bilateria has not reached a stable state (see Chapter 12). The position of *Trichoplax* as sister group of the Bilateria is becoming the more accepted position (Chapter 11), but especially the ctenophores are apparently problematic (Chapter 15).

Interesting subjects for future research

1. A reinvestigation of hydrozoan gap junctions
2. Ultrastructure of cubozoan and scyphozoan cnidocytes
3. Origin of the secondary body axis
4. Extracellular matrix

References

Bridge, D., Cunningham, C.W., Schierwater, B., DeSalle, R. and Buss, L.W. 1992. Class-level relationships in the phylum Cnidaria: evidence from mitochondrial genome structure. *Proc. Natl. Acad. Sci. USA* **89**: 8750–8753.

Burton, P.M. 2007. Insights from diploblasts; the evolution of mesoderm and muscle. *J. Exp. Zool. (Mol. Dev. Evol.)* **310B**: 5–14.

Cartwright, P., Halgedahl, S.L., Hendricks, J.R., *et al.* 2007. Exceptionally preserved jellyfishes from the Middle Cambrian. *PLoS ONE* **2(10)**: e1121.

Chapman, D.M. 1978. Microanatomy of the cubopolyp, *Tripedalia cystophora* (Class Cubozoa). *Helgol. Wiss. Meeresunters.* **31**: 128–168.

Chapman, J.A., Kirkness, E.F., Simakov, O., *et al.* 2010. The dynamic genome of *Hydra*. *Nature* **464**: 592–596.

Chia, F.S., Amerongen, H.M. and Peteya, D.J. 1984. Ultrastructure of the neuromuscular system of the polyp of *Aurelia aurita* L., 1758 (Cnidaria, Scyphozoa). *J. Morphol.* **180**: 69–79.

Collins, A.G., Schuchert, P., Marques, A.C., *et al.* 2006. Medusozoan phylogeny and character evolution clarified by new large and small subunit rDNA dana and assessment of the utility of phylogenetic mixture methods. *Syst. Biol.* **55**: 97–115.

Darling, J.A., Reitzel, A.M., Burton, P.M., *et al.* 2005. Rising starlet: the starlet sea anemone, *Nematostella vectensis*. *BioEssays* **27**: 211–221.

David, C.N., Özbek, S., Adamczyk, P., *et al.* 2008. Evolution of complex structures: minicollagens shape the cnidarian nematocyst. *Trends Genet.* **24**: 431–436.

Ehlers, U. 1993. Ultrastructure of the spermatozoa of *Halammohydra schulzei* (Cnidaria, Hydrozoa): the significance of acrosomal structures for the systematization of the Eumetazoa. *Microfauna Mar.* **8**: 115–130.

Evans, N.M., Lindner, A., Raikova, E.V., Collins, A.G. and Cartwright, P. 2008. Phylogenetic placement of the enigmatic parasite, *Polypodium hydriforme*, within the phylum Cnidaria. *BMC Biology* **8**: 139.

Fadlallah, Y.H. 1983. Sexual reproduction, development and larval biology in scleractinian corals. *Coral Reefs* **2**: 129–150.

Fautin, D.G. and Mariscal, R.N. 1991. Cnidaria: Anthozoa. In F.W. Harrison (ed.): *Microscopic Anatomy of Invertebrates*, vol. 2, pp. 267–358. Wiley-Liss, New York.

Freeman, G. 1990. The establishment and role of polarity during embryogenesis in hydrozoans. In D.L. Stocum (ed.): *The Cellular and Molecular Biology of Pattern Formation*, pp. 3–30. Oxford Univ. Press, Oxford.

Fritzenwanker, J.H., Genikhovich, G., Kraus, Y. and Technau, U. 2007. Early development and axis specification in the sea anemone *Nematostella vectensis*. *Dev. Biol.* **310**: 264–279.

Galliot, B., Quiquand, M., Ghila, L., *et al.* 2009. Origins of neurogenesis, a cnidarian view. *Dev. Biol.* **332**: 2–24.

Garm, A., Ekström, P., Boudes, M. and Nilsson, D.E. 2006. Rhopalia are integrated parts of the central nervous system in box jellyfish. *Cell Tissue Res.* **325**: 333–343.

Grimmelikhuijzen, C.J.P. 1985. Antisera to the sequence Arg-Phe-amide visualize neuronal centralization in hydroid polyps. *Cell Tissue Res.* **241**: 171–182.

Grimmelikhuijzen, C.J.P. and Westfall, J.A. 1995. The nervous system of cnidarians. In O. Breidbach (ed.): *The Nervous Systems of Invertebrates: An Evolutionary and Comparative Approach*, pp. 7–24. Birkhäuser Verlag, Basel.

Holland, J.W., Okamura, B., Hartikainen, H. and Secombes, C.J. 2011. A novel minicollagen gene links cnidarians and myxozoans. *Proc. R. Soc. Lond. B.* **278**: 546–553.

Holstein, T. 1981. The morphogenesis of nematocytes in *Hydra* and *Forskålia*: an ultrastructural study. *J. Ultrastruct. Res.* **75**: 276–290.

Hou, X.-G., Aldridge, R.J., Bergström, J., *et al.* 2004. *The Cambrian Fossils of Chengjiang, China*. Blackwell, Malden, MA.

Hwang, J.S., Nagai, S., Hayakawa, S., Takaku, Y. and Gojobori, T. 2008. The search for the origin of cnidarian nematocysts in dinoflagellates. In P. Pontarotti (ed.): *Evolutionary Biology from Concept to Application*, pp. 135–152. Springer, Berlin.

Jiménez-Guri, E., Philippe, H., Okamura, B. and Holland, P.W.H. 2007. *Buddenbrockia* is a cnidarian worm. *Science* **317**: 116–118.

Kent, M.L., Andree, K.B., Bartholomew, J.L., *et al.* 2001. Recent advances in our knowledge of the Myxozoa. *J. Eukaryot. Microbiol.* **48**: 395–413.

Lesh-Laurie, G.E. and Suchy, P.E. 1991. Cnidaria: Scyphozoa and Cubozoa. In F.W. Harrison (ed.): *Microscopic Anatomy*

of *Invertebrates*, vol. 2, pp. 185–266. Wiley-Liss, New York.

Lichtneckert, R. and Reichert, H. 2007. Origin and evolution of the first nervous system. In J.H. Kaas (ed.): *Evolution of Nervous Systems. A Comprehensive Reference*, vol. 1, pp. 289–315. Academic Press, Amsterdam.

Lom, J. 1990. Phylum Myxozoa. In L. Margulis (ed.): *Handbook of Protoctista*, pp. 36–52. Jones and Bartlett Publishers, Boston.

Mackie, G.O., Nielsen, C. and Singla, C.L. 1989. The tentacle cilia of *Aglantha digitale* (Hydrozoa: Trachylina) and their control. *Acta Zool. (Stockh.)* **70**: 133–141.

Marlow, H.Q., Srivastava, M., Matus, D.Q., Rokhsar, D. and Martindale, M.Q. 2009. Anatomy and development of the nervous system of *Nematostella vectensis*, an anthozoan cnidarian. *Dev. Neurobiol.* **69**: 235–254.

Nakanishi, N., Yuan, D., Jacobs, D.K. and Hartenstein, V. 2008. Early development, pattern, and reorganization of the planula nervous system in *Aurelia* (Cnidaria, Scyphozoa). *Dev. Genes Evol.* **218**: 511–524.

Nielsen, C. 1984. Notes on a *Zoanthina*-larva (Cnidaria) from Phuket, Thailand. *Vidensk. Medd. Dan. Naturhist. Foren.* **145**: 53–60.

Nielsen, C. 1998. Origin and evolution of animal life cycles. *Biol. Rev.* **73**: 125–155.

Özbek, S., Balasubramanian, P.G. and Holstein, T.W. 2009. Cnidocyst structure and the biomechanics of discharge. *Toxicon* **54**: 1038–1045.

Petersen, K.W. 1990. Evolution and taxonomy in capitate hydroids and medusae (Cnidaria: Hydrozoa). *Zool. J. Linn. Soc.* **100**: 101–231.

Phelan, P. 2005. Innexins: members of an evolutionarily conserved family of gap-junction proteins. *Biochim. Biophys. Acta* **1711**: 225–245.

Pick, K., Philippe, H., Schreiber, F., *et al.* 2010. Improved phylogenomic taxon sampling noticeably affects non-bilaterian relationships. *Mol. Biol. Evol.* **27**: 1983–1987.

Putnam, N.H., Srivastava, M., Hellsten, U., *et al.* 2007. Sea anemone genome reveals ancestral eumetazoan gene repertoire and genomic organization. *Science* **317**: 86–94.

Raikova, E.V. 1994. Life cycle, cytology, and morphology of *Polypodium hydriforme*, a coelenterate parasite of the eggs of the acipenseriform fishes. *J. Parasitol.* **80**: 1–22.

Schlegel, M., Lom, J., Stechmann, A., *et al.* 1996. Phylogenetic analysis of complete small subunit ribosomal RNA coding region of *Myxidium lieberkuehni*: evidence that Myxozoa are Metazoa and related to the Bilateria. *Arch. Protistenkd.* **147**: 1–9.

Seipel, K. and Schmid, V. 2005. Evolution of striated muscle: jellyfish and the origin of triploblasty. *Dev. Biol.* **282**: 14–26.

Siddall, M.E. and Whiting, M.F. 1999. Long-branch abstractions. *Cladistics* **15**: 9–24.

Siddall, M.E., Martin, D.S., Bridge, D., Desser, S.S. and Cone, D.K. 1995. The demise of a phylum of protists: phylogeny of Myxozoa and other parasitic Cnidaria. *J. Parasitol.* **81**: 961–967.

Sun, W.-G. and Hou, X.-g. 1987. Early Cambrian medusae from Chengjiang, Yunnan, China. *Acta Palaeontol. Sin.* **26**: 257–271.

Tardent, P. 1978. Coelenterata, Cnidaria. In F. Seidel (ed.): *Morphogenese der Tiere, Deskriptive Morphogenese, 1. Lieferung*, pp. 69–415. VEB Gustav Fischer, Jena.

Tardent, P. 1995. The cnidarian cnidocyte, a high-tech cellular weaponry. *BioEssays* **17**: 351–362.

Tardent, P. and Holstein, T. 1982. Morphology and morphodynamics of the stenothele nematocyst of *Hydra attenuata* Pall. (Hydrozoa, Cnidaria). *Cell Tissue Res.* **224**: 269–290.

Thomas, M.B. and Edwards, N.C. 1991. Cnidaria: Hydrozoa. In F.W. Harrison (ed.): *Microscopic Anatomy of Invertebrates*, vol. 2, pp. 91–183. Wiley-Liss, New York.

Tiemann, H. and Jarms, G. 2010. Organ-like gonads, complex oocyte formation, and long-term spawning in *Periphylla periphylla* (Cnidaria, Scyphozoa, Coronatae). *Mar. Biol.* **157**: 527–535.

Trembley, A. 1744. *Mémoires, pour servir à l'histoire d'un genre de polypes d'eau douce, à bras en forme de cornes*. Verbeek, Leiden.

Watanabe, H., Fujisawa, T. and Holstein, T.W. 2009. Cnidarians and the evolutionary origin of the nervous system. *Dev. Growth Differ.* **51**: 167–183.

Werner, B. 1973. New investigations on systematics and evolution of the class Scyphozoa and the phylum Cnidaria. *Publ. Seto Mar. Biol. Lab.* **20**: 35–61.

Werner, B. 1984. Stamm Cnidaria, Nesseltiere. In H.E. Gruner (ed.): *A. Kaestner's Lehrbuch der Speziellen Zoologie*, 4th ed., 1. Band, 2. Teil, pp. 11–305. Gustav Fischer, Stuttgart.

Westfall, J.A. 1996. Ultrastructure of synapses in the first-evolved nervous systems. *J. Neurocytol.* **25**: 735–746.

Yuan, D., Nakanishi, N., Jacobs, D.K. and Hartenstein, V. 2008. Embryonic development and metamorphosis of the scyphozoan *Aurelia*. *Dev. Genes Evol.* **218**: 525–539.

Zhao, Y. and Bengtson, S. 1999. Embryonic and post-embryonic development of the early Cambrain cnidarian *Olivooides*. *Lethaia* **32**: 181–195.

Zrzavý, J., Mihulka, S., Kepka, P., Bezděk, A. and Tietz, D. 1998. Phylogeny of the Metazoa based on morphological and 18S ribosomal DNA evidence. *Cladistics* **14**: 249–285.

14

TRIPLOBLASTICA

Triploblasts are built of three germ layers: the primary ectoderm and endoderm, and the secondary mesoderm. The mesoderm is surrounded by the basal membranes of ectoderm and endoderm. The term 'Diploblastica' has been used for its 'sister taxon' with various combinations of non-bilaterian groups. The group was called Coelenterata by Hatschek (1888), and this concept has survived in various combinations. It has been recovered in a few molecular analyses (Dellaporta *et al.* 2006; Wang and Lavrov 2007; Ruiz-Trillo *et al.* 2008; Wang and Lavrov 2008; studies based on mitochondrial genomes that quite often give unexpected trees, and Schierwater *et al.* 2009; study with low support). Morphology does not show any synapomorphies of this group, all the 'defining characters' being plesiomorphies; it will not be discussed further.

Mesoderm is often organized as epithelia, covering, for example, muscles and other mesodermal organs or lining cavities, such as coeloms, or in a few cases as compact, apolar, mesenchymatous tissue (Rieger 1986). Epithelia show a pronounced polarity with a basal membrane at the basal surface, septate/tight cell junctions, and Golgi apparatus in the apical part of the cell and in some cases one or more cilia at the apical pole. Mesodermal tissues are easily distinguished from the rather isolated cells in the mesogloea of the cnidarians (Chapter 13). Well-defined mesoderm is found in ctenophores (Chapter 15).

Mesoderm originates through ingression or invagination from ectoderm or endoderm, or from the blastopore lips where the two primary germ layers are in contact. Four main types of mesoderm formation can be recognized: (1) ingression of ectodermal cells; (2) ingression of one or a few cells at the blastopore rim; (3) compact, hollow pockets, or egressions from the endoderm; and (4) neural crest cells of the vertebrates (see Chapter 65).

Mesoderm formed through the first-mentioned process is called ectomesoderm, and has been reported from many of the spiralian phyla, where it usually originates from the a-c cells of the 2nd and 3rd micromere quartet (Boyer *et al.* 1996; Hejnol *et al.* 2007). Mesoderm from the d cells of these quartets has been reported in old studies of echiurans and nemertines, but Henry and Martindale (1998) did not find ectomesoderm originating from these cells in the nemertine *Cerebratulus*. Ectomesoderm is usually not reported from ecdysozoans, but as the cleavage of several nematodes has now been described in great detail, it is possible that some of the cells that form muscles could be classified as ectomesoderm (Chapter 49). There are only a few reports of ectomesoderm in the deuterostomes (Salvini-Plawen and Spelchtna 1979), and most of the reports of normal ectodermal origin of parts of coeloms are connected with coelomoducts. Organs that are normally formed by mesoderm and endoderm may develop from the ectoderm under regeneration or budding; these special cases are discussed under the respective phyla.

Mesoderm formed from the blastoporal lips, or from one cell located at the posterior side of the blastopore (the 4d cell in the spiral cleavage), is found in most spiralians (Chapter 23), but it has not been observed in deuterostomes. This type of mesoderm mixes freely with ectomesoderm so that the origin of various mesodermal structures from these two sources can only be distinguished by cell-lineage studies.

Mesoderm formed from various parts of the archenteron is characteristic of deuterostomes, but also the chaetognaths (which are now classified as protostomes) have mesoderm originating from the archenteron (Chapter 55).

It is obvious that the cells that give rise to mesoderm have different origin in different phyla, and nothing indicates that the mesoderm of all phyla is homologous. Endoderm is formed in many different ways in cnidarians (Fig. 13.2), so the different types of mesoderm formation may just represent specializations of an unspecified ancestral ability to proliferate cells from the primary germ layers.

Cnidarians have epitheliomuscular cells, i.e. the muscle cells either have a small part that is situated in the epithelium, or they lie at the base of the epithelium above the basement membrane. Ctenophores and bilaterians have true myocytes that are situated between the basal membranes of ectoderm and endoderm; epitheliomuscular cells are only found in certain organs. The mesoderm of the ctenophores is discussed in Chapter 15. A long series of genes are known to be involved in endodermal specification in bilaterians, and some of these genes have been identified in the endodermal tissue of the sea anemone *Nematostella*, indicating that the ancestral origin of mesoderm is from the endoderm (Martindale *et al.* 2004).

Other apomorphies of the Triploblastica are more difficult to point out.

The group was recognized already by Ax (1995), who used the name Acrosomata for this group, but an acrosome is now known to be present in the homoscleromorphs (Chapter 9).

Molecular phylogenies usually place ctenophores at or near the base of the metazoan tree, but their position as triploblasts seems strongly supported by morphology. The molecular analyses are discussed in Chapter 15.

The adaptational value of the triploblast organization seems obvious: many different types of organs can grow and differentiate when their tissues are isolated from the primary germ layers, and this permits evolution of larger and more complex body plans.

References

Ax, P. 1995. *Das System der Metazoa I*. Gustav Fischer, Stuttgart.

Boyer, B.C., Henry, J.Q. and Martindale, M.Q. 1996. Dual origins of mesoderm in a basal spiralian: cell lineage analyses in the polyclad turbellarian *Hoploplana inquilina*. *Dev. Biol.* **179**: 329–338.

Dellaporta, S.L., Xu, A., Sagasser, S., *et al.* 2006. Mitochondrial genome of *Trichoplax adhaerens* supports Placozoa as the basal lower metazoan phylum. *Proc. Natl. Acad. Sci. USA* **103**: 8751–8756.

Hatschek, B. 1888. *Lehrbuch der Zoologie, 1. Lieferung*, pp. 1–144. Gustav Fischer, Jena.

Hejnol, A., Martindale, M.Q. and Henry, J.Q. 2007. High-resolution fate map of the snail *Crepidula fornicata*: the origins of ciliary bands, nervous system, and muscular elements. *Dev. Biol.* **305**: 63–76.

Henry, J.J. and Martindale, M.Q. 1998. Conservation of the spiralian developmental program: cell lineage of the nemertean, *Cerebratulus lacteus*. *Dev. Biol.* **201**: 253–269.

Martindale, M.Q., Pang, K. and Finnerty, J.R. 2004. Investigating the origins of triploblasty: 'mesodermal' gene expression in a diploblastic animal, the sea anemone *Nematostella vectensis* (Phylum, Cnidaria; class, Anthozoa). *Development* **131**: 2463–2472.

Rieger, R.M. 1986. Über den Ursprung der Bilateria: die Bedeutung der Ultrastrukturforschung für ein neues Verstehen der Metazoenevolution. *Verh. Dtsch. Zool. Ges.* **79**: 31–50.

Ruiz-Trillo, I., Roger, A.J., Burger, G., Gray, M.W. and Lang, B.F. 2008. A phylogenomic investigation into the origin of Metazoa. *Mol. Biol. Evol.* **25**: 664–672.

Salvini-Plawen, L.v. and Spelchtna, H. 1979. Zur Homologie der Keimblätter. *Z. Zool. Syst. Evolutionsforsch.* **17**: 10–30.

Schierwater, B., Eitel, M., Jakob, W., *et al.* 2009. Concatenated analysis sheds light on early metazoan evolution and fuels a modern 'urmetazoan' hypothesis. *PLoS Biol.* **7(1)**: e1000020.

Wang, X. and Lavrov, D.V. 2007. Mitochondrial genome of the homoscleromorph *Oscarella carmela* (Porifera, Demospongiae) reveals unexpected complexity in the common ancestor of sponges and other animals. *Mol. Biol. Evol.* **24**: 363–373.

Wang, X. and Lavrov, D.V. 2008. Seventeen new complete mtDNA sequences reveal extensive mitochondrial genome evolution within the Demospongiae. *PLoS ONE* **3(7)**: e2723.

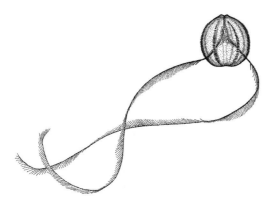

15

Phylum Ctenophora

Ctenophora, or comb jellies, is a small phylum of about 150 living, marine species. Most species are holopelagic, transparent, and sometimes too fragile to be collected with any of the more conventional methods. The few creeping, benthic species are more compact, and the sessile *Tjalfiella* is almost leathery. *Lampetia* has a rather undifferentiated larval stage (called *Gastrodes*) that parasitizes salps (Mortensen 1912; Komai 1922). Surprisingly well-preserved specimens have been reported from Cambrian deposits from Chengjiang (Chen and Zhou 1997). Recent reviews of the phylum based on morphological and 18S sequence studies (Harbison 1985; Podar *et al.* 2001) give no clear picture, but it seems that the atentaculate *Beroe* is not the sister group of the remaining families as believed previously. The many new types that are currently being discovered, for example during bluewater diving, indicate that our knowledge of the phylum is still quite incomplete. Considerable morphological variation is found within the phylum, but, nevertheless, it is very well delimited.

The ctenophores have the structure of a gastrula, with the blastopore remaining as the mouth–anus, and the archenteron as the sac-shaped adult gut, but the presence of a mesodermal (mesogloeal and muscular) layer between the ectoderm and endoderm indicates the higher level of organization. The apical–blastoporal axis is retained throughout life as the main

axis. A pair of tentacles and the main axis define the tentacular plane, and the perpendicular plane is called oral (or sagittal) because the mouth and stomodaeum are flattened in this plane. This type of symmetry is called biradial. The almost spherical *Pleurobrachia* appears to be close to the ancestral type, with retractable tentacles with tentillae with colloblasts. Most of the more 'advanced' types, such as *Mnemiopsis*, go through a juvenile stage of a similar type, called the cydippid stage, but the later stages lose the tentacles and develop lobes at the oral side; these lobes are used in feeding. Also the creeping and sessile forms have cydippid juveniles, but *Beroe* lacks tentacles in all stages. The microscopical anatomy was reviewed by Hernandez-Nicaise (1991).

The body is spherical in the supposedly primitive, cydippid forms, but various parts of the body may be expanded into folds or lappets, or the whole body may be band-shaped with extreme flattening in the tentacular plane (*Cestus*). The benthic forms creep or are attached with the oral side (Harbison and Madin 1982). The cylindrical tentacles can usually be retracted into tentacle sheaths and have specialized side branches (tentillae) in many species (Mackie *et al.* 1988). The adhesive colloblasts are specialized ectodermal cells of the tentacles (see below). Eight meridional rows of comb plates, which are very large compound cilia, are the main locomotory organs in most pelagic forms,

Chapter vignette: *Pleurobrachia pileus.* (Based on Brusca and Brusca 1990.)

and can also be recognized in developmental stages of the benthic species, which lack the comb plates in the adult phase. Some pelagic forms have short, apically located comb rows, and the expanded oral lobes or the tentacles appear to be more important in locomotion. The benthic forms creep by means of cilia on the expanded pharyngeal or oral epithelium.

The ectoderm is monolayered in the early developmental stages, but adults have both an external epithelium, with ciliated cells, glandular cells and 'supporting' cells, and nerve cells and ribbon-shaped, smooth parietal muscles at the base of the ectoderm of the body and the pharynx. The epithelial cells are joined by spot desmosomes, zonula adherens, and series of punctate contacts, resembling the vertebrate zonula occludens (Tyler 2003). The ectoderm is underlain by a conspicuous basal membrane. The ciliated cells are multiciliate (except in some sense organs) and several specialized types can be recognized. The comb plates consist of many aligned cilia from several cells and show an orthoplectic beat pattern; their structure is unique, with compartmentalizing lamellae between the lateral doublets (numbers 3 and 8) of the axoneme and the cell membrane. The comb plates are used in swimming, which is normally with the apical pole in front, i.e. the effective stroke is towards the oral pole, but their beat can be reversed locally, or on the whole animal so that oriented swimming, for example associated with feeding, is possible (Tamm and Moss 1985). Unique macrocilia with several hundred axonemes are found at the mouth of beroids. The cells at the base of the apical organ and the polar fields, and ciliated furrows, that extend from the apical organ, have separate cilia.

The apical organ is a statocyst with four compound cilia, called balancers, carrying a compound statolith. The balancer cells are monociliate, and the individual otoliths are formed as specialized cells from a region adjacent to the balancer cells. The whole structure is enclosed in a dome-shaped cap consisting of cilia from cells at the periphery of the organ. The organ protrudes from the apical pole in most species, but is situated in an invagination in *Coeloplana* (Abbott 1907).

The very characteristic colloblasts (Fig. 15.1) are formed continuously from undifferentiated ectodermal cells of the basal growth zone of the tentacles. Fully-grown colloblasts have a very characteristic structure, with a spirally coiled thread around the stalk, and a head with numerous small peripheral granules with a mucous substance that is released by contact with a prey (Franc 1978). They bear no resemblance to cnidarian nematocysts, which are intracellular organelles (compare with Fig. 13.1).

The tentacles of *Haeckelia* lack colloblasts, but contain nematocysts (cleptocnidia), which originate from ingested medusae (Mills and Miller 1984). They are enclosed in a vacuole in an innervated cell that lacks the cnidocil and the other structures associated with the nematocyst when it is in the normal position in a cnidarian nematocyte (Carré and Carré 1989). The

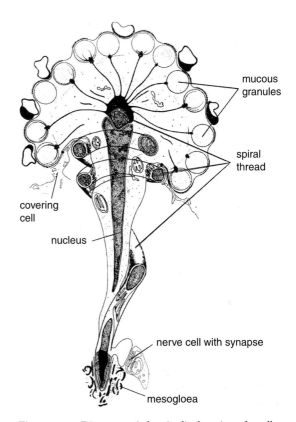

mucous granules

spiral thread

covering cell

nucleus

nerve cell with synapse

mesogloea

Figure 15.1. Diagrammatic longitudinal section of a colloblast. (Modified from Franc 1978.)

tentacles also have many cells of a type called pseudo-colloblasts, which are used in prey capture before the cleptocnidia have become functional, but their homology with colloblasts seems uncertain.

The nervous system consists of a nerve net at the base of the ectoderm, with conspicuous concentrations along the comb rows and at the mouth opening, and one at the base of the endoderm; nerve nets occur also in the mesoderm, especially concentrated in the tentacles. The synapses have a unique structure, with a 'presynaptic triad', and a thickened postsynaptic membrane. Synapses are both chemical, with acetylcholine and FRMFamide, and gap junctions. There is a large concentration of nerve cells in the epithelium below the apical organ, and this concentration serves many of the functions of a brain.

The gut or gastrovascular cavity is a complicated system of branched canals with eight major, meridional canals along the comb-plate rows. The pharynx should probably be interpreted as a stomodaeum, as indicated by its origin from the apical micromeres (see below), and by its innervation and parietal musculature. A narrow apical extension of the gut reaches to the underside of the apical organ, where it forms a pair of Y-shaped canals in the oral plane. One branch on each side ends in a small ampulla while the other forms a small pore to the outside; these pores may function as anal openings (Main 1928), but the undigested remains of the prey are usually egested through the mouth (Bumann and Puls 1997). Prey may be captured with the tentacles or the large lobes, but smaller organisms may be wafted to the mouth by ciliary currents; the beroids capture other gelatinous prey by use of the macrocilia.

The walls of peripheral parts of the gastrovascular system show 'ciliated rosettes', consisting of a double ring of endodermal cells surrounding a pore, that can be constricted by the ring of cells in plane with the gut wall. This ring of cells has a conical tuft of cilia protruding into the gut cavity. A similar ring of ciliated cells protrudes into the mesogloea. Experiments indicate that the rosettes can transport water between the gut and mesogloea, but their function remains unknown.

The mesoderm, or mesogloea, is a hyaline, gelatinous extracellular matrix with muscle cells, nerve cells, and mesenchyme cells; epithelia are not formed. The matrix contains a meshwork of fibrils, which are banded like collagen in certain areas, especially in the tentillae, but it is now assumed that collagen is a major component of the whole meshwork. The matrix appears to be secreted mainly by ectodermal cells, but some smooth muscle cells of the mesoderm also secrete collagen. The muscle cells in the body are very large, branched, and smooth. The smooth, longitudinal muscle cells of the tentacles are arranged around a core of matrix with nerves (Fig. 15.2). The tentillae of *Euplokamis* contain striated muscle 'cells', which lack a nucleus. The ultrastructure of their sarcomeres indicates that they contract but do not relax, and this indicates that they only contract once, and that they cannot be considered homologous of the eubilaterian striated muscle cells (Mackie *et al.* 1988; Burton 2007). The mesenchymal cells are of two types, but their functions are unknown.

The gonads differentiate from the endoderm of the eight meridional gastrovascular canals. The developing oocyte is connected with three clusters of nurse cells through intercellular bridges, and the polar bodies are given off in a constant relation to these bridges (Martindale and Henry 1997a). The sperm has an acrosome (Franc 1973). The gametes are shed through pores in the epidermis above the gonads (Pianka 1974).

Fertilization takes place at spawning. As a unique feature of ctenophores, as many as 20 spermatozoa enter the egg (in *Beroe*; Sardet *et al.* 1990). The female pronucleus moves through the cytoplasm and 'selects' one male pronucleus for syngamy. The apical blastoporal axis is apparently not fixed in the oocyte, but the position of the 'selected' male pronucleus appears to determine the position of the blastoporal pole; the polar bodies are often but not always situated in this region.

Development is highly determined (Martindale 1986; Martindale and Henry 1997a,b; Henry and Martindale 2004); isolated blastomeres of the 2-cell stage develop into half-larvae, and blastomeres of the

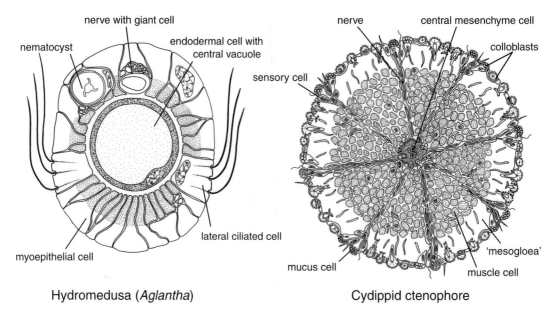

Figure 15.2. Cross sections of a tentacle of the trachyline hydromedusa *Aglantha digitale* (redrawn from Mackie *et al.* 1989) and a cydippid ctenophore (based on Hernandez-Nicaise 1973).

Table 15.1. Cell lineage of one quadrant of the embryo of the ctenophore *Mnemiopsis leidyi*. (Based on Martindale and Henry 1997a.) Capital letters: macromeres, lower-case letters: apical micromeres. The 2M macromeres divide once more before giving off the oral micromeres from which the mesodermal elements develop.

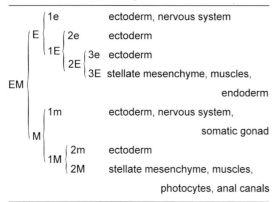

4-cell stage each develop into one quadrant of an animal. The first cleavage is in the oral plane and the second in the tentacular plane. The first cleavage begins at the blastoporal pole, and the cleavage furrow cuts towards the apical pole, so that a peculiar stage with the two blastomeres connected by an apical 'handle' is

formed before the blastomeres finally separate; similar shapes are seen in the following two cleavages (Freeman 1977). The third cleavage is in the oral plane, but slightly shifted towards a radial pattern; the embryo now consists of four median (M) cells and four external (E) cells, with highly determined fates (Table 15.1). At the fourth cleavage each large cell gives off a small cell at the apical pole, and this is repeated once in the M cells and twice in the E cells. During the following cleavages, the micromeres first form an oval on top of the eight macromeres; later on the oval develops into a sheet of cells covering the apical side of the macromeres (Fig. 15.3). The macromeres then produce some micromeres at the blastoporal pole. The embryo goes through an embolic gastrulation through the spreading of the apical micromeres over the macromeres to the blastoporal pole, where an invagination forms so that the blastoporal micromeres become situated at the bottom of an archenteron, with the lower part of the invagination (the pharynx-stomodaeum) covered by apical micromeres (Fig. 15.3). *Brachyury* is expressed in the stomodaeum, as in several deuterostomes (Yamada *et al.* 2010) and cnidarians

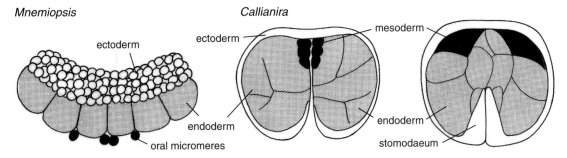

Figure 15.3. Ctenophore embryology: 128-cell stage (the 4E cells have not divided yet) of *Mnemiopsis leidyi* (based on Freeman and Reynolds 1973), and early and late gastrula of *Callianira bialata* (redrawn from Metschnikoff 1885).

(Fritzenwanker *et al.* 2004). The apical micromeres remain in their respective quadrants, whereas the macromeres and blastoporal micromeres mix freely. The apical micromeres form the ectoderm, including the apical organ, comb rows, tentacle epidermis with colloblasts, pharyngeal epithelium, and nervous system; the macromeres form the endoderm. The fate of the oral micromeres has not been followed directly, but has been inferred from experiments with markings of the remaining macromeres; they form stellate mesenchyme, muscles, and photocytes. The four quadrants are almost identical, except for the four M cells, two of which give rise to the anal canals.

The tentacles develop just after gastrulation as ectodermal thickenings, which invaginate to form the tentacle sheaths; the tentacles sprout from the bottom of these invaginations. Metschnikoff (1885) reported similar epithelial thickenings in *Beroe*, which lacks tentacles, but this should be reinvestigated.

Dissogony, i.e. sexual maturity, both in early larval and adult stages separated by a period with reduced gonads, has been observed, for example, in *Pleurobrachia* and *Mnemiopsis* (Remane 1956; Martindale 1987) (see Fig. 10.2).

The monophyly of the Ctenophora can hardly be questioned. The comb plates and the apical organ are unique, and further apomorphies can be pointed out, for example the very unusual oogenesis and embryology, and the naked extracellular bundles of tubulin structures resembling ciliary axonemes, found in grooves along the smooth muscle cells

(Tamm and Tamm 1991). The ancestral ctenophore probably resembled a cydippid, with tentacles with colloblasts.

A sister-group relationship between cnidarians and ctenophores is indicated by the unusual shape of the first embryonic cleavages and the position of the polar bodies. Cnidarians have remained at the gastrula stage with an acellular mesogloea, whereas ctenophores have an extracellular matrix with mesodermal cells, and mesodermal muscles in the tentacles developing from micromeres of the invaginated blastoporal area of the early embryo (Fig. 15.2). The latter character links the ctenophores to the bilaterians. The intracellular nematocysts of cnidarians are completely different from the colloblasts of ctenophores (compare Figs. 13.1 and 15.1).

One *Hox1* gene was found by Finnerty *et al.* (1996), but no Hox genes were found in the study of Pang and Martindale (2008).

Older studies of molecular phylogeny using ribosomal RNA have shown ctenophores in many different positions (review in Nielsen 2008), and even the newest publications disagree strongly about the position of the group. Many studies do not include the ctenophores. The expressed sequence tag study of Dunn *et al.* (2008) and the phylogenomic study of Hejnol *et al.* (2009) showed them as sister group to the remaining metazoans. The analyses of nearly complete mitochondrial RNA by Mallatt *et al.* (2010) placed them as sister group of the Calcarea. Analyses of nuclear-encoded proteins and mitochondrial genes

(Philippe *et al.* 2009; Schierwater *et al.* 2009) showed them as the sister group of the Cnidaria (in the traditional 'Coelenterata'). An extensive phylogenomic study placed the Ctenophora as sister group of Cnidaria + Placozoa + Bilateria (Pick *et al.* 2010). However, to a traditional morphologist it seems obvious that the Ctenophora belong to the Neuralia, as the sister group of the Bilateria. Any position further 'down' on the tree would require either massive losses of complex characters in 'sponges' and/or cnidarians, such as nervous system, mesoderm, and epithelia, or massive homoplasies in ctenophores, cnidarians, and some bilaterians. It can only be hoped that future studies, perhaps of whole genomes, can produce more acceptable results.

Interesting subjects for future studies

1. Development of the nervous system
2. A whole genome

References

Abbott, J.F. 1907. The morphology of *Coeloplana. Zool. Jahrb., Anat.* **24**: 41–70.

Brusca, R.C. and Brusca, G.J. 1990. *Invertebrates.* Sinauer Associates, Sunderland, MA.

Bumann, D. and Puls, G. 1997. The ctenophore *Mnemiopsis leidyi* has a flow-through system for digestion with three consecutive phases of extracellular digestion. *Physiol. Zool.* **70**: 1–6.

Burton, P.M. 2007. Insights from diploblasts; the evolution of mesoderm and muscle. *J. Exp. Zool. (Mol. Dev. Evol.)* **310B**: 5–14.

Carré, D. and Carré, C. 1989. Acquisition de cnidocystes et différenciation de pseudocolloblastes chez les larves et les adultes de deux cténophores du genre *Haeckelia* Carus, 1863. *Can. J. Zool.* **67**: 2169–2179.

Chen, J. and Zhou, G. 1997. Biology of the Chengjiang fauna. *Bull. Natl. Mus. Nat. Sci.* **10**: 11–105.

Dunn, C.W., Hejnol, A., Matus, D.Q., *et al.* 2008. Broad phylogenomic sampling improves resolution of the animal tree of life. *Nature* **452**: 745–749.

Finnerty, J.R., Master, V.A., Irvine, S., *et al.* 1996. Homeobox genes in the Ctenophora: identification of paired-type and Hox homologues in the atentaculate ctenophore, *Beroë ovata. Mol. Mar. Biol. Biotechnol.* **5**: 249–258.

Franc, J.M. 1978. Organization and function of ctenophore colloblasts: an ultrastructural study. *Biol. Bull.* **155**: 527–541.

Franc, S. 1973. Etude ultrastructurale de la spermatogénèse du Cténaire *Beroe ovata. J. Ultrastruct. Res.* **42**: 255–267.

Freeman, G. 1977. The establishment of the oral-blastoporal axis in the ctenophore embryo. *J. Embryol. Exp. Morphol.* **42**: 237–260.

Freeman, G. and Reynolds, G.T. 1973. The development of bioluminescence in the ctenophore *Mnemiopsis leidyi. Dev. Biol.* **31**: 61–100.

Fritzenwanker, J.H., Saina, M. and Technau, U. 2004. Analysis of *forkhead* and *snail* expression reveals epithelial-mesenchymal transitions during embryonic and larval development of *Nematostella vectensis. Dev. Biol.* **275**: 389–402.

Harbison, G.R. 1985. On the classification and evolution of the Ctenophora. In S. Conway Morris (ed.): *The Origins and Relationships of Lower Invertebrates*, pp. 78–100. Oxford Univ. Press, Oxford.

Harbison, G.R. and Madin, L.P. 1982. Ctenophora. In S.P. Parker (ed.): *Synopsis and Classification of Living Organisms*, pp. 707–715. McGraw-Hill, New York.

Hejnol, A., Obst, M., Stamatakis, A., *et al.* 2009. Assessing the root of bilaterian animals with scalable phylogenomic methods. *Proc. R. Soc. Lond. B* **276**: 4261–4270.

Henry, J.Q. and Martindale, M.Q. 2004. Inductive interactions and embryonic equivalence groups in a basal metazoan, the ctenophore *Mnemiopsis leidyi. Evol. Dev.* **6**: 17–24.

Hernandez-Nicaise, M.L. 1973. Le systéme nerveux des Cténaires. I. Structure et ultrastructure des réseaux épithéliaux. *Z. Zellforsch.* **143**: 117–133.

Hernandez-Nicaise, M.-L. 1991. Ctenophora. In F.W. Harrison (ed.): *Microscopic Anatomy of Invertebrates*, vol. 2, pp. 359–418. Wiley-Liss, New York.

Komai, T. 1922. *Studies on two aberrant Ctenophores Coeloplana and Gastrodes.* Published by the Author, Kyoto.

Mackie, G.O., Mills, C.E. and Singla, C.L. 1988. Structure and function of the prehensile tentilla of *Euplokamis* (Ctenophora, Cydippida). *Zoomorphology* **107**: 319–337.

Mackie, G.O., Nielsen, C. and Singla, C.L. 1989. The tentacle cilia of *Aglantha digitale* (Hydrozoa: Trachylina) and their control. *Acta Zool. (Stockh.)* **70**: 133–141.

Main, R.J. 1928. Observations of the feeding mechanism of a ctenophore, *Mnemiopsis leidyi. Biol. Bull.* **55**: 69–78.

Mallatt, J., Craig, C.W. and Yoder, M.J. 2010. Nearly complete rRNA genes assembled from across the metazoan animals: Effects of more taxa, a structure-based alignment, and paired-sites evolutionary models on phylogeny reconstruction. *Mol. Phylogenet. Evol.* **55**: 1–17.

Martindale, M.Q. 1986. The ontogeny and maintenance of adult symmetry properties in the ctenophore, *Mnemiopsis mccradyi. Dev. Biol.* **118**: 556–576.

Martindale, M.Q. 1987. Larval reproduction in the ctenophore *Mnemiopsis mccradyi* (order Lobata). *Mar. Biol.* **94**: 409–414.

Martindale, M.Q. and Henry, J. 1997a. Ctenophorans, the comb jellies. In S.F. Gilbert and A.M. Raunio (eds): *Embryology. Constructing the Organism*, pp. 87–111. Sinauer Associates, Sunderland, MA.

Martindale, M.Q. and Henry, J.Q. 1997b. Reassessing embryogenesis in the Ctenophora: the inductive role of e₁ micromeres in organizing ctene row formation in the 'mosaic' embryo, *Mnemiopsis leidyi*. *Development* **124**: 1999–2006.

Metschnikoff, E. 1885. Vergleichend-embryologische Studien. *Z. Wiss. Zool.* **42**: 648–673.

Mills, C.E. and Miller, R.L. 1984. Ingestion of a medusa (*Aegina citrea*) by the nematocyst-containing ctenophore *Haeckelia rubra* (formerly *Euchlora rubra*): phylogenetic implications. *Mar. Biol.* **78**: 215–221.

Mortensen, T. 1912. Ctenophora. *The Danish Ingolf Expedition* **5A(2)**: 1–96.

Nielsen, C. 2008. Six major steps in animal evolution - are we derived sponge larvae? *Evol. Dev.* **10**: 241–257.

Pang, K. and Martindale, M.Q. 2008. Developmental expression of homeobox genes in the ctenophore *Mnemiopsis leidyi*. *Dev. Genes Evol.* **218**: 307–319.

Philippe, H., Derelle, R., Lopez, P., *et al.* 2009. Phylogenomics revives traditional views on deep animal relationships. *Curr. Biol.* **19**: 706–712.

Pianka, H.D. 1974. Ctenophora. In A.C. Giese and J.S. Pearse (eds): *Reproduction of Marine Invertebrates*, vol. 1, pp. 201–265. Academic Press, New York.

Pick, K.S., Philippe, H., Schreiber, F., *et al.* 2010. Improved phylogenomic taxon sampling noticeably affects nonbilaterian relationships. *Mol. Biol. Evol.* **27**: 1983–1987.

Podar, M., Haddock, S.H.D., Sogin, M.L. and Harbison, G.R. 2001. A molecular phylogenetic framework for the phylum Ctenophora using 18S rRNA genes. *Mol. Phylogenet. Evol.* **21**: 218–230.

Remane, A. 1956. Zur Biologie des Jugendstadiums der Ctenophore *Pleurobrachia pileus* O. Müller. *Kieler Meeresforsch.* **12**: 72–75.

Sardet, C., Carr, D. and Rouvière, C. 1990. Reproduction and development in ctenophores. *NATO ASI Ser., Life Sci.* **195**: 83–94.

Schierwater, B., Eitel, M., Jakob, W., *et al.* 2009. Concatenated analysis sheds light on early metazoan evolution and fuels a modern 'urmetazoan' hypothesis. *PLoS Biol.* **7(1)**: e1000020.

Tamm, S. and Tamm, S.L. 1991. Extracellular ciliary axonemes associated with the surface of smooth muscle cells of ctenophores. *J. Cell Sci.* **94**: 713–724.

Tamm, S.L. and Moss, A.G. 1985. Unilateral ciliary reversal and motor responses during prey capture by the ctenophore *Pleurobrachia*. *J. Exp. Biol.* **114**: 443–461.

Tyler, S. 2003. Epithelium – the primary building block for metazoan complexity. *Integr. Comp. Biol.* **43**: 55–63.

Yamada, A., Martindale, M.Q., Fukui, A. and Tochinai, S. 2010. Highly conserved functions of the Brachyury gene on morphogenetic movements: insights from the early-diverging phylum Ctenophora. *Dev. Biol.* **339**: 212–222.

16

BILATERIA

For more than a century it has been customary to contrast two main groups within the Eumetazoa, viz. Coelenterata and Bilateria (Hatschek 1888). The first-mentioned group has previously been rejected as non-monophyletic (Chapters 8 and 10), but the monophyly of the Bilateria has been accepted in practically all recent papers, whether based on morphological, molecular, or developmental analyses (Hejnol *et al.* 2009; Paps *et al.* 2009a; Paps *et al.* 2009b), although the Acoelomorpha has only been included in more recent analyses. The most conspicuous characteristic of the group is of course the bilaterality, with an anteroposterior axis forming an angle with the primary, apical-blastoporal axis, but there are not many other morphological apomorphies.

The origin of bilaterality with the establishment of the new anteroposterior axis appears intimately linked with a change of life style, from holopelagic or sessile to creeping. It was probably the adults that took up the creeping, benthic life style, whereas the larval stages remained pelagic (see below and Fig. 10.3). Adult sponges do not retain the larval primary body axis, which is the only body axis in cnidarians and ctenophores. Bilateral symmetry is seen in anthozoans and a few hydrozoan polyps (Chapter 8), but they are sessile and none of them have an anterior pole with a brain. The anthozoan bilaterality, with a 'directive axis' defined by the siphonoglyph(s) perpendicular to the primary, apical-blastoporal axis, has been interpreted as homologous to bilaterian homology, based on similar expression of some genes along this and the bilaterian dorsal-ventral axis (Finnerty *et al.* 2004; Matus

et al. 2006). It seems clear that Hox genes are involved in the patterning along the primary (anterior-posterior) axis in *Nematostella* (Ryan *et al.* 2007), although the expression pattern is different in hydrozoans (Raible and Steinmetz 2010), and it is also clear that they are active in the patterning along the directive axis. However, newer studies show that both the expression of bone morphongenetic protein genes and Hox genes have evolved along separate lines in cnidarians and bilaterians, so 'true bilaterality' is a bilaterian apomorphy (Raible and Steinmetz 2010).

Bilateria comprises two well-separated groups, Acoelomorpha and Eubilateria (Fig. 16.1). The acoelomorphs have the structure of a bilateral gastraea, whereas the eubilaterians have a tube-shaped gut. The Hox gene cluster of the acoelomorphs is short, significantly lacking the *Hox3* gene, whereas the eubilaterians have a long cluster, which indicates that the platyhelminthes that lack an anus, are derived from an ancestor with an anus (Fig. 21.3). A recent study that placed the acoelomorphs as the sister group of the ambulacrarians (Philippe *et al.* 2011) implies massive gene losses for which there is no evidence.

References

Finnerty, J.R., Pang, K., Burton, P., Paulson, D. and Martindale, M.Q. 2004. Origins of bilateral symmetry: *Hox* and *Dpp* expression in a sea anemone. *Science* **304**: 1335–1337.

Hatschek, B. 1888. *Lehrbuch der Zoologie, 1. Lieferung* (pp 1–144). Gustav Fischer, Jena.

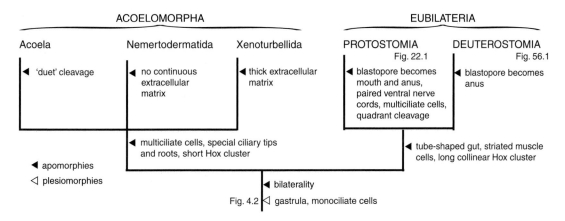

Figure 16.1. Phylogeny of the Bilateria.

Hejnol, A., Obst, M., Stamatakis, A., *et al.* 2009. Assessing the root of bilaterian animals with scalable phylogenomic methods. *Proc. R. Soc. Lond. B* **276**: 4261–4270.

Matus, D.Q., Pang, K., Marlow, H., *et al.* 2006. Molecular evidence for deep evolutionary roots of bilaterality in animal development. *Proc. Natl. Acad. Sci. USA* **103**: 11195–11200.

Paps, J., Baguñá, J. and Riutort, M. 2009a. Lophotrochozoa internal phylogeny: new insights from an up-to-date analysis of nuclear ribosomal genes. *Proc. R. Soc. Lond. B* **276**: 1245–1254.

Paps, J., Baguñà, J. and Riutort, M. 2009b. Bilaterian phylogeny: A broad sampling of 13 nuclear genes provides a new Lophotrochozoa phylogeny and supports a paraphyletic basal Acoelomorpha. *Mol. Biol. Evol.* **26**: 2397–2406.

Philippe, H., Brinkmann, H., Copley, R.R., *et al.* 2011. Acoelomorph flatworms are deuterostomes related to *Xenoturbella*. *Nature* **470**: 255–258.

Raible, F. and Steinmetz, P.R.H. 2010. Metazoan Complexity. In J.M. Cock, K. Tessmar-Raible, C. Boyen and F. Viard (eds): *Introduction to Marine Genomics*, pp. 143–178. Springer, Dordrecht.

Ryan, J.F., Mazza, M.E., Pang, K., *et al.* 2007. Pre-bilaterian origins of the Hox cluster and the Hox code: evidence from the sea anemone, *Nematostella vectensis*. *PLoS ONE* **2(1)**: e153.

17

ACOELOMORPHA

Acoels, nemertodermatids, and *Xenoturbella* have been regarded as closely related 'primitive turbellarians' already from the first description of *Xenoturbella* (Westblad 1950), and several subsequent studies have described morphological details supporting the monophyly of the Acoelomorpha (review in Nielsen 2010). Molecular phylogenetics has conclusively removed Acoela and Nemertodermatida from the Platyhelminthes (Philippe *et al.* 2007, 2011; Egger *et al.* 2009; Mallatt *et al.* 2010), and placed them either as sister groups in the Acoelomorpha (Ruiz-Trillo *et al.* 2002), or separately as subsequent early off-shots of the line to the Eubilateria (Jondelius *et al.* 2002; Wallberg *et al.* 2007; Hejnol and Martindale 2009; Paps *et al.* 2009a,b). *Xenoturbella* was taken on a long detour, via molluscs and deuterostomes, to end up as a group belonging to the Acoelomorpha, with support from phylogenomics and Hox genes (Fritsch *et al.* 2008; Hejnol *et al.* 2009; Nielsen 2010). The recent analysis of Philippe *et al.* (2011) places the acoelomorphs as the sister group of the ambulacrarians, whereas that of Edgecombe *et al.* (2011) supports the position favoured here.

In addition to the similar general organization of the three phyla, a small number of morphological synapomorphies can be pointed out. The epithelium of the three groups shows several specializations in the ultrastructure of the ciliary tips and similar, complicated root systems (Franzén and Afzelius 1987; Pedersen and Pedersen 1988; Rohde *et al.* 1988; Lundin 1997, 1998). An unusual feature is that cells can be withdrawn from the epithelium and digested (Lundin 2001). The rather diffuse nervous system with no brain is another similarity between the groups, but specific synapomorphies are difficult to point out. The Hox genes (see below) may show other important characters, but the nemertodermatids are poorly studied. Acoels and nemertodermatids have unique frontal organs with glandular and sensory cells (Ehlers 1992).

The embryology of the acoels has been studied in great detail, whereas that of the nemertodermatids is poorly known, and that of *Xenoturbella* is virtually unknown, so nothing can be said about the ancestral developmental type of the acoelomorphs.

Acoelomorphs can be characterized as bilateral, triploblastic gastrulae. In contrast to the cnidarians, the acoelomorphs have a pronounced anteroposterior axis different from the primary axis, often with a statocyst-like sensory organ in the anterior end. The circumoral nerve ring of the cnidarians is not found, and there is no well-defined brain. The sack-shaped gut could be interpreted as the result of a loss of the anus as that seen in the Platyhelminthes (Chapter 29). However, the acoelomorphs have genes belonging to most of the groups characteristic of the bilaterian Hox cluster, but they have few genes of the central and posterior groups and lack *Hox3*, whereas the platyhelminths, which have lost the anus secondarily, have retained an almost full cluster (Fig. 21.3); this indicates that the sack-shaped gut of the acoelomorphs is ancestral. So Hyman (1951) was actually right in pointing to the acoel organization as being ancestral of the Bilateria, although the derivation from a compact planula appears highly unlikely (Chapter 10).

The basal position of the acoelomorphs is further supported by the paucity of microRNA in the acoel *Symsagittifera* (Sempere *et al.* 2007).

The three phyla are treated here as the sister group of the Eubilateria, but it cannot be excluded that they represent subsequent side branches on the line leading to the Eubilateria.

References

Edgecombe, G.D., Giribet, G., Dunn, C.W., *et al.* 2011. Higher-level metazoan relationships: recent progress and remaining questions. *Org. Divers. Evol.* **11**: 151–172.

Egger, B., Steinke, D., Tarui, H., *et al.* 2009. To be or not to be a flatworm: the acoel controversy. *PLoS ONE* **4(5)**: e5502.

Ehlers, U. 1992. Frontal glandular and sensory structures in *Nemertoderma* (Nemertodermatida) and *Paratomella* (Acoela): ultrastructure and phylogenetic implications for the monophyly of the Euplathelminthes (Plathelminthes). *Zoomorphology* **112**: 227–236.

Franzén, Å. and Afzelius, B.A. 1987. The ciliated epidermis of *Xenoturbella bocki* (Platyhelminthes, Xenoturbellida) with some phylogenetic considerations. *Zool. Scr.* **16**: 9–17.

Fritsch, G., Böhme, M.U., Thorndyke, M., *et al.* 2008. PCR survey of *Xenoturbella bocki* Hox genes. *J. Exp. Zool. (Mol. Dev. Evol.)* **310B**: 278–284.

Hejnol, A. and Martindale, M.Q. 2009. Coordinated spatial and temporal expression of *Hox* genes during embryogenesis in the acoel *Convolutriloba longifissura*. *BMC Biology* **7**: **65**.

Hejnol, A., Obst, M., Stamatakis, A., *et al.* 2009. Assessing the root of bilaterian animals with scalable phylogenomic methods. *Proc. R. Soc. Lond. B* **276**: 4261–4270.

Hyman, L.H. 1951. *The Invertebrates, vol. 2. Platyhelminthes and Rhynchocoela. The Acoelomate Bilateria.* McGraw-Hill, New York.

Jondelius, U., Ruiz-Trillo, I., Baguñá, J. and Riutort, M. 2002. The Nemertodermatida are basal bilaterians and not members of the Platyhelminthes. *Zool. Scr.* **31**: 201–215.

Lundin, K. 1997. Comparative ultrastructure of the epidermal ciliary rootlets and associated structures of the Nemertodermatida and Acoela (Plathelminthes). *Zoomorphology* **117**: 81–92.

Lundin, K. 1998. The epidermal ciliary rootlets of *Xenoturbella bocki* (Xenoturbellida) revisited: new support for a possible kinship with the Acoelomorpha (Platyhelminthes). *Zool. Scr.* **27**: 263–270.

Lundin, K. 2001. Degenerating epidermal cells in *Xenoturbella bocki* (phylum uncertain), Nemertodermatida and Acoela (Platyhelminthes). *Belg. J. Zool.* **131** (Supplement): 153–157.

Mallatt, J., Craig, C.W. and Yoder, M.J. 2010. Nearly complete rRNA genes assembled from across the metazoan animals: Effects of more taxa, a structure-based alignment, and paired-sites evolutionary models on phylogeny reconstruction. *Mol. Phylogenet. Evol.* **55**: 1–17.

Nielsen, C. 2010. After all: *Xenoturbella* is an acoelomorph! *Evol. Dev.* **12**: 241–243.

Paps, J., Baguñá, J. and Riutort, M. 2009a. Lophotrochozoa internal phylogeny: new insights from an up-to-date analysis of nuclear ribosomal genes. *Proc. R. Soc. Lond. B* **276**: 1245–1254.

Paps, J., Baguñà, J. and Riutort, M. 2009b. Bilaterian phylogeny: A broad sampling of 13 nuclear genes provides a new Lophotrochozoa phylogeny and supports a paraphyletic basal Acoelomorpha. *Mol. Biol. Evol.* **26**: 2397–2406.

Pedersen, K.J. and Pedersen, L.R. 1988. Ultrastructural observations on the epidermis of *Xenoturbella bocki* Westblad, 1949; with a discussion of epidermal cytoplasmic filament systems of invertebrates. *Acta Zool. (Stockh.)* **69**: 231–246.

Philippe, H., Brinkmann, H., Copley, R.R., *et al.* 2011. Acoelomorph flatworms are deuterostomes related to *Xenoturbella*. *Nature* **470**: 255–258.

Philippe, H., Brinkmann, H., Martinez, P., Riutort, M. and Baguñá, J. 2007. Acoel flatworms are not Platyhelminthes: evidence from phylogenomics. *PLoS ONE* **2(8)**: e717.

Rohde, K., Watson, N. and Cannon, L.R.G. 1988. Ultrastructure of epidermal cilia of *Pseudactinoposthia* sp. (Platyhelminthes, Acoela); implications for the phylogenetic status of the Xenoturbellida and Acoelomorpha. *J. Submicrosc. Cytol. Pathol.* **20**: 759–767.

Ruiz-Trillo, I., Paps, J., Loukota, M., *et al.* 2002. A phylogenetic analysis of myosin heavy chain type II sequences corroborates that Acoela and Nemertodermatida are basal bilaterians. *Proc. Natl. Acad. Sci. USA* **99**: 11246–11251.

Sempere, L.F., Martinez, P., Cole, C., Baguñá, J. and Peterson, K.J. 2007. Phylogenetic distribution of microRNAs supports the basal position of acoel flatworms and the polyphyly of Platyhelminthes. *Evol. Dev.* **9**: 409–415.

Wallberg, A., Curini-Galletti, M., Ahmadzadeh, A. and Jondelius, U. 2007. Dismissal of Acoelomorpha: Acoela and Nemertodermatida are separate early bilaterian clades. *Zool. Scr.* **36**: 509–523.

Westblad, E. 1950. *Xenoturbella bocki* n.g., n.sp. a peculiar, primitive turbellarian type. *Ark. Zool.*, 2. Ser. **1**: 11–29.

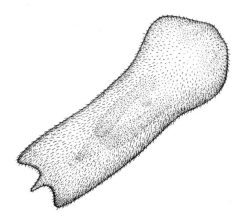

18

Phylum **Acoela**

The Acoela is a small phylum of about 400 species of small, completely ciliated, mostly marine 'worms'. Its monophyly has not been questioned.

Until recently, the acoels were regarded as 'primitive' platyhelminths, but almost all of the newer molecular studies place the acoels at the base of the Eumetazoa (together with the nemertodermatids, and sometimes also with *Xenoturbella*, see Chapter 17) as the sister group of the Eubilateria (Ruiz-Trillo *et al.* 1999; Egger *et al.* 2009; Hejnol *et al.* 2009; Paps *et al.* 2009). Only a few studies are more undecided between a position together with the Platyhelminthes and the 'basal' position (Egger *et al.* 2009). The basal position is supported by the low number of microRNAs (Sempere *et al.* 2007).

The acoels are organized as bilateral, creeping gastrulae with mesoderm. The microscopic anatomy was reviewed by Rieger *et al.* (1991). The cilia of the monolayered epithelium have microtubular doublets numbers 4–7 ending below the tip, which has a lamellar cap; their basal system is unusually complicated, with a posterior rootlet with a brush of microtubules (Rohde *et al.* 1988; Lundin 1997; Tyler and Hooge 2004). The basal membrane is almost absent, only represented by small 'islands' of extracellular matrix (Tyler and Hooge 2004). The statocyst is surrounded by a proteinaceous capsule (Ferrero 1973), but whether this represents a basal membrane remains unknown.

The mesoderm comprises a subepidermal layer of thin muscle cells showing a highly variable pattern between species (Tyler and Hyra 1998; Tyler and Rieger 1999). The muscles are characterized as smooth or pseudo-striated (Todt and Tyler 2006).

The midventral mouth opens directly to a central digestive syncytium, surrounded by a layer of parenchymal cells, but the central syncytium is apparently shed after digestion of a prey in some species (Smith and Tyler 1985). Only *Paratomella* has an epithelium-like organization of the endoderm, but without cell junctions. Protonephridia are absent.

The nervous system consists of a ring-shaped anterior ganglion surrounding a statocyst with one lithocyte and various patterns of longitudinal nerve cords with transverse commissures (Kotikova and Raikova 2008; Semmler *et al.* 2010). Unique ocelli with a pigment cell and sensory cells without cilia or rhabdomeres are found in several species (Yamasu 1991; Nilsson 2009). There is a frontal gland complex consisting of ciliated sensory cells and glandular cells, and with a common apical pore (Ehlers 1992).

The gonads are groups of germ cells in the parenchyma surrounded by various types of accessory cells. Some species have more organized male and female gonads, with various copulatory structures.

The spermatozoa are biciliate, but show much variation with cortical or axial microtubules in the tail

Chapter vignette: *Convolutriloba longifissura*. (Based on an illustration by Eric Rottinger, Kahikai.)

(resembling some rhabditophoran platyhelminths), and various numbers of central microtubules in the axoneme (Raikova and Justine 1994; Raikova *et al.* 2001; Tekle *et al.* 2007).

Cleavage shows a characteristic pattern that was earlier interpreted as a modified spiral pattern, with duets instead of quartets, but this idea has now been abandoned. However, the nomenclature with micromeres, macromeres, and 'duets' is still used. Earlier studies recorded the cleavage pattern of a number of genera (e.g. Bresslau 1909; Apelt 1969), and Henry *et al.* (2000) documented the early cell lineage through marking of blastomeres (Table 18.1). The first cleavage is equal, and divides the embryo into a left-dorsal part and a right-ventral part. The following three cleavages are unequal so that three duets of micromeres are given off all through laeotropic cleavages. Gastrulation is embolic, and descendants of all three micromere duets form the ectoderm with the nervous system of the juvenile. The third duet macromeres both give rise to endoderm and mesoderm. The following development was followed by Ramachandra *et al.* (2002). Gastrulation is embolic and the ectodermal cells become ciliated. A nervous concentration develops at the anterior pole, and muscle cells differentiate at the periphery of the invaginated endo- and mesoderm. A concentration of neurons differentiate into an unpaired brain ganglion, with perikarya surrounding a neuropile and paired longitudinal nerve cords (see also Bery *et al.* 2010). The mesoderm differentiates into a complicated system of fine muscles (Ladurner and Rieger 2000), and the endodermal cells fuse into the digestive syncytium.

Table 18.1. Cell lineage of *Neochildia fusca*. (Modified from Henry *et al.* 2000.)

Z	A	1a		ectoderm (neurons)
		1A	2a	ectoderm (neurons)
			2A	3a	ectoderm (neurons)
				3A	endoderm, mesoderm
	B	1b		ectoderm (neurons)
		1B	2b	ectoderm (neurons)
			2B	3b	ectoderm (neurons)
				3B	endoderm, mesoderm

The acoel *Childia* is able to form normal larvae when one blastomere of the 2-cell stage is deleted (Boyer 1971).

Hox1, *Hox4/5* and a posterior Hox gene have been identified (Cook *et al.* 2004; Hejnol and Martindale 2009) (Fig. 21.3).

The phylogenetic position of the Acoela as one of the 'basal' metazoan groups is now well-documented (see Chapter 17). A sister-group relationship with the Nemertodermatids is indicated by the possession in both groups of the characteristic frontal organ. The relationships of the three acoelomorph phyla are discussed in Chapter 17.

Interesting subjects for future research

1. Nature of the statocyst capsule

References

Apelt, G. 1969. Fortpflanzungsbiologie, Entwicklungszyklen und vergleichende Frühentwicklung acoeler Turbellarien. *Mar. Biol.* **4**: 267–325.

Bery, A., Cardona, A., Martinez, P. and Hartenstein, V. 2010. Structure of the central nervous system of a juvenile acoel, *Symsagittifera roscoffensis*. *Dev. Genes Evol.* **220**: 61–76.

Boyer, B.C. 1971. Regulative development in a spiralian embryo as shown by cell deletion experiments on the acoel. *Childia. J. Exp. Zool.* **176**: 97–105.

Bresslau, E. 1909. Die Entwicklung der Acoelen. *Verh. Dtsch. Zool. Ges.* **19**: 314–324.

Cook, C.E., Jiménez, E., Akam, M. and Saló, E. 2004. The Hox gene complement of acoel flatworms, a basal bilaterian clade. *Evol. Dev.* **6**: 154–163.

Egger, B., Steinke, D., Tarui, H., *et al.* 2009. To be or not to be a flatworm: the acoel controversy. *PLoS ONE* **4(5)**: e5502.

Ehlers, U. 1992. Frontal glandular and sensory structures in *Nemertoderma* (Nemertodermatida) and *Paratomella* (Acoela): ultrastructure and phylogenetic implications for the monophyly of the Euplathelminthes (Plathelminthes). *Zoomorphology* **112**: 227–236.

Ferrero, E. 1973. A fine structural analysis of the statocyst in Turbellaria Acoela. *Zool. Scr.* **2**: 5–16.

Hejnol, A. and Martindale, M.Q. 2009. Coordinated spatial and temporal expression of *Hox* genes during embryogenesis in the acoel *Convolutriloba longifissura*. *BMC Biology* **7**: 65.

Hejnol, A., Obst, M., Stamatakis, A., *et al.* 2009. Assessing the root of bilaterian animals with scalable phylogenomic methods. *Proc. R. Soc. Lond. B* **276**: 4261–4270.

Henry, J.Q., Martindale, M.Q. and Boyer, B.C. 2000. The unique developmental program of the acoel flatworm, *Neochildia fusca. Dev. Biol.* **220**: 285–295.

Kotikova, E.A. and Raikova, O.I. 2008. Architectonics of the central nervous system of Acoela, Platyhelminthes, and Rotifera. *J. Evol. Biochem. Physiol.* **44**: 95–108.

Ladurner, P. and Rieger, R. 2000. Embryonic muscle development of *Convoluta pulchra* (Turbellaria-Acoelomorpha, Platyhelminthes). *Dev. Biol.* **222**: 359–375.

Lundin, K. 1997. Comparative ultrastructure of the epidermal ciliary rootlets and associated structures of the Nemertodermatida and Acoela (Plathelminthes). *Zoomorphology* **117**: 81–92.

Nilsson, D.-E. 2009. The evolution of eyes and visually guided behaviour. *Phil. Trans. R. Soc. Lond. B* **364**: 2833–2847.

Paps, J., Baguñà, J. and Riutort, M. 2009. Bilaterian phylogeny: A broad sampling of 13 nuclear genes provides a new Lophotrochozoa phylogeny and supports a paraphyletic basal Acoelomorpha. *Mol. Biol. Evol.* **26**: 2397–2406.

Raikova, O., Reuter, M. and Justine, J.L. 2001. Contributions to the phylogeny and systematics of the Acoelomorpha. In D.T.J. Littlewood and R.A. Bray (eds): *Interrelationships of the Platyhelminthes (Systematics Association Special Volume 60)*, pp. 13–23. Taylor and Francis, London.

Raikova, O.I. and Justine, J.L. 1994. Ultrastructure of spermiogenesis and spermatozoa in three acoels (Platyhelminthes). *Ann. Sci. Nat., Zool., 13. Ser.* **15**: 63–75.

Ramachandra, N.B., Gates, R.D., Ladurner, P., Jacobs, D.K. and Hartenstein, V. 2002. Embryonic development in the primitive bilaterian *Neochildia fusca*: normal morphogenesis and isolation of POU genes *Brn-1* and *Brn-2. Dev. Genes Evol.* **212**: 55–69.

Rieger, R.M., Tyler, S., Smith, J.P.S., III and Rieger, G.E. 1991. Platyhelminthes: Turbellaria. In F.W. Harrison (ed.): *Microscopic Anatomy of Invertebrates*, vol. 3, pp. 7–140. Wiley-Liss, New York.

Rohde, K., Watson, N. and Cannon, L.R.G. 1988. Ultrastructure of epidermal cilia of *Pseudactinoposthia* sp.

(Platyhelminthes, Acoela); implications for the phylogenetic status of the Xenoturbellida and Acoelomorpha. *J. Submicrosc. Cytol. Pathol.* **20**: 759–767.

Ruiz-Trillo, I., Riutort, M., Littlewood, D.T.J., Herniou, E.A. and Baguñá, J. 1999. Acoel flatworms: earliest extant bilaterian metazoans, not members of Platyhelminthes. *Science* **283**: 1919–1923.

Semmler, H., Chiodin, M., Bailly, X., Martinez, P. and Wanninger, A. 2010. Steps towards a centralized nervous system in basal bilaterians: Insights from neurogenesis of the acoel *Symsagittifera roscoffensis. Dev. Growth Differ.* **52**: 701–713.

Sempere, L.F., Martinez, P., Cole, C., Baguñá, J. and Peterson, K.J. 2007. Phylogenetic distribution of microRNAs supports the basal position of acoel flatworms and the polyphyly of Platyhelminthes. *Evol. Dev.* **9**: 409–415.

Smith, J.P.S., III and Tyler, S. 1985. The acoel turbellarians: kingpins of metazoan evolution or a specialized offshot? In S. Conway Morris (ed.): *The Origins and Relationships of Lower Invertebrates*, pp. 123–142. Oxford University Press, Oxford.

Tekle, Y.I., Raikova, O.I., Justine, J.-L., Hendelberg, J. and Jondelius, U. 2007. Ultrastructural and immunocytochemical investigation of acoel sperms with 9 + 1 axoneme structure: new sperm characters for unraveling phylogeny in Acoela. *Zoomorphology* **126**: 1–16.

Todt, C. and Tyler, S. 2006. Morphology and ultrastructure of the pharynx in Solenofilomorphidae (Acoela). *J. Morphol.* **267**: 776–792.

Tyler, S. and Hooge, M. 2004. Comparative morphology of the body wall in flatworms (Platyhelminthes). *Can. J. Zool.* **82**: 194–210.

Tyler, S. and Hyra, G.S. 1998. Patterns of musculature as taxonomic characters for the Turbellaria Acoela. *Hydrobiologia* **383**: 51–59.

Tyler, S. and Rieger, R.M. 1999. Functional morphology of musculature in the acoelomate worm, *Convoluta pulchra* (Plathelminthes). *Zoomorphology* **119**: 127–141.

Yamasu, T. 1991. Fine structure and function of ocelli and sagittocysts of acoel flatworms. *Hydrobiologia* **227**: 273–282.

19

Phylum **Nemertodermatida**

Nemertodermatida is a very small phylum, comprising about ten species of millimetre-sized 'turbellariform', mostly interstitial worms; *Meara* is a commensal in the pharynx of holothurians (Sterrer 1998). The microscopic anatomy was reviewed in Rieger *et al.* (1991; see also Lundin and Sterrer 2001).

The monolayered, multiciliary epithelium shows a complex system of connecting ciliary roots. The cilia have characteristic tips with microtubular doublets 4–7, terminating at a distance from the tip and a terminal disc (Lundin 1997). Worn ciliated cells become internalized and digested (Lundin 2001). There is no continuous extracellular matrix (Tyler and Hooge 2004).

In most species, the midventral mouth opens into a short ciliated pharynx. The mouth opening is absent in *Ascoparia*, but digestive glands at the expected position of the mouth may indicate the formation of a temporary opening. The gut has a narrow lumen and lacks cilia. There are no protonephridia

The nervous system is basiepithelial, with an annular concentration at the anterior end and one or two pairs of weak longitudinal nerves; there is no stomogastric nerve system (Gustafsson *et al.* 2002). A small statocyst has two statoliths (Westblad 1937; Raikova *et al.* 2004). A 'frontal organ' consists of a group of gland cells and ciliated, putative sensory cells (Ehlers 1992).

Both ovaries and testes are situated between the dorsal-body wall and the gut. The mature sperm is collected in a vesiculum seminalis, and is probably transferred to the female through hypodermal injection. The sperm is fusiform with a slender nuclear head and a long mid piece, with a pair of spirally coiled mitochondria and a long cilium (Lundin and Hendelberg 1998).

The embryology is not well known. The first three cleavages of *Nemertoderma* are duet cleavages, with micro- and macromeres, but the 16-cell stage consists of a ring of micromeres and a ring of macromeres; the following development has not been described (Jondelius *et al.* 2004).

Two central and one posterior Hox gene have been identified (Jiménez-Guri *et al.* 2006).

Nemertodermatida appears always to be regarded as a monophyletic group, closely related to Acoela, and a number of papers also point to a close relationship with *Xenoturbella*. Possible synapomorphies include several characters of the cilia. Readers are referred to Chapter 17 for further discussion.

Interesting subjects for future research

1. Embryology
2. Hox genes

Chapter vignette: *Ascoparia neglecta*. (Redrawn from Sterrer 1998.)

References

Ehlers, U. 1992. Frontal glandular and sensory structures in *Nemertoderma* (Nemertodermatida) and *Paratomella* (Acoela): ultrastructure and phylogenetic implications for the monophyly of the Euplathelminthes (Plathelminthes). *Zoomorphology* **112**: 227–236.

Gustafsson, M.K.S., Halton, D.W., Kreshchenko, N.D., *et al.* 2002. Neuropeptides in flatworms. *Peptides* **23**: 2053–2061.

Jiménez-Guri, E., Paps, J., García-Fernàndez, J. and Saló, E. 2006. Hox and ParaHox genes in Nemertodermatida, a basal bilaterian clade. *Int. J. Dev. Biol.* **50**: 675–679.

Jondelius, U., Larsson, K. and Raikova, O. 2004. Cleavage in *Nemertoderma westbladi* (Nemertodermatida) and its phylogenetic significance. *Zoomorphology* **123**: 221–225.

Lundin, K. 1997. Comparative ultrastructure of the epidermal ciliary rootlets and associated structures of the Nemertodermatida and Acoela (Plathelminthes). *Zoomorphology* **117**: 81–92.

Lundin, K. 2001. Degenerating epidermal cells in *Xenoturbella bocki* (phylum uncertain), Nemertodermatida and Acoela (Platyhelminthes). *Belg. J. Zool.* **131** (**Supplement**): 153–157.

Lundin, K. and Hendelberg, J. 1998. Is the sperm type of the Nemertodermatida close to that of the ancestral Platyhelminthes? *Hydrobiologia* **383**: 197–205.

Lundin, K. and Sterrer, W. 2001. The Nemertodermatida. In D.T.J. Littlewood and R.A. Bray (eds): *Interrelationships of the Platyhelminthes*, pp. 24–27. Taylor & Francis, London.

Raikova, O.I., Reuter, M., Gustafsson, M.K.S., *et al.* 2004. Basiepithelial nervous system in *Nemertoderma westbladi* (Nemertodermatida): GYIRFamide immunoreactivity. *Zoology* **107**: 75–86.

Rieger, R.M., Tyler, S., Smith, J.P.S., III and Rieger, G.E. 1991. Platyhelminthes: Turbellaria. In F.W. Harrison (ed.): *Microscopic Anatomy of Invertebrates*, vol. 3, pp. 7–140. Wiley-Liss, New York.

Sterrer, W. 1998. New and known Nemertodermatida (Platyhelminthes—Acoelomorpha)—a revision. *Belg. J. Zool.* **128**: 55–92.

Tyler, S. and Hooge, M. 2004. Comparative morphology of the body wall in flatworms (Platyhelminthes). *Can. J. Zool.* **82**: 194–210.

Westblad, E. 1937. Die Turbellarien-Gattung *Nemertoderma* Steinböck. *Acta Soc.Fauna Flora Fenn.* **60**: 45–89.

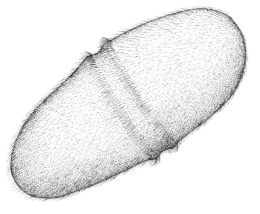

Phylum **Xenoturbellida**

The two described species of *Xenoturbella* are centimetre-long, 'turbellariform', marine worms, with a thick ciliated epidermis, a thick extracellular matrix, a layer of strong muscles, and a sac-shaped gut with a mid-ventral mouth opening (Westblad 1950).

The ectoderm is heavily ciliated and shows openings of numerous mucus glands (Pedersen and Pedersen 1988). The cilia have the characteristic acoelomorph morphology, with the axonemal doublets numbers 4–7, ending at a small plateau below the tip, which has an intracellular cap. Their basal structure comprises double, striated roots and a basal foot with a brush of microtubules (Franzén and Afzelius 1987). An unusual process of withdrawing and resorbing epithelial cells was also observed (Lundin 2001). The extracellular matrix comprises a thick layer of striated filaments, probably collagen (Pedersen and Pedersen 1986). The strong musculature consists of an outer layer of annular muscles, an inner layer of longitudinal muscles, and a system of radial muscles (Ehlers and Sopott-Ehlers 1997; Raikova *et al.* 2000). Characters of the musculature led Ehlers and Sopott-Ehlers (1997) to place *Xenoturbella* as the sister taxon of all other bilaterians.

There is a diffuse, intraepithelial nerve net without any special concentrations (Raikova *et al.* 2000). An anterior 'statocyst' has a number of mobile, monociliated cells (Ehlers 1991), but its function as a statocyst was doubted by Israelsson (2007).

The midventral mouth opens directly into a gut that Westblad (1950) described as syncytial, but this has turned out not to be correct (personal communication, Dr Kenneth Lundin, Göteborg Natural History Museum). The food consists almost exclusively of bivalves, especially *Nucula* (Bourlat *et al.* 2008).

Some of the developmental stages found inside *Xenoturbella* are probably of bivalve origin (Bourlat *et al.* 2003), and the embryology is unknown. The sperm is of the 'primitive' type, with a rounded head with an acrosomal vesicle, four mitochondria in a long cilium (Obst *et al.* 2011), but sperm resembling that of the bivalve *Nucula* have been observed too (Claus Nielsen, unpublished observations).

Both the general morphology and most of the detailed observations of the structure of *Xenoturbella* indicate that it is a basal bilaterian close to the Acoela and Nemertodermatida. Only the transmission electron microscopy studies of the extracellular matrix (Pedersen and Pedersen 1986) and immune reactions of the nervous system (Stach *et al.* 2005) hint to a closer relationship to enteropneusts. A formal cladistic analysis of morphological characters (Haszprunar 1996) placed *Xenoturbella* with the acoelomorphs.

Chapter vignette: *Xenoturbella bocki* (based on various sources.)

The first molecular studies indicated that *Xenoturbella* is a bivalve (Norén and Jondelius 1997), but it was subsequently demonstrated that the molluscan RNA was from *Xenoturbella*'s food (mostly bivalves of the genus *Nucula*), and placed it as sister group to the Ambulacraria (Bourlat *et al.* 2003). This position was subsequently recovered in a series of analyses (see Nielsen 2010), and this finds some morphological support for a relationship with the enteropneusts (Rieger *et al.* 1991). However, a few newer analyses show *Xenoturbella* in a more basal position, as the sister group to all other deuterostomes (Bourlat *et al.* 2009; Perseke *et al.* 2007). Finally, the phylogenomic study of Hejnol *et al.* (2009) showed reasonably good support for a position of *Xenoturbella* as the sister group of Acoela + Nemertodermatida together, forming the Acoelomorpha at the base of the eubilaterian tree, in good agreement with the morphological data. Further support for this position is found in the structure of the short Hox cluster (Fritsch *et al.* 2008), which lacks the *Hox3* gene characteristic of the eubilaterians (see Fig. 21.3). Analyses of the mitochondrial genome excludes a position of *Xenoturbella* within the Aceolomorpha, but its position is very uncertain (Mwinyi *et al.* 2010).

Acoela, Nemertodermatida, and Xenoturbellida may constitute the monophyletic sister group of the Bilateria, as suggested by Hejnol *et al.* (2009), or represent one or two side branches on the line leading to the Eubilateria. The recent study of (Philippe *et al.* 2011) supports the position of *Xenoturbella* within the Acoelomorpha, but places this group as the sister group of the Ambulacraria (see Chapter 17).

Interesting subjects for future studies

1. Embryology

References

Bourlat, S.J., Nakano, H., Åkerman, M., *et al.* 2008. Feeding biology of *Xenoturbella bocki* (phylum Xenoturbellida) revealed by genetic barcoding. *Mol. Ecol. Resour.* **8**: 18–22.

Bourlat, S.J., Nielsen, C., Lockyer, A.E., Littlewood, D.T.J. and Telford, M.J. 2003. *Xenoturbella* is a deuterostome that eats molluscs. *Nature* **424**: 925–928.

Bourlat, S.J., Rota-Stabelli, O., Lanfear, R. and Telford, M.J. 2009. The mitochondrial genome structure of *Xenoturbella bocki* (phylum Xenoturbellida) is ancestral within the deuterostomes. *BMC Evol. Biol.* **9**: 107.

Ehlers, U. 1991. Comparative morphology of statocysts in the Plathelminthes and the Xenoturbellida. *Hydrobiologia* **227**: 263–271.

Ehlers, U. and Sopott-Ehlers, B. 1997. Ultrastructure of the subepidermal musculature of *Xenoturbella bocki*, the adelphotaxon of the Bilateria. *Zoomorphology* **117**: 71–79.

Franzén, Å. and Afzelius, B.A. 1987. The ciliated epidermis of *Xenoturbella bocki* (Platyhelminthes, Xenoturbellida) with some phylogenetic considerations. *Zool. Scr.* **16**: 9–17.

Fritsch, G., Böhme, M.U., Thorndyke, M., *et al.* 2008. PCR survey of *Xenoturbella bocki* Hox genes. *J. Exp. Zool. (Mol. Dev. Evol.)* **310B**: 278–284.

Haszprunar, G. 1996. Plathelminthes and Plathelminthomorpha—paraphyletic taxa. *J. Zool. Syst. Evol. Res.* **34**: 41–48.

Hejnol, A., Obst, M., Stamatakis, A., *et al.* 2009. Assessing the root of bilaterian animals with scalable phylogenomic methods. *Proc. R. Soc. Lond. B* **276**: 4261-4270.

Israelsson, O. 2007. Ultrastructural aspects of the 'statocyst' of *Xenoturbella* (Deuterostomia) cast doubt on its function as a georeceptor. *Tissue Cell* **39**: 171–177.

Lundin, K. 2001. Degenerating epidermal cells in *Xenoturbella bocki* (phylum uncertain), Nemertodermatida and Acoela (Platyhelminthes). *Belg. J. Zool.* **131(Supplement)**: 153–157.

Mwinyi, A., Bailly, X., Bourlat, S.J., *et al.* 2010. The phylogenetic position of Acoela as revealed by the complete mitochondrial genome of *Symsagittifera roscoffensis*. *BMC Evol. Biol.* **10**: 309.

Nielsen, C. 2010. After all: *Xenoturbella* is an acoelomorph! *Evol. Dev.* **12**: 241–243.

Norén, M. and Jondelius, U. 1997. *Xenoturbella*'s molluscan relatives. *Nature* **390**: 31–32.

Obst, M., Nakano, H., Bourlat, S.J., *et al.* 2011. Spermatozoon ultrastructure of *Xenoturbella bocki* (Westblad 1949). *Acta Zool. (Stockh.)* **92**: 109–115.

Pedersen, K.J. and Pedersen, L.R. 1986. Fine structural observations on the extracellular matrix (ECM) of *Xenoturbella bocki* Westblad, 1949. *Acta Zool. (Stockh.)* **67**: 103–113.

Pedersen, K.J. and Pedersen, L.R. 1988. Ultrastructural observations on the epidermis of *Xenoturbella bocki* Westblad, 1949; with a discussion of epidermal cytoplasmic filament systems of invertebrates. *Acta Zool. (Stockh.)* **69**: 231–246.

Perseke, M., Hankeln, T., Weich, B., *et al.* 2007. The mitochondrial DNA of *Xenoturbella bocki*: genomic architecture and phylogenetic analysis. *Theory Biosci.* **126**: 35–42.

Philippe, H., Brinkmann, H., Copley, R.R., *et al.* 2011. Acoelomorph flatworms are deuterostomes related to *Xenoturbella*. *Nature* **470**: 255–258.

Raikova, O.I., Reuter, M., Jondelius, U. and Gustafsson, M.K.S. 2000. An immunocytochemical and ultrastructural study of the nervous and muscular systems of *Xenoturbella westbladi* (Bilateria inc. sed.). *Zoomorphology* **120**: 107–118.

Rieger, R.M., Tyler, S., Smith, J.P.S., III and Rieger, G.E. 1991. Platyhelminthes: Turbellaria. In F.W. Harrison (ed.): *Microscopic Anatomy of Invertebrates*, vol. 3, pp. 7–140. Wiley-Liss, New York.

Stach, T., Dupont, S., Israelsson, O., *et al.* 2005. Nerve cells of *Xenoturbella bocki* (phylum uncertain) and *Harrimania kupfferi* (Enteropneusta) are positively immunoreactive to antibodies raised against echinoderm neuropeptides. *J. Mar. Biol. Assoc. U.K.* **85**: 1519–1524.

Westblad, E. 1950. *Xenoturbella bocki* n.g., n.sp. a peculiar, primitive turbellarian type. *Ark. Zool.*, 2. *Ser.* **1**: 11–29.

21

EUBILATERIA

Eubilateria is one of the most well-defined metazoan groups. It shows a whole suite of morphological apomorphies, and practically all molecular phylogenies and studies of Hox genes agree on the monophyly. The most important apomorphies of the clade comprise the presence of a tubular gut with mouth and anus, a well-defined anterior brain, and a long Hox-cluster collinear with an anteroposterior axis and functioning as an organizer of the body parts along this axis. The Hox cluster comprises the eubilaterian signature (a group 3 gene and the two most anterior central genes), plus varying numbers of more posterior central and posterior genes (see below). Not many other morphological apomorphies can be pointed out; protonephridia have been mentioned, and this character has inspired the alternative name Nephrozoa (Jondelius *et al.* 2002), but protonephridia are apparently only found in the protostomes. These and a number of other characters will be discussed below. The tubular gut has made a rapid radiation possible, probably in the Ediacaran (see for example Nielsen and Parker 2010), because this new organization allowed for occupation of a number of new ecospaces (Xiao and Laflamme 2008).

Bilateral symmetry is seen in anthozoans, a few hydrozoan polyps (Chapter 13), and in acoelomorphs (Chapters 17–20), but they lack all the eubilaterian characteristics listed above. All bilaterians have a secondary, anteroposterior axis (usually with anterior brain and mouth), which forms an angle to the primary, apical–blastoporal (animal–vegetal) axis, and this defines the bilateral plane of symmetry characteristic of the Bilateria. If one accepts that the eubilaterian ancestor was organized as a gastraea, then the tubular gut could have evolved in two ways: either by division of the blastopore into two openings, mouth and anus, or through the establishing of a new opening, where the blastopore was retained as the anus, and the new opening became the mouth or vice versa. This is discussed below and in more detail in Chapters 22 (Protostomia) and 56 (Deuterostomia).

Fossils resembling compact embryos with large cells, indicating lecithotrophy, have been described from the Neoproterozoic (Xiao and Knoll 2000; Hagadorn *et al.* 2006; Chen *et al.* 2009), and other fossils from the same period strongly resemble thin-walled gastrulae of deuterostomes and protostomes, indicating the presence of planktotrophic larval stages (Chen *et al.* 2000). However, the very nature of both types of fossils has been questioned by a number of authors (for example, Bailey *et al.* 2007; Bengtson and Budd 2004), so the evidence is inconclusive.

Unfortunately, the fossil record of adult organisms tells very little about early animal radiation. The early ancestors were small and without skeletons or tough cuticles. Most of the Vendian organisms interpreted as cnidarians, annelids, or arthropods are highly questionable (McCall 2006). The 'Ediacaran–Cambrian explosion' with rich fossil faunas, for example, from Chiengjiang and Sirius Passet (Early Cambrian; Hou *et al.* 2004; Conway Morris and Peel 2008), Burgess Shale (Middle Cambrian; Briggs *et al.* 1994), and the

Orsten fauna (mainly Upper Cambrian; Maas *et al.* 2006) are 'windows' showing that most of the groups that we call phyla today, or their stem lineages, were established already at that time together with a number of forms that are difficult to relate to existing phyla. It seems likely that a period of radiation preceded the faunal 'explosion' we see in the Early Cambrian, and this is also indicated by studies of the 'molecular clock' (Peterson *et al.* 2008; Nielsen and Parker 2010). The Cambrian faunas have given us good information about the diversity and morphological complexity of the animal life at that time, but they do not give us much information about animal radiation.

The morphology of the ancestral eubilaterian is difficult to visualize because the protostomes and deuterostomes show such different body plans and embryology, but it was probably a pelago–benthic bilateral gastraea with an apical brain and a nervous concentration on the ventral side (facing the substratum). Evolution of an early protostome from this common ancestor seems easy to reconstruct in a series of steps that can be recognized in ontogenetic stages of several protostomes, and that shows functional continuity (the trochaea theory, see Chapter 22). On the contrary, the evolution of the deuterostome ancestor is difficult to infer, because the two sister clades, Ambulacraria and Chordata, are so different, both in development and in adult structure.

The traditional division of the Eubilateria into Protostomia and Deuterostomia is based on several morphological characters, and the two groups are recovered in almost all molecular phylogenies, including studies of microRNAs (Wheeler *et al.* 2009).

The two groups have been given various names by different authors to underline the differences each author considered most important. The name pairs introduced by some important classical authors are: Zygoneura/Ambulacralia + Chordonia (Hatschek 1888); Protostomia/Deuterostomia (Grobben 1908); Ecterocoelia/Enterocoelia (Hatschek 1911); Hyponeuralia/Epineuralia (Cuénot 1940); and Gastroneuralia/Notoneuralia (Ulrich 1951). Hatschek's first system was based on his trochophora theory, which emphasized the unity of the Zygoneura with a paired ventral nerve chord; Grobben's division was based on the differences in the fate of the blastopore (see below); Hatschek's second scheme was based on differences in the formation of the mesoderm and coelom; and Cuénot's and Ulrich's names refer to the position of the main parts of the central nervous systems. There are no nomenclatural rules for the highest systematic categories in zoology, but Protostomia and Deuterostomia have become the preferred names. An alternative scheme that until recently has been followed by a number of college textbooks (even as recent as Margulis and Chapman 2009) is Hyman's (1940) arrangement of her chapters on bilaterians under three headings: acoelomates, pseudocoelomates, and coelomates. This scheme has unfortunately been used by a number of molecular phylogeneticists under the misleading name 'the traditional morphology-based phylogeny' (see Nielsen 2010). However, Hyman did not consider this as a phylogenetic scheme (Jenner 2004), and the names are not found in her 'hypothetical diagram of the relationships of the phyla' (Hyman 1940: fig. 5.1). Luckily, this three-group classification is now almost totally abandoned.

A number of characters and organ systems of phylogenetic interest are discussed below in an attempt to evaluate information about their origin and phylogenetic importance. More detailed discussions can be found in Schmidt-Rhaesa (2007).

Tube-shaped gut with mouth and anus This is one of the defining apomorphies of the Eubilateria. The exclusively parasitic Nematomorpha have nonfunctioning rudiments of a gut, and completely gutless groups, such as acanthocephalans, cestodes, and 'pogonophorans', occur within a number of phyla where the lack of a gut can be interpreted with confidence as a specialization. Free-living eubilaterians without an anus are found within phyla, where the presence of an anus is the 'normal' condition, for example in rotifers, such as *Asplanchna* (Chapter 34), the articulate brachiopods (Chapter 41), and the ophiuroids (Chapter 58). The platyhelminths all lack an anus, but both embryological and molecular data support the interpretation of this as a derived character (Chapter 29).

A more difficult question is whether gut, mouth, and anus of all eubilaterias are homologous. The position of the gut, its general structure and function, and its origin from the archenteron of a gastraea appear similar, so homology of the guts (endoderm) appears well founded. However, as mentioned above, there appears to be two fundamentally different ways of deriving a tube-shaped gut from the sac-shaped gut of a gastraea: (1) the lateral lips of the blastopore may fuse so that the blastopore becomes divided into mouth and anus (as observed in many protostomes, see Fig. 22.2); and (2) the blastopore may become the anus and a new mouth formed as an opening from the bottom of the archenteron (as observed in some deuterostomes). This demonstrates that the mouth openings of the two main groups of bilateral animals, viz. Protostomia and Deuterostomia, cannot *a priori* be regarded as homologous. Blastopore fate and the development of the adult gut in the two groups are discussed in Chapters 22 and 56.

Monociliate/multiciliate epithelial cells Monociliated epithelia are the rule in all the non-triploblastic phyla and this must be interpreted as a plesiomorphy (Chapter 4).

The multiciliate condition has obviously evolved more than once outside the animal kingdom (for example in polymastigine flagellates, ciliates, and sperm of ferns) and it is therefore not unexpected that it has evolved more than once within the animals. Multiciliary cells are found in hexactinellid larvae (Chapter 5), in few cnidarians (Chapter 13), in ctenophores (Chapter 15), and in the acoelomorphs (Chapters 17–20). It appears that the protostomian ancestor evolved the multiciliate condition, and that this has then been lost in certain clades. The monociliate annelid *Owenia* (belonging to a family with multiciliate species) is a good example of this (Chapter 25). All gnathostomulids and some gastrotrich clades are monociliate (Chapters 30, 32), and this is best interpreted as reversals to the ancestral type of ciliation. The brachiozoans (Chapters 39–41) are exclusively monociliate. Multiciliarity appears to have evolved independently in enteropneusts, which have monociliate ciliary bands in the larvae (Chapter 60) and in urochordates + vertebrates (Chapters 64–65).

Ciliary bands Ciliated epithelia are used in locomotion in many aquatic metazoans, either in creeping or in swimming. A special type of ciliated epithelia are the narrow ciliary bands with diaplectic metachronism (Knight-Jones 1954) used in swimming and feeding. Such bands are characteristic of practically all types of ciliated planktotrophic invertebrate larvae, and are also found in some adult rotifers and annelids, and in all adult entoprocts, bryozoans, cycliophorans, phoronids, brachiopods, and pterobranchs. The special type found on the gills of some autobranch bivalves (called cirral trapping, see Riisgård and Larsen 2001) is not discussed here.

The ciliary bands fall in well-defined groups according to structure and function. The ciliated cells may be monociliate or multiciliate, and the cilia may be separate or compound. The bands may function in downstream collecting, ciliary sieving, or upstream collecting.

Planktotrophic trochophora larvae of annelids, molluscs, and entoprocts, and adult entoprocts, some rotifers, and annelids have downstream collecting bands composed of compound cilia on multiciliate cells (Nielsen 1987) (Fig. 22.6). The filtering systems are organized with two bands of compound cilia, with opposed orientation of their effective strokes (this type is also called the opposed band or double-band system). The compound cilia create a water current and 'catch up' with suspended particles that are then pushed onto a band of separate cilia, which transports the particles to the mouth (Strathmann *et al.* 1972; Riisgård *et al.* 2000).

Adult bryozoans and cyphonautes larvae capture particles by setting up a water current with a lateral band of separate cilia on multiciliate cells, and filtering particles from the water current by a row of stiff laterofrontal cilia; the captured particles are then transported further through ciliary or muscular movements of the filtering structure (Chapter 38). Phoronids and brachiopods show a similar system, but the cells are all monociliate (Chapters 39–41).

Planktotrophic larvae of enteropneusts and echinoderms use ciliary bands of separate cilia on monociliate cells in an upstream collecting system (Strathmann *et al.* 1972; Nielsen 1987; Strathmann 2007). The ciliary bands both create the water current and strain the food particles, apparently by reversal of the ciliary beat. The pterobranch tentacles have been described as having laterofrontal cilia as those of the bryozoans and brachiozoans, but there are no laterofrontal cells in the tentacles of *Cephalodiscus* (Chapter 61), so their ciliary bands have the same structure and probably also function as those of the ambulacrarian larvae (Chapter 57).

The chordates use mucociliary filtering systems where the water currents are created by cilia and the particles caught by a mucous filter (Chapter 62).

Nervous systems Most ciliated bilaterian larvae have a ciliated apical organ like that of the cnidarians (Fig. 10.1) that is lost before or at metamorphosis (Chapter 10). The larval spiralians have an apical organ with an apical ganglion that is lost at metamorphosis, and a pair of cerebral ganglia that is retained in the adult brain (Chapter 23), but the ecdysozoan brains are of different types. However, all protostomes appear to have a dorsal/circumoral brain and a pair of longitudinal ventral nerve cords (Chapter 22, Fig. 22.3). The deuterostomes show two quite different types of nervous systems. The ambulacrarians lack a centralized brain (Chapter 57), whereas the chordates have the characteristic neural tube (Chapter 62). Gene-expression patterns in the ventral nerve cords of annelids and arthropods show strong similarities with those of the chordate neural tube, see for example Denes *et al.* (2007). It has therefore been suggested that the ventral cord of the protostomes, which originates from the blastopore margin, is homologous with the neural tube of the chordates; see further discussion in Chapter 62.

At present, it seems almost impossible to envisage the morphology of the ancestral eubilaterian nervous system.

Muscle cells Striated muscle cells have been reported from certain hydromedusae (Chapter 13) and the very peculiar, anucleate tentillae of a ctenophore (Chapter 15), but it is generally believed that striated muscle cells are an apomorphy of the Eubilateria (Burton 2007).

Coelom A coelom is defined as a cavity surrounded by mesoderm, which is usually a peritoneum with basal membrane and apical cell junctions, and with the apical side facing the cavity. The discussions about the origin and homologies of the mesodermal cavities in the various phyla have been very extensive in the past, and the phylogenetic significance of the coeloms has definitely been grossly exaggerated. Here it may suffice to say that the cavities that fall under this definition are of quite different morphology and ontogeny. Only where the coeloms have an identical origin and an identical, complex structure, as for example the endodermally derived proto-, meso-, and metacoel of the deuterostomes, can one speak of well-founded homology.

The origin of the coelomic compartments has been related to the locomotory habits of the organisms, with the coeloms functioning as hydrostatic skeletons (Clark 1964), and this correlation appears very well founded in annelids. However, the fluid-filled body cavity of priapulans that is used as a hydrostatic skeleton in the large burrowing species is not a coelom (Chapter 52). Fluid-filled body cavities function as hydrostatic skeletons in protrusion and retraction of anterior parts of the body in many types of animals. The cavity is a primary body cavity in rotifers and pelagic larvae of annelids and molluscs, and a large coelom in the adult bryozoans and sipunculans. It appears that fluid-filled cavities have evolved independently in several phyla. There is nothing to indicate that the various coeloms are homologous, and this opinion is now shared by most authors.

The function of the coelomic cavities as hydrostatic organs does not necessitate the presence of coelomoducts, and the association between gonads and coeloms (see below) is probably secondary. Coelomoducts may have originated in connection with the association between gonads and coelom, so that the original function of coelomoducts was that of gonoducts.

Circulatory systems Special fluid transport systems are absent in cycloneuralians, in microscopic animals, and in larvae of many of the spiralian and deuterostome phyla where diffusion and circulation of the fluid in the primary body cavity appears sufficient for the transport of gases and metabolites. In adults of larger compact and coelomate organisms these methods are usually supplemented by special transport systems, which fall into two main types: coelomic and haemal (Ruppert and Carle 1983).

Coelomic circulatory systems are lined by coelomic epithelia and the circulation is caused by the cilia of the peritoneum, which have now been observed in a number of phyla (Ruppert and Carle 1983), or by muscles. The coelomic cavities of many polychaetes are segmental and cannot transport substances along the body, whereas other annelids have extensive coelomic cavities formed by fusion of the segmental coelomic sacs so that the coelomic fluid, which contains respiratory pigments in many forms, can circulate through most of the animal. In the leeches, the coelomic cavities have been transformed into a system of narrow canals that are continuous throughout the body and that function as a circulatory system (Chapter 25). Coelomic specializations such as the tentacle coelom, with one or two tubular, contractile 'compensation sacs' of the sipunculans (Chapter 26), and the water-vascular system of the echinoderms (Chapter 58), have respiratory functions too.

Haemal systems, also called blood vascular systems, are cavities between the basal membranes of the epithelia, for example blood vessels in the dorsal and ventral mesenteries in annelids (Ruppert and Carle 1983; Ruppert *et al.* 2004). This position could indicate that the blood spaces are remnants of the blastocoel, but the blood vessels arise *de novo* between cell layers in most cases (Ruppert and Carle 1983). The only major deviations from this structure appear in vertebrates that have blood vessels with endothelial walls, and in cephalopods that have endothelium in some vessels (Yoshida *et al.* 2010).

The blood vessels may be well defined in the whole organism, for example in some annelids, or there may be smaller or larger blood sinuses or lacunae, in addition to a heart and a few larger vessels, for example in molluscs. These two types are sometimes called closed and open haemal systems, respectively, but the phylogenetic value of this distinction is dubious.

Arthropods and onychophorans show a very peculiar organization of coelom and haemal system. During ontogeny, the contractile dorsal vessel is formed by the dorsal parts of the coelomic sacs, but the 'coelomic parts' of the coelomic cavities fuse with the haemal system so that a large mixocoel is formed (Chapters 44 and 45).

Contractile blood vessels are found in many animals, and their muscular walls are, in all cases, derivatives of the surrounding mesoderm, which in most cases can be recognized as coelomic epithelia. The well-defined hearts consist of usually paired coelomic pouches that surround a blood vessel; each has a muscular inner wall and a thin outer wall, separated by a coelomic space, the pericardial cavity, that facilitates the movements of the heart.

Unique circulatory systems of various types are found in a few phyla. Many platyhelminths have lacunae between the mesenchymal cells, and some digenean trematodes have a well-defined system of a right and a left channel lined by mesodermal syncytia surrounded by muscle cells (Strong and Bogitsch 1973). Carle and Ruppert (1983) interpreted the funiculus of the bryozoans as a haemal system, because it usually consists of one or more hollow strands of mesoderm where the lumen is sometimes lined by a basement membrane. It is possible that this structure transports nutrients from the gut to the testes, which are usually located on the funiculus (and to the developing statoblasts of phylactolaemates), but neither position, structure, or function of the funiculus bears resemblance to blood vessels of other metazoans, and it may simply be a highly specialized mesentery (Chapter 38). Also chaetognaths have a small basement-membrane-lined cavity that has been interpreted as a haemal space, but a circulatory function is not obvious (Chapter 55). The lateral channels of the nemertines are discussed in Chapter 28.

The two types of circulatory system are usually not found in the same organism, but the echinoderms are a remarkable exception in that they have both a haemal system and a coelomic circulatory system - the water vascular system clearly transports, for example, oxygen from the podia (Chapter 58).

There seems to be no morphological indication of an ancestral eubilaterian circulatory system. The orthologous regulatory genes *tinman* in *Drosophila* and *Nkx2.5* in the mouse specify the development of the mesodermal heart, and this has led some authors (for example De Robertis and Sasai 1996) to postulate that the eubilaterian ancestor had a circulatory system with a mesodermal heart. However, the homologous gene *ceh-22* regulates differentiation of the ectodermal myoepithelial pharynx in *Caenorhabditis* (Lam *et al.* 2006), where a homology seems very far fetched. Davidson (2006) interpreted the highly complex heart-gene network common to *Drosophila* and the mouse as a kernel specifying a pulsating tissue, and proposed that this ancestral network has lost some of its components in *Caenorhabditis*. It could be interesting to look for similar genes in other metazoans.

Nephridia/coelomoducts/gonoducts Half a century ago Goodrich (1946) summarized the available knowledge about nephridia/coelomoducts/gonoducts and concluded that there are two main types of nephridia: protonephridia with a closed inner part; and metanephridia with a ciliated funnel opening into a coelom. He further concluded that the coelomoducts appear to have originated as gonoducts, and that these canals may secondarily have become engaged in excretion, and eventually have lost the primary function as gonoducts. The different types of organs may fuse so that compound organs (nephromixia) are formed.

Protonephridia (Ruppert and Smith 1988; Bartolomaeus and Ax 1992; Bartolomaeus and Quast 2005; Schmidt-Rhaesa 2007) are ectodermally derived canals with inner (terminal) cells with narrow, slit-like gaps or interdigitations with a canal-cell, usually covered by a basal membrane that functions as an ultrafilter in the formation of the primary urine. The filtration is usually from the blastocoel or from interstices between the mesodermal cells, but protonephridia draining coelomic spaces are found in some annelids (Bartolomaeus and Quast 2005). The primary urine is modified during the passage through the canal cells, and the duct opens to the exterior through a nephridiopore cell. The definition based on the ectodermal origin is more narrow than the usual one and excludes, for example the cyrtopodocytes of amphioxus, which develop from mesodermal cells (Chapter 63).

Protonephridia are probably an ancestral character in spiralian larvae and are found in a number of adults too. The ecdysozoans lack primary larvae, but protonephridia are found in a number of adult cycloneuralians, so protonephridia may have been present in the protostomian ancestor. There is considerable variation in the morphology of these organs, but it is difficult to see any evolutionary tendencies (Schmidt-Rhaesa 2007). The nematodes have an excretory organ of one or a few cells without a filtration weir (Chapter 49), and its homology is uncertain. Protonephridia are found in phoronids (Chapter 40), and this is one of the characters that indicate their spiralian relationship.

The driving force for filtration is usually said to come from the movements of the cilium/cilia, but I find it difficult to see how this should work.

Metanephridia/gonoducts (Ruppert and Smith 1988; Bartolomaeus 1999; Schmidt-Rhaesa 2007) are mesodermal organs that transport and modify coelomic fluid (and sometimes also gametes) through a canal that often has an ectodermal distal part (for example in clitellates) (Weisblat and Shankland 1985; Gustavsson and Erséus 1997). The coelomic fluid can be regarded as the primary urine in most organisms and it is formed by filtration from the blood through ultrafilters, formed by cells of blood vessels specialized as podocytes. There is, however, one important exception to this, namely the sipunculans, which have large metanephridia but lack a haemal system (Chapter 26). The coelomic compartment may be large and have retained the primary functions (as hydrostatic skeleton or gonocoel), or be very restricted so that the blood vessel(s), podocytes, coelomic cavity, and sometimes also the modifying nephridial canal are united into a complex organ, for example the antennal gland of the crustaceans (Chapter 44), the axial complex of echinoderms (Chapter 58) and enteropneusts (Chapter 60), and the vertebrate nephron. Annelid metanephridia (Bartolomaeus 1999) often develop from a pair of intersegmentally situated mesodermal cells that divide and develop a lumen with cilia; this primordial structure may either form a weir and become one or more cyrtocytes, or open to become a ciliated funnel (a traditional metanephridium). Genital ducts in the form of ciliated mesodermal funnels develop independently or in association with the nephridia.

Segmentation Repeated units of anatomical structures are found in a number of eubilaterians, and a complex type, comprising both mesodermal and ectodermal elements, such as coelomic sacs, ganglia, and excretory organs, sequentially added from a posterior growth zone, are called segments or somites (Tautz 2004). Such 'true' segments are characteristic of Annelida (Chapter 25), Panarthropoda (Chapter 43), and Chordata (Chapter 62). The three 'segments' of the ambulacrarians, i.e. prosome, mesosome, and metasome, are excluded here because there is no terminal addition of segments.

Several authors have pointed out that homologous genes are involved in organizing segmentation in all the three 'segmented clades', and this has led to the hypothesis that the common ancestor of the eubilaterians, 'Urbilateria', was segmented (Balavoine and Adoutte 2003; De Robertis 2008; Saudemont *et al.* 2008; Couso 2009). This would imply that the bilaterian ancestor was a coelomate, and that both coeloms and segmentation have been lost numerous times during the bilaterian radiation. Other authors point to the differences in the segmentation mechanisms (Blair 2008), and the question is far from settled.

The patterning of protostomian-type segmentation has been studied extensively in annelids and especially in arthropods, both by use of direct observation of cleavage patterns and by observations of gene expression.

Arthropods, especially malacostracans (Dohle and Scholtz 1997; Browne *et al.* 2005), show teloblastic growth of both ectoderm and endoderm (Chapter 44). Each ring of cells given off from the ectoteloblasts divides twice forming a parasegment with four rows of cells. The anteriormost row of cells in each parasegment and the three posterior rows of the anterior parasegment fuse into a 'true' segment with a ring of sclerites. This seems necessary for the function of the mesodermal muscles connecting the sclerites of two

segments. The gene *engrailed* is expressed in the anterior row of cells in the parasegment and *wingless* in the posterior row of cells, so that both of these two stripes become located in the posterior part of each segment. Other arthropods do not show teloblasts, but the same pattern of gene expression is seen in insects and spiders (Damen 2002; Tautz 2004).

In the annelid *Platynereis* (Prud'homme *et al.* 2003; Dray *et al.* 2010), a similar striped gene expression pattern is found, but the *engrailed* stripe remains in the anterior side of the (para)segment, so that the border between two segments is between the *wingless/Wnt1* and *engrailed/hedgehog* stripes. In contrast to arthropod musculature, the annelid musculature functions primarily within each (para)segment, possibly because the segmentation ancestrally was functioning as rows of hydrostatic compartments

The annelid parapodia develop from the middle of the segments, whereas the arthropod limbs develop at the border between parasegments, so the two structures cannot be homologous (Prud'homme *et al.* 2003; Prpic 2008; Dray *et al.* 2010).

It appears that the segmentation genes of annelids and arthropods could be an extreme example of co-option of genes into similar function in two separate lineages (Chipman 2010).

It should be mentioned that two hypotheses for the evolution of the protostomes exist. The 'classical' Articulata hypothesis, which emphasizes the similarities in morphology of segmentation in annelids and arthropods and in the similarity of gene expression of segmentation genes (see for example Saudemont *et al.* 2008; Dray *et al.* 2010), and the Ecdysozoa hypothesis, which is supported by the very consistent tree topology, with Spiralia and Ecdysozoa as sister groups seen in molecular phylogenetic analysis and in the Hox-gene signatures (see below and Fig. 21.3). I have chosen to follow the Ecdysozoa hypothesis, but this leaves the problem of explaining the gene expression patterns. Two possibilities have been proposed. One is that the ancestral bilaterian was segmented and that this character was subsequently lost in all the non-segmented phyla, and the other is that the segmentation with the genes evolved convergently in annelids,

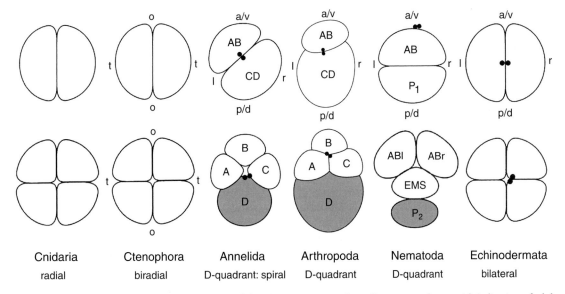

Cnidaria
radial

Ctenophora
biradial

Annelida
D-quadrant: spiral

Arthropoda
D-quadrant

Nematoda
D-quadrant

Echinodermata
bilateral

Figure 21.1. Examples of cleavage types in major eubilaterian groups; 2- and 4-cell stages are shown with indication of adult body orientations. The cells giving rise to germ cells are shaded. Ctenophores have biradial cleavage with strictly defined cleavage planes; o-o, the oral plane, t-t, the tentacle plane. Spiralians have spiral cleavage with oblique cleavage planes, so that four quadrants are formed: A, left; B, anteroventral; C, right; D, posterodorsal. Arthropods (exemplified by *Semibalanus*) with similar quadrants. Cycloneuralians (exemplified by the nematode *Caenorhabditis*) first have a transverse cleavage; at the second cleavage the anterior cell (AB) divides medially and one of the resulting cells immediately slides to the dorsal side (see Fig. 37.2), whereas the posterior cell (P_1) divides transversely into the anterior EMS and the posterior P_2. Deuterostomes have a bilateral cleavage pattern where the first cleavage is median and the second transverse.

arthropods, and chordates (see for example Balavoine and Adoutte 2003). Much more information is obviously needed to resolve the discrepancy between the various types of information.

The chordates show a probably non-homologous segmentation organized from the mesodermal somites, which is patterned by segment polarity genes of other types (Holley 2007) (see Chapter 62).

It should of course be mentioned once again that expression of a gene in structures or functions in different lineages does not prove homology (Nielsen and Martinez 2003). A good example is the presence of the microRNA-183 family in ciliated sensory cells of various sense organs in both protostomes and deuterostomes (Pierce *et al.* 2008). Nobody would regard nematode sensilla, fruit-fly sensilla, sensory cells at the sea-urchin tube feet, and the ciliated cells of the vertebrate ear as homologous organs. The presence of segmentation genes in non-segmented bilaterians is

possibly another example of 'preadaptation' (Marshall and Valentine 2010).

Gonads The origin of the germ cells has been studied in many bilaterian phyla, and it appears that it is always mesodermal, but there are differences in the timing of the segregation of the germ cells. In protostomes with spiral cleavage, the germ cells differentiate from the 4d cell at a late stage of development, and this seems also to be the case in many arthropods. In the cycloneuralians, the germ cells can be traced back to the 2- and 4-cell stages. This is discussed further below and in connection with cleavage patterns of protostomes (Chapter 22, Table 22.1). Deuterostomes generally have a late segregation of the germ cells, but this is not the case in the vertebrates (Juliano and Wessel 2010).

Almost all bilaterians studied so far have the polar bodies situated, at least initially, at the pole where the apical organ will be formed, opposite to the position of the blastopore at the point where the two first cleavage furrows intersect. The polar bodies are situated at the blastoporal pole in cnidarians and ctenophores (Chapters 13 and 15). Oocyte polarity has only been studied in a few groups, but the apical face of the oocyte in the ovarial epithelium becomes the apical pole of the embryo in nemertines (Chapter 28), brachiopods (Chapter 41), and amphioxus (where there is even a small cilium; Chapter 63).

Spawning may have been through rupture of the external body wall in the earliest triploblastic organisms, but this method is rare among extant organisms (amphioxus is one example; see Chapter 63), and the isolated examples must be interpreted as apomorphies. Gonoducts of acoelomate organisms are typically formed by fusion of an ectodermal invagination, with an extension of the gonadal wall (for example in nemertines; Chapter 28). In coelomate groups, the gametes are usually shed via the coelom and the gonoducts are therefore coelomoducts. These structures are usually formed through fusion between an ectodermal invagination and an extension from the coelom.

Cleavage patterns Early metazoan development shows a high degree of variation, both intraphyletic

and between phyla. Only development with total cleavage will be discussed here.

The 'sponges' (Chapters 5,7,9) do not show any characteristic pattern. Cnidarians show much variation (see Fig. 13.1), but a radial pattern can be recognized in most species. There is only the primary, apical-blastoporal axis, and this is apparently the primitive cleavage pattern in the eumetazoans. Ctenophores (Chapter 15) have a highly determined biradial cleavage pattern, with the first cleavage through the oral plane and the second through the tentacle plane, reflecting the symmetry of the adults. This is definitely an apomorphy of the phylum. Acoels show a highly specialized cleavage pattern (Chapter 18), but the embryology of Nemertodermatida is poorly known and that of *Xenoturbella* unknown.

Eubilaterian cleavage patterns show much variation too, but it appears that three main types of total cleavage can be recognized (Fig. 21.1). Many protostomes with total cleavage show unequal cleavage with the two first cleavage planes cutting along the primary axis, but being oblique, so that three smaller cells (left, anteroventral, and right) and a larger, posterodorsal cell are formed. These four blastomeres

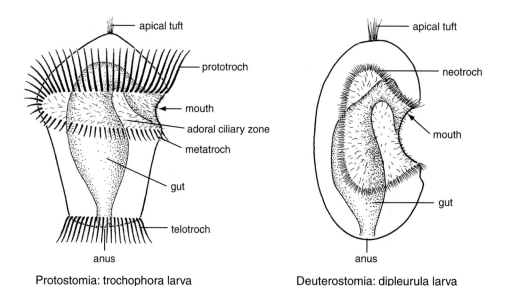

Protostomia: trochophora larva   Deuterostomia: dipleurula larva

Figure 21.2. Diagrams of the two main types of larvae of bilaterians, the protostomian trochophora and the deuterostomian (ambulacrarian) dipleurula.

give rise to quartets of cells with more or less fixed fates. The three smaller cells, called A-C, have rather similar fates, whereas the fourth cell, called D, usually contributes to other tissues, such as endomesoderm and germ cells. This pattern, which is here called quadrant cleavage, is considered a protostomian apomorphy. A special, well-known example of this type of cleavage is the spiral cleavage found in most of the spiralian phyla (Chapter 23). Among the ecdysozoans, some arthropods with small eggs and total cleavage and most nematodes show quadrant cleavage.

Deuterostomes have a bilateral cleavage pattern (Chapter 56), where the first cleavage usually divides the embryo medially and the second cleavage is transverse, resulting in anterior and posterior blastomeres.

Life cycles and larval types Metazoan life cycles show a bewildering variation, but biphasic, pelago-benthic life cycles with a ciliated larva are known for the 'sponges' and in a number of eubilaterian groups. From the phylogeny presented here it seems clear that the eumetazoan ancestor was a gastraea, and that an adult benthic stage was added independently in cni-

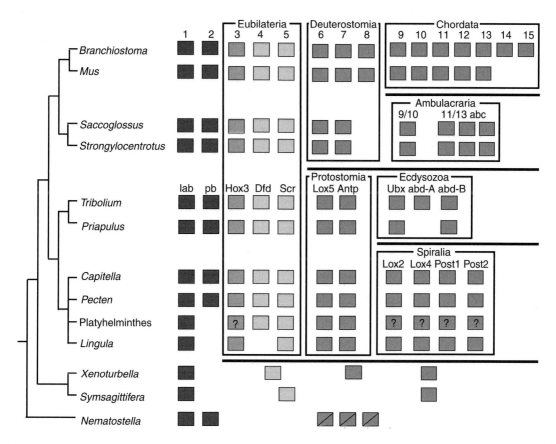

Figure 21.3. Hox genes of a cnidarian and selected representatives of the major bilaterian groups, demonstrating the existence of series of 'Hox signatures' (see the text). The Eubilaterian Hox signature consists of a group 3 gene and the two anteriormost central genes. Genes belonging to the same orthology group have the same colour: anterior genes, red; group three, orange; central group, green; and posterior group, blue. Duplications are not shown. Diagram based on: *Branchiostoma*: Holland *et al.* (2008), *Mus*: Carroll *et al.* (2005), *Saccoglossus*: Aronowicz and Lowe (2006), *Strongylocentrotus*: Cameron *et al.* (2006), *Tribolium*: Shippy *et al.* (2008), *Priapulus*: de Rosa *et al.* (1999), *Capitella*: Fröbius *et al.* (2008), *Pecten*: Canapa *et al.* (2005), Platyhelminthes: Olson (2008); Badets and Verneau (2009), *Lingula*: de Rosa *et al.* (1999), *Xenoturbella*: Fritsch *et al.* (2008), *Symsagittifera*: Moreno *et al.* (2009), *Nematostella*: Ryan *et al.* (2007). The platyhelminth sequences are rather divergent so the genes have been inserted with question marks. See Colour Plate 2.

darians and once or twice in the bilaterians (Fig. 10.3; Nielsen 1998, 2009).

It seems obvious that the cnidarians have evolved from a planktotrophic gastraea thay has added a sessile adult stage (Chapter 13). The acoelomorphs are creeping bilateral gastraeae (Chapters 17–20).

The eubilaterian sister groups, Protostomia and Deuterostomia, both comprise two lineages one of which has retained the primary, ciliated larvae (Fig. 21.2), whereas the other has lost it.

Among the protostomes, the spiralians have the trochophore as the typical larva. Planktotrophic trochophores are known from annelids, molluscs, and entoprocts, and some of the adult rotifers have similar ciliary bands and can be interpreted as progenetic. This larval type is characterized by a downstream-collecting ciliary system consisting of proto- and metatroch of compound cilia on multiciliate cells and an adoral ciliary zone (see Chapter 23). The ecdysozoans lack the primary, ciliated larvae.

Among the deuterostomes, the ambulacrarians have the dipleurula as the typical larva. Planktotrophic larvae of this type are known from enteropneusts and echinoderms. It is characterized by the upstream-collecting ciliary band of single cilia on monociliate cells called the neotroch (see Chapter 57).

The two larval types are so fundamentally different that it seems impossible to derive one from the other. It appears easy to derive the trochophore from a gastraea (Chapter 23), whereas the origin of the deuterostome life cycle is very difficult to visualize.

Hox genes The Hox genes are instrumental in the patterning of organs along the anteroposterior axis of the bilaterians. The full complement of 10–15 Hox genes is only found in the eubilaterians (Fig. 21.3). Acoelomorphs and cnidarians have much shorter Hox clusters and lack the full complement of the Hox3–5 genes, and the homologies of their central and posterior genes are uncertain. It appears that both Protostomia and Deuterostomia have specific signatures, which are also found both in Spiralia and Ecdysozoa, and in Ambulacraria and Chordata (de Rosa *et al.* 1999; Telford 2000; Ogishima and Tanaka 2007; Thomas-Chollier *et al.* 2010).

Interesting subjects for future research

1. Function of protonephridia

References

Aronowicz, J. and Lowe, C.J. 2006. *Hox* gene expression in the hemichordate *Saccoglossus kowalevskii* and the evolution of deuterostome nervous system. *Integr. Comp. Biol.* **46**: 890–901.

Badets, M. and Verneau, O. 2009. Hox genes from the Polystomatidae (Platyhelminthes, Monogenea). *Int. J. Parasitol.* **39**: 1517–1523.

Bailey, J.V., Joye, S.B., Kalanetra, K., Flood, B.E. and Corsetti, F.A. 2007. Evidence of giant sulphur bacteria in Neoproterozoic phosphorites. *Nature* **445**: 198–201.

Balavoine, G. and Adoutte, A. 2003. The segmented *Urbilateria*: a testable hypothesis. *Integr. Comp. Biol.* **43**: 137–147.

Bartolomaeus, T. 1999. Structure, function and development of segmental organs in Annelida. *Hydrobiologia* **402**: 21–37.

Bartolomaeus, T. and Ax, P. 1992. Protonephridia and metanephridia - their relation within the Bilateria. *Z. Zool. Syst. Evolutionsforsch.* **30**: 21–45.

Bartolomaeus, T. and Quast, B. 2005. Structure and development of nephridia in Annelida and related taxa. *Hydrobiologia* **535/536**: 139–165.

Bengtson, S. and Budd, G. 2004. Comment on 'Small bilaterian fossils from 40 to 55 million years before the Cambrian'. *Science* **306**: 1291a.

Blair, S.S. 2008. Segmentation in animals. *Curr. Biol.* **18**: R991–R995.

Briggs, D.E.G., Erwin, D.H. and Collier, F.J. 1994. *The Fossils of the Burgess Shale.* Smithsonian Institution Press, Washington.

Browne, W.E., Price, A.L., Gerberding, M. and Patel, M.N. 2005. Stages of embryonic development in the amphipod crustacean, *Parhyale hawaiensis. Genesis* **42**: 124–149.

Burton, P.M. 2007. Insights from diploblasts; the evolution of mesoderm and muscle. *J. Exp. Zool. (Mol. Dev. Evol.)* **310B**: 5–14.

Cameron, A.R., Bowen, L., Nesbitt, R., *et al.* 2006. Unusual gene order and organization of the sea urchin Hox cluster. *J. Exp. Zool. (Mol. Dev. Evol.)* **306B**: 45–58.

Canapa, A., Biscotti, M.A., Olmo, E. and Barucca, M. 2005. Isolation of Hox and ParaHox genes in the bivalve *Pecten maximus. Gene* **348**: 83–88.

Carle, K.J. and Ruppert, E.E. 1983. Comparative ultrastructure of the bryozoan funiculus: a blood vessel homologue. *Z. Zool. Syst. Evolutionsforsch.* **21**: 181–193.

Carroll, S.B., Grenier, J.K. and Weatherbee, S.D. 2005. *From DNA to Diversity. Molecular Genetics and the Evolution of Animal Design*, 2nd ed. Blackwell Publishing, Malden, MA.

Chen, J.-Y., Bottjer, D.J., Li, G., *et al.* 2009. Complex embryos displaying bilaterian characters from Precambrian Doushantuo phosphate deposits, Weng'an, Guizhou, China. *Proc. Natl. Acad. Sci. USA* **106**: 19056–19060.

Chen, J.Y., Oliveri, P., Li, C.W., *et al.* 2000. Precambrian animal diversity: putative phosphatized embryos from the Doushanto Formation of China. *Proc. Natl. Acad. Sci. USA* **97**: 4457–4462.

Chipman, A.D. 2010. Parallel evolution of segmentation by co-option of ancestral gene regulatory networks. *BioEssays* **32**: 60–70.

Clark, R.B. 1964. *Dynamics in Metazoan Evolution.* Clarendon Press, Oxford.

Conway Morris, S. and Peel, J.S. 2008. The earliest annelids: Lower Cambrian polychaetes from the Sirius Passet Lagerstätte, Peary Land, North Greenland. *Acta Palaeontol. Pol.* **51**: 137–148.

Couso, J.P. 2009. Segmentation, metamerism and the Cambrian explosion. *Int. J. Dev. Biol.* **53**: 1305–1316.

Cuénot, L. 1940. Essai d'arbre généalogique du règne animal. *C. R. Hebd. Seanc. Acad. Sci. Paris* **210**: 196–199.

Damen, W.G.M. 2002. Parasegmental organization of the spider embryo implies that the parasegment is an evolutionary conserved entity in arthropod embryogenesis. *Development* **129**: 1239–1250.

Davidson, E.H. 2006. *The Regulatory Genome. Gene Regulatory Networks in Development and Evolution.* Academic Press, Amsterdam.

De Robertis, E.M. 2008. Evo-devo: variations on ancestral themes. *Cell* **132**: 185–195.

De Robertis, E.M. and Sasai, Y. 1996. A common plan for dorsoventral patterning in Bilateria. *Nature* **380**: 37–40.

de Rosa, R., Grenier, J.K., Andreeva, T., *et al.* 1999. Hox genes in brachiopods and priapulids and protostome evolution. *Nature* **399**: 772–776.

Denes, A.S., Jékely, G., Steinmetz, P.R.H., *et al.* 2007. Molecular architecture of annelid nerve cord supports common origin of nervous system centralization in Bilateria. *Cell* **129**: 277–288.

Dohle, W. and Scholtz, G. 1997. How far does cell lineage influence cell fate specification in crustacean embryos? *Semin. Cell Dev. Biol.* **8**: 379–390.

Dray, N., Tessmar-Raible, K., Le Gouar, M., *et al.* 2010. Hedgehog signaling regulates segment formation in the annelid *Platynereis. Science* **329**: 339–342.

Fritsch, G., Böhme, M.U., Thorndyke, M., *et al.* 2008. PCR survey of *Xenoturbella bocki* Hox genes. *J. Exp. Zool. (Mol. Dev. Evol.)* **310B**: 278–284.

Fröbius, A.C., Matus, D.Q. and Seaver, E.C. 2008. Genomic organization and expression demonstrate spatial and temporal Hox gene colinearity in the lophotrochozoan *Capitella* sp. 1. *PLoS ONE* **3(12)**: e4004.

Goodrich, E.S. 1946. The study of nephridia and genital ducts since 1895. *Q. J. Microsc. Sci., N. S.* **86**: 113–392.

Grobben, K. 1908. Die systematische Einteilung des Tierreichs. *Verh. Zool.-Bot. Ges. Wien* **58**: 491–511.

Gustavsson, L.M. and Erséus, C. 1997. Morphogenesis of the genital ducts and spermathecae in *Clitellio arenarius, Heterochaeta costata, Tubificoides benedii* (Tubificidae) and *Stylaria lacustris* (Naididae) (Annelida, Oligochaeta). *Acta Zool. (Stockh.)* **78**: 9–31.

Hagadorn, J.W., Xiao, S., Donoghue, P.C.J., *et al.* 2006. Cellular and subcellular structure of Neoproterozoic animal embryos. *Science* **314**: 291–294.

Hatschek, B. 1888. *Lehrbuch der Zoologie, 1. Lieferung* (pp. 1–144). Gustav Fischer, Jena.

Hatschek, B. 1911. *Das neue zoologische System.* W. Engelmann, Leipzig.

Holland, L.Z., Albalat, R., Azumi, K., *et al.* 2008. The amphioxus genome illuminates vertebrate origins and cephalochordate biology. *Genome Res.* **18**: 1100–1111.

Holley, S.A. 2007. The genetics and embryology of zebrafish metamerism. *Dev. Dyn.* **236**: 1422–1449.

Hou, X.-G., Aldridge, R.J., Bergström, J., *et al.* 2004. *The Cambrian Fossils of Chengjiang, China.* Blackwell, Malden, MA.

Hyman, L.H. 1940. *The Invertebrates, vol. 1. Protozoa through Ctenophora.* McGraw-Hill, New York.

Jenner, R.A. 2004. Libbie Henrietta Hyman (1888–1969): from developmental mechanisms to the evolution of animal body plans. *J. Exp. Zool. (Mol. Dev. Evol.)* **302B**: 413–423.

Jondelius, U., Ruiz-Trillo, I., Baguñá, J. and Riutort, M. 2002. The Nemertodermatida are basal bilaterians and not members of the Platyhelminthes. *Zool. Scr.* **31**: 201–215.

Juliano, C. and Wessel, G. 2010. Versatile germline genes. *Science* **329**: 649–641.

Knight-Jones, E.W. 1954. Relations between metachronism and the direction of ciliary beat in Metazoa. *Q. J. Microsc. Sci., N. S.* **95**: 503–521.

Lam, N., Chesney, M.A. and Kimble, J. 2006. Wnt signaling and CEH-22/tinman/Nkx2.5 specify a stem cell niche in *C. elegans. Curr. Biol.* **16**: 287–295.

Margulis, L. and Chapman, M.J. 2009. *Kingdoms & Domains. An Illustrated Guide to the Phyla of Life on Earth.* Academic Press, Amsterdam.

Marshall, C.R. and Valentine, J.W. 2010. The importance of preadapted genomes on the origin of the animal bodyplans and the Cambrian explosion. *Evolution* **64**: 1189–1201.

McCall, G.J.H. 2006. The Vendian (Ediacaran) in the geological record: Enigmas in geology's prelude to the Cambrian explosion. *Earth Sci.Rev.* **77**: 1–229.

Moreno, E., Nadal, M., Baguñà, J. and Martínez, P. 2009. Tracking the origins of the bilaterian *Hox* patterning system: insights from the acoel flatworm *Symsagittifera roscoffensis. Evol. Dev.* **11**: 574–581.

Maas, A., Braun, A., Dong, X.-P., *et al.* 2006. The 'Orsten'—More than a Cambrian Konservat-Lagerstätte yielding exceptional preservation. *Palaeoworld* **15**: 266–282.

Nielsen, C. 1987. Structure and function of metazoan ciliary bands and their phylogenetic significance. *Acta Zool. (Stockh.)* **68**: 205–262.

Nielsen, C. 1998. Origin and evolution of animal life cycles. *Biol. Rev.* **73**: 125–155.

Nielsen, C. 2009. How did indirect development with planktotrophic larvae evolve? *Biol. Bull.* **216**: 203–215.

Nielsen, C. 2010. The 'new phylogeny'. What is new about it? *Palaeodiversity* **3 (Suppl.)**: 149–150.

Nielsen, C. and Martinez, P. 2003. Patterns of gene expression: homology or homocracy? *Dev. Genes Evol.* **213**: 149–154.

Nielsen, C. and Parker, A. 2010. Morphological novelties detonated the Ediacaran-Cambrian 'explosion'. *Evol. Dev.* **12**: 345–346.

Ogishima, S. and Tanaka, H. 2007. Missing link in the evolution of Hox clusters. *Gene* **387**: 21–30.

Olson, P.D. 2008. Hox genes and the parasitic flatworms: new opportunities, challenges and lessons from the free-living. *Parasitol. Int.* **57**: 8–17.

Peterson, K.J., Cotton, J.A., Gehling, J.G. and Pisani, D. 2008. The Ediacaran emergence of bilaterians: congruence between the genetic and the geological fossil records. *Phil. Trans. R. Soc. Lond. B* **363**: 1435–1443.

Pierce, M.L., Weston, M.D., Fritsch, B., *et al.* 2008. MicroRNA-183 family conservation and ciliated neurosensory organ expression. *Evol. Dev.* **10**: 106–113.

Prpic, N.-M. 2008. Parasegmental appendage allocation in annelids and arthropods and the homology of parapodia and arthropodia. *Front. Zool.* **5**: 17.

Prud'homme, B., de Rosa, R., Arendt, D., *et al.* 2003. Arthropod-like expression patterns of *engrailed* and *wingless* in the annelid *Platynereis dumerilii* suggest a role in segment formation. *Curr. Biol.* **13**: 1876–1881.

Riisgård, H.U. and Larsen, P.S. 2001. Minireview: Ciliary filter feeding and bio-fluid mechanics—present understanding and unsolved problems. *Limnol. Oceanogr.* **46**: 882–891.

Riisgård, H.U., Nielsen, C. and Larsen, P.S. 2000. Downstream collecting in ciliary suspension feeders: the catch-up principle. *Mar. Ecol. Prog. Ser.* **207**: 33–51.

Ruppert, E.E. and Carle, K.J. 1983. Morphology of metazoan circulatory systems. *Zoomorphology* **103**: 193–208.

Ruppert, E.E. and Smith, P.R. 1988. The functional organization of filtration nephridia. *Biol. Rev.* **63**: 231–258.

Ruppert, E.E., Fox, R.S. and Barnes, R.D. 2004. *Invertebrate Zoology: a Functional Evolutionary Approach* (7th ed. of R.D. Barnes' *Invertebrate Zoology*). Brooks/Cole, Belmont, CA.

Ryan, J.F., Mazza, M.E., Pang, K., *et al.* 2007. Pre-bilaterian origins of the Hox cluster and the Hox code: evidence from the sea anemone, *Nematostella vectensis*. *PLoS ONE* **2(1)**: e 153.

Saudemont, A., Dray, N., Hudry, B., *et al.* 2008. Complementary striped expression patterns of *NK* homeobox genes during segment formation in the annelid *Platynereis*. *Dev. Biol.* **317**: 430–443.

Schmidt-Rhaesa, A. 2007. *The Evolution of Organ Systems.* Oxford University Press, Oxford.

Shippy, T.L., Ronshaugen, M., Cande, J., *et al.* 2008. Analysis of the *Tribolium* homeotic complex: insights into mechanisms constraining insect Hox clusters. *Dev. Genes Evol.* **218**: 127–139.

Strathmann, R.R. 2007. Time and extent of ciliary response to particles in a non-filtering feeding mechanism. *Biol. Bull.* **212**: 93–103.

Strathmann, R.R., Jahn, T.L. and Fonseca, J.R. 1972. Suspension feeding by marine invertebrate larvae: Clearance of particles by ciliated bands of a rotifer, pluteus, and trochophore. *Biol. Bull.* **142**: 505–519.

Strong, P.A. and Bogitsch, B.J. 1973. Ultrastructure of the lymph system of the trematode *Megalodiscus temperatus*. *Trans. Am. Microsc. Soc.* **92**: 570–578.

Tautz, D. 2004. Segmentation. *Dev. Cell* **7**: 301–312.

Telford, M.J. 2000. Turning Hox 'signatures' into synapomorphies. *Evol. Dev.* **2**: 360–364.

Thomas-Chollier, M., Ledent, V., L., L. and Vervoort, M. 2010. A non-tree-based comprehensive study of metazoan Hox and ParaHox genes prompts new insights into their origin and evolution. *BMC Evol. Biol.* **10**: 73.

Ulrich, W. 1951. Vorschläge zu einer Revision der Grosseinteilung des Tierreichs. *Zool. Anz. Suppl.* **15**: 244–271.

Weisblat, D.A. and Shankland, M. 1985. Cell lineage and segmentation on the leech. *Phil. Trans. R. Soc. Lond. B* **312**: 39–56.

Wheeler, B., Heimberg, A.M., Moy, V.N., *et al.* 2009. The deep evolution of metazoan microRNAs. *Evol. Dev.* **11**: 50–68.

Xiao, S. and Knoll, A.H. 2000. Phosphatized animal embryos from the Neoproterozoic Doushanto Formation at Weng'an, Guizhou, South China. *J. Paleont.* **74**: 767–788.

Xiao, S. and Laflamme, M. 2008. On the eve of animal radiation: phylogeny, ecology and evolution of the Ediacara biota. *Trends Ecol. Evol.* **24**: 31–40.

Yoshida, M.-A., Shigeno, S., Tsuneki, K. and Furuya, H. 2010. Squid vascular endothelial growth factor receptor: a shared molecular signature in the convergent evolution of closed circulatory systems. *Evol. Dev.* **12**: 25–33.

22

PROTOSTOMIA

Protostomia is a large clade of bilateral animals characterized by a mosaic of features, but few groups exhibit them all (Table 22.1). Hatschek (1888) named the group Zygoneura with reference to the characteristic paired ventral nerve cords. However, Grobben (1908) introduced the name Protostomia for the group, based on the fate of the blastopore, which should become the mouth (as opposed to it becoming the anus as in the Deuterostomia). Unfortunately, Grobben's name has prevailed although the fate of the blastopore is highly variable even within phyla (see below), and this has inspired very many fruitless phylogenetic discussions. The protostomes are considered monophyletic in almost all morphological and molecular studies.

Modern studies divide the group into Spiralia (Lophotrochozoa) and Ecdysozoa (Fig. 22.1) as this is shown by all the newer molecular phylogenies. Ecdysozoa is well characterized morphologically by a suite of characters related to the lack of ciliated ectoderm, such as the predominantly chitinous (ancestral?) cuticle, and the lack of primary larvae (Chapter 42). Unfortunately, Spiralia is mostly characterized by plesiomorphies. The presence of the ciliated epithelium (found in all the more 'basal' metazoans and in the deuterostomes) and the primary larvae are clearly plesiomorphies, and it is quite difficult to find good apomorphies. Adult morphology and ontogeny of the central nervous system indicate that the evolution has proceeded from a gastraea-like ancestor (trochaea) with a lateral blastopore closure, leaving mouth and anus (amphistomy, see Arendt and Nübler-Jung 1997)

as explained in the trochaea theory (see below). The cilia and the primary larvae have then been lost in the ecdysozoans.

The Chaetognatha (Chapter 55) has always been a problematic group, but the morphology and ontogeny of the nervous system leaves no doubt about their protostomian affinities, and this is now supported by most molecular analyses. A closer relationship with one of the protostome phyla has not been indicated, and also the Hox genes indicate their isolated position. They are here placed in a trichotomy with Spiralia and Ecdysozoa (Fig. 22.1).

All molecular analyses place the Brachiozoa (Phoronida + Brachiopoda) in the Spiralia, but they lack almost all of the characteristics emphasized here as apomorphies of the Protostomia, except the protonephridia of the phoronid larva (Table 22.1). To facilitate the following discussion, they are discussed separately in Chapters 39–41.

Classification has often been based on one character only, but the following complex of characters should be considered apomorphic of the protostomes: (1) the blastopore becomes divided by the fusing lateral lips leaving only mouth and anus (Fig. 22.2); (2) the nervous system includes a dorsal-circumpharyngeal brain, and a pair of ventral longitudinal nerve cords (sometimes fused) formed from longitudinal zones along the fused blastopore lips (Fig. 22.3); (3) the primary larvae are trochophore-types with a downstream-collecting ciliary system consisting of bands or compound cilia on multiciliate cells (Figs. 22.4-7) (primary larvae are lacking in the

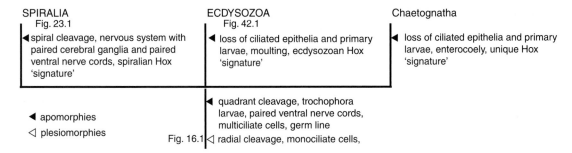

Figure 22.1. Phylogeny of the Protostomia.

Table 22.1. Phylogenetically informative characters of spiralians, ecdysozoans, chaetognaths, and deuterostomes. The rotifers can be interpreted as progenetic trochophores. The non–ciliated Ecdysozoa lack an apical organ, but their brain is situated in a position similar to that of the spiralians. Nemertines and platyhelminths lack the ventral part of the nervous system.

	Blastopore becomes mouth + anus	Trochophora larvae	Circumoral brain + ventral nerve cords	D–quadrant cleavage	Spiral cleavage	Protonephridia
Annelida	+	+	+	+	+	+
Sipuncula	−	+	+	+	+	−
Mollusca	+	+	+	+	+	+
Nemertini	−	(+)	−	+	+	+
Platyhelminthes	−	(+)	(+)	+	+	+
Gastrotricha	−	−	+	(+)	−	+
Gnathostomulida	−	?	+	(+)	(+)	+
Micrognathozoa	−	?	?	?	?	+
Rotifera	−	(+)	(+)	+	−	+
Entoprocta	−	+	−	+	+	+
Cycliophora	−	?	(+)	?	?	+
Ectoprocta	−	−	−	−	−	−
Phoronida	−	−	−	−	−	+
Brachiopoda	−	−	−	−	−	−
Onychophora	+	−	+	−	−	−
Arthropoda	−	−	+	(+)	−	−
Tardigrada	−	−	+	?	−	−
Nematoda	+	−	+	+	−	−
Nematomorpha	−	−	+	?	−	−
Priapula	?	−	+	−	−	+
Kinorhyncha	?	−	+	?	?	+
Loricifera	?	−	+	?	?	+
Chaetognatha	−	−	+	+	−	−
Enteropneusta	−	−	−	−	−	−
Pterobranchia	−	−	−	−	−	−
Echinodermata	−	−	−	−	−	−
Cephalochordata	−	−	−	−	−	−
Urochordata	−	−	−	−	−	−
Vertebrata	−	−	−	−	−	−

Ecdysozoa, which lack the ciliated ectoderm); and (4) the first two cleavages separate one cell that has a special fate, usually contributing a large part of the body ectoderm, endomesoderm, and germ cells (Fig. 21.1). The spiral pattern is a special case of this pattern and is perhaps the most important spiralian apomorphy, although it could of course have been lost together with the ciliation in the ecdysozoans.

Only some polychaetes show both the life cycle, and the larval and adult structures, that are considered ancestral. However, Table 22.1 shows that a number of these characters occur in every protostome phylum (except the Bryozoa (Ectoprocta) that are discussed in Chapter 38).

The fate of the blastopore has, for almost a century, been used as the characterizing feature of the protostomes (see Chapter 21), usually with the understanding that the blastopore should become the mouth (as indicated by the name). Unfortunately, this character is highly variable even within phyla, so it is of little value in phylogenetic analyses. The amount of yolk or an evolution of placental nourish-

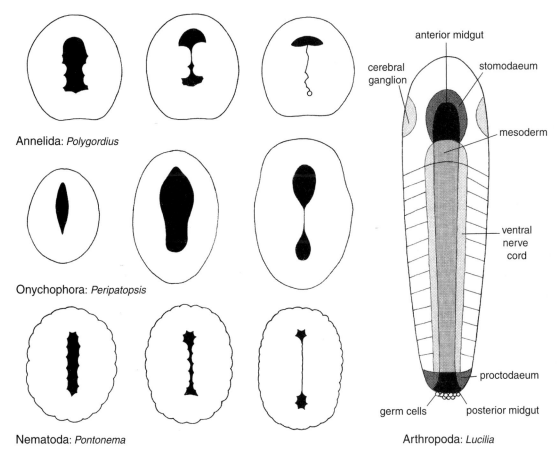

Figure 22.2. The lateral blastopore closure is considered ancestral in the Protostomia; embryos of some annelids, onychophorans, and nematodes with embolic gastrulation directly show the fusion of the lateral blastopore lips, leaving only mouth and anus; embryos with epibolic gastrulation or more derived types have fate maps that reflect the ancestral pattern with the ventral nerve cords developing from the ectoderm along the line between mouth and anus. Annelida: *Polygordius* sp.; the white circle shows the position of the future anus. (Based on Woltereck 1904.) Onychophora: *Peripatopsis capensis*. (Redrawn from Manton 1949.). Nematoda: *Pontonema vulgare*. (Redrawn from Malakhov 1986.) Arthropoda: fate map of the fly *Lucilia sericata*. (Based on Davis 1967 and Anderson 1973.)

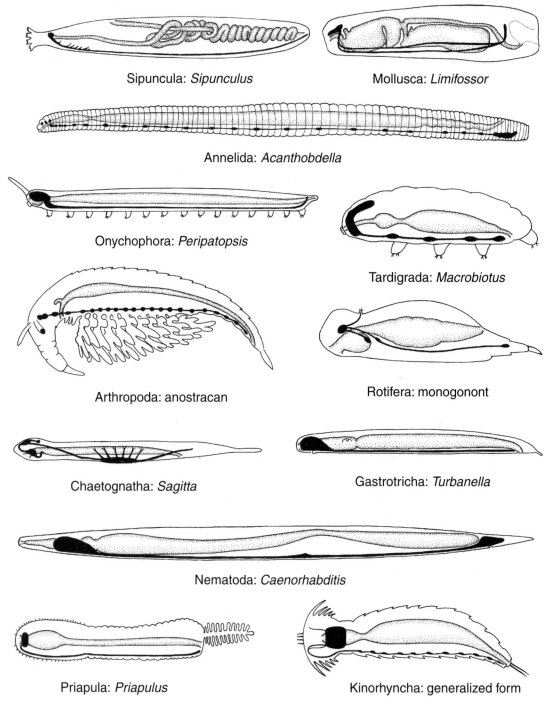

Sipuncula: *Sipunculus*

Mollusca: *Limifossor*

Annelida: *Acanthobdella*

Onychophora: *Peripatopsis*

Tardigrada: *Macrobiotus*

Arthropoda: anostracan

Rotifera: monogonont

Chaetognatha: *Sagitta*

Gastrotricha: *Turbanella*

Nematoda: *Caenorhabditis*

Priapula: *Priapulus*

Kinorhyncha: generalized form

Figure 22.3. Left views of central nervous systems of protostomes. Sipuncula: *Sipunculus nudus*. (Based on Metalnikoff 1900.) Mollusca: *Limifossor talpoideus*. (After an unpublished drawing by Drs A.H. Scheltema and M.P. Morse, based on Heath 1905.) Annelida: the hirudinean *Acanthobdella peledina*. (Based on Storch and Welsch 1991.) Onychophora: *Peripatoides novaezelandiae*. (Based on Snodgrass 1938.) Tardigrada: *Macrobiotus hufelandi*. (Modified from Cuénot 1949.) Arthropoda: generalized anostracan. (Redrawn from Storch and Welsch 1991.) Rotifera: generalized monogonont. (Redrawn from Hennig 1984.) Chaetognatha: *Sagitta crassa*. (Based on Goto and Yoshida 1987.) Gastrotricha: *Turbanella cornuta*. (Based on Teuchert 1977.) Nematoda: *Caenorhabditis elegans* (see Fig. 35.1). Priapula: *Priapulus caudatus*. (Based on Apel 1885.) Kinorhyncha: generalized kinorhynch. (Redrawn from Hennig 1984.)

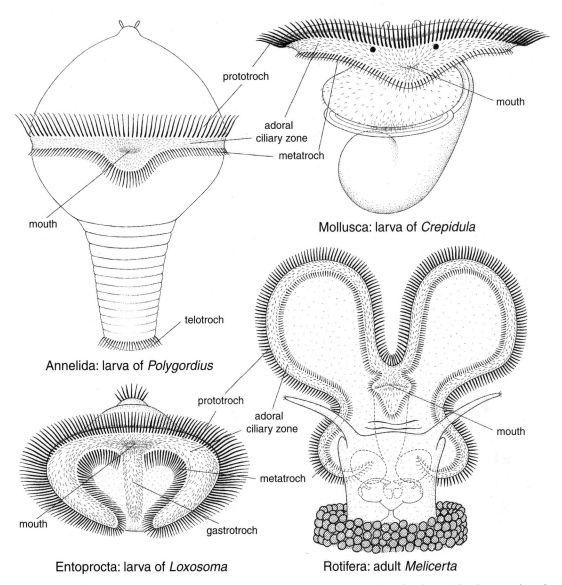

Figure 22.4. Planktotrophic spiralian larvae and an adult rotifer, showing the ciliary bands considered ancestral in the Protostomia. (*Polygordius appendiculatus* based on Hatschek 1878; the other three drawings from Nielsen 1987.)

ment of the embryos have decisive influence both on the cleavage patterns and on gastrulation. In many species a normal gastrulation can hardly be recognized, and a blastopore is not recognized as an opening to an archenteron. Many examples of epibolic gastrulation, superficial cleavage, discoidal cleavage, etc. will be mentioned in the discussions of the various phyla, but most of these cases can be interpreted as variations on the general theme of holoblastic cleavage, followed by embolic gastrulation. Fusion of the lateral blastoporal lips is only observed directly in few species representing few groups, but the trochaea theory (see below) proposes that the protostomian ancestor had a nerve ring along the blastopore rim, with an archaeotroch, and the development of the ventral nerve cords from areas

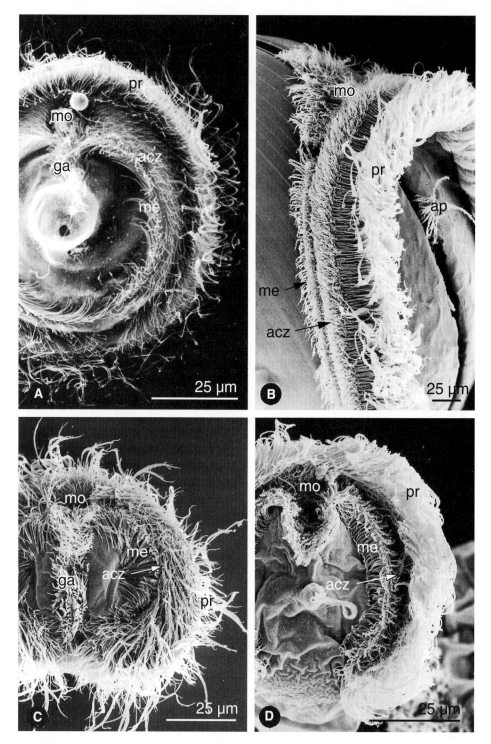

Figure 22.5. Oral area with prototroch, adoral ciliary zone, metatroch, and gastrotroch of larval and adult protostomes (SEM). (A) Larva of the polychaete *Serpula oregonensis* (Friday Harbor Laboratories, WA, USA, July 1980). (B) Larva of the bivalve *Barnea candida* (plankton, off Frederikshavn, Denmark, August 1984). (C) Larva of the entoproct *Loxosoma pectinaricola* (Øresund, Denmark, October 1981). (D) Adult of the rotifer *Conochilus unicornis* (Almind Lake, Denmark, May 1983). acz, adoral ciliary zone; ap, apical organ; ga, gastrotroch; me, metatroch; mo, mouth; pr, prototroch. (See also Nielsen 1987.)

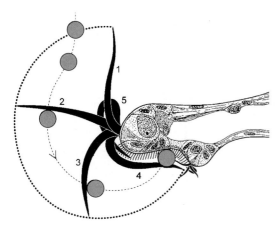

Figure 22.6. Diagram of the downstream-collecting ciliary system at the edge of the velum of a gastropod. The large compound cilia of the prototroch catch up with the particles in the water and push them into the adoral ciliary zone; the opposed beat of the metatroch apparently prevents the particles escaping from the ciliary band. (After Riisgård *et al.* 2000.)

corresponding to the lateral blastopore lips can be followed in most of the protostomian groups, also in groups with epibolic gastrulation, and direct development. In annelids, the two longitudinal cords develop from the ectoderm along the lateral blastopore lips or corresponding lateral areas, for example in 'polychaetes' with lecithotrophic larvae (Fig. 25.4) and in the leeches, which have yolk-rich eggs and direct development (Fig. 25.8; Table 25.3). In the nematodes, which have epibolic gastrulation with a few endodermal cells, the lateral blastopore closure can nevertheless clearly be recognized, and the development of the median nerve cord from the ectoderm of the lateral blastopore lips has been documented in every detail (Fig. 49.3). A further example is the chaetognaths, where the ventral 'ganglion' develops from paired ectodermal bands that fuse in the midline (Fig. 55.1).

The lack of an anus in the Platyhelminthes is discussed in Chapter 29.

The paired or fused ventral, longitudinal nerve cord is perhaps the most stable character of the protostomes (Table 22.1). As mentioned above, the cords differentiate from areas along the lateral blastopore lips of the

embryo, or from comparable areas in species with modified embryology. This is well known from many studies of various spiralians, whereas the embryology of many of the ecdysozoan phyla is poorly studied or completely unknown. However, the detailed studies of the nematode *Caenorhabditis* have shown that the ventral nervous cells originate from the lateral blastoporal lips (Fig. 49.3). The position of the ventral cord(s) is intraepithelial in the early stages, and it remains so in some phyla, whereas it becomes internalized in some types within other phyla (Fig. 25.4). A ventral cord in the primitive, intraepithelial/basiepithelial position, i.e. between the ventral epithelium and its basal lamina, is found in adults of annelids (both in several of the primitive 'polychaete' families and in some oligochaetes), arthropods, chaetognaths, gastrotrichs, nematodes, kinorhynchs, loriciferans, and priapulans. In other groups, for example nematomorphs, the cord is folded in and almost detached from the epithelium, but it is surrounded by the peritoneum and a basal membrane, which is continuous with that of the ventral epithelium. A completely detached nerve cord surrounded by peritoneum and the basal membrane is found in sipunculans, molluscs, many annelids, onychophorans, tardigrades, and arthropods.

Nemertines and platyhelminths appear to lack the ventral component of the protostomian nervous system (Chapters 28, 29). Their main nervous system develops entirely from the cerebral ganglia, and there is no indication of ventral nerves originating from a blastopore closure. Their larvae have diminutive hypospheres that can be interpreted as a reduction of the ventral part of the body.

Polyzoans are sessile and lack cerebral ganglia and the ventral nerve cord (Chapters 35–38).

The main part of the spiralian brain, the cerebral ganglia, develop from paired areas of the larval episphere, but a small part originates from the anterior side of the circumblastoporal nerve ring (observed in annelids, see below). A similar organization of the ecdysozoan brains can perhaps be recognized in the arthropod nervous system (Chapter 45).

The 'pioneer cells' (cells located at the posterior pole of the embryo, and that are believed to guide the

Planktotrophic larvae	Pericalymma larvae type 1	Pericalymma larvae type 2	Pericalymma larvae type 3

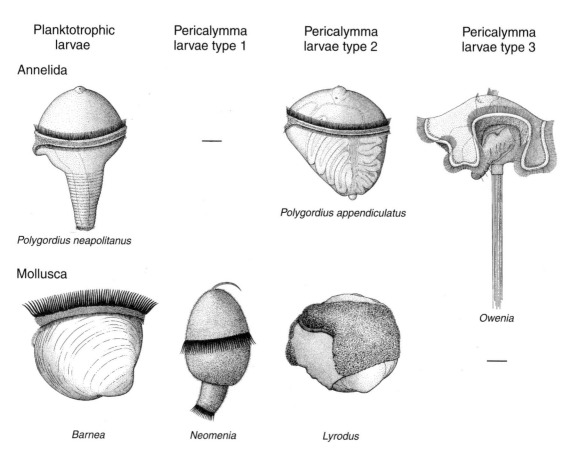

Annelida

Polygordius neapolitanus

Polygordius appendiculatus

Owenia

Mollusca

Barnea   *Neomenia*   *Lyrodus*

Figure 22.7. Normal trochophore larvae and different types of pericalymma larvae of molluscs and annelids. Annelida: planktotrophic trochophore of *Polygordius neapolitanus*. (based on Woltereck 1902); pericalymma larva type 2: endolarva of *Polygordius appendiculatus* (based on Herrmann 1986); pericalymma larva type 3: endolarva of *Owenia fusiformis* (after Wilson 1932). Mollusca: planktotrophic veliger larva of the bivalve *Barnea candida* (based on Nielsen 1987); pericalymma larva type 1: lecithotrophic larva of the solenogaster *Neomenia carinata* (based on Thompson 1960); pericalymma larva type 2: brooded larval stage of the bivalve *Lyrodus pedicellatus*. (Long Beach, CA, USA; after a scanning micrograph by Drs C.B. Calloway and R.D. Turner, Museum of Comparative Zoology, Harvard University, MA, USA.)

neurons during the establishment of the ventral nerve cords) may be another apomorphy of the protostomes. They have been reported in annelids (*Platytnereis*; Dorresteijn *et al.* 1993), molluscs (*Mytilus*; Raineri 1995), arthropods (*Artemia* and *Gonodactylus*; Fischer and Scholtz 2010), and possibly in platyhelminths (*Maritigrella*; Rawlinson 2010).

The trochophora larva was pointed out as a protostomian characteristic already by Hatschek (1891). As stated above, the lack of primary larvae in the ecdysozoans must be a derived character. The ancestral trochophore is believed to have been planktotrophic with

a tube-shaped gut, and the five ciliary bands found, for example in the planktotrophic larvae of the annelids *Polygordius* and *Echiurus*, are: (1) a prototroch of compound cilia anterior to the mouth (a small dorsal break in this band is observed in early stages of several spiralians; see Chapter 23); (2) an adoral ciliary zone of separate cilia surrounding the mouth and with lateral extensions between proto- and metatroch; (3) a metatroch of compound cilia with a break behind the mouth; (4) a gastrotroch that is a midventral extension of the adoral from the mouth to the anus; and (5) a telotroch of compound cilia surrounding the anus, except for a break at the ventral side. All these bands are formed from multiciliate cells. The prototroch is the main locomotory organ, together with the telotroch when this is present. Prototroch, adoral ciliary zone, and metatroch form the feeding organ, which is a downstream-collecting system in that the large compound cilia of the opposed prototroch and metatroch cut through the water and catch up with particles, which are then pushed onto the adoral ciliary zone, this in turn carries them to the mouth (Fig. 22.6). This is called the 'catch-up principle' (Riisgård *et al.* 2000), and it has only been found in spiralian phyla, viz. annelids, sipunculans, molluscs, rotifers, cycliophorans, and entoprocts (Nielsen 1987). The particles may be rejected at the mouth and transported along the gastrotroch to the anal region where they leave the water currents of the larva. Other characteristics of trochophore larvae include the presence of an apical ganglion (probably a eumetazoan character) and a pair of protonephridia.

The larvae of platyhelminths and nemertines may be interpreted as derived trochophores, and this is supported by the identical cell lineage of the prototrochs (Chapters 28, 29).

However, only the planktotrophic larvae have the metatroch, and other of the characteristic ciliary bands are absent in various types of non-feeding larvae. Among the phyla, here referred to the Protostomia, only the bryozoans (ectoprocts) have larvae with a different ciliary filter system, which is of a unique structure (Chapter 38). Some rotifers have a ciliary feeding organ that is of exactly the same structure and function as that of the *Polygordius* and *Echiurus* larvae, and they have been interpreted as progenetic trochophores (Chapter 34).

None of the deuterostome larvae have downstream-collecting ciliary bands, and the upstream-collecting ciliary bands found in the larval (and some adult) ambulacrarians are in all cases made up of separate cilia on monociliate cells in contrast to the compound cilia on multiciliate cells observed in the protostomes.

The original concept of the trochophora larva (Hatschek 1891) comprised also the actinotroch larva of *Phoronis*. The sharpened definition of structure and function of the ciliary bands given above (see also Nielsen 1985, 1987) excludes the actinotroch because it has an upstream-collecting ciliary system and ciliary bands consisting of separate cilia on monociliate cells.

The trochophora concept has been attacked several times (Salvini-Plawen 1980; Rouse 1999, 2000a,b), but I find it useful to maintain the term for the hypothetical ancestral planktotrophic larva, and to use it for actual larvae that have all or most of the ciliary bands described above; it provides a set of names for the various ciliary bands, which are very useful in comparisons. The cell-lineage studies of several spiralian embryos have clearly demonstrated the homology of the prototrochs in the various phyla (see for example Henry *et al.* 2007).

Some authors believe that the original protostomian (and bilaterian in general) life cycle was a direct one, and that the larval stages, especially the planktotrophic trochophore, were 'intercalated' in several lineages (see for example Sly *et al.* 2003). I have discussed this in some detail (Nielsen 2009) and come to the conclusions that a non-feeding planula must be a very unlikely ancestor, and that the characteristic planktotrophic ciliated larvae found in cnidarians, spiralians, and ambulacrarians most probably have originated from a common holopelagic planktotrophic ancestor, gastraea, that gave rise to pelago-benthic descendents through addition of adult benthic stages, with retention of the planktotrophic stage as a larva (Fig. 10.3).

A convergent evolution of the morphologically and functionally identical filter-feeding ciliary systems, with identical cell lineages in planktotrophic larvae from uniformly ciliated lecithotrophic larvae in several lineages, appears highly unlikely. An origin of a locomotory prototroch (and a telotroch) could well be envisaged as an adaptation to more efficient swimming (Emlet 1991). However, the evolution of a metatroch, which is apparently a necessary component in the downstream or opposed-band system, from a uniform post-prototrochal ciliation, as proposed by various authors (for example Salvini-Plawen 1980; Ivanova-Kazas 1987; Ivanova-Kazas and Ivanov 1988), appears highly unlikely. A developing metatroch, which is without any function before it is fully developed (and that may even hamper the swimming because the direction of its beat is opposite to that of the prototroch), appears without adaptational value, and a driving force for the evolution is therefore lacking.

Hejnol *et al.* (2007) and Henry *et al.* (2007) proposed that the row of secondary trochoblasts of the *Crepidula* larva in some way has become separated from the primary trochoblasts, and become the metatroch. This would not only require an unexplained development of the adoral ciliary zone separating the two bands of compound cilia, but also a reversal of the beat of the metatroch cells. It appears impossible to imagine the intermediate stages in this evolution and the necessary adaptational advantages leading through this process.

Miner *et al.* (1999) observed that small larvae of the polychaete *Armandia* collect particles using their downstream bands, and that both the length of the bands and of the prototroch cilia increase with growth, thus increasing the filtration capacity. However, this increase cannot catch up with the strong growth of the adult, segmented body, and some of the oral cilia increase in size and become specialized for capturing large particles, which are handled individually. This indicates that the downstream system is phylogenetically older, and that the various systems of large oral cilia found in several other polychaete larvae, for example the ciliary 'brush' found on the left side of the mouth in polynoid larvae, are derived.

Salvini-Plawen (1980) restricted the term 'trochophora' for the larvae of annelids and echiurans, and created new names for larvae of other phyla. The mollusc larvae, for example, were stated generally to lack both a metatroch and protonephridia, but a metatroch is present in almost all planktotrophic larvae of gastropods and bivalves, and protonephridia are now considered to be an ancestral character of mollusc larvae (Chapter 27).

A useful term introduced by Salvini-Plawen (1972, 1980) is the 'pericalymma' larva, which comprises trochophore-like larvae with most of the hyposphere covered by a usually ciliated expansion (often called serosa)

from an anterior zone (these larvae have also been called 'Hüllglocken', test-cell, or serosa larvae). However, it is important to note that these expansions originate from different areas in different larvae (Fig. 22.7): In the larvae of molluscs such as the protobranch bivalves *Yoldia* and *Acila* (Drew 1899; Zardus and Morse 1998), the solenogaster *Neomenia*, and in the sipunculan *Sipunculus* (Hatschek 1883), the expansion originates from the prototroch area (or from the episphere) (type 1); the three bands of compound cilia in the mollusc larvae correspond to the three rows of cells that form the prototroch in scaphopods (van Dongen and Geilenkirchen 1974). In the annelids *Polygordius lacteus* and the phyllodocids (Pernet *et al.* 2002), and in the bivalve *Lyrodus*, the expansion originates from the zone just behind the mouth (type 2). In the oweniid polychaetes, the serosa is formed from an area further away from the ciliary bands, viz. a zone between the two first segments with parapodia, so that the very long setae of the first segment are exposed (type 3). It should be clear that the serosae of these three types of pericalymma larvae are non-homologous. The fact that variations from the more usual planktotrophic trochophore larvae to pericalymma types can be observed even within a genus, makes it very unlikely that the evolution has gone from various lecithotrophic pericalymma larvae to the structurally and functionally complicated planktotrophic trochophores, which are very similar among the phyla. The pericalymma larvae must thus be interpreted as independent specializations in different phylogenetic lines.

Quadrant cleavage and blastomere fates. In most of the studied protostomes, one cell of the 4-cell stage is different from the other three (Fig. 21.1) and gives rise to the endomesoderm and germ cells, and I propose the term 'quadrant cleavage' for this type of cleavage.

In the spiralians, the two first-cleavage furrows separate an anterior, a pair of lateral, and a posterior blastomere (called D). A germ line has been demonstrated in a few spiralian phyla, but it is generally believed that the germ cells differentiate from the endomesoderm, which derives from the 4d cell. Direct observations are few, but the presence of a cell lineage

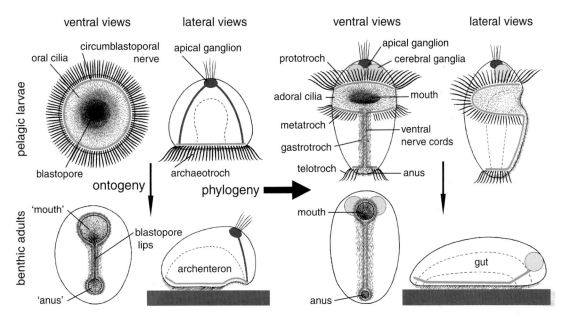

Figure 22.8. The trochaea theory. Evolution of the pelago-benthic protostomian ancestor gastroneuron (right) from a radially symmetric, holopelagic trochaea (upper left), through an advanced, bilaterally symmetric pelago-benthic trochaea (left). Short thin lines, separate cilia; long thick lines, compound cilia; red, apical organ; yellow, cerebral ganglia; green, circumblastoporal nerve ring. (See Nielsen and Nørrevang 1985; Nielsen 2010.) See Colour Plate 3.

is well documented in the mollusc *Sphaerium* (Woods 1932) (Chapter 27). In the rotifer *Asplanchna*, the 6D cell is the primordial germ cell (Table 34.1). Phoronids and brachiopods have cleavage patterns and late blastomere specification, as seen in the deuterostomes (see Chapters 39–41).

In the ecdysozoans, the existence of a special cell of the 4-cell stage has been demonstrated in a few arthropods (Chapter 44). A large D cell is seen in the arthropod *Semibalanus* (Fig. 44.3; Table 44.2), and a corresponding, but not large cell has been observed in the decapod *Sicyonia* (Table 44.3). One of the blastomeres of the 4-cell stages of *Parhyale* and *Holopedium* is a germinal cell, and corresponds to the D cell of the spiralians (Baldass 1937; Gerberding *et al.* 2002). The lineage of the germinal cells is well documented in the nematodes *Caenorhabditis* and *Romanomermis*, where the P_4 cell of the 4-cell stage is the primordial germ cell (Tables 49.1 and 49.2). The other cycloneuralian phyla have not been studied.

The chaetognath *Sagitta* has a 'germ-cell determinant' that has been followed from the zygote to the germ cells (Fig. 55.1).

The deuterostomes generally show the two first cleavages separating anterior right and left, and posterior right and left blastomeres (Fig. 21.1). The deuterostomes do not show a germ line.

Blastomere specification It has been known for more than a century that isolated blastomeres of spiralian embryos in some species develop as if they were still in their place in the embryo, as shown for example for *Patella* (Wilson 1904) and *Nereis* (Costello 1945). This type of development with very limited powers of regulation has been called mosaic development. It has been extensively studied in annelids (Chapter 25) and molluscs (Chapter 27), but studies of other phyla are scarce. The nemertine *Cerebratulus* is reported to regulate after isolation of blastomeres of the 4-cell stage, but the larvae have not been followed to metamorphosis (Hörstadius 1937). The platyhelminth *Hoploplana*

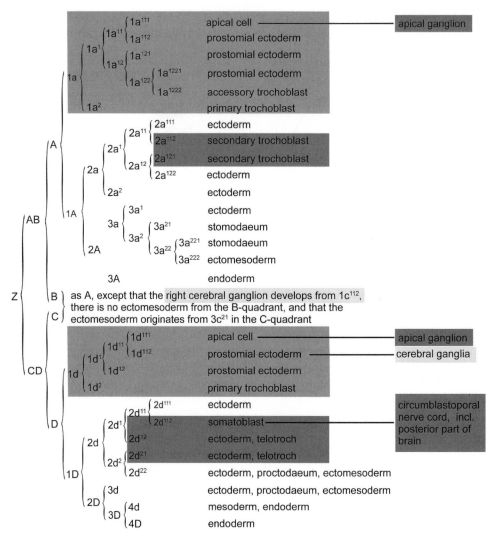

Figure 22.9. Generalized cell lineage of an annelid. (Based on *Podarke* and *Arenicola*; Child 1900; Treadwell 1901.) See Colour Plate 4.

showed only partial regulation after deletion of one blastomere at the 2- and 4-cell stages (Boyer 1989). The rotifer *Asplanchna* shows a high degree of determination already at the earliest cleavage stages (Lechner 1966). Zimmer (1973) reported 'full regulation' of isolated blastomeres of 2-cell stages in the bryozoan *Membranipora*, but without further details. Nematodes show very limited powers of regulation (Chapter 49).

The trochaea theory (Fig. 22.8–10) The protostomian characteristics discussed above have led me to propose

the 'trochaea theory' (Nielsen 1979, 1985; Nielsen and Nørrevang 1985), which proposed the trochaea as the ancestor of all the bilaterians, but I have subsequently restricted it to the Protostomia (Nielsen 2001, 2009).

The theory proposes that an early protostomian ancestor was a uniformly ciliated, planktotrophic gastraea that developed a ring of special locomotory cilia around the blastopore (the archaeotroch). It appears that the multiciliate cells evolved at this stage, and that the cilia of the archaeotroch became organized as

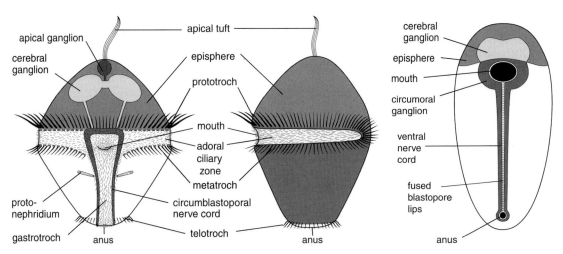

Figure 22.10. Larval regions of epidermis and nervous systems of a larva and an adult of the ancestral spiralian, gastroneuron, with a trochophora larva and a benthic adult. The two left diagrams show the trochophore from the ventral and the left side, and the right diagram shows the adult from the ventral side. The right and left diagrams show the development of the nervous system, and the middle diagram shows the embryological origin of the main epithelial areas. The blue areas in the middle diagram indicate the secondary trochoblasts originating from the second micromere quartet, from the cells $2a^2$–$2d^2$, and the whole area covered by the descendents of the 2d cell, which have spread enormously to cover almost the whole post-trochal region. See Colour Plate 5.

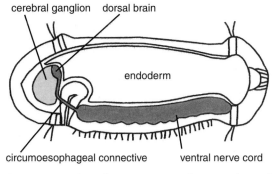

Figure 22.11. Central nervous system of a stage-6 larva of *Capitella telata* in left view. The left cerebral ganglion (yellow) is derived from 1c cells, and the circumblastoporal nerves (green, posterodorsal brain, circumoesophageal connectives, and ventral nerve cord) are derived from 2d cells. (Based on Meyer and Seaver 2010.) See Colour Plate 6.

compound cilia, which can grow longer and carry out the usual effective stroke with increased force. The archaeotroch became specialized as a downstream-collecting system, based on the catch-up principle (Riisgård *et al.* 2000) to strain particles from the water and transfer them to the cilia of the area surrounding the mouth. A functional ring-shaped 'archaeotroch' is found in the cycliophoran *Symbion* (Chapter 37). This hypothetical holopelagic stage is called trochaea.

If a trochaea went down to the bottom and stopped the ciliary beat of the archaeotroch at the end of the recovery stroke, it could collect deposited detritus particles from the bottom with the blastoporal cilia. However, if the detritus on the bottom should be exploited efficiently it would be advantageous for the organism to be able to move along the bottom, and a preferred direction of creeping became established. The blastopore elongated along the new anteroposterior axis perpendicular to the primary, apical-blastoporal axis. This created one-way traffic along the archenteron, from the anterior to the posterior end, and this movement and the digestion of food particles could be enhanced by pressing the lateral blastopore lips together, creating a functionally tubular gut. The lateral blastopore closure could later become permanent by a fusion of the lips, with the anterior mouth and the posterior anus as the only remains of the blastopore.

This evolution is perhaps best illustrated by the annelid development, where the cell lineage is so well known (Fig. 22.9; see also Table 25.2). The elongation of the body along the new axis is caused by a disproportionate growth of the D quadrant (Wilson 1892; Mead 1897; Shankland and Seaver 2000), especially of the 2d cell, the somatoblast (Fig. 22.10). The micromeres of the first quartet form the epithelium of the prostomium, and the descendants of the other micromeres of the A–C quadrants are restricted to the head and mouth region. The descendants of the 2d cell expand enormously over the whole post-oral part of the body as the somatic plate, finally covering almost the whole post-trochal region and fusing along the ventral midline (Fig. 22.10). Its posterior edge carries the telotroch surrounding the anus. Thus, the 'blastoporal edge' of the second micromere quartet carries the ancestral archaeotroch, which is retained as the secondary prototroch cells, the metatroch (see Chapter 25), and the telotroch. The circumblastoporal nerve is retained as a small posterior brain region, a pair of circumoesophageal connectives, the paired ventral nerve cords, and a loop around the anus (Fig. 22.11). The band of compound cilia had lost its functions in the benthic adult and disappeared.

However, the pelagic stage was retained as a dispersal stage, and its structure, including that of the ciliary bands, changed in connection with the changes in the adult morphology (adultation; see Jägersten 1972). With the lateral blastopore closure, the parts of the archaeotroch along the fusing blastopore lips disappeared, whereas the anterior part, now surrounding the mouth, developed lateral loops so that its pre-oral part became the large prototroch used also for swimming and the post-oral part became the smaller metatroch; the posterior part surrounding the new anus became the telotroch. The circumblastoporal field of separate cilia became the adoral ciliary field around the mouth, with extensions between proto- and metatroch, and the gastrotroch along the ventral side. This pattern of ciliary bands can be recognized in several protostome larvae and in some adult rotifers (Figs 22.4 and 22.5). This ancestral protostome, with a pelago-benthic life cycle, is called gastroneuron (Nielsen 1985; Nielsen and Nørrevang 1985).

The trochaea had a small apical ganglion with connections to a ring nerve along the archaeotroch. This ganglion is found in the trochophore, where it disappears at metamorphosis; it is a plesiomorphy, found also in cnidarian and deuterostome larvae (Fig. 10.1). A pair of cerebral ganglia developed near the new anterior pole. The ring nerve along the archaeotroch became deformed by the blastopore closure to attain the shape of a loop around the mouth, a pair of longitudinal nerves along the blastopore lips, and a small loop around the anus. This corresponds to the idea that the whole bilaterian central nervous system is homologous to the circumblastoporal nerve ring of the cnidarians (Koizumi 2007). The cerebral ganglia fused with the anterior loop forming the brain. This is the central nervous system seen in most protostomes (Fig. 22.3). The morphology of the arthropod nervous system indicated that the protocerebrum is the homologue of the cerebral ganglia, and the deutocerebrum (surrounding the oesophagus) represents the anterior part of the circumblastoporal loop (Anderson 1969; Harzsch et al. 2005).

It appears that the trochaea theory is the only available theory that explains the origin and evolution of the complex of the tubular gut, the characteristic trochophora larva, and the characteristic nervous system through continuous modifications of functioning organisms. All other theories propose changes without explaining the structure and function of the intermediary stages, and none of the theories explain the adaptational advantages of the changes. Both of these requirements should, when possible, be fulfilled in acceptable evolutionary theories (Frazzetta 1975).

An alternative theory of the origin and evolution of the nervous system of protostomes is the orthogon theory (Reisinger 1925; Hanström 1928; Reisinger 1972). It envisages a transformation series, from the diffuse nervous net of the cnidarians via an orthogonal nervous system with a circumoral brain, and typically eight longitudinal nerves and a series of annular nerves, to a nervous system with only a pair of midventral longitudinal nerves. However, a typical orthogon is only found in a few platyhelminths; it is definitely absent in nematodes (Chapter 49) (and other cycloneuralians), and its presence in annelids and molluscs is in the eye of the beholder.

A number of other characters have been used in characterizing major bilaterian groups.

Mesoderm and coelom have played major roles in phylogenetic discussions. It has often been stated that one of the major differences between protostomes and deuterostomes is that the protostomes form coelomic cavities through schizocoely, and the deuterostomes through enterocoely. There is some truth in the statement, but when the variation within the two groups is considered it turns out that the situation is much more complex, and that the origin of the mesoderm, which by definition surrounds the coelomic cavities, is perhaps more significant. The mesoderm and coelom of the deuterostomes are discussed in Chapter 56, and here it should suffice to mention that their mesoderm in all cases originates from the walls of the archenteron (and from the neural crest in the chordates); the coelomic pouches are often formed through enterocoely, but even among, for example, enteropneusts there is a variation ranging from typical schizocoely to typical enterocoely (Chapter 60; see also Fig. 57.2).

In the protostomes, the mesoderm originates from the blastopore lips and as ectomesoderm, and the coelomic cavities originate through schizocoely; only the chaetognaths form an exception, with coelomic pouches originating from the archenteron (Chapter 55). Well-defined coeloms are found in Schizoecoelia (Annelida, Sipuncula, and Mollusca), Ectoprocta, Arthropoda, Onychophora, and Chaetognatha, but their morphology shows much variation. Many of the schizocoelians have metanephridia that drain and modify the primary urine in the coelomic sacs and in most cases function as gonoducts (see Chapter 21). There is an enormous literature discussing the phylogenetic importance of coeloms and metanephridia, but it appears that the structure of coeloms is of such low complexity that its phylogenetic significance is very limited.

Protonephridia are probably an ancestral character of the trochophora larva (see Chapter 21). This type of excretory organ is found in adults of most of the protostome phyla. With the brachiozoan phyla included in the Protostomia, the protonephridia appears to be a synapomorphy.

Segmentation, i.e. repetition of structural units comprising a pair of coelomic sacs and associated structures, such as metanephridia and ganglia, is seen in annelids and arthropods, and the several significant similarities of segmentation of these groups has been emphasized by many authors, arguing for the 'Articulata hypothesis' (Scholtz 2002), which considers annelids and arthropods as sister groups. This is discussed further in Chapter 21.

Special genes Specific, rarely changing amino acids in the mitochondrial gene *nad5* have been identified only in protostomian phyla, viz. annelids, molluscs, platyhelminths, brachiopods, chaetognaths, priapulids, nematodes, and arthropods (Papillon *et al.* 2004; Telford *et al.* 2008). This seems to be an important marker for this large clade.

Hox genes show a characteristic protostomian signature (Fig. 21.3), with the genes *ftz* and *Antp* found only in the protostomian phyla. Also the clades Spiralia and Ecdysozoa are characterized by special signatures.

As emphasized above, the protostomes fall into two morphologically rather well-separated groups: Spiralia (Lophotrochozoa) (Chapter 23) and Ecdysozoa (Chapter 42), and it appears that the spiralians have retained a number of the ancestral characters that have been lost in the ecdysozoans. This could indicate that the Ecdysozoa is an ingroup of the Spiralia, but the molecular analyses, including the analyses of Hox genes, are almost unanimous about the monophyly of the Spiralia. The interpretation proposed here, viz. that the protostomian ancestor had a pelago-benthic life cycle with a trochophora larva, and that the ecdysozoans secondarily lost the ciliation, the trochophore, and perhaps even the spiral cleavage, is in full accordance with the modern 'tree thinking' (Baum *et al.* 2005).

The reasons for including the Brachiozoa (Phoronida + Brachiopoda) in the Spiralia are discussed in Chapter 39.

Interesting subjects for future research

1. Are the cerebral ganglia of the annelids and the protocerebrum of the arthropods homologous?

References

Anderson, D.T. 1969. On the embryology of the cirripede crustaceans *Tetraclita rosea* (Krauss), *Tetraclita purpurascens* (Wood), *Chthamalus antennatus* (Darwin) and *Chamaesipho columna* (Spengler) and some considerations of crustacean phylogenetic relationships. *Phil. Trans. R. Soc. Lond. B* **256**: 183–235.

Anderson, D.T. 1973. *Embryology and Phylogeny of Annelids and Arthropods (International Series of Monographs in pure and applied Biology, Zoology 50)*. Pergamon Press, Oxford.

Apel, W. 1885. Beitrag zur Anatomie und Histologie des *Priapulus caudatus* Lam und des *Halicryptus spinulosus* (v. Sieb.). *Z. Wiss. Zool.* **42**: 459–529.

Arendt, D. and Nübler-Jung, K. 1997. Dorsal or ventral: similarities in fate maps and gastrulation patterns in annelids, arthropods and chordates. *Mech. Dev.* **61**: 7–21.

Baldass, F. 1937. Entwicklung von *Holopedium gibberum*. *Zool. Jahrb. Anat.* **63**: 399–454.

Baum, D.A., Smith, S.D. and Donovan, S.S. 2005. The tree-thinking challenge. *Science* **310**: 979–980.

Boyer, B.C. 1989. The role of the first quartet micromeres in the development of the polyclad *Hoploplana inquilina*. *Biol. Bull.* **177**: 338–343.

Child, C.M. 1900. The early development of *Arenicola* and *Sternaspis*. *Arch Entwicklungsmech. Org.* **9**: 587–723, pls 521–525.

Costello, D.P. 1945. Experimental studies of germinal localization in *Nereis*. *J. Exp. Zool.* **100**: 19–66.

Cuénot, L. 1949. Les Tardigrades. *Traité de Zoologie*, vol. 6, pp. 39–59. Masson, Paris.

Davis, C.W.C. 1967. A comparative study of larval embryogenesis in the mosquito *Culex fatigans* Wiedemann (Diptera: Culicidae) and the sheep fly *Lucilia sericata* Meigen (Diptera: Calliphoridae). *Aust. J. Zool.* **15**: 547–579.

Dorresteijn, A.W.C., O'Grady, B., Fischer, A., Porchet-Henner, E. and Boilly-Marer, Y. 1993. Molecular specification of cell lines in the embryo of *Platynereis* (Annelida). *Roux's Arch. Dev. Biol.* **202**: 260–269.

Drew, G.A. 1899. Some observations on the habits, anatomy and embryology of members of the Protobranchia. *Anat. Anz.* **15**: 493–519.

Emlet, R.B. 1991. Functional constraints on the evolution of larval forms of marine invertebrates: experimental and comparative evidence. *Am. Zool.* **31**: 707–725.

Fischer, A.H.L. and Scholtz, G. 2010. Axogenesis in the stomatopod crustacean *Gonodactylus falcatus* (Malacostraca). *Invert. Biol.* **129**: 59–76.

Frazzetta, T.H. 1975. *Complex Adaptations in Evolving Populations*. Sinauer Associates, Sunderland, MA.

Gerberding, M., Browne, W.E. and Tatel, N.H. 2002. Cell lineage analysis of the amphiopod crustacean *Parahyale hawaiensis* reveals an early restriction of cell fates. *Development* **129**: 5789–5801.

Goto, T. and Yoshida, M. 1987. Nervous system in Chaetognatha. In M.A. Ali (ed.): *Nervous Systems in Invertebrates* (NATO ASI, Ser. A 141), pp. 461–481. Plenum Press, New York.

Grobben, K. 1908. Die systematische Einteilung des Tierreichs. *Verh. Zool.-Bot. Ges. Wien* **58**: 491–511.

Hanström, B. 1928. *Vergleichende Anatomie des Nervensystems der wirbellosen Tiere*. Springer, Berlin.

Harzsch, S., Wildt, M., Battelle, B. and Waloszek, D. 2005. Immunohistochemical localization of neurotransmitters in the nervous system of larval *Limulus polyphemus* (Chelicerata, Xiphosura): evidence for a conserved protocerebral architecture in Euarthropoda. *Arthropod Struct. Dev.* **34**: 327–342.

Hatschek, B. 1878. Studien über Entwicklungsgeschichte der Anneliden. *Arb. Zool. Inst. University Wien.* **1**: 277–404.

Hatschek, B. 1883. Über Entwicklung von *Sipunculus nudus*. *Arb. Zool. Inst. University Wien.* **5**: 61–140.

Hatschek, B. 1888. *Lehrbuch der Zoologie, 1. Lieferung* (pp. 1–144). Gustav Fischer, Jena.

Hatschek, B. 1891. *Lehrbuch der Zoologie, 3. Lieferung* (pp. 305–432). Gustav Fischer, Jena.

Hejnol, A., Martindale, M.Q. and Henry, J.Q. 2007. High-resolution fate map of the snail *Crepidula fornicata*: the origins of ciliary bands, nervous system, and muscular elements. *Dev. Biol.* **305**: 63–76.

Hennig, W. 1984. *Taschenbuch der Zoologie, Band 2, Wirbellose I*. Gustav Fischer, Jena.

Henry, J.Q., Hejnol, A., Perry, K.J. and Martindale, M.Q. 2007. Homology of ciliary bands in spiralian trochophores. *Integr. Comp. Biol.* **47**: 865–871.

Herrmann, K. 1986. *Polygordius appendiculatus* (Archiannelida)—Metamorphose. Publ. Wiss. Film., Biol., 18. Ser. **36**: 1–15.

Hörstadius, S. 1937. Experiments on determination in the early development of *Cerebratulus lacteus*. *Biol. Bull.* **73**: 317–342.

Ivanova-Kazas, O.M. 1987. Origin, evolution and phylogenetic significance of ciliated larvae. *Zool. Zh.* **66**: 325–338.

Ivanova-Kazas, O.M. and Ivanov, A.V. 1988. Trochaea theory and phylogenetic significance of ciliate larvae. *Sov. J. Mar. Biol.* **13**: 67–80.

Jägersten, G. 1972. *Evolution of the Metazoan Life Cycle*. Academic Press, London.

Koizumi, O. 2007. Nerve ring of the hypostome in *Hydra*: is it an origin of the central nervous system of bilaterian animals? *Brain Behav. Evol.* **69**: 151–159.

Lechner, M. 1966. Untersuchungen zur Embryonalentwicklung des Rädertieres *Asplanchna girodi* de Guerne. *Wilhelm Roux' Arch. Entwicklungmech. Org.* **157**: 117–173.

Malakhov, V.V. 1986. *Nematodes. Anatomy, Development, Systematics and Phylogeny*. Nauka, Moskva (In Russian).

Manton, S.F. 1949. Studies on the Onychophora VII. The early embryonic stages of *Peripatopsis*, and some general considerations concerning the morphology and phylogeny of the Arthropoda. *Phil. Trans. R. Soc. Lond. B* **233**: 483–580.

Mead, A.D. 1897. The early development of marine annelids. *J. Morphol.* **13**: 227–326.

Metalnikoff, S. 1900. *Sipunculus nudus*. *Z. Wiss. Zool.* **68**: 261–322.

Meyer, N.P. and Seaver, E. 2010. Cell lineage and fate map of the primary somatoblast of the polychaete annelid *Capitella teleta*. *Integr. Comp. Biol.* **50**: 756–767.

Miner, B.G., Sanford, E., Strathmann, R.R., Pernet, B. and Emlet, R.E. 1999. Functional and evolutionary implications of opposed bands, big mouths, and extensive oral ciliation in larval opheliids and echiurids (Annelida). *Biol. Bull.* **197**: 14–25.

Nielsen, C. 1979. Larval ciliary bands and metazoan phylogeny. *Fortschr. Zool. Syst. Evolutionsforsch.* **1**: 178–184.

Nielsen, C. 1985. Animal phylogeny in the light of the trochaea theory. *Biol. J. Linn. Soc.* **25**: 243–299.

Nielsen, C. 1987. Structure and function of metazoan ciliary bands and their phylogenetic significance. *Acta Zool. (Stockh.)* **68**: 205–262.

Nielsen, C. 2001. *Animal Evolution: Interrelationships of the Living Phyla*, 2nd ed. Oxford University Press, Oxford.

Nielsen, C. 2009. How did indirect development with planktotrophic larvae evolve? *Biol. Bull.* **216**: 203–215.

Nielsen, C. 2010. Some aspects of spiralian development. *Acta Zool. (Stockh.)* **91**: 20–28.

Nielsen, C. and Nørrevang, A. 1985. The trochaea theory: an example of life cycle phylogeny. In S. Conway Morris, J.D. George, R. Gibson and H.M. Platt (eds): *The Origins and Relationships of Lower Invertebrates*, pp. 28–41. Oxford University Press, Oxford.

Papillon, D., Perez, Y., Caubit, X. and Le Parco, Y. 2004. Identification of chaetognaths as protostomes is supported by the analysis of their mitochondrial genome. *Mol. Biol. Evol.* **21**: 2122–2129.

Pernet, B., Qian, P.Y., Rouse, G., Young, C.M. and Eckelbarger, K.J. 2002. Phylum Annelida: Polychaeta. In C.M. Young (ed.): *Atlas of Marine Invertebrate Larvae*, pp. 209–243. Academic Press, San Diego.

Raineri, M. 1995. Is a mollusc an evolved bent metatrochophore? A histochemical investigation of neurogenesis in *Mytilus* (Mollusca: Bivalvia). *J. Mar. Biol. Assoc. U.K.* **75**: 571–592.

Rawlinson, K.A. 2010. Embryonic and post-embryonic development of the polyclad flatworm *Maritigrella crozieri*; implications for the evolution of spiralian life history traits. *Front. Zool.* **7**: 12.

Reisinger, E. 1925. Untersuchungen am Nervensystem der *Bothrioplana semperi* Braun. *Z. Morphol. Oekol. Tiere* **5**: 119–149.

Reisinger, E. 1972. Die Evolution des Orthogons der Spiralier und das Archicoelomatenproblem. *Z. Zool. Syst. Evolutionsforsch.* **10**: 1–43.

Riisgård, H.U., Nielsen, C. and Larsen, P.S. 2000. Downstream collecting in ciliary suspension feeders: the catch-up principle. *Mar. Ecol. Prog. Ser.* **207**: 33–51.

Rouse, G. 2000a. Bias? What bias? The evolution of downstream larval-feeding in animals. *Zool. Scr.* **29**: 213–236.

Rouse, G.W. 1999. Trochophore concepts: ciliary bands and the evolution of larvae in spiralian Metazoa. *Biol. J. Linn. Soc.* **66**: 411–464.

Rouse, G.W. 2000b. The epitome of hand waving? Larval feeding and hypotheses of metazoan phylogeny. *Evol. Dev.* **2**: 222–233.

Salvini-Plawen, L.V. 1972. Zur Morphologie und Phylogenie der Mollusken: Die Beziehungen der Caudofoveata und der Solenogastres als Aculifera, als Mollusca und als Spiralia. *Z. Wiss. Zool.* **184**: 205–394.

Salvini-Plawen, L.V. 1980. Was ist eine Trochophora? Eine Analyse der Larventypen mariner Protostomier. *Zool. Jahrb., Anat.* **103**: 389–423.

Scholtz, G. 2002. The Articulata hypothesis—or what is a segment? *Org. Divers. Evol.* **2**: 197–215.

Shankland, M. and Seaver, E.C. 2000. Evolution of the bilaterian body plan: what have we learned from annelids? *Proc. Natl. Acad. Sci. USA* **97**: 4434–4437.

Sly, B.J., Snoke, M.S. and Raff, R.A. 2003. Who came first—larvae or adults? Origins of bilaterian metazoan larvae. *Int. J. Dev. Biol.* **47**: 623–632.

Snodgrass, R.E. 1938. Evolution of the Annelida, Onychophora, and Arthropoda. *Smithson. Misc. Collect.* **97(6)**: 1–159.

Storch, V. and Welsch, U. 1991. *Systematische Zoologie*, 4th ed. Gustav Fischer, Stuttgart.

Telford, M.J., Bourlat, S.J., Economou, A., Papillon, D. and Rota-Stabelli, O. 2008. The evolution of the Ecdysozoa. *Phil. Trans. R. Soc. Lond. B* **363**: 1529–1537.

Teuchert, G. 1977. The ultrastructure of the marine gastrotrich *Turbanella cornuta* Remane (Macrodasyoidea) and its functional and phylogenetic importance. *Zoomorphologie* **88**: 189–246.

Thompson, T.E. 1960. The development of *Neomenia carinata* Tullberg (Mollusca Aplacophora). *Proc. R. Soc. Lond. B* **153**: 263–278.

Treadwell, A.L. 1901. Cytogeny of *Podarke obscura* Verrill. *J. Morphol.* **17**: 399–486.

van Dongen, C.A.M. and Geilenkirchen, W.L.M. 1974. The development of *Dentalium* with special reference to the significance of the polar lobe. I–III. Division chronology and development of the cell pattern in *Dentalium dentale* (Scaphopoda). *Proc. K. Ned. Akad. Wet., Ser. C* **77**: 57–100.

Wilson, D.P. 1932. On the mitraria larva of *Owenia fusiformis* Delle Chiaje. *Phil. Trans. R. Soc. Lond. B* **221**: 231–334.

Wilson, E.B. 1892. The cell lineage of *Nereis*. *J. Morphol.* **6**: 361–480.

Wilson, E.B. 1904. Experimental studies in germinal localization. II. Experiments on cleavage-mosaic in *Patella* and *Dentalium*. *J. Exp. Zool.* **1**: 197–268.

Woltereck, R. 1902. Trochophora-Studien I. Histologie der Larve und die Entstehung des Annelids bei den *Polygordius*-Arten der Nordsee. *Zoologica (Stuttg.)* **13**: 1–71.

Woltereck, R. 1904. Wurm 'kopf', Wurmrumpf und Trochophora. *Zool. Anz.* **28**: 273–322.

Woods, F.H. 1932. Keimbahn determinants and continuity of the germ cells in *Sphaerium striatinum* (Lam.). *J. Morphol.* **53**: 345–365.

Zardus, J.D. and Morse, M.P. 1998. Embryogenesis, morphology and ultrastructure of the pericalymma larva of *Acila castrensis* (Bivalvia: Protobranchia: Nuculoida). *Invert. Biol.* **117**: 221–244.

Zimmer, R.L. 1973. Morphological and developmental affinities of the lophophorates. In G.P. Larwood (ed.): *Living and Fossil Bryozoa*, pp. 593–599. Academic Press, London.

23

SPIRALIA (LOPHOTROCHOZOA)

It seems obvious that the presence of ciliated ectoderm and primary larvae in the Spiralia is plesiomorphic, because all the outgroups of the Protostomia are ciliated, and the lack of these two characters and the presence of moulting in the Ecdysozoa must therefore be apomorphic. This makes it easy to define the Ecdysozoa, whereas the Spiralia show many plesiomorphies, which of course cannot be used as defining characters. However, molecular analyses almost unanimously recover the Spiralia as a clade (Dunn *et al.* 2008; Hejnol *et al.* 2009; Grande 2010; Hejnol 2010; Mallatt *et al.* 2010; Pick *et al.* 2010).

The spiral cleavage, which has given the name to the clade, is observed in representatives of most of the phyla included here (Fig. 23.1). It may be a spiralian apomorphy, a specialization of the D-quadrant cleavage (Fig. 21.1), but it cannot be excluded that the protostomian ancestor had spiral cleavage that was then lost in the ecdysozoan lineage together with the ciliated larva. It is proposed here that the protostomian ancestor had ciliated epithelia and a trochophora larva. This is in full accord with the morphology of the nervous systems in adult spiralians and ecdysozoans (see Chapter 22, Fig. 22.3).

Spiralian apomorphies, independent of cleavage patterns and the trochophora complex, are difficult to identify, but a few will be discussed at the end of this chapter.

The spiral cleavage is a highly characteristic developmental type, which exhibits not only very conspicuous cleavage patterns (Fig. 23.2), but also strongly conserved blastomere fates (van den Biggelaar *et al.* 1997; Boyer and Henry 1998; Henry and Martindale 1999; Nielsen 2004, 2005; Lambert 2010). The generally accepted notation for the blastomeres (first used by Conklin 1897 for the cleavage of the gastropod *Crepidula*) is shown in the generalized cell-lineage diagram in Table 23.1 (see also Fig. 22.9). This cleavage type is easily recognized in many of the phyla included in the group, not only in many annelids, molluscs, sipunculans, and entoprocts, which have typical trochophora larvae, but also in platyhelminths and nemertines, which have larvae of a derived type (Fig. 23.3). On the other hand, the pattern has obviously been lost in certain lineages within phyla that otherwise have spiral cleavage, for example in the cephalopods within the Mollusca (Hejnol 2010). It is therefore not unexpected that this cleavage type can have been lost in whole phyla (Fig. 23.1).

The main (apical-blastoporal) axis of the egg/embryo is fixed during oogenesis (Huebner and Anderson 1976; Nielsen 2010). The orientation is related to the position of the oocyte in the ovarial epithelium, with the apical (animal) pole facing the coelomic cavity, for example in several molluscs with intraovarian oogenesis (Raven 1976) and in nemertines (Wilson 1903); in many polychaetes the oocytes mature floating in the coelomic fluid so that this relationship cannot be recognized (Eckelbarger 1988). The polar bodies are given off at the apical (animal) pole, and as fertilization usually takes place before the meiotic divisions have been completed, the polar bodies are

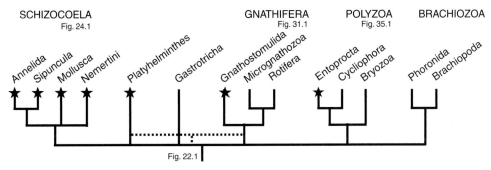

Figure 23.1. Phylogeny of the Spiralia. The phyla with representatives showing spiral cleavage are indicated by asterisks. The Platyzoa hypothesis is indicated by stippled lines.

Table 23.1. Cell lineage of an annelid (*Arenicola*) and a mollusc (*Trochus*) with lecithotrophic development. (Based on Siewing 1969.)

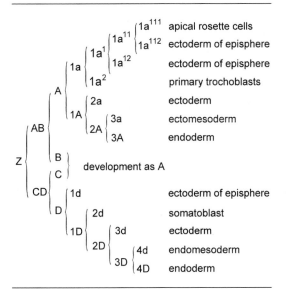

				$1a^{111}$	apical rosette cells
			$1a^{11}$	$1a^{112}$	ectoderm of episphere
		$1a^1$	$1a^{12}$		ectoderm of episphere
	$1a$	$1a^2$			primary trochoblasts
A	$1A$	$2a$			ectoderm
		$2A$	$3a$		ectomesoderm
			$3A$		endoderm
AB	B				development as A
	C				
CD	D	$1d$			ectoderm of episphere
Z		$1D$	$2d$		somatoblast
		$2D$	$3d$		ectoderm
			$3D$	$4d$	endomesoderm
				$4D$	endoderm

retained inside the fertilization membrane and can be used as markers for the orientation of the embryos.

Fertilization may take place anywhere on the egg, for example in some polychaetes. The entry point of the sperm determines the secondary main axis of the embryo (the dorsoventral 'axis') in some species because the first-cleavage furrow forms through this point and the apical pole (Guerrier 1970; van den

Biggelaar and Guerrier 1983; Luetjens and Dorresteijn 1998). The axes of the embryo and the adult may therefore be determined very early. There is an asymmetry already in the first cleavage in unequally dividing embryos, and this becomes conspicuous during the spiral cleavages with the shifts between dextral and sinistral cleavages. The mechanism behind this chirality is unknown, but the handedness of adult snails is related to that of the spiral cleavage (Kuroda *et al.* 2009), and depends on a single locus in the maternal gene (Freeman and Lundelius 1982). The axes can be changed experimentally (Arnolds *et al.* 1983; Goldstein and Freeman 1997).

The first two cleavages are through the main axis and usually oriented so that the embryo becomes divided into four quadrants, with the quadrants A and C being left and right, respectively, and B and D being anterior/ventral and posterior/dorsal, respectively (Fig. 21.1). The sizes of the first four blastomeres may be equal so that it is difficult to orient the early embryo, but size differences or other characters make it possible to name the individual blastomeres as early as at the 2-cell stage of many species. When size differences are apparent, the D-quadrant is larger than the others (Figs 21.1 and 23.2). The 4-cell stage is radially symmetrical or bilateral, but the following cleavages show chirality, with oblique mitotic spindles resulting in the characteristic pattern, with blastomere quartets given off in an alternating clockwise and counterclockwise pattern (Fig. 23.2) (see below).

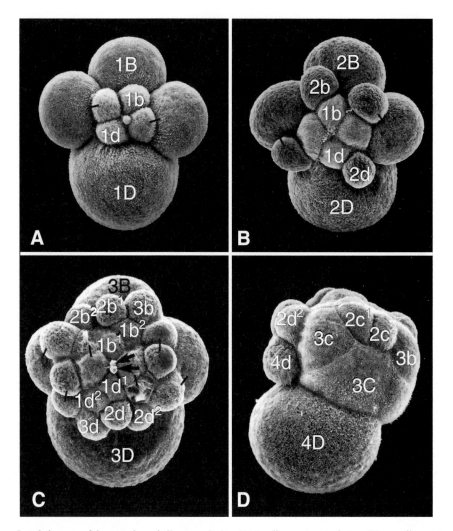

Figure 23.2. Spiral cleavage of the prosobranch *Ilyanassa obsoleta*. (A) 8-cell stage in apical view. (B) 12-cell stage in apical view. (C) 24-cell stage in apical view. (D) 28-cell stage in right view. In A–C the cells of the B and D quadrants are numbered, and the daughter blastomeres of the last cleavage are united by short lines in the A and C quadrants; the polar bodies are indicated by arrows. (SEM courtesy of Dr M.M. Craig, Southwest Missouri State University, Springfield, MO, USA; see Craig and Morrill 1986.)

Some embryos, especially of annelids and molluscs, show cross-like patterns of more conspicuous blastomeres, but this seems to be of no phylogenetic significance (see Chapter 24).

The 'typical' spiral cleavage, with four micromere quartets and mesoderm-formation from the 4d cell, has been found in several phyla and is described in most text-books. The cleavage and the normal fate of the resulting blastomeres can be correlated in details in embryos of several phyla (compare Fig. 22.9 and Tables 23.1, 25.1–3, 27.1–3, 28.1–2, and 29.1), whereas the cleavage of other spiralian phyla have not been studied in sufficient detail to allow similar comparisons.

The cleavage spindles of the early cleavages form an angle with the main (apical-blastoporal) axis, so that the A and C blastomeres of the 4-cell stage are often in wide contact at the apical pole, and the B and D

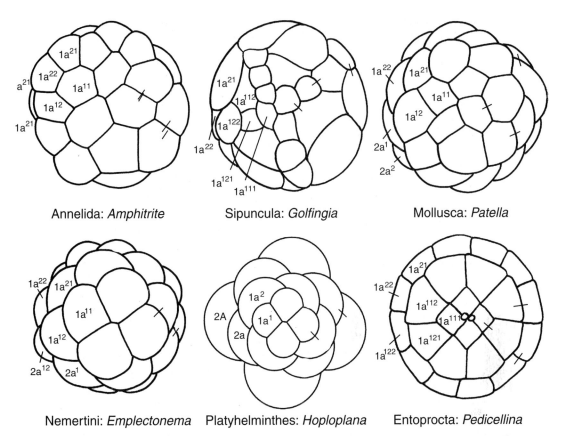

Figure 23.3. Spiral cleavage patterns in spiralian phyla; embryos seen from the apical pole. Annelida: 32-cell stage of *Amphitrite ornate*. (Redrawn from Mead 1897.) Sipuncula: 48-cell stage of *Golfingia vulgare*. (Redrawn from Gerould 1906.) Mollusca: 32-cell stage of *Patella vulgate*. (Redrawn from Damen and Dictus 1994.) Nemertini: 44-cell stage of *Emplectonema gracile*. (Redrawn from Delsman 1915.) Platyhelminthes: 32-cell stage of *Hoploplana inquilina*. (Redrawn from Boyer *et al.* 1998.) Entoprocta: 48-cell stage of *Pedicellina cernua*. (Redrawn from Marcus 1939.) The blastomeres of the A quadrant are numbered and the blastomere pairs of the latest cleavage are indicated by small lines in the C quadrant.

blastomeres in wide contact at the blastoporal pole. The future anteroposterior (ventrodorsal) axis is usually almost in line with the B–D axis. At the following cleavages, small apical micromeres are given off from the (usually) large blastoporal macromeres, with the spindles alternately twisted, so that the micromeres are given off alternately clockwise (dexiotropic) and counterclockwise (laeotropic) when seen from the apical pole (Figs. 23.2 and 23.3). The first micromere quartet may be twisted 45° so that the micromeres a and b (and c and d) are situated in a bilateral pattern. In species where the whole larval ectoderm is formed from these cells, the fate map of the ectoderm shows a bilateral pattern (for example the direct developing nemertine *Nemertopsis bivittata*: Martindale and Henry 1995) and the same holds true for the episphere of indirect-developing species (for example the nemertine *Cerebratulus lacteus*: Henry and Martindale 1994, and the polychaete *Polygordius*: Woltereck 1904). The same pattern has been reported from the direct-developing leeches (see Table 25.3 and Fig. 25.6), but a certain obliqueness is observed in the position of the mesoteloblasts. The rotation of the micromeres may be quite small and a number of intermediate patterns have been described.

Nemertines have the 'micromeres' that are larger than the 'macromeres', and the 4A–D cells, which become endoderm in most annelids and molluscs, are very small and become yolk granules in polyclad turbellarians; the cytoplasm has been shifted so much towards the micromeres that the first cleavage appears to be median, but the following development reveals the spiral nature (see Chapter 29).

Cell-lineage studies have shown that the prototroch of annelids, molluscs, sipunculans, and entoprocts, and the conspicuous band of long cilia of planktotrophic nemertine end platyhelminth larvae develop from the same cells, viz. primary trochoblasts from $1a^2$–$1d^2$, variously supplemented by accessory trochoblasts, from $1a^{1222}$–$1c^{1222}$, and secondary trochoblasts from $2a^1$–$2c^1$(Henry $et\ al.$ 2007). Trochoblasts are cleavage-arrested in a number of species, but there are many exceptions, for example in larvae with large prototrochs. The trochoblasts always degenerate at metamorphosis, even in species where these cells do not show a special ciliation (Maslakova $et\ al.$ 2004). Early stages of several species show a small temporary break at the dorsal side of the prototroch, between the blastomeres $1c^2$ and $1d^2$, corresponding to the break postulated in the ancestral trochophora larva (Chapter 22). Mesoderm develops from different micromere groups, but may apparently mix rather freely in later stages. Endomesoderm develops from the 4d cell, whereas ectomesoderm develops from micromeres of the second and third quartet (Boyer and Henry 1998; Hejnol $et\ al.$ 2007). Neither of these patterns resembles mesoderm development in cycloneuralians or deuterostomes.

The spiral pattern is easily recognized in species with small eggs (typically 50–300 μm in diameter), but many types of modifications occur for example in species with large, yolky eggs, and in species with placentally nourished embryos. There is, however, no direct correlation between the amount of yolk and cleavage pattern. The eggs of the prosobranch $Busycon\ carica$ are about 1.7 mm in diameter, but the cleavage is nevertheless holoblastic and the spiral pattern can easily be followed (Conklin 1907). The very large eggs of the cephalopods (0.6–17 mm in diameter) alternatively all show discoidal cleavage (Fioroni 1978). Large systematic groups, such as classes, may have uniform developmental features, but in other groups considerable variation may occur even within genera. An example of this is the hardly recognizable spiral pattern in the embryos of the entoproct $Loxosomella\ vivipara$, which has very small eggs and placentally nourished embryos, whereas many other species of the same genus have a normal spiral pattern (Chapter 36). It is generally accepted that spiral cleavage has been lost in many groups under the influence of large amounts of yolk or placental nourishment of the embryos, so it should not be controversial to include classes, or even phyla now classified in the Spiralia (Hejnol 2010).

The high degree of determination in the development is demonstrated by the fact that isolated blastomeres or blastomere groups are generally not able to regulate and form normal embryos. This does not mean that the spiralian embryo lacks regulative powers, as shown in annelids by Dorresteijn $et\ al.$ (1987) and molluscs by van den Biggelaar and Guerrier (1979). Separations of blastomeres of 2- and 4-cell stages of the nemertine $Cerebratulus$ have given at least partially normal embryos (Chapter 28), but the small larvae were not followed to metamorphosis. The brachiozoans resemble the echinoderms in their ability to regenerate, with complete totipotentiality of the blastomeres of the 4-cell stage and the ability to regenerate completely from half-gastrulae (see Chapters 54 and 55).

Apical organs are described in ciliated larvae of almost all eumetazoans, but there has been some confusion about the terminology. In the spiralians, an apical ganglion, usually with up to eight cells, some of which are flask-shaped, can be recognized in most groups (Richter $et\ al.$ 2010; note that this review does not distinguish between the apical and cerebral ganglia). A pair of cerebral ganglia develop from the episphere lateral to the apical ganglion, and often become tightly apposed to the apical ganglion, so that a composite apical organ is formed (Figs. 22.8–10), but the pilidium larva of the nemertines shows the cerebral ganglia developing far removed from the apical ganglion (Chapter 28). The apical ganglion is lost at metamorphosis, whereas the cerebral ganglia become the main part of the adult brain (see also Fig. 27.3). The spiralian apical organ is often onion-shaped with a pair of anterior nerves to the ventral nervous system (or to the prototroch nerve), and a pair of muscles to the region of the mouth (Fig. 23.4). The molluscs have very small apical organs with the cerebral ganglia developing from cells of the episphere just lateral to the apical cells; ventral extensions from these ganglia go to the anterior part of the foot, where a pair of sta-

tocysts are formed (Conklin 1897). A muscle from the apical organ to the mouth region is seen in larvae of the platyhelminth *Hoploplana* (Reiter *et al.* 1996). The brachiozoans (Chapter 39) have apical ganglia of a somewhat different morphology. The ecdysozoans lack the ciliated apical ganglion. The apical organs of cnidarian larvae are usually a rather thin ectodermal-

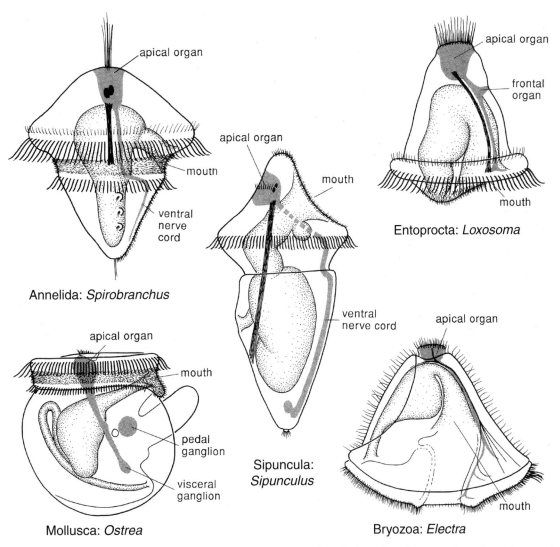

Figure 23.4. Apical organs of larval or juvenile spiralians showing apical cerebral ganglia and nerves and muscles to the ventral side; in entoprocts and ectoprocts the nerves and muscles extend only to the prototroch zone; all specimens are seen from the right side. Sipuncula: *Sipunculus nudus* larva ready for metamorphosis. (Redrawn from Hatschek 1883.) Mollusca: larva of *Ostrea edulis.* (Based on Erdmann 1935.) Annelida: early metatrochophore of *Spirobranchus polycerus.* (Based on Lacalli 1984.) Entoprocta: larva of *Loxosoma pectinaricola.* (Based on Nielsen 1971.) Ectoprocta: larva of *Electra pilosa.* (Based on Nielsen 1971.)

cell group, with a tuft of long cilia and a basiepithelial nerve plexus. Among the deuterostomes, only the tornaria larva (Chapter 60) has a more onion-shaped apical organ in some species, even with a pair of eyes, but it lacks nerves to ventral nerve cords and also muscles to the mouth region.

The Spiralia comprises five phyla or groups of phyla: Schizocoelia, Platyhelminthes, Gastrotricha, Gnathifera, and Polyzoa, but I have found it impossible to resolve their interrelationships and they have therefore been left in a polytomy (Fig. 23.1). Some molecular analyses indicate the existence of the clade Platyzoa, comprising Platyhelminthes, Gastrotricha, and Gnathifera (Passamaneck and Halanych 2006; Giribet *et al.* 2009). Computer analyses can of course force a resolution, but this will depend on subjective choices and interpretations of characters, and will result in a false impression of the state of knowledge.

References

Arnolds, W.J.A., van den Biggelaar, J.A.M. and Verdonk, N.H. 1983. Spatial aspects of cell interactions involved in the determination of dorsoventral polarity in equally cleaving gastropods and regulative abilities of their embryos, as studied by micromere deletions in *Lymnaea* and *Patella*. *Roux's Arch. Dev. Biol.* **192**: 75–85.

Boyer, B.C. and Henry, J.Q. 1998. Evolutionary modifications of the spiralian developmental program. *Am. Zool.* **38**: 621–633.

Boyer, B.C., Henry, J.J. and Martindale, M.Q. 1998. The cell lineage of a polyclad turbellarian embryo reveals close similarity to coelomate spiralians. *Dev. Biol.* **204**: 111–123.

Conklin, E.G. 1897. The embryology of *Crepidula*. *J. Morphol.* **13**: 1–226.

Conklin, E.G. 1907. The embryology of *Fulgur*: a study of the influence of yolk on development. *Proc. Acad. Sci. Nat. Phila.* **59**: 320–359.

Craig, M.M. and Morrill, J.B. 1986. Cellular arrangements and surface topography during early development in embryos of *Ilyanassa obsoleta*. *Int. J. Invertebr. Reprod. Dev.* **9**: 209–228.

Damen, P. and Dictus, W.J.A.G. 1994. Cell-lineage analysis of the prototroch of the gastropod mollusc *Patella vulgata* shows conditional specification of some trochoblasts. *Roux's Arch. Dev. Biol.* **203**: 187–198.

Delsman, H.C. 1915. Eifurchung und Gastrulation bei *Emplectonema gracile* Stimpson. *Tijdschr. Ned. Dierkd. Ver.* **14**: 68–114.

Dorresteijn, A.W.C., Bornewasser, H. and Fischer, A. 1987. A correlative study of experimentally changed first cleavage and Janus development in the trunk of *Platynereis dumerilii* (Annelida, Polychaeta). *Roux's Arch. Dev. Biol.* **196**: 51–58.

Dunn, C.W., Hejnol, A., Matus, D.Q., *et al.* 2008. Broad phylogenomic sampling improves resolution of the animal tree of life. *Nature* **452**: 745–749.

Eckelbarger, K.J. 1988. Oogenesis and female gametes. *Microfauna Mar.* **4**: 281–307.

Erdmann, W. 1935. Untersuchungen über die Lebensgeschichte der Auster. Nr. 5. Über die Entwicklung und die Anatomie der 'ansatzreifen' Larve von *Ostrea edulis* mit Bemerkungen über die Lebensgeschichte der Auster. *Wiss. Meeresunters., Helgol., N. F.* **19(6)**: 1–25.

Fioroni, P. 1978. Cephalopoda, Tintenfische. In F. Seidel (ed.): *Morphogenese der Tiere, Deskriptive Morphogenese*, 2. Lieferung, pp. 1–181. VEB Gustav Fischer, Jena.

Freeman, G. and Lundelius, J.W. 1982. The developmental genetics of dextrality and sinistrality in the gastropod *Lymnaea peregra*. *Roux's Arch. Dev. Biol.* **191**: 69–83.

Gerould, J.H. 1906. The development of *Phascolosoma*. *Zool. Jahrb., Anat.* **23**: 77–162.

Giribet, G., Dunn, C.W., Edgecombe, G.D., *et al.* 2009. Assembling the spiralian tree of life. In M.J. Telford and D.T.J. Littlewood (eds): *Animal Evolution. Genomes, Fossils, and Trees*, pp. 52–64. Oxford University Press, Oxford.

Goldstein, B. and Freeman, G. 1997. Axis specification in animal development. *BioEssays* **19**: 105–116.

Grande, C. 2010. Left–right asymmetries in Spiralia. *Integr. Comp. Biol.* **50**: 744–755.

Guerrier, P. 1970. Les caractères de la segmentation et de la détermination de la polarité dorsoventrale dans le développement de quelques Spiralia. III. *Pholas dactylus* et *Spisula subtruncata* (Mollusques, Lamellibranches). *J. Embryol. Exp. Morphol.* **23**: 667–692.

Hatschek, B. 1883. Über Entwicklung von *Sipunculus nudus*. *Arb. Zool. Inst. University Wien.* **5**: 61–140.

Hejnol, A. 2010. A twist in time—The evolution of spiral cleavage in the light of animal phylogeny. *Integr. Comp. Biol.* **50**: 695–706.

Hejnol, A., Martindale, M.Q. and Henry, J.Q. 2007. High-resolution fate map of the snail *Crepidula fornicata*: the origins of ciliary bands, nervous system, and muscular elements. *Dev. Biol.* **305**: 63–76.

Hejnol, A., Obst, M., Stamatakis, A., *et al.* 2009. Assessing the root of bilaterian animals with scalable phylogenomic methods. *Proc. R. Soc. Lond. B* **276**: 4261–4270.

Henry, J.J. and Martindale, M.Q. 1999. Conservation and innovation in spiralian development. *Hydrobiologia* **402**: 255–265.

Henry, J.Q., Hejnol, A., Perry, K.J. and Martindale, M.Q. 2007. Homology of ciliary bands in spiralian trochophores. *Integr. Comp. Biol.* **47**: 865–871.

Henry, J.Q. and Martindale, M.Q. 1994. Establishment of the dorsoventral axis in nemertean embryos: evolutionary considerations of spiralian development. *Dev. Genet.* **15**: 64–78.

Huebner, E. and Anderson, E. 1976. Comparative spiralian oogenesis—structural aspects: an overview. *Am. Zool.* **16**: 315–343.

Kuroda, R., Endo, B., Abe, M. and Shimizu, M. 2009. Chiral blastomere arrangement dictates zygotic left–right asymmetry pathway in snails. *Nature* **462**: 790–794.

Lacalli, T.C. 1984. Structure and organization of the nervous system in the trochophore larva of *Spirobranchus*. *Phil. Trans. R. Soc. Lond. B* **306**: 79–135.

Lambert, J.D. 2010. Developmental patterns in spiralian embryos. *Curr. Biol.* **20**: R72-R77.

Luetjens, C.M. and Dorresteijn, A.W.C. 1998. The site of fertilisation determines dorsoventral polarity but not chirality in the zebra mussel embryo. *Zygote* **6**: 125–135.

Mallatt, J., Craig, C.W. and Yoder, M.J. 2010. Nearly complete rRNA genes assembled from across the metazoan animals: Effects of more taxa, a structure-based alignment, and paired-sites evolutionary models on phylogeny reconstruction. *Mol. Phylogenet. Evol.* **55**: 1–17.

Marcus, E. 1939. Briozoários marinhos brasileiros III. *Bol. Fac. Filos. Cienc. Let. Univ S. Paulo, Zool.* **3**: 111–354.

Martindale, M.Q. and Henry, J.Q. 1995. Modifications of cell fate specification in equal-cleaving nemertean embryos: alternate patterns of spiralian development. *Development* **121**: 3175–3185.

Maslakova, S.A., Martindale, M.Q. and Norenburg, J.L. 2004. Fundamental properties of the spiralian development program are displayed by the basal nemertean *Carinoma tremaphoros* (Palaeonemertea, Nemertea). *Dev. Biol.* **267**: 342–360.

Mead, A.D. 1897. The early development of marine annelids. *J. Morphol.* **13**: 227–326.

Nielsen, C. 1971. Entoproct life-cycles and the entoproct/ectoproct relationship. *Ophelia* **9**: 209–341.

Nielsen, C. 2004. Trochophora larvae: cell-lineages, ciliary bands and body regions. 1. Annelida and Mollusca. *J. Exp. Zool. (Mol. Dev. Evol.)* **302**B: 35–68.

Nielsen, C. 2005. Trochophora larvae: cell-lineages, ciliary bands and body regions. 2. Other groups and general discussion. *J. Exp. Zool. (Mol. Dev. Evol.)* **304**B: 401–447.

Nielsen, C. 2010. Some aspects of spiralian development. *Acta Zool. (Stockh.)* **91**: 20–28.

Passamaneck, Y. and Halanych, K.M. 2006. Lophotrochozoan phylogeny assessed with LSU and SSU data: evidence of lophophorate polyphyly. *Mol. Phylogenet. Evol.* **40**: 20–28.

Pick, K.S., Philippe, H., Schreiber, F., *et al.* 2010. Improved phylogenomic taxon sampling noticeably affects nonbilaterian relationships. *Mol. Biol. Evol.* **27**: 1983–1987.

Raven, C.P. 1976. Morphogenetic analysis of spiralian development. *Am. Zool.* **16**: 395–403.

Reiter, D., Boyer, B., Ladurner, P., *et al.* 1996. Differentiation of the body wall musculature in *Macrostomum hystricum marinum* and *Hoploplana inquilina* (Plathelminthes), as models for muscle development in lower Spiralia. *Roux's Arch. Dev. Biol.* **205**: 410–423.

Richter, S., Loesel, L., Purschke, G., *et al.* 2010. Invertebrate neurophylogeny: suggested terms and definitions for a neuroanatomical glossary. *Front. Zool.* **7**: 29.

Siewing, R. 1969. *Lehrbuch der vergleichenden Entwicklungsgeschichte der Tiere*. Paul Parey, Hamburg.

van den Biggelaar, J.A.M. and Guerrier, P. 1979. Dorsoventral polarity and mesentoblast determination as concomitant results of cellular interactions in the mollusk *Patella vulgata*. *Dev. Biol.* **68**: 462–471.

van den Biggelaar, J.A.M. and Guerrier, P. 1983. Origin of spatial organization. In K.M. Wilbur (ed.): *The Mollusca*, vol. 3, pp. 179–213. Academic Press, New York.

van den Biggelaar, J.A.M., Dictus, W.J.A.G. and van Loon, A.E. 1997. Cleavage patterns, cell-lineages and cell specification are clues to phyletic lineages in Spiralia. *Semin. Cell Dev. Biol.* **8**: 367–378.

Wilson, E.B. 1903. Experiments on cleavage and localization in the nemertine-egg. *Arch Entwicklungsmech. Org.* **16**: 411–460.

Woltereck, R. 1904. Beiträge zur praktischen Analyse der *Polygordius*-Entwicklung nach dem 'Nordsee-' und dem 'Mittelmeer-Typus'. *Arch Entwicklungsmech. Org.* **18**: 377–403.

24

SCHIZOCOELIA

The new division of Protostomia into Spiralia (Lophotrochozoa) and Ecdysozoa splits my earlier concept of the Schizocoelia into two widely separated groups, but I have chosen to retain the name of the old group for the spiralian phyla, i.e. Annelida, Sipuncula, and Mollusca, and to add the Nemertini (Fig. 24.1). This makes the name synonymous with Eutrochozoa, in the sense used, for example, by Peterson and Eernisse (2001). Within the Spiralia, these groups are characterized by the lateral coelomic sacs developing through schizocoely in the mesoderm formed from the two primary mesoblasts, which are descendants of the 4d cell in the spiral cleavage. This mode of coelom formation is unique among spiralians. Mesodermal cavities are found in Bryozoa (Chapter 38) but, although these cavities must be classified as coeloms, i.e. cavities surrounded by a mesodermal epithelium, their ontogeny gives no indication of homology with the coelomic cavities of

the Schizocoelia. The Brachiozoa (Phoronida + Brachiopoda) develop paired coelomic cavities in lateral mesodermal tissues, but the mesoderm does not originate from special mesoblasts (Chapters 39–41). The presence of metanephridia is of course related to the presence of a coelom. A haemal system is found in most of these phyla, but it is peculiarly absent in Sipuncula (Chapter 26).

A concept called the spiralian cross, with the two types annelid and molluscan cross, has haunted the systematic/cladistic literature for many decades. It is based on the fact that in blastula stages of some annelids, the cells $1a^{112}$–$1d^{112}$ and their descendants are rather large and form a cross-shaped figure, whereas a cross-shaped figure formed by large $1a^{12}$–$1d^{12}$ cells is seen in some molluscs and sipunculans. It is of course always possible to identify corresponding cells of the spiral cleavage in various embryos, but it is without meaning to speak about two different types of crosses if the cells have no other characteristics

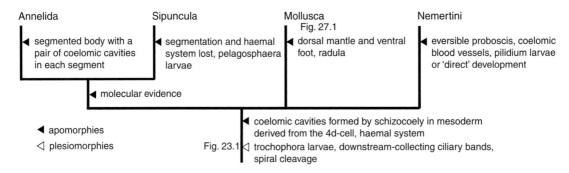

Figure 24.1. Phylogeny of the Schizocoelia.

than their size or shape. This is clearly demonstrated in the study by Maslakova *et al.* (2004). Analyses based on this concept must be rejected.

An overwhelming majority of the molecular phylogenetic analyses now supports the monophyly of the Schizocoelia comprising Annelida, Sipuncula, Mollusca, and Nemertini (Dunn *et al.* 2008; Helmkampf *et al.* 2008; Struck and Fisse 2008; Bleidorn *et al.* 2009; Hejnol *et al.* 2009; Paps *et al.* 2009; Hausdorf *et al.* 2010; Mallatt *et al.* 2010), usually with the Brachiozoa included as the sister group of the nemertines (see Chapter 39). Only the studies of mitochondrial genes give erratic results (Chen *et al.* 2009; Mwinyi *et al.* 2009; Podsiadlowski *et al.* 2009).

References

Bleidorn, C., Podsiadlowski, L., Zhong, M., *et al.* 2009. On the phylogenetic position of Myzostomida: can 77 genes get it wrong? *BMC Evol. Biol.* **9**: 150.

Chen, H.-X., Sundberg, P., Norenburg, J.L. and Sun, S.-C. 2009. The complete mitochondrial genome of *Cephalothrix simula* (Iwata)(Nemertea: Palaeonemertini). *Gene* **442**: 8–17.

Dunn, C.W., Hejnol, A., Matus, D.Q., *et al.* 2008. Broad phylogenomic sampling improves resolution of the animal tree of life. *Nature* **452**: 745–749.

Hausdorf, B., Helmkampf, M., Nesnidal, M.P. and Bruchhaus, I. 2010. Phylogenetic relationships within the lophophorate lineages (Ectoprocta, Brachiopoda and Phoronida). *Mol. Phylogenet. Evol.* **55**: 1121–1127.

Hejnol, A., Obst, M., Stamatakis, A., *et al.* 2009. Assessing the root of bilaterian animals with scalable phylogenomic methods. *Proc. R. Soc. Lond. B* **276**: 4261–4270.

Helmkampf, M., Bruchhaus, I. and Hausdorf, B. 2008. Phylogenomic analyses of lophophorates (brachiopods, phoronids and bryozoans) confirm the Lophotrochozoa concept. *Proc. R. Soc. Lond. B* **275**: 1927–1933.

Mallatt, J., Craig, C.W. and Yoder, M.J. 2010. Nearly complete rRNA genes assembled from across the metazoan animals: Effects of more taxa, a structure-based alignment, and paired-sites evolutionary models on phylogeny reconstruction. *Mol. Phylogenet. Evol.* **55**: 1–17.

Maslakova, S.A., Martindale, M.Q. and Norenburg, J.L. 2004. Fundamental properties of the spiralian development program are displayed by the basal nemertean *Carinoma tremaphoros* (Palaeonemertea, Nemertea). *Dev. Biol.* **267**: 342–360.

Mwinyi, A., Meyer, A., Bleidorn, C., *et al.* 2009. Mitochondrial genome sequences and gene order of *Sipunculus nudus* give additional support for an inclusion of Sipuncula into Annelida. *BMC Genomics* **10**: 27.

Paps, J., Baguñà, J. and Riutort, M. 2009. Bilaterian phylogeny: A broad sampling of 13 nuclear genes provides a new Lophotrochozoa phylogeny and supports a paraphyletic basal Acoelomorpha. *Mol. Biol. Evol.* **26**: 2397–2406.

Peterson, K.J. and Eernisse, D.J. 2001. Animal phylogeny and the ancestry of bilaterians: inferences from morphology and 18S rDNA sequences. *Evol. Dev.* **3**: 170–205.

Podsiadlowski, L., Braband, A., Struck, T.H., von Döhren, J. and Bartolomaeus, T. 2009. Phylogeny and mitochondrial gene order variation in Lophotrochozoa in the light of new mitogenomic data from Nemertea. *BMC Genomics* **10**: 364.

Struck, T. and Fisse, F. 2008. Phylogenetic position of Nemertea derived from phylogenomic data. *Mol. Biol. Evol.* **25**: 728–736.

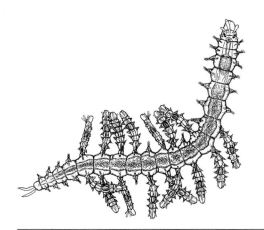

25

Phylum Annelida

Annelida is an ecologically and systematically important phylum of aquatic or terrestrial animals, comprising more than 16 500 described, living species. The fossil record is meagre: the Precambrian *Dickinsonia*, which has been interpreted as an annelid, is now assigned to the Placozoa (Chapter 11). *Spriggina* has been interpreted as a cnidarian, an annelid and an arthropod (McCall 2006), and even its animal nature appears uncertain. The earliest unquestionable annelid fossil appears to be *Phragmochaeta* from the Lower Cambrian of Sirius Passet (Conway Morris and Peel 2008). Annelids can generally be recognized by their segments with coelomic compartments and bundles of chitinous chaetae, but both characters show variation (Purschke 2002).

Annelid evolution is still poorly understood. The traditional classification with Polychaeta, Oligochaeta, and Hirudinea has now been abandoned because almost all morphological and molecular analyses show that Hirudinea is an ingroup of Clitellata, which in turn is an ingroup of the Polychaeta. This makes the name Polychaeta synonymous with Annelida, but the term 'polychaetes' is still used for the non-oligochaete, mainly marine groups, in parallel to the likewise paraphyletic 'turbellarians' and 'invertebrates'. The archiannelids, which were earlier considered a separate class, are now regarded as specialized interstitial forms, and are integrated in various polychaete orders or families with larger forms or as separate orders (Westheide 1990; Rouse 1999; Worsaae *et al.* 2005). *Diurodrilus* is discussed in the chapter Problematica (Chapter 66). Fauchald and Rouse (1997) and Rouse and Fauchald (1997) made cladistic analyses based on a large database of morphological characters, Rousset *et al.* (2007) analyzed a large dataset of ribosomal genes, and Zrzavý *et al.* (2009) made combined analyses of morphological and molecular characters. The Clitellata was recovered in all three studies, and some orders and families were recognized in two or all three studies, but the overall picture is very inconsistent. A new phylogenomic study supports the traditional Errantia-Sedentaria classification, but with the clitellates as an in-group of the Sedentaria (Struck *et al.* 2011).

A number of groups that have previously been regarded as separate phyla are now more or less firmly nested within the Annelida.

Echiura resembles annelids in most features of anatomy (Pilger 1993) and embryology (Hatschek 1880; Newby 1940), except that they seem to lack any trace of segmentation. However, new observations on the development of the nervous system (Hessling 2002; Hessling and Westheide 2002) indicate a segmentation of the ventral nerve cord in early developmental stages. Almost all studies of molecular phylogeny show the Echiura as an ingroup of the

---

Chapter vignette: The polychaete *Exogone gemmifera* with attached juveniles. (After Rasmussen 1973.)

Annelida, probably related to the Capitellidae (Bleidorn *et al.* 2003; Hall *et al.* 2004; Rousset *et al.* 2007; Struck *et al.* 2007; Hejnol *et al.* 2009; Mwinyi *et al.* 2009).

Pogonophora and Vestimentifera (Southward *et al.* 2005) have earlier been regarded as separate phyla (or one phylum) belonging to the Deuterostomia, but new knowledge about their ontogeny has revealed their true systematic position as an ingroup of the Annelida. Cleavage has now been shown to be spiral; the larvae have prototroch and telotroch, and the juveniles have a through gut, from which the trophosome with the symbiotic bacteria develops; also an apical organ and a pair of protonephridia lateral to the mouth have been described (Jones and Gardiner 1988,1989; Southward 1988; Callsen-Cencic and Flügel 1995; Young *et al.* 1996; Malakhov *et al.* 1997) (Fig. 25.1). Studies on molecular phylogeny almost unanimously support this interpretation (Rousset *et al.* 2007; Struck *et al.* 2007; Bleidorn *et al.* 2009; Mwinyi *et al.* 2009) and the two groups are now often treated as one family, the Siboglinidae. They may be close to the 'bone-eating' *Osedax* (Worsaae and Rouse 2010).

Myzostomida (Eeckhaut and Lanterbecq 2005) is a small group of parasites on crinoids (external or gall forming), and endoparasites of asteroids and ophiuroids. They show remarkable co-evolution with their crinoid hosts (Lanterbecq *et al.* 2010). Their pelagic larvae (Jägersten 1939; Eeckhaut *et al.* 2003) (Fig. 25.2) are typical polychaete nectochaetes with larval chaetae, whereas the parasitic adults, not unexpectedly, are quite modified. The adults show five segments having neuropodia with chaetae, acicula, and protonephridia (Pietsch and Westheide 1987). Molecular studies generally support their inclusion in the Annelida (Bleidorn *et al.* 2007,2009; Mwinyi *et al.* 2009; Dordel *et al.* 2010).

Lobatocerebrum comprises a few small, interstitial, unsegmented, completely ciliated 'worms', which are usually regarded as very specialized annelids (Rieger 1980, 1981, 1988, 1991b; Smith *et al.* 1986). They resemble small turbellarians and some of the interstitial polychaetes. The ultrastructure shows no sign of segmentation of the mesoderm or of coelomic cavities;

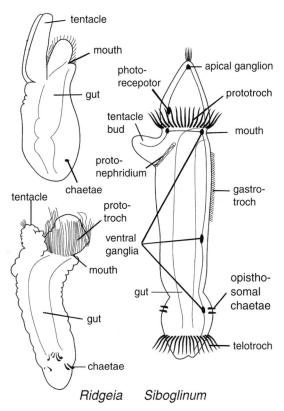

Figure 25.1. Larval stages of pogonophores showing mouth and gut. *Ridgeia* sp., newly settled individual with a pair of tentacle buds and juvenile with two longer tentacles. (Redrawn from Southward 1988; not to scale.) *Siboglinum poseidoni*, young bottom stage. (Based on Callsen-Cencic and Flügel 1995.)

similar 'acoelomate' conditions have been described from small species belonging to a number of polychaete families that show a segmentation (Fransen 1980). Segmentation is one of the most conspicuous characters of annelids, but the tiny neotenic dwarf male of the polychaete *Dinophilus gyrociliatus* is unsegmented (Westheide 1988), so *Lobatocerebrum* may well be a highly specialized annelid. To my knowledge they have not been included in any molecular analysis.

Jennaria pulchra is another small interstitial 'worm' described by Rieger (1991a,b). It appears unsegmented and acoelomate, and chaetae are absent. The gut is complete with an anus, and there is a series of protonephridia. It was interpreted as a specialized annelid, and until further information becomes available, no further

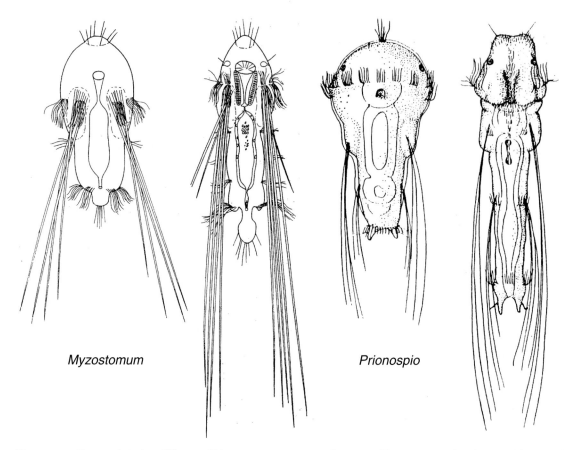

Myzostomum   *Prionospio*

Figure 25.2. Four- and six-day-old larvae of *Myzostomum parasiticum* and young and late 2-segment larval stage of the spionid *Prionospio malmgreni*; both types of larvae have chaetae that are shed in later stages. (From Jägersten 1939 and Hannerz 1956.)

conclusions can be drawn. It has apparently not been seen since the original sampling.

The sipunculans are morphologically quite distinct from the annelids, but they show up as an ingroup in Annelida in many molecular phylogenies (Struck *et al.* 2007; Dunn *et al.* 2008; Bleidorn *et al.* 2009; Hejnol *et al.* 2009; Zrzavý *et al.* 2009; Dordel *et al.* 2010), whereas other studies show the two as sister groups (Mwinyi *et al.* 2009; Sperling *et al.* 2009). For clarity, I have chosen to treat them as a separate phylum (Chapter 26).

Annelids are typically segmented, with groups of chaetae (setae) on both sides of each segment (lacking in the hirudineans and a few interstitial forms). The segmentation is almost complete, with septa and mesenteries separating a row of paired coelomic sacs in many 'errant' and tubicolous forms, such as nerei-

dids, spionids, and sabellids, but both septa and mesenteries are lacking in the anterior portion of the body in many forms, with a large, eversible pharynx, such as polynoids and glycerids, and the inner partitions are also strongly reduced in burrowing forms, such as arenicolids and scalibregmids, in tubicolous forms, such as pectinariids, and in the pelagic *Poeobius* (Robbins 1965). On the other hand, the coelomic cavities are completely absent in interstitial forms, such as *Protodrilus* and *Psammodriloides* (Fransen 1980).

The head region consists of the pre-oral, presegmental prostomium and the peri-oral peristomium (Fig. 25.3); postoral parapodia may move forwards, lose their chaetae, and become incorporated in the head complex, and even the peristomial cirri may have originated in this way (Dorresteijn *et al.* 1987; Fischer

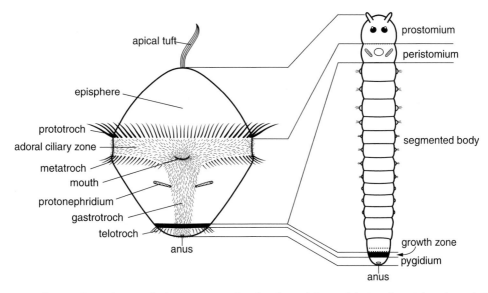

Figure 25.3. The correlation between body regions in a larval and an adult annelid. (Based on Schroeder and Hermans 1975.)

1999). The prostomium may carry eyes, nuchal organs (Purschke 2005), and various structures called antennae and palps. The peristomium carries the mouth and originates in front of the posterior growth zone, which proliferates the ectoderm of the body segments, and it seems that the ectoderm of the mouth region is topographically and phylogenetically presegmental (Schroeder and Hermans 1975); this is supported by clitellate embryology (see below). The peristomium may carry various types of tentacles, and both nuchal organs and palps from the prostomium may have moved to the peristomium. The mesoderm of prostomium and peristomium is less well described, but mesoderm in the shape of one or more pairs of coelomic sacs may be present; it appears that the mesoderm initially originates from ectomesodermal cells, and that additional mesoderm may be formed by proliferation from the anterior coelomic sacs formed from the mesoteloblasts (Anderson 1966, 1973).

The microscopic anatomy was reviewed in the edited volume by Harrison (1992). In the following, emphasis is on the more 'primitive' polychaetes.

The ectoderm (Hausen 2005b) is monolayered except in *Travisia*, which has a stratified epithelium with extensive intercellular spaces. Myoepithelial cells have

been observed in tentacular cirri of nereidids and syllids. Regular moulting has been observed in leeches, where it is correlated with cyclical changes in the concentration of 20-hydroxyecdysone (Sauber *et al.* 1983). Moulting has also been observed several times in the Flabelligeridae (*Brada* and *Pherusa*) by Dr K.W. Ockelmann (University Copenhagen), and casual observations of what appears to be moulting have been made in the Opheliidae by Dr D. Eibye-Jacobsen (University Copenhagen). Moulting of the jaw apparatus in the onuphid *Diopatra* has been reported by Paxton (2005).

The ciliated epithelia of annelids consist of multiciliate cells, but a few exceptions have been described. *Owenia* has only monociliate cells both as adults and larvae—even the prototroch and metatroch arise from monociliate cells. This was interpreted as a plesiomorphic feature, and the oweniids were accordingly regarded as located at the base of the polychaetes (Smith *et al.* 1987); but the larvae of another oweniid, *Myriochele*, have multiciliate cells, so it appears more plausible that it is an advanced character—a reversal to the original monociliate stage through loss of the 'additional cilia' (see Chapter 21).

The cuticle consists of several layers of parallel collagen fibrils with alternating orientation and with microvilli extending between the fibrils to the surface. The microvilli often terminate in a small knob, and there is in many cases an electron-dense epicuticle at the surface. Chitin is generally absent in the cuticle, including the jaws of several species, but Bubel *et al.* (1983) demonstrated the presence of α-chitin in the opercular filament cuticle of the serpulid *Pomatoceros*.

The chaetae (setae) (Hausen 2005a) consist mainly of β-chitin associated with protein, each chaeta being secreted by a chaetoblast with a bundle of long parallel microvilli; the chaetae have characteristic longitudinal channels corresponding to these microvilli. Some polychaetes, such as capitellids and oweniids, have the chaetae projecting directly from the cylindrical body and are burrowing or tubicolous, and have segments with longitudinal and circular muscles functioning as hydrostatic units (Clark 1964). Others are creeping on ciliary fields, for example many of the interstitial types. However, most polychaetes have protruding muscular appendages, parapodia, with chaetae on a dorsal and a ventral branch and with an elaborate musculature that makes the parapodia suited for various types of creeping or swimming; the circular segmental muscles may be rather weak.

The chaetae of β-chitin are here considered one of the most important apomorphies of the Annelida, although similar structures are known from the mantle edge of brachiopods (Chapter 41). The annelid chaetae are formed in lateral groups along the body; the larval brachiopod chaetae occur in a similar pattern, while those of the adults are situated along the mantle edge. It appears that Remane's homology criterion of position (Remane 1952) is not fulfilled, and it is therefore not probable that the various chaeta-like structures are homologous. Some cephalopod embryos and juveniles have numerous organs called Kölliker's organs (Brocco *et al.* 1974) scattered over the body, each comprising a cell with many microvilli, each secreting a chitinous tubule; the tubules may lie close together resembling an annelid chaeta, but they may also spread out completely; this structure is obviously a cephalopod apomorphy. Some polyplacophorans have hair-like structures on the girdle, but their cuticular part is secreted by a number of epithelial cells (Leise 1988).

It has been customary to regard types with large parapodia, such as *Nereis*, as the typical polychaetes, but Fauchald (1974) and Fauchald and Rouse (1997) proposed that the ancestral polychaete was a burrowing form with chaetae but without parapodia, resembling a capitellid or an oligochaete. The parapodia should then be seen as locomotory appendages that enabled the more advanced polychaetes to crawl in soft, flocculent substrates, such as the rich detritus layer at the surface of the sediment. The swimming and tube-building types should be more advanced. This interpretation does not agree with the generally accepted evolution of the annelids (see above and Westheide 1997).

The pharynx is a stomodaeum lined with a cuticle similar to that of the outer body wall, but variously specialized in connection with different feeding strategies. Microphagous annelids have heavily ciliated regions of the stomodeum. Most species of the orders Phyllodocida and Eunicida have jaws that are heavily sclerotized parts of the cuticle. Collagen is an important constituent and quinone tanning has been demonstrated in some species; chitin has not been found. The basal layer of most jaws shows short canals with microvilli.

Many polychaetes build tubes from secretions of epidermal glands, with or without incorporated mud, shells, sand grains, or other foreign objects; the tubes of serpulids are heavily calcified. The composition of the organic material is not well known, but both carbohydrates and proteins, in some species in a keratin-like form, are present. Chitin is generally not present, but it is a main component of the tubes of siboglinids, with the tubes of *Siboglinum* containing about 30% chitin.

The central nervous system consists of a brain with paired, often more or less fused cerebral ganglia, connectives on each side of the pharynx united in a suboesophageal ganglion behind the mouth, and a pair of ventral longitudinal nerves. Other ganglia are found around the mouth and segmentally arranged along the ventral nerves (Orrhage and Müller 2005). These nerves are situated within the epithelium in early developmental stages and also in many adult forms, but in some of the larger forms the cords sink in from the epithelium during ontogeny, surrounded by the

  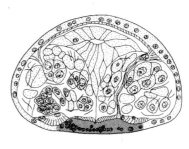

embryo of *Scoloplos*   adult *Scoloplos*   adult *Hesionides*

Figure 25.4. Transverse sections of polychaetes showing the position of the ventral longitudinal nerve cords (dark shading). Two-chaetiger embryo of *Scoloplos armiger* with intraepithelial nerve cords. (Redrawn from Anderson 1959.) Adult *Scoloplos armiger* with completely internalized and fused nerve cords. (Redrawn from Mau 1881.) Adult *Hesionides arenaria* with intraepithelial ventral nerve cords. (Redrawn from Westheide 1987.)

basal membrane (Fig. 25.4). The perikarya become arranged in paired ganglia connected by transverse and paired longitudinal nerves. There is usually one pair of ganglia (sometimes fused) per segment, but the ganglia may be less well defined, and two or three pairs of ganglion-like swellings with lateral nerves are observed in each parapodial segment, for example, in *Pectinaria*. Many species have one or three thin, median ventral nerves. Various numbers of lateral and dorsal, longitudinal nerves are found in the different species, but they are completely absent, e.g. in *Neanthes arenaceodentata* (Winchell *et al.* 2010), and there seems to be no support for the orthogon theory, which proposes an ancestral nervous system with longitudinal and circular nerves. Special parapodial ganglia usually connected by lateral nerves are found in species with well-developed parapodia. The innervation of the various anterior appendages, such as palps and antennae, give strong indications of homologies.

Photoreceptors are found in a number of adult annelids (Purschke *et al.* 2006). Cerebrally innervated eyes are rhabdomeric and vary from a simple type consisting of a pigment cell and a receptor cell, for example in *Protodrilus*, to large eyes with primary and secondary retina, a lens, and several types of accessory cells in alciopids. Segmental eyes are found in pairs along the sides of the body, for example in some opheliids, pygidial eyes are known from sabellids, and branchial, ciliary eyes are found on the tentacles of several sabellids; some of the last-mentioned eyes are compound (Nilsson 1994). It is obvious that these eyes are not all homologous (Nilsson 2009).

The midgut is a straight tube in most smaller forms, but more complicated shapes, for example with lateral diverticula, are found in, for example, aphroditids. Juvenile pogonophores and vestimentiferans have a normal gut and feed on bacteria, but after a short period, one type of the ingested bacteria becomes incorporated in the midgut epithelium that becomes transformed into a voluminous trophosome with bacteria, whereas the remainder of the gut degenerates (Southward 1988; Gardiner and Jones 1993; Callsen-Cencic and Flügel 1995).

The mesoderm lines the coelomic cavities (Rieger and Purschke 2005), which are restricted to one pair in each segment in many species, but the coelomic sacs become confluent in the anterior region in species with a large proboscis and in the whole body, for example, in *Pectinaria* and *Poeobius*. The coelomic lining varies from a simple myoepithelium to a thick multilayered tissue, with the myocytes covered by a peritoneum. Monociliate myoepithelial cells have been reported from *Owenia* and *Magelona*. Small species representing various families have a very narrow coelom and some lack cavities totally, for example adults of *Microphthalmus* and juveniles of *Drilonereis*.

Most annelids have a haemal system that consists of more or less well-defined vessels surrounded by basal membranes of the various epithelia; there is no endothelium. The main, dorsal vessel is contractile and pumps the blood anteriorly. Some blood vessels consist of podocytes and are the site of ultrafiltration of the primary urine to the coelom. Capitellids and glycerids lack blood vessels, and the coelomic fluid, which may contain respiratory pigment, functions as a circulatory system. Hirudineans have a highly specialized vascular system consisting of confluent segmental coelomic cavities that have subsequently become modified into narrow canals (Bürger 1891; Sawyer 1986).

The excretory organs (Bartolomaeus and Quast 2005) show enormous variation. Protonephridia of several types are found in both larvae and adults, and metanephridia occur in adults of many families. Larval protonephridia (sometimes called head kidneys) are situated in the peristome and show considerable vari-ation, from simple, monociliated terminal cells, to complicated organs with several multiciliate terminal cells of various types. The various types of protone-phridia are clearly of ectodermal origin and are gener-ally known to be surrounded by a basal membrane. The nephridial sacs of the mitraria larva of *Owenia* have complicated, podocyte-like fenestrated areas, but are, in principle, like protonephridia. Metanephridia are usually thought to be modified coelomoducts and to originate from the mesoderm (and this has been shown to be the case in oligochaetes, see below), but the metanephrida of a number of polychaetes develop from protonephridia that open up into the coelom.

Gonads of mesodermal origin are found in a large number of segments in many families, but, for exam-ple, some capitellids and the clitellates have the gonads restricted to a small number of segments; in some spe-cies, the germ cells become liberated to the coelom, where the final maturation takes place. Spawning is through the ciliated metanephridia or gonoducts, or by rupture of the body wall.

Table 25.1. Cell lineage of *Polygordius*. (Based on Woltereck 1904.) * Interpretation according to Nielsen (2004).

The sperm shows much variation, from 'primitive' types, with a rounded to conical head with the nucleus and four mitochondrial spheres, to highly specialized types with almost filiform head with spirally coiled acrosome (Rouse 2000).

The embryology was reviewed by Nielsen (2004, 2005). Polychaetes exhibit a wide variation in developmental types, whereas the clitellates have direct development. A number of polychaetes are free spawners, and the zygote develops into a planktotrophic larva that metamorphoses into a benthic adult; this is considered the ancestral developmental type in the annelids (Nielsen 1998). Other forms have large yolky eggs that develop into lecithotrophic larvae, or the development may be direct without a larval stage (see the chapter vignette).

Meiosis is usually halted in the prophase of the first division and becomes reactivated at fertilization. The apical-blastoporal axis is apparently fixed already during maturation, and the entrance of the spermatozoon may determine the position of the first cleavage and thereby the orientation of the anteroposterior axis. The egg is surrounded by a vitelline membrane, which in some species becomes incorporated in the larval cuticle through which the cilia penetrate; in other species this membrane forms a protecting envelope from which the larva hatches.

Cleavage is total and spiral (Tables 23.1, 25.1–2), and as the polar bodies are retained at the apical (animal) pole and the D-cell is often larger than the other three, it has been possible to follow the lineage of many of the important cells from the 2-cell stage. Polar lobes have been observed in a few species (Dorresteijn 2005).

Cell-lineage studies are available for *Polygordius* with a planktotrophic larva (Woltereck 1904) (Table

Table 25.2. Cell lineage of *Capitella*. The contribution from the 2d cell to the central nervous system is in bold. (Based on Meyer *et al.* 2010.)

Lineage					Fate
Z	AB	A	1A	1a	1a^1 — episphere, left cerebral ganglion, accessory trochoblasts
					1a^2 — primary trochoblasts
				2A	2a — secondary trochoblasts, stomodaeum, proctodaeum, ectomesoderm
					3a — stomodaeum, neurons, ectomesoderm
					3A — endoderm
		B	1B	1b	1b^1 — episphere, right cerebral ganglion, accessory trochoblasts
					1b^2 — primary trochoblasts
				2B	2b — secondary trochoblasts, stomodaeum
					3b — stomodaeum
					3B — endoderm
	CD	C	1C	1c	1c^1 — episphere, right cerebral ganglion, accessory trochoblasts, trunk ectoderm
					1c^2 — primary trochoblasts
				2C	2c — secondary trochoblasts, stomodaeum, proctodaeum, ectomesoderm
					3c — right mesodermal band, stomodaeum, neurotroch
					3C — endoderm
		D	1D	1d	1d^1 — episphere, left cerebral ganglion, accessory trochoblasts, trunk ectoderm
					1d^2 — primary trochoblasts
				2D	2d — trunk and pygidial ectoderm, **dorsal brain, circumoesophageal connectives, ventral nerve cord,** neurotroch, telotroch
				3D	3d — left mesodermal band, stomodaeum, neurotroch
					4d — muscle cells, germ cells
					4D — endoderm

Table 25.3. Cell lineage of *Helobdella*; l, left; r, right. (Based on Shankland and Savage 1997.)

		1a			micromeres: prostomial ectoderm, cerebral ganglion and provisional integument
	A	1A { 2a			
AB			2A { 3a		
			3A		endoderm
	B } development as A				
Z	C				
		1d			micromeres: prostomial ectoderm, cerebral ganglion and provisional integument
CD	D	NOPQ	NOPQ^l { OPQ^l { N^l, OP, Q^l }		ectoteloblasts
	1D		NOPQ^r development as NOPQ^l		
		M { M^l, M^r }			mesoteloblasts

25.1), for a number of species with lecithotrophic larvae, for example *Nereis* (Wilson 1892), *Amphitrite* (Mead 1897), and *Podarke* (Treadwell 1901) (Table 23.1), and from the recent study using blastomere marking of *Platynereis* (Ackermann *et al.* 2005) and *Capitella* (Meyer and Seaver 2010) (Table 25.2), and for the direct-developing hirudineans (Table 25.3).

In *Polygordius* (Table 25.1), the early development is an equal spiral cleavage with the $1a^2$–$1d^2$ cells becoming the primary trochoblasts; there is no accessory or secondary trochoblasts. The blastula becomes very flat, so that the cell fates in the hyposphere become easy to observe. Gastrulation is embolic with the blastopore becoming divided through the fusion of the lateral blastopore lips (from 3c and 3d cells). The anterior part of the blastopore remains open and sinks in with the stomodeum, whereas the posterior opening closes; after some time an anus breaks through between descendants of the 4d cell. Larval protonephridia develop from 3c and 3d cells. Woltereck (1904) stated that the ciliated lower lip consists of descendents of 3c and 3d cells. He interpreted a pair of cells (named 3c/1/post and 3d/1/post) as precursors of the metatroch, but these cells are sister cells of the lateral cells of the lower lip and their ciliation forms latero-posterior extensions of the ciliation of the mouth, so they are more probably the precursors of the adoral ciliary zone. The two cells on the median side of these cells (named 2d/2/1/1/2 and 2d/1/2/2) could then be the precursors of the metatroch. The larvae (Fig. 25.5) become very large and were the first trochophores described (Lovén 1840). In *P. lacteus/appendiculatus* the segmented body develops retracted in an accordion-like manner, covered by an extension of the epithelium below the metatroch (type 2 Pericalymma larva, see Fig. 22.7). This extension is shed together with the ciliated bands at metamorphosis, whereas the apical region with the brain and tentacles is retained as the adult anterior end.

The cell lineage of the lecithotrophic *Nereis*, *Amphitrite* and *Podarke* (Table 23.1), studied by traditional observations of the developing embryo, shows general agreement with that of *Polygordius*, and with those of *Platynereis* and *Capitella* studied with blastomere marking.

Fate maps of 64-cell stage embryos of lecithotrophic species, with the positions of the cells indicated in the notation of the spiral cleavage, have been

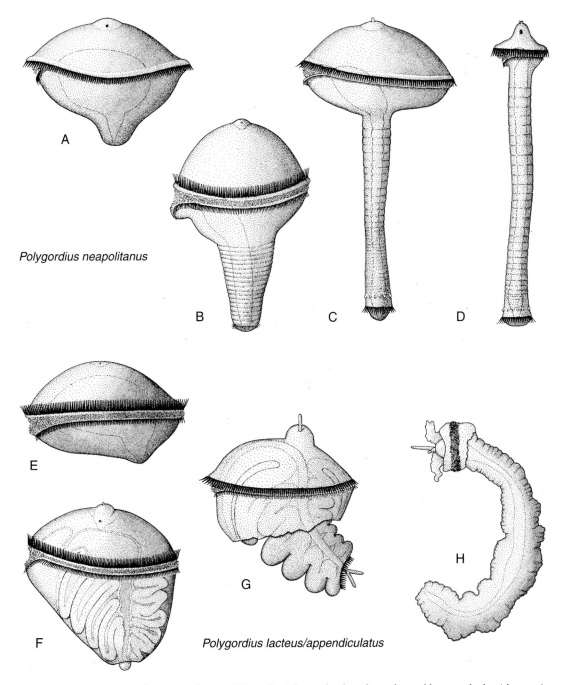

Polygordius neapolitanus

Polygordius lacteus/appendiculatus

Figure 25.5. Larval types and metamorphoses in *Polygordius*. The trochophora larvae have a blown-up body with a spacious blastocoel in both main types, but the segmented body of the metatrochophores develops along two main lines: the exolarva, which has the body as a posterior appendage (*P. neapolitanus*, redrawn from Hatschek 1878), or the endolarva (a type 2 pericalymma), in which the segmented body is contracted like an accordion and retracted into an extension of the post-metatrochal zone (serosa) of the trochophore (*P. lacteus* and *P. appendiculatus*, redrawn from Woltereck 1902, 1926, and Herrmann 1986). (A) A young trochophore. (B,C) The segmented body develops. (D) The larval body has contracted strongly and the larva is ready for settling. (E) A young trochophore. (F) The segmented body develops inside the serosa. (G) The serosa ruptures when the segmented body stretches. (H) Most of the larval organs degenerate and the juvenile is ready for the benthic life.

constructed for a number of lecithotrophic species based on classical studies (Anderson 1973); the map of *Podarke* can be taken as an example (Fig. 25.6). The apical cells give rise to the apical ganglion with a tuft of long cilia. An almost equatorial circle of cells, with a posterior break, gives rise to the prototroch; the number of cells varies somewhat between species, but the prototroch cells are always descendants of the primary trochoblasts, $1a^2$–$1d^2$, the accessory trochoblasts, $1a^{12}$–$1c^{12}$, and the secondary trochoblasts, $2a^{11}$–$2c^{11}$; the posterior break usually closes at a later stage. The cells of the 1st micromere quartet, the episphere, are presumptive ectoderm cells of the head, as are the cells in the posterior break of the prototroch. A narrow posterior ring of ectodermal cells, the so-called ectoteloblasts (descendants of 2d, (3a ?), 3c and 3d cells in species with small eggs), form the growth zone where segments become added at the anterior side. In species with large D-cells (Table 23.1) the 2d cell, the somatoblast, proliferates profusely, spreading from the dorsal side to the ventral midline and finally forming the ectoderm of the whole segmented body region, with a posterior ring of cells, descendants of $2d^{1222}$ and $2d^{2121}$, forming the telotroch (sometimes with a ventral gap). A narrow mid-ventral zone becomes the cells of the gastrotroch. The lecitho-

trophic species lack the metatroch. The blastoporal pole of the blastula is occupied by the future endoderm cells and posteriorly by the mesoderm originating from the 4d cell, or from the 3c-d cells in *Capitella* (Meyer *et al.* 2010). At each side there is a narrow area of future ectomesoderm just above the endoderm, originating from the cells 3a-d. After gastrulation the cell areas lie in the positions characteristic of the trochophore. The 4d cell divides into a right and a left cell, which give off a few small cells to the endoderm before becoming the mesoteloblasts.

The cell-lineage studies of *Platynereis* (Ackermann *et al.* 2005) and *Capitella* (Meyer and Seaver 2010) confirm a number of the above-mentioned observations, and demonstrate that in this species the ectoderm of the whole trunk, including ventral nerve cords and telotroch, develops from the 2d cell. The posterior part of the brain, just in front of the oesophagus, is formed from the 2d cell (Fig. 22.11).

There is a good deal of variation in the extent and shape of the areas in the fate maps, and the areas are not strictly related to identical cells, but the maps can all be seen as modifications of the general pattern described above. The spiral pattern gives rise to some characteristic symmetries. The macromeres B and D in a bilateral position, and the spiral cleavage shifts the

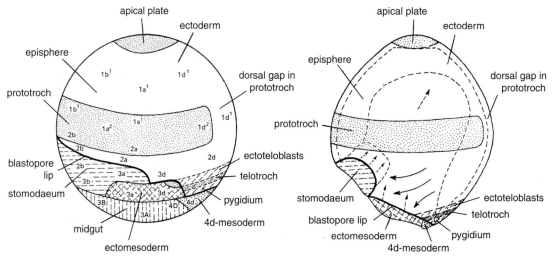

Figure 25.6. Fate maps of blastula (left) and gastrula (right) stages of *Podarke obscura* seen from the left side. The heavy lines (marked blastopore lip) separate ectoderm from mesoderm + endoderm. (Modified from Anderson 1973.)

micromeres 1a-d relative to the midline, so that a more or less conspicuous bilateral symmetry is established with the a and d cells on the left side, and b and c on the right side (in the dexiotropic cleavage). This is clearly seen both in *Polygordius* (Table 25.1) and in *Capitella* (Table 25.2).

Gastrulation is embolic in species with small eggs and a blastocoel, such as *Podarke*, *Eupomatus*, and *Polygordius*. The blastopore becomes laterally compressed leaving the adult mouth and anus, and a tube-shaped gut (Fig. 22.2). The formation of a deep stomodeum makes the mouth sink into a deep funnel and the anus closes temporarily, but reopens in the same region later on. More commonly, the blastopore closes from behind so that only the mouth remains, while the anus breaks through to the sac-shaped archenteron at a later stage (for example in *Eupomatus*). In *Eunice*, gastrulation is embolic and the blastopore constricts completely in the area where the anus develops at a later stage; the stomodeum develops from an area isolated from the blastopore by a wide band of ectoderm. The development of *Owenia* shows an invagination gastrula and the blastopore directly becomes the anus (Smart and Von Dassow 2009). In types with epibolic gastrulation, the lumen of the gut forms as a slit between the endodermal cells, while the two ectodermal invaginations, stomodeum and proctodeum, break through to the gut to form mouth and anus, respectively (as in *Arenicola*). The two types of mesoderm give rise to different structures in most forms, the ectomesoderm, developing into muscles traversing the blastocoel in the episphere of the larva and musculature in the prostomium-peristomium of the adult (see for example Åkesson 1968; Anderson 1973), and the 4d mesoderm (or 3c-d in *Capitella*), developing into the mesoderm of the true segments. The development of the paired mesodermal segments from the pygidial growth zone and the development of the coelomic spaces through schizocoely have been documented for several species (for example *Owenia*: Wilson 1932, and *Scoloplos*: Anderson 1959). A number of Hox genes are expressed in the early development of *Nereis* (Kulakova *et al.* 2007) and *Capitella* (Fröbius *et al.* 2008). *Hox1-5* are expressed mainly in the segmental ectoderm, and

later on in the ganglia, in an anteroposterior sequence as that observed in arthropods (Chapter 41); the more posterior Hox genes are expressed in the posterior growth zone.

The trochophore with an apical tuft, a prototroch and a metatroch of compound cilia, functioning in a downstream-collecting system using the catch-up principle (Riisgård *et al.* 2000) (Fig. 22.6), an adoral zone of single cilia, transporting the captured particles to the mouth, a gastrotroch of single cilia, and a telotroch of compound cilia around the anus (as in *Polygordius*, which lacks the gastrotroch) are the types from which all the other types may have developed through losses of one or more of the ciliary bands. The telotroch is absent in many planktotrophic larvae, which therefore swim only by means of the prototroch (for example *Serpula*, see Fig. 25.7). The metatroch is absent in all lecithotrophic larvae, which may have a telotroch (as in the larvae of spionids and terebellids) or that may only have the prototroch (as in many phyllodocids). Accessory rings of single or compound cilia are found in many species, and some specialized types of feeding in the plankton are also observed. The unusual mitraria larva of *Owenia* shows no cleavage-arrested trochoblasts, and both prototroch and metatroch are from monociliate cells (Smart and Von Dassow 2009).

Annelid ontogeny can be interpreted in terms of the trochophore theory as follows: the primary and accessory trochoblasts of the first micromere quartet form the prototroch (sometimes with contributions from the second micromere quartet). The dorsal cells of the second micromere quartet proliferate enormously, spreading from the dorsal side to cover the whole body of the worm. The edge of this field of cells carries the just-mentioned secondary trochoblasts including the cells of the telotroch; in my interpretation they also form the compound cilia of the metatroch. The general ciliation of the blastopore is retained as the ciliation of the mouth, with lateral extensions as the adoral ciliary zone, and the cilia of the lateral blastopore lips as the gastrotroch (neurotroch).

The trochophore is in principle unsegmented, and the true segments of the body become added from the growth zone in front of the telotroch/pygidium in the

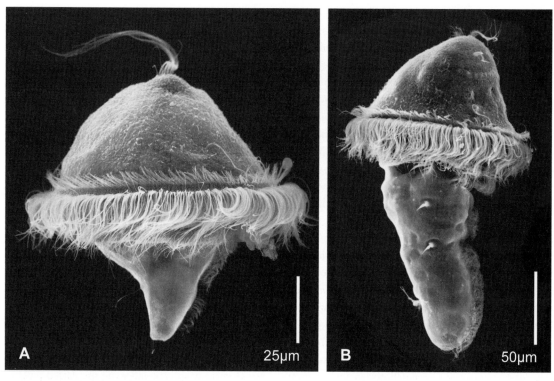

Figure 25.7. Trochophora (A) and three-pair chaetiger (B) larva of *Serpula columbiana*. (Friday Harbor Laboratories, WA, USA, July 1980; species identification after Kupriyanova (1999).)

metatrochophora stages (Fig. 25.3). The segment borders are first recognized as epithelial grooves, but the segmental groups of chaetae soon develop. Several species (such as *Nereis* and *Serpula*, see Fig. 25.7) show simultaneous development of three or four anterior segments with chaetae, followed somewhat later by the addition of segments from the posterior growth zone. The significance of this is uncertain. Expression of the *engrailed* gene has been observed in the cells of the prototroch, i.e. along the posterior border of the prostomium, along the posterior border of the peristomium (possibly indicating the cells that form the metatroch in planktotrophic species), and in the posterior borders of the developing body segments (Dorresteijn *et al.* 1993). The parapodia develop from one segment, as opposed to the limbs of the arthropods, where the *engrailed*-expressing row of cells of one 'parasegment' fuses with the segment at its posterior side forming a segment; the limb then develops

from the *engrailed* cell row and cells of the posterior segment (Chapter 41) (Prud'homme *et al.* 2003; Prpic 2008; Dray *et al.* 2010).

The development of the nervous system has been studied in a few species. The apical ganglion and the cerebral ganglia, which are often closely apposed, have usually been treated together as the apical organ, so the fate of the two types of ganglia are impossible to trace in the older studies. Lacalli (1981, 1984, 1986) studied the larval nervous systems of *Phyllodoce* and *Spirobranchus* by use of serial transmission-electron-micrography sections, and identified eight cells in the apical ganglion of *Spirobranchus*, two ciliated cells, three plexus cells, and three capsular cells. Two groups of 'reticular cells' probably represent the cerebral ganglia connected by a commissure. A prototroch nerve and a metatroch nerve were also documented. A pair of rhabdomeric photoreceptor cells situated laterally on the episphere without connection to the apical organ innervate the prototroch directly (Jékely *et al.* 2008). Newer studies used immunocytochemical methods to show the developing nervous system. Hay-Schmidt (1995) studied *Polygordius*, but did not distinguish the closely apposed apical and cerebral ganglia. Voronezhskaya *et al.* (2003) studied *Phyllodoce* and found four FMRFamidergic cells in the apical ganglion of early embryos and, slightly later, a number of serotonergic cells. McDougall *et al.* (2006) studied *Pomatoceros* and found a few FMRFamidergic cells in the early apical ganglion, and a number of cells in the cerebral ganglia developing later on in ectodermal pockets lateroposterior to the apical ganglion. The apical ganglion disappeared at metamorphosis and the cerebral ganglia became connected to a pair of ventral nerves with segmental commissures. The serotonergic system develops slightly later and comprises a prototroch and a metatroch nerve, in addition to nerves following the FMRFamidergic system. In later larvae of *Sabellaria*, Brinkmann and Wanninger (2008) observed serotonergic and FRMFamidergic nervous systems with cerebral ganglia connected to paired ventral nerves, ending in a peri-anal loop below the telotroch. The serotonergic system of the young larvae consisted of three cell bodies in the apical organ, two at the oesophagus, and two ventrally below the metatroch, and a number of axonal bundles, i.e. a pair from the apical organ along the lateral sides of the blown-up larval body to a midventral pair along the slender posterior part, two rings along the prototroch and one along the metatroch, and a pair of small nerves from the apical organ to the oesophageal cells. Later stages showed additional cells forming a small ventral ganglion. The FMRFamidergic system of the young larvae consisted of eight perikarya at the periphery of the apical organ and a system of axons following that of the serotonergic system; late larvae showed additional cells in the apical zone and also in the ventral ganglion. Immunocytochemical studies of *Platynereis* nectochaetes (Dorresteijn *et al.* 1993; Ackermann *et al.* 2005) have revealed a pair of 'pioneer' nerve cells, 2d^{12} and 2d^2, in the pygidium with neural projections extending forwards along the ventral side; the following stages show an increase of nerve fibres along the

two first ones, and segmental commissures begin to develop. Similar pioneer cells have been observed in *Pomatoceros* (McDougall *et al.* 2006); in *Phyllodoce*, Voronezhskaya *et al.* (2003) observed one posterior and two anterior cells located at the prototroch. The trochaea theory (Figs. 22.8, 22.10) implies that the protostomian brain should consist of components from the cerebral ganglia, with connectives around the oesophagus, and a ventral component originating from the anterior part of the circumblastoporal nerve. The observations of Ackermann *et al.* (2005: fig. 31) demonstrate the presence of a streak of 2d cells along the posterior side of the cerebral ganglia (from the 1c and 1d cells), and this must be interpreted as the 'blastoporal' part of the brain. The cell-lineage study of *Capitella* (Meyer *et al.* 2010) elegantly demonstrated the origin of a posterodorsal brain region, the circumoesophageal connectives, and the ventral nerve cord from the circumblastoporal descendants of the 2d cell (Table 25.2).

The apical organ is usually interpreted as a sense organ, but this has, to my knowledge, never been proved. The cerebral ganglia often develop rhabdomeric eyes with pigment cells, and sometimes with a vestigial cilium (Arendt *et al.* 2009) and innervate appendages of the prostomium. The ventral chain of ganglia develop from the ectoderm along the fused blastopore lips (Anderson 1959, 1973), probably guided by the axons from the pioneer cells. The ventral cords remain intraepithelial, for example in many of the small, interstitial species (Westheide 1990), but sink in and become situated along the ventral attachment of the mesentery, surrounded by its basal membrane in many of the larger forms (Fig. 25.4). The two pioneer nerve cells resemble the first nerve cells at the posterior end of early bivalve and gastropod larvae (Chapter 27), and one could speculate that they represent a 'pre-segmented' evolutionary stage.

The trochophore larvae have developing coelomic sacs and chaetae already in later planktonic stages (nectochaetes), sometimes with long, special larval chaetae functioning as protection towards predators. Coelomic sacs extend from the lateral position around the gut to meet middorsally; the dorsal blood vessel develops in the dorsal mesenterium where the two epithelia, with their basement membranes, separate creating a longitudinal haemal space. Metamorphosis may be rather gradual, as in many *Nereis* species with lecithotrophic larvae, or more abrupt, as in sabellariid larvae that shift from planktotrophic larvae to sessile adults with a new feeding apparatus; metamorphosis may even be 'catastrophic', as in *Owenia* and *Polygordius*, where the larval organs used in feeding are cast off (see below).

Special larval types called pericalymma (or serosa) larvae (Fig. 25.5) are found in some species of *Polygordius*, in *Owenia* and *Myriochele* (family Oweniidae), and in some phyllodocids. All *Polygordius* larvae have more or less 'blown-up' bodies with the normal trochophore, prototroch, and metatroch at the equator. *P. neapolitanus* has a rather 'normal' nectochaete, which metamorphoses rather gradually into the adult, as in most other polychaete larvae, whereas *P. appendiculatus* and *P. lacteus* have larvae in which the segmented body develops strongly retracted (like an accordion), and covered by a circular fold of the region behind the metatroch (pericalymma type 2 larvae; Fig. 22.7); at metamorphosis the body stretches out and the larger part of the spherical larval body with the ciliary feeding apparatus is shed and engulfed (Woltereck 1902; Herrmann 1986). The pericalymma type 2 larvae of some phyllodocids are morphologically similar to the *Polygordius* pericalymma larvae (Tzetlin 1998), but some larvae are able to retract the long expanded body into the serosa again, in an irregular, knot-like shape (own observations from Hawaii plankton). The oweniid larvae (called mitraria) have ciliary bands that form wide lobes, and the chaetae of the first segment develop early and become very long; the following segments have short chaetae and are pulled up into a deep circular fold behind the long chaetae (pericalymma type 3 larvae; Fig. 22.7). At metamorphosis, the parts of the hyposphere carrying the ciliary feeding structures are cast off together with the long larval chaetae, and the body stretches out so that a small worm resembling an adult emerges in less than an hour (Wilson 1932). The pericalymma larvae of *Polygordius*, phyllodocids and oweniids, are thus only superficially similar, and they have obviously evolved

independently, so they can be of no importance at the higher phylogenetic level.

Clitellates are characterized by the reduction of the number of segments with reproductive organs, and by development of special structures of the fertile region (the clitellum) connected with copulation and formation of protective cocoons for the eggs. All species have eggs with considerable amounts of yolk and direct development. A few oligochaetes, such as *Criodrilus*, have a coeloblastula with no size difference between presumptive endodermal and ectodermal cells, and gastrulation that is described as a type between emboly and epiboly. The mesoteloblasts were very clearly seen, but ectoteloblasts were not reported.

Other clitellates have a highly unequal cleavage and epibolic gastrulation; the ectodermal and mesodermal cells form a micromere cap that spreads ventrally over the large endodermal cells (Anderson 1966). The cell lineages of the glossiphoniid leeches *Helobdella* and *Theromyzon* have been studied in great detail (see review in Shankland and Savage 1997; Fig. 25.8; Table 25.3).

After the two first cleavages, the D cell is larger than the A-C cells, with the B and D cells lying in the midline of the developing embryo. The A-C cells each give off three small micromeres and the D cell gives off one. The micromeres give rise to the ectoderm of the cephalic region, the anterior and lateral parts of the cephalic ganglion, and the stomodeum (foregut). The macromeres 3A-C give rise to the midgut. The 1D cell divides into a cell called DNOPQ situated in the midline, and a cell called DM, which is displaced to the left; DNOPQ gives off three small micromeres and DM gives off two. The two remaining large cells, NOPQ and M, divide transversely and give rise to right and left ectodermal and mesodermal germinal bands respectively. The ectodermal bands are named n, o, p, and q, with the n bands situated along the ventral midline. Rather small cells are given off in a fixed sequence from the large teloblasts, with two cells per segment in the p and q bands and one cell in the other bands; the arrangement of the cells does not follow the boundaries of the adult segments exactly and the descendant cells intermingle in a fixed pattern.

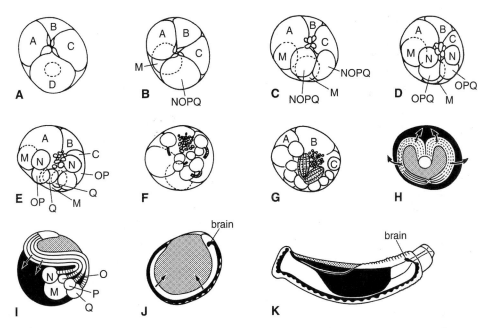

Figure 25.8. Embryology of *Helobdella triserialis*. (A–G) Embryos seen from the apical pole, the letters A–C indicate quadrants not blastomeres; compare with Table 25.3 (H–I) Embryos with germ bands (white), provisional integument formed by apical micromeres (stippled), and macromeres that will form endoderm (black). (J) Late embryo with ventrally fused germ bands, which have formed the ventral nerve cord (black) and provisional integument (stippled). (K) Almost fully formed embryo with endoderm:midgut (black), nerve cord (black), and the provisional integument restricted to a narrow, posterodorsal area (stippled). (Combined after Weisblat *et al.* 1980, 1984, and Shankland and Savage 1997.)

However, a segmental pattern can be found in the ectoderm with the N cells giving off blast cells with alternating fates, each pair giving rise to one parasegment. At a later stage a row of cells from each anterior blast cell expresses *engrailed* and the adult segment border develops just behind these cells (Ramírez *et al.* 1995). Neurons differentiate from all the ectodermal bands and a few from the mesodermal bands. The ectodermal cell bands become S shaped, and fuse midventrally from the anterior end leaving the posterior anus. The dorsal side of the embryo is at first unsegmented, covered by cells from the second micromere quartet (called provisional integument or temporary yolk-sac ectoderm), but this gradually becomes covered by the lateral parts of the ectodermal bands that extend dorsally. The two M-cells in each segment give rise to all of the musculature, nephridia, connective tissue, coelomic epithelia, and some neurons, and the ectodermal bands give rise to

ectoderm and nervous system. The fate of individual cells in the germinal bands is determined by cell-cell interactions.

Other oligochaetes and hirudineans have more aberrant cleavage and cell lineage (Anderson 1966). The midgut originates solely from the D quadrant in *Erpobdella*, and in *Stylaria*, descendants of the first micromeres, and the macromeres 1A and 1B become an embryonic envelope that is shed at a later stage (Dawydoff 1941).

Annelid embryology shows most of the characteristics predicted by the trochaea theory, so the phylum is firmly rooted in the Spiralia. This finds almost unanimous support from the molecular analyses. The radiation of the spiralians is not resolved convincingly either by morphological or by molecular analyses. Annelida and Sipuncula are clearly either sister groups, or the Sipuncula is an ingroup of the Annelida. Annelida + Sipuncula, Mollusca, and

Nemertini share the schizocoelic mode of forming coelomic compartments with metanephridia and a haemal system, and this indicates a close relationship, as indicated in the name Schizocoelia. This finds support from several of the molecular analyses (see Chapter 24).

Interesting subjects for future research

1. Segment development and differentiation in polychaetes: parasegments, ganglia, nephridia
2. Moulting and ecdysteroids
3. Ontogeny of the haemal coeloms of the leeches

References

Ackermann, C., Dorresteijn, A. and Fischer, A. 2005. Clonal domains in postlarval *Platynereis dumerilii* (Annelida: Polychaeta). *J. Morphol.* **266**: 258–280.

Åkesson, B. 1968. The ontogeny of the glycerid prostomium (Annelida; Polychaeta). *Acta Zool. (Stockh.)* **49**: 203–217.

Anderson, D.T. 1959. The embryology of the polychaete *Scoloplos armiger*. *Q. J. Microsc. Sci., N. S.* **100**: 89–166.

Anderson, D.T. 1966. The comparative early embryology of the Oligochaeta, Hirudinea and Onychophora. *Proc. Linn. Soc. N. S. W.* **91**: 11–43.

Anderson, D.T. 1973. *Embryology and Phylogeny of Annelids and Arthropods (International Series of Monographs in pure and applied Biology, Zoology 50)*. Pergamon Press, Oxford.

Arendt, D., Hausen, H. and Purschke, G. 2009. The 'division of labour' model of eye evolution. *Proc. R. Soc. Lond. B* **364**: 2809–2817.

Bartolomaeus, T. and Quast, B. 2005. Structure and development of nephridia in Annelida and related taxa. *Hydrobiologia* **535/536**: 139–165.

Bleidorn, C., Eeckhaut, I., Podsiadlowski, L., *et al.* 2007. Mitochondrial genome and nuclear sequence data support Myzostomida as part of the annelid radiation. *Mol. Biol. Evol.* **24**: 1690–1701.

Bleidorn, C., Podsiadlowski, L., Zhong, M., *et al.* 2009. On the phylogenetic position of Myzostomida: can 77 genes get it wrong? *BMC Evol. Biol.* **9**: 150.

Bleidorn, C., Vogt, L. and Bartolomaeus, T. 2003. New insights into polychaete phylogeny (Annelida) inferred from 18S rDNA sequences. *Mol. Phylogenet. Evol.* **29**: 279–288.

Brinkmann, N. and Wanninger, A. 2008. Larval neurogenesis in *Sabellaria alveolata* reveals plasticity in polychaete neural patterning. *Evol. Dev.* **10**: 606–616.

Brocco, S.L., O'Clair, R. and Cloney, R.A. 1974. Cephalopod integument: The ultrastructure of Kölliker's organs and their relationships to setae. *Cell Tissue Res.* **151**: 293–308.

Bubel, A., Stephens, R.M., Fenn, R.H. and Fieth, P. 1983. An electron microscope, X-ray diffraction and amino acid analysis study of the opercular filament cuticle, calcareous opercular plate and habitation tube of *Pomatoceros lamarckii* Quatrefages (Polychaeta: Serpulidae). *Comp. Biochem. Physiol.* **74B**: 837–850.

Bürger, O. 1891. Beiträge zur Entwicklungsgeschiche der Hurudineen. Zur Embryologie von *Nephelis*. *Zool. Jahrb., Anat.* **4**: 697–738.

Callsen-Cencic, P. and Flügel, H.J. 1995. Larval development and the formation of the gut of *Siboglinum poseidoni* Flügel & Langhof (Pogonophora, Perviata). Evidence of protostomian affinity. *Sarsia* **80**: 73–89.

Clark, R.B. 1964. *Dynamics in Metazoan Evolution*. Clarendon Press, Oxford.

Conway Morris, S. and Peel, J.S. 2008. The earliest annelids: Lower Cambrian polychaetes from the Sirius Passet Lagerstätte, Peary Land, North Greenland. *Acta Palaeontol. Pol.* **51**: 137–148.

Dawydoff, C. 1941. Etudes sur l'embryologie des Naïdidae Indochinoises. *Arch. Zool. Exp. Gen.* **81(Notes et Revue)**: 173–194.

Dordel, J., Fisse, F., Purschke, G. and Struck, T. 2010. Phylogenetic position of Sipuncula derived from multigene and phylogenomic data and its implication for the evolution of segmentation. *J. Zool. Syst. Evol. Res.* **48**: 197–207.

Dorresteijn, A. 2005. Cell lineage and gene expression in the development of polychaetes. *Hydrobiologia* **235/236**: 1–22.

Dorresteijn, A.W.C., Bornewasser, H. and Fischer, A. 1987. A correlative study of experimentally changed first cleavage and Janus development in the trunk of *Platynereis dumerilii* (Annelida, Polychaeta). *Roux's Arch. Dev. Biol.* **196**: 51–58.

Dorresteijn, A.W.C., O'Grady, B., Fischer, A., Porchet-Henner, E. and Boilly-Marer, Y. 1993. Molecular specification of cell lines in the embryo of *Platynereis* (Annelida). *Roux's Arch. Dev. Biol.* **202**: 260–269.

Dray, N., Tessmar-Raible, K., Le Gouar, M., *et al.* 2010. Hedgehog signaling regulates segment formation in the annelid *Platynereis*. *Science* **329**: 339–342.

Dunn, C.W., Hejnol, A., Matus, D.Q., *et al.* 2008. Broad phylogenomic sampling improves resolution of the animal tree of life. *Nature* **452**: 745–749.

Eeckhaut, I. and Lanterbecq, D. 2005. Myzostomida: A review of the phylogeny and ultrastructure. *Hydrobiologia* **535/536**: 253–275.

Eeckhaut, I., Fievez, L. and Müller, M.C.M. 2003. Larval development of *Myzostomum cirriferum* (Myzostomida). *J. Morphol.* **258**: 269–283.

Fauchald, C. and Rouse, G. 1997. Polychaete systematics: past and present. *Zool. Scr.* **26**: 71–138.

Fauchald, K. 1974. Polychaete phylogeny: a problem in protostome evolution. *Syst. Zool.* **24**: 493–506.

Fischer, A. 1999. Reproductive and developmental phenomena in annelids: a source of exemplary research problems. *Hydrobiologia* **402**: 1–20.

Fransen, M.E. 1980. Ultrastructure of coelomic organization in annelids. I. Archiannelids and other small polychaetes. *Zoomorphology* **95**: 235–249.

Fröbius, A.C., Matus, D.Q. and Seaver, E.C. 2008. Genomic organization and expression demonstrate spatial and temporal *Hox* gene colinearity in the lophotrochozoan *Capitella* sp. 1. *PLoS ONE* **3(12)**: e4004.

Gardiner, S.L. and Jones, M.L. 1993. Vestimentifera. In F.W. Harrison (ed.): *Microscopic Anatomy of Invertebrates*, vol. 12, pp. 371–460. Wiley-Liss, New York.

Hall, K.A., Hutchings, P. and Colgan, D.J. 2004. Further phylogenetic studies of the Polychaeta using 18S rDNA sequence data. *J. Mar. Biol. Assoc. U.K.* **84**: 949–960.

Hannerz, L. 1956. Larval development of the polychaete families Spionidae Sars, Disomidae Mesnil, and Poecilochaetidae n.fam. in the Gullmar Fjord. *Zool. Bidr. Upps.* **31**: 1–204.

Harrison, F.W. (ed.) 1992. *Microscopic Anatomy of Invertebrates*, vol. 7. *Annelida*. Wiley-Liss, New York.

Hatschek, B. 1878. Studien über Entwicklungsgeschichte der Anneliden. *Arb. Zool. Inst. University Wien.* **1**: 277–404.

Hatschek, B. 1880. Ueber Entwicklungsgeschichte von *Echiurus* und die systematische Stellung der Echiuridae (Gephyrei chaetiferi). *Arb. Zool. Inst. University Wien.* **3**: 45–78.

Hausen, H. 2005a. Chaetae and chaetogenesis in polychaetes (Annelida). *Hydrobiologia* **535/536**: 37–52.

Hausen, H. 2005b. Comparative structure of the epidermis in polychaetes (Annelida). *Hydrobiologia* **535/536**: 25–35.

Hay-Schmidt, A. 1995. The larval nervous system of *Polygordius lacteus* Schneider, 1868 (Polygordiidae, Polychaeta): immunocytochemical data. *Acta Zool. (Stockh.)* **76**: 121–140.

Hejnol, A., Obst, M., Stamatakis, A., *et al.* 2009. Assessing the root of bilaterian animals with scalable phylogenomic methods. *Proc. R. Soc. Lond. B* **276**: 4261–4270.

Herrmann, K. 1986. *Polygordius appendiculatus* (Archiannelida)—Metamorphose. *Publ. Wiss. Film., Biol.,* 18. Ser. **36**: 1–15.

Hessling, R. 2002. Metameric organisation of the nervous system in developmental stages of *Urechis caupo* (Echiura) and its phylogenetic implications. *Zoomorphology* **121**: 221–234.

Hessling, R. and Westheide, W. 2002. Are Echiura derived from a segmented ancestor? Immunohistochemical analysis of the nervous system in developmental stages of *Bonellia viridis*. *J. Morphol.* **252**: 100–113.

Jägersten, G. 1939. Zur Kenntniss der Larvenentwicklung bei *Myzostomum*. *Ark. Zool.* **31A**: 1–21.

Jékely, G., Colombelli, J., Hausen, H., *et al.* 2008. Mechanism of phototaxis in marine zooplankton. *Nature* **456**: 395–399.

Jones, M.L. and Gardiner, S.L. 1988. Evidence for a transient digestive tract in Vestimentifera. *Proc. Biol. Soc. Wash.* **101**: 423–433.

Jones, M.L. and Gardiner, S.L. 1989. On the early development of the vestimentiferan tube worm *Ridgeia* sp. and observations on the nervous system and trophosome of *Ridgeia* sp. and *Riftia pachyptila*. *Biol. Bull.* **177**: 254–276.

Kulakova, M., Bakalenko, N., Novikova, E., *et al.* 2007. Hox gene expression in larval development of the polychaetes *Nereis virens* and *Platynereis dumerilii* (Annelida, Lophotrochozoa). *Dev. Genes Evol.* **217**: 39–54.

Kupriyanova, E.K. 1999. The taxonomic status of *Serpula* cf. *columbiana* Johnson, 1901 from the American and Asian coasts of the North Pacific Ocean (Polychaeta: Serpulidae). *Ophelia* **50**: 21–34.

Lacalli, T.C. 1981. Structure and development of the apical organ in trochophores of *Spirobranchus polycerus*, *Phyllodoce maculata* and *Phyllodoce mucosa* (Polychaeta). *Proc. R. Soc. Lond. B* **212**: 381–402.

Lacalli, T.C. 1984. Structure and organization of the nervous system in the trochophore larva of *Spirobranchus*. *Phil. Trans. R. Soc. Lond. B* **306**: 79–135.

Lacalli, T.C. 1986. Prototroch structure and innervation in the trochophore larva of *Phyllodoce* (Polychaeta). *Can. J. Zool.* **64**: 176–184.

Lanterbecq, D., Rouse, G. and Eeckhaut, I. 2010. Evidence for cospeciation events in the host–symbiont system involving crinoids (Echinodermata) and their obligate associates, the myzostomids (Myzostomida, Annelida). *Mol. Phylogenet. Evol.* **54**: 357–371.

Leise, E.M. 1988. Sensory organs in the hairy girdles of some mopaliid chitons. *Am. Malacol. Bull.* **6**: 141–151.

Lovén, S. 1840. Iakttagelse öfver metamorfos hos en Annelid. *K. Sven. Naturvetenskapsakad. Handl.* **1840**: 93–98.

Malakhov, V.V., Popelyaev, I.S. and Galkin, S.V. 1997. Organization of Vestimentifera. *Zool. Zh.* **76**: 1308–1335.

Mau, W. 1881. Über *Scoloplos armiger* O.F. Müller. *Z. Wiss. Zool.* **36**: 389–432.

McCall, G.J.H. 2006. The Vendian (Ediacaran) in the geological record: Enigmas in geology's prelude to the Cambrian explosion. *Earth Sci.Rev.* **77**: 1–229.

McDougall, C., Chen, W.C., Shimeld, S.M. and Ferrier, D.E.K. 2006. The development of the larval nervous system, musculature and ciliary bands of Pomatoceros lamarcki (Annelida): heterochrony in polychaetes. *Front. Zool.* **3**: 16.

Mead, A.D. 1897. The early development of marine annelids. *J. Morphol.* **13**: 227–326.

Meyer, N.P. and Seaver, E. 2010. Cell lineage and fate map of the primary somatoblast of the polychaete annelid *Capitella teleta*. *Integr. Comp. Biol.* **50**: 756–767.

Meyer, N.P., Boyle, M.J., Martindale, M.Q. and Seaver, E.C. 2010. A comprehensive fate map by intracellular injection of identified blastomeres in the marine polychaete *Capitella teleta*. *EvoDevo* **1**: 8.

Mwinyi, A., Meyer, A., Bleidorn, C., *et al.* 2009. Mitochondrial genome sequences and gene order of *Sipunculus nudus* give additional support for an inclusion of Sipuncula into Annelida. *BMC Genomics* **10**: 27.

Newby, W.W. 1940. The embryology of the echiuroid worm *Urechis caupo*. *Mem. Am. Philos. Soc.* **16**: 1–219.

Nielsen, C. 1998. Origin and evolution of animal life cycles. *Biol. Rev.* **73**: 125–155.

Nielsen, C. 2004. Trochophora larvae: cell-lineages, ciliary bands and body regions. 1. Annelida and Mollusca. *J. Exp. Zool. (Mol. Dev. Evol.)* **302B**: 35–68.

Nielsen, C. 2005. Trochophora larvae: cell-lineages, ciliary bands and body regions. 2. Other groups and general discussion. *J. Exp. Zool. (Mol. Dev. Evol.)* **304B**: 401–447.

Nilsson, D.-E. 1994. Eyes as optical alarm systems in fan worms and ark clams. *Phil. Trans. R. Soc. Lond. B* **346**: 195–212.

Nilsson, D.-E. 2009. The evolution of eyes and visually guided behaviour. *Phil. Trans. R. Soc. Lond. B* **364**: 2833–2847.

Orrhage, L. and Müller, M.C.M. 2005. Morphology of the nervous system of the Polychaeta (Annelida). *Hydrobiologia* **535/536**: 97–111.

Paxton, J. 2005. Molting polychaete jaws—ecdysozoans are not the only molting animals. *Evol. Dev.* **7**: 337–340.

Pietsch, A. and Westheide, W. 1987. Protonephridial organs in *Myzostoma cirriferum* (Myzostomida). *Acta Zool. (Stockh.)* **68**: 195–203.

Pilger, J.F. 1993. Echiura. In F.W. Harrison (ed.): *Microscopic Anatomy of Invertebrates*, vol. 12, pp. 185–236. Wiley-Liss, New York.

Prpic, N.-M. 2008. Parasegmental appendage allocation in annelids and arthropods and the homology of parapodia and arthropodia. *Front. Zool.* **5**: 17.

Prud'homme, B., de Rosa, R., Arendt, D., *et al.* 2003. Arthropod-like expression patterns of *engrailed* and *wingless* in the annelid *Platynereis dumerilii* suggest a role in segment formation. *Curr. Biol.* **13**: 1876–1881.

Purschke, G. 2002. On the ground pattern of Annelida. *Org. Divers. Evol.* **2**: 181–196.

Purschke, G. 2005. Sense organs in polychaetes (Annelida). *Hydrobiologia* **535/536**: 53–78.

Purschke, G., Arendt, D., Hausen, H. and Müller, M.C.M. 2006. Photoreceptor cells and eyes in Annelida. *Arthropod Struct. Dev.* **35**: 211–230.

Ramírez, F.A., Wedeen, C.J., Stuart, D.K., Lans, D. and Weisblat, D.A. 1995. Identification of a neurogenic sublineage required for CNS segmentation in an annelid. *Development* **121**: 2091–2097.

Rasmussen, E. 1973. Systematics and ecology of the Isefjord marine fauna. *Ophelia* **11**: 1–495.

Remane, A. 1952. *Die Grundlagen des natürlichen Systems, der vergleichenden Anatomie und der Phylogenetik*. Akademische Verlagsgesellschaft, Leipzig.

Rieger, R.M. 1980. A new group of interstitial worms, Lobatocerebridae nov. fam. (Annelida) and its significance for metazoan phylogeny. *Zoomorphologie* **95**: 41–84.

Rieger, R.M. 1981. Fine structure of the body wall, nervous system, and digestive tract in the Lobatocerebridae Rieger and the organization of the gliointerstitial system in Annelida. *J. Morphol.* **167**: 139–165.

Rieger, R.M. 1988. Comparative ultrastructure and the Lobatocerebridae: Keys to understand the phylogenetic relationship of Annelida and the Acoelomates. *Microfauna Mar.* **4**: 373–382.

Rieger, R.M. 1991a. *Jennaria pulchra*, nov. gen. nov. spec., eine den psammobionten Anneliden nahestehende Gattung aus dem Küstengrundwasser von North Carolina. *Ber. Nat.-Med. Ver. Innsbruck* **78**: 203–215.

Rieger, R.M. 1991b. Neue Organisationstypen aus der Sandlückenfauna: die Lobatocerebriden und *Jennaria pulchra*. *Verh. Dtsch. Zool. Ges.* **84**: 247–259.

Rieger, R.M. and Purschke, G. 2005. The coelom and the origin of the annelid body plan. *Hydrobiologia* **535/536**: 127–137.

Riisgård, H.U., Nielsen, C. and Larsen, P.S. 2000. Downstream collecting in ciliary suspension feeders: the catch-up principle. *Mar. Ecol. Prog. Ser.* **207**: 33–51.

Robbins, D.E. 1965. The biology and morphology of the pelagic annelid *Poeobius meseres* Heath. *J. Zool.* **146**: 197–212.

Rouse, G.W. 1999. Trochophore concepts: ciliary bands and the evolution of larvae in spiralian Metazoa. *Biol. J. Linn. Soc.* **66**: 411–464.

Rouse, G.W. 2000. Polychaeta, including Pogonophora and Myzostomida. In K.G. Adiyodi and R.G. Adiyodi (eds): *Reproductive Biology of Invertebrates*, vol 9B, pp. 81–124. Wiley, Chichester.

Rouse, G.W. and Fauchald, K. 1997. Cladistics and polychaetes. *Zool. Scr.* **26**: 139–204.

Rousset, V., Pleijel, F., Rouse, G.W., Erséus, C. and Siddall, M.E. 2007. A molecular phylogeny of annelids. *Cladistics* **23**: 41–63.

Sauber, F., Reuland, M., Berchtold, J.P., *et al.* 1983. Cycle de mue et ecdystéroïdes chez une Sangsue, *Hirudo medicinalis*. *C. R. Seanc. Acad. Sci., Sci. Vie* **296**: 413–418.

Sawyer, R.T. 1986. *Leech Biology and Behaviour. Vol. 1. Anatomy, Physiology, and Behaviour*. Oxford University Press, Oxford.

Schroeder, P.C. and Hermans, C.O. 1975. Annelida: Polychaeta. In A.C. Giese and J.S. Pearse (eds): *Reproduction of Marine Invertebrates*, vol. 3, pp. 1–213. Academic Press, New York.

Shankland, M. and Savage, R.M. 1997. Annelids, the segmented worms. In S.F. Gilbert and A.M. Raunio (eds): *Embryology. Constructing the Organism*, pp. 219–235. Sinauer Associates, Sunderland, MA.

Smart, T.I. and Von Dassow, G. 2009. Unusual development of the mitraria larva in the polychaete *Owenia collaris*. *Biol. Bull.* **217**: 253–268.

Smith, P.R., Lombardi, J. and Rieger, R.M. 1986. Ultrastructure of the body cavity lining in a secondary acoelomate, *Microphthalmus* cf. *listensis* Westheide (Polychaeta: Hesionidae). *J. Morphol.* **188**: 257–271.

Smith, P.R., Ruppert, E.E. and Gardiner, S.L. 1987. A deuterostome-like nephridium in the mitraria larva of *Owenia fusiformis* (Polychaeta, Annelida). *Biol. Bull.* **172**: 315–323.

Southward, E.C. 1988. Development of the gut and segmentation of newly settled stages of *Ridgeia* (Vestimentifera):

implications for relationships between Vestimentifera and Pogonophora. *J. Mar. Biol. Assoc. U.K.* **68**: 465–487.

Southward, E.C., Schulze, A. and Gardiner, S.L. 2005. Pogonophora (Annelida): form and function. *Hydrobiologia* **535/536**: 227–251.

Sperling, E.A., Vinther, J., Moy, V.N., *et al.* 2009. MicroRNAs resolve an apparent conflict between annelid systematics and their fossil record. *Proc. R. Soc. Lond. B* **276**: 4315–4322.

Struck, T., Schult, N., Kusen, T., *et al.* 2007. Annelid phylogeny and the status of Sipuncula and Echiura. *BMC Evol. Biol.* **7**: 57.

Struck, T.H., Paul, C., Hill, N., *et al.* 2011. Phylogenomic analyses unravel annelid evolution. *Nature* **471**: 95–98.

Treadwell, A.L. 1901. Cytogeny of *Podarke obscura* Verrill. *J. Morphol.* **17**: 399–486.

Tzetlin, A.B. 1998. Giant pelagic larvae of Phyllodocidae (Polychaeta, Annelida). *J. Morphol.* **238**: 93–107.

Voronezhskaya, E.E., Tsitrin, E.B. and Nezlin, L.P. 2003. Neuronal development in larval polychaete *Phyllodoce maculata* (Phyllodocidae). *J. Comp. Neurol.* **455**: 299–309.

Weisblat, D.A., Kim, S.Y. and Stent, G.S. 1984. Embryonic origins of cells in the leech *Helobdella triserialis*. *Dev. Biol.* **104**: 65–85.

Weisblat, D.A., Harper, G., Stent, G.S. and Sawyer, R.T. 1980. Embryonic cell lineages in the nervous system of the glossiphoniid leech *Helobdella triserialis*. *Dev. Biol.* **76**: 58–78.

Westheide, W. 1987. Progenesis as a principle in meiofauna evolution. *J. Nat. Hist.* **21**: 843–854.

Westheide, W. 1988. The nervous system of the male *Dinophilus gyrociliatus* (Annelida: Polychaeta). I. Number, types and distribution pattern of sensory cells. *Acta Zool. (Stockh.)* **69**: 55–64.

Westheide, W. 1990. Polychaetes: Interstitial families. *Synop. Br. Fauna, N.S.* **44**: 1–152.

Westheide, W. 1997. The direction of evolution within the Polychaeta. *J. Nat. Hist.* **31**: 1–15.

Wilson, D.P. 1932. On the mitraria larva of *Owenia fusiformis* Delle Chiaje. *Phil. Trans. R. Soc. Lond. B* **221**: 231–334.

Wilson, E.B. 1892. The cell lineage of *Nereis*. *J. Morphol.* **6**: 361–480.

Winchell, C.J., Valencia, J.E. and Jacobs, D.K. 2010. Concfocal analysis of nervous system architecture in direct-developing juveniles of *Neanthes arenaceodentata* (Annelida, Nereididae). *Front. Zool.* **7**: 17.

Woltereck, R. 1902. Trochophora-Studien I. Histologie der Larve und die Entstehung des Annelids bei den *Polygordius*-Arten der Nordsee. *Zoologica (Stuttg.)* **13**: 1–71.

Woltereck, R. 1904. Beiträge zur praktischen Analyse der *Polygordius*-Entwicklung nach dem 'Nordsee-' und dem 'Mittelmeer-Typus'. *Arch Entwicklungsmech. Org.* **18**: 377–403.

Woltereck, R. 1926. Neue und alte Beobachtungen zur Metamorphose der Endolarve von *Polygordius*. *Zool. Anz.* **65**: 49–60.

Worsaae, K. and Rouse, G.W. 2010. The simplicity of males: dwarf males of four species of *Osedax* (Siboglinidae; Annelida) investigated by confocal laser scanning microscopy. *J. Morphol.* **271**: 127–142.

Worsaae, K., Nygren, A., Rouse, G.W., *et al.* 2005. Phylogenetic position of Nerillidae and Aberranta (Polychaeta, Annelida), analysed by direct optimization of combined molecular and morphological data. *Zool. Scr.* **34**: 313–328.

Young, C., Výzquez, E., Metaxas, A. and Tyler, P.A. 1996. Embryology of vestimentiferan tube worms from deep-sea methane/sulphide seeps. *Nature* **381**: 514–516.

Zrzavý, J., Říha, P., Piálek, L. and Janouškovec, J. 2009. Phylogeny of Annelida (Lophotrochozoa): total-evidence analysis of morphology and six genes. *BMC Evol. Biol.* **9**: 189.

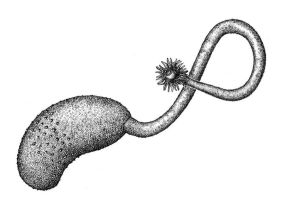

26

Phylum Sipuncula

Sipuncula is a small, well-defined phylum of about 150 benthic, marine species. A number of strikingly 'modern' sipunculans have been described from the Lower Cambrian (Huang *et al.* 2004).

The almost cylindrical organisms are divided into a trunk and a more slender, anterior, retractable introvert. The terminal mouth is usually surrounded by a ring of ciliated tentacles. The gut forms a loop and the anus is situated dorsally at the base of the introvert. The microscopical anatomy was reviewed by Rice (1993; see also Schulze *et al.* 2005).

The monolayered ectoderm is covered by a cuticle with cross-layered collagen fibres between microvilli; the ectoderm of the oral area and the oral side of the tentacles consists of multiciliated cells and the cuticle is quite thin. Hooks of a proteinaceous material are found on various parts of the introvert and body. Chitin appears to be completely absent.

The nervous system (Fig. 22.3) comprises a dorsal, bilobed brain that is connected with a ventral nerve cord by a pair of circumoesophageal connectives. Ultrastructure and innervation of a paired ciliated organ, situated on the dorsal side of the mouth, indicates that it is not homologous with the nuchal organs of the polychaetes (Purschke *et al.* 1997). Ocelli situated at the bottom of tubular invaginations into the brain are found in some species; they resemble ocelli of certain polychaetes (Åkesson 1958). The ventral nerve

cord is surrounded by a thin peritoneum and lies free in the body cavity only attached in the ventral midline by a number of fine nerves covered by the peritoneum; the nerve cord is accompanied by a pair of lateral muscles that cover the whole periphery in some zones.

There are two body cavities, a small, anterior tentacle coelom and a large, posterior body coelom. Both cavities contain various types of coelomocytes, which are formed from the mesothelia, and which under certain circumstances may move between the cavities; nevertheless, the two cavities are normally isolated from each other as shown by the different properties of the haemerythrins in the two cavities (Manwell 1963).

The tentacle coelom is circumoesophageal and sends canals into the branched tentacles, and one or two median canals (compensation sacs) along the gut. The peritoneum is ciliated especially in the tentacle canals, where the cilia are presumed to create a circulation of the coelomic fluid with haemocytes through a median and two lateral canals in each tentacle. Some types have two rather short compensation sacs without diverticula, for example *Sipunculus*, while the dorsal sac is greatly expanded posteriorly (called the contractile vessel), with numerous long, thin diverticula in others, such as *Themiste*. This system clearly functions both as a hydrostatic skeleton expanding the tentacles, and as a respiratory system that can transport oxygen from the expanded tentacles to the body coelom (Ruppert and Rice 1995). It has been described as a system of blood vessels, but the ciliated epithelial lining shows that it is a true coelom.

The main body coelom is a spacious cavity lined by a peritoneum overlying longitudinal and circular muscles and containing a fluid with haemocytes and other cell types. The muscle layers are more or less continuous in the genera considered primitive, and divided into separate muscles in the more advanced genera. The coelomic cavity extends into partially ciliated, longitudinal canals or sacs between the muscles in the advanced genera with small tentacles, where it is believed to be important for circulation (Ruppert and Rice 1995). The body coelom functions as a hydrostatic skeleton, both in eversion of the introvert and in burrowing.

A haemal system is absent, part of its functions apparently being carried out by the tentacle coelom and the coelomic canals of the body wall.

A pair of large metanephridia is found in the body coelom with the nephridiopores situated near the anus. The funnel is very large and ciliated, and has a special function in separating the ripe eggs from the several other cell types in the coelomic fluid. In most metazoans, the primary urine is filtered from a haemal system to the coelom, but as pointed out by Ruppert and Smith (1988) and Bartolomaeus and Ax (1992), the sipunculan excretory system with a metanephridium is exceptional in lacking a haemal system from which the primary urine can be filtered. The suggestion of a filtration from the compensation sac of the tentacle coelom is not supported by experiments, and appears unlikely as the tentacle coelom is so restricted and the body coelom is in direct contact with most of the muscles through the coelomic canals of the body wall. Podocytes have been observed in the trunk peritoneum of the contractile vessel, but not in the apposed peritoneum of the tentacle coelom. Podocytes obviously without a function in ultrafiltration, have been reported from crustaceans (Wägele and Walter 1990) and enteropneusts (Chapter 50), so the mere presence of podocytes is not proof of production of primary urine. An unusually complicated metanephridium was described in *Thysanocardia*, which shows podocytes on

the coelomic side of the funnel and protruding groups of excretory cells of two types on the inner side (Adrianov *et al.* 2002).

The gonad is a ventral, lobed organ surrounded by peritoneum and suspended in a mesentery. The oocytes develop to the first meiotic prophase in the ovary and are then released into the coelom, where vitellogenesis takes place. The ripe egg is surrounded by a thick envelope with many pores. Some species have spherical eggs, while others have spindle-shaped to flattened eggs with a shallow depression at the apical pole, indicating that the polarity of the egg is determined before spawning. The sperm may penetrate the egg envelope everywhere except at the apical pole. Most species are free spawners, and the polar bodies are given off at the apical pole soon after fertilization (Rice 1989).

The embryology was reviewed by Nielsen (2005). Cleavage is spiral (Fig. 23.2; Table 26.1) with a cell lineage closely resembling that of annelids and molluscs. An apical ganglion develops at the position of the polar bodies. The early prototroch is formed by descendants of the four primary trochoblasts, $1a^2$–$1d^2$, and three secondary trochoblasts, $1a^{122}$–$1c^{122}$, so that there is a narrow dorsal gap, which closes at a later stage when the two ends of the band fuse. Gastrulation is embolic to epibolic according to the amount of yolk. Mesoderm is formed from the 4d cell, and a pair of mesoteloblasts has been observed both in *Sipunculus*, *Phascolopsis*, and *Phascolosoma*. Gerould (1906) described and illustrated three to four small coelomic pouches in early trochophores of *Phascolopsis*, but according to Hyman (1959: p. 657), he later changed his interpretation and explained that the apparent metamerism was caused by contraction and buckling. The compact endoderm hollows out to the tubular gut that becomes connected to a stomodeum and a small dorsal proctodeum. The ventral zone between mouth and anus forms a growing bulge, which contains the gut, and the whole body thus becomes elongate perpendicular to the original anteroposterior axis. A pair of coelomic cavities develop through schizocoely in the paired mesodermal bands, but the two coelomic sacs fuse completely at a later stage. In *Phascolion* (Wanninger *et al.* 2005), two FMRFamidergic cells can be recognized in the apical ganglion, and paired cerebral ganglia develop lateral to these. Nerves develop from the cerebral ganglia and extend posteriorly around the mouth and along the ventral side. In *Phascolosoma* (Kristof *et al.* 2008; Wanninger *et al.* 2009), perikarya of the anterior part of the ventral nerve cords are arranged in pairs, suggesting traces of segmentation. The paired ventral nerves subsequently fuse in early larvae of both genera. In *Sipunculus*, the ventral nerve cord develops as a median longitudinal thickening of the ectoderm that later splits off from the ectoderm and sinks into the body cavity covered by the mesothelium.

Hatschek (1883) illustrated an early metamorphosis stage of *Sipunculus* with an undivided coelomic cavity extending to the anterior end of the larva in front of the mouth, and the tentacles were described as developing from the rim of the mouth; the origin of the tentacle coelom was not mentioned, but his observations indicate that it becomes pinched off from the

body coelom after metamorphosis. Protonephridia have not been observed at any stage, but a pair of metanephridia develops in the early pelagosphaera stage. A yellowish cell was found completely embedded in each of the lateral groups of mesodermal cells. These cells developed into U-shaped cell groups with a narrow canal, which came into contact with the ectoderm and formed the nephridiopore, while the opposite end broke through the peritoneum, which formed the ciliated funnel of a typical metanephridium. Rice (1973) observed a similar development in *Phascolosoma* with ectodermal cells giving rise to the pore region. However, Gerould (1906) believed that the main part of the metanephridia in *Golfingia* develops from the ectoderm, with only the ciliated funnel originating from the mesoderm. It cannot be excluded that variations occur in the development of metanephridia, but Gerould's report appears less well documented.

A few species have direct development, but most species have lecithotrophic trochophores that swim with the cilia of the prototroch protruding through pores in the egg envelope. In some of these species, the trochophores metamorphose directly into the juveniles, but most species go through a planktonic, planktotrophic or lecithotrophic stage called pelagosphaera (Rice 1973, 1981; Fig. 26.1). This larval type is characterized by a prominent ring of compound cilia behind the mouth, and a prototroch that has become overgrown more or less completely by surrounding ectodermal cells. The fully developed pelagosphaera larva has an extended ciliated lower lip, with a buccal organ similar to that found in many polychaetes (Tzetlin and Purschke 2006) and a lip gland. The buccal organ can be protruded from a deep, transverse ectodermal fold. A pair of ocelli consist of a pair of rhabdomeric sensory cells and a pigment-cup cell, structurally identical to the ocelli of polychaete trochophores (Blumer 1997). A pair of muscles develop between the apical organ and the body wall behind the anus (Fig. 23.3), and a further pair develop from the oral region to the posterior part of the body wall. These muscles can retract the whole anterior part of the body, including the ciliary ring, into the posterior part of the body, and the retracted

Table 26.1. Cell lineage of *Phascolopsis* and *Golfingia*. (Based on Gerould 1906).

				$1a^{111}$	apical cell
			$1a^{11}$	$1a^{112}$	ectoderm
		$1a^1$		$1a^{121}$	ectoderm
			$1a^{12}$	$1a^{122}$	accessory trochoblast
	$1a$	$1a^2$			primary trochoblast
	A	$2a$			ectoderm
AB	$1A$		$3a$		ectoderm
		$2A$	$3A$	$4a$	ectoderm
Z				$4A$	ectoderm
	B				
	C	development as A			
CD				$1d^{111}$	apical cell
			$1d^{11}$	$1d^{112}$	ectoderm
		$1d^1$	$1d^{12}$		ectoderm
	$1d$	$1d^2$			primary trochoblast
	D	$2d$	$2d^1$		somatoblast
			$2d^2$		ectoderm
	$1D$	$2D$	$3d$		ectoderm
			$3D$	$4d$	mesoderm, endoderm
				$4D$	endoderm

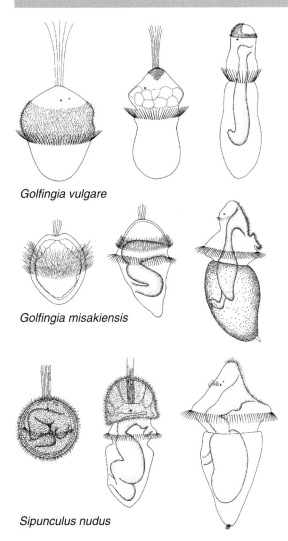

Golfingia vulgare

Golfingia misakiensis

Sipunculus nudus

Figure 26.1. Pelagic developmental types of sipunculans. *Golfingia vulgare* has a completely lecithotrophic development; the first stage is a roundish trochophore that swims with a wide band of cilia anterior to and around the mouth; the second stage is more elongate, the prototroch cells have become infolded and degenerate, and the larva swims with a ring of compound cilia behind the mouth; the third stage is cylindrical and the gut is developing. (Based on Gerould 1906.) *Golfingia misakiensis* goes through a similar trochophora stage, but the gut becomes functional at an early stage; the fully-grown pelagosphaera larva swims with the post-oral ring of compound cilia and is able to retract the anterior part of the body, including the ciliary ring, into the posterior part. (Redrawn from Rice 1978.) *Sipunculus nudus* has a pericalymma larva (type 1) with the hyposphere completely covered by an extension of the prototrochal area; the hyposphere breaks out through the posterior end of the serosa, which is for a short time carried as a helmet over the episphere and then cast off; the fully-grown larva is a normal pelagosphaera. (Based on Hatschek 1883.)

part can be enclosed by a strong constrictor muscle. A characteristic retractile terminal organ contains both sensory and secretory cells (Ruppert and Rice 1983).

Sipunculus has a special pericalymma larva with the prototrochal epithelium extended posteriorly, covering mouth and hyposphere (type 1 pericalymma; Fig. 22.7); this thin extension, called the serosa, is shed at hatching and the larva becomes a normal pelagosphaera.

Gerould (1906) observed the development of circular muscles from ectomesodermal cells in the zone just behind the prototroch. The two pairs of retractor muscles are believed to be ectomesodermal (Rice 1973).

The planktotrophic pelagosphaera larvae do not use the ciliary bands in filter feeding, but the nature of the feeding mechanism is unknown. Jägersten (1963) observed swallowing of large particles, such as fragments of other larvae, and found copepods in the gut of freshly caught larvae, so the larvae may be carnivorous.

Monophyly of the Sipuncula seems unquestioned. The U-shaped gut indicates that the unsegmented body has not evolved simply by loss of the segmentation along the anteroposterior axis. The ontogeny resembles the metamorphosis of the phoronids (Chapter 54), but this must be a homoplasy. The presence of a typical spiral cleavage with mesoderm originating from the 4d cell and the development and morphology of the central nervous system clearly places the Sipuncula in the spiralian line of the Protostomia (the significance of a 'molluscan cross' is discussed in Chapter 24). The teloblastic proliferation of the mesoblasts, and the schizocoelic formation of paired coelomic pouches, indicate a close relationship with the molluscs and annelids, and the group is here called Schizocoelia. The complete absence of a haemal system is enigmatic, so much more as organs with the same function formed by coelomic compartments are found in several species (see above), and the close relationship with the annelids and molluscs indicate that the common ancestor had a haemal system. The pelagosphera larva develops after the trochophora stage in some species and should perhaps be interpreted as a secondary larva; its prominent ring of compound cilia behind the mouth is usually interpreted as a metatroch, but the direction of the effective stroke is

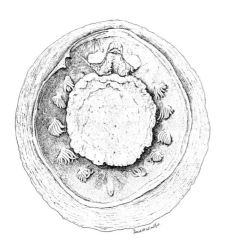

27

Phylum **Mollusca**

Mollusca is one of the largest animal phyla, with about 200 000 described living species and a very extensive fossil record. The number of described living species is somewhat uncertain because malacologists, from amateur shell collectors to professional biologists, during the last centuries have given names to the same species several times, often with rather chaotic results. The eight living classes are predominantly marine; only gastropods and bivalves have spread to freshwater, and only the gastropods have entered the terrestrial habitat. Some families of snails and mussels comprise commensalistic species and one family of snails contains endoparasitic species, which can only be recognized as molluscs by their larvae.

The living molluscs are a morphologically very well-defined, monophyletic group, comprising clearly delimited classes but, quite surprisingly, this is only very rarely shown even in the most recent studies on molecular phylogeny. Only studies with a limited number of molluscan taxa show the monophyly (Pick *et al.* 2010). The study of Wilson *et al.* (2010), based on eight ribosomal and mitochondrial genes from 91 species, representing all eight classes and a few outgroup species, failed to show the monophyly. The study of Meyer *et al.* (2010), based on 18S rRNA from 111 molluscs, representing all classes and a few outgroup species, showed molluscan monophyly, but a very surprising arrangement of the molluscan classes with a

diphyletic Bivalvia. The study of Mallatt *et al.* (2010), based on nearly complete rRNA genes from 197 metazoan taxa, showed a mix of molluscan classes, with annelids, sipunculans, nemertines, phoronids, and brachiopods. The study of Chen *et al.* (2009), based on complete mitochondrial genomes of 34 lophotrochozoans, showed the molluscs mixed up with rotifers and phoronids. Similar results are found in many other studies using ribosomal and mitochondrial genes, which are apparently not suited for studies on lophotrochozoan interrelationships. Recent studies based on 'expressed sequence tag' data and phylogenomic methods show molluscan monophyly, but with similarly surprising relationships between the classes (Dunn *et al.* 2008; Hejnol *et al.* 2009). Only on the lower levels, such as the Bivalvia (Campbell 2000; Giribet and Wheeler 2002; Giribet 2008), has reasonable agreement between morphology and molecules been attained.

Early Cambrian faunas comprise a diversity of 'small shelly fossils' (Parkhaev 2008), larger unquestionable molluscs, such as *Halkieria* (Vinther and Nielsen 2005), and several more questionable forms. *Halkieria* and *Sinosachites* (Vinther 2009) have an elongate body, with the dorsal side covered by a layer of calcareous spicules resembling the polyplacophoran perinotum, and an anterior and a posterior dorsal shell. They probably represent an extinct molluscan

---

Chapter vignette: The monoplacophoran *Neopilina galatheae*. (From Lemche and Wingstrand 1959.)

clade and bear no resemblance to annelids. The naked Middle Cambrian *Odontogriphus* shows radular teeth and a general outline resembling that of *Halkieria* and polyplacophorans (Caron *et al.* 2006, 2007). The Late Precambrian *Kimberella* has been interpreted as a mollusc (Fedonkin and Waggoner 1997), but the reconstruction shows a periphery with a 'crenellated zone' that does not resemble any known structure from living molluscs, and there is no sign of foot, mouth, or radula. Long scratches in the microbial mat near the anterior end of some specimens have been interpreted as radular marks (Fedonkin *et al.* 2007), but the relationships of *Kimberella* still seem uncertain. The Hyolithida has been interpreted as a group of molluscs, but this seems unfounded. Some of the well-preserved shells with distinct muscle scars have given the study of the early radiation of the shelled groups, especially the Conchifera (see below), a firm basis. However, the unshelled forms are almost unknown as fossils, so the early phylogeny must be based on soft-part anatomy and embryology of living forms. The surprisingly well-preserved Silurian *Acaenoplax* (Sutton *et al.* 2001, 2004) resembles an 'aplacophoran with shells', but it is difficult to fit into a phylogeny.

There is a wide variation in morphology of both adults and larvae, but a number of characters can be recognized in almost all molluscs, and can be interpreted as ancestral of the phylum. The latest molluscan ancestor probably had three characters.

1. The mantle: a large area of the dorsal epithelium with a thickened cuticle with calcareous spicules. The mantle is usually expanded into a peripheral fold that overarches a mantle cavity. The mantle is easily recognized in representatives of all the eight living classes.
2. The foot: a flat, ciliated, postoral, ventral expansion used in creeping or modified for other types of locomotion. A series of muscles from mantle to foot can pull the protective mantle towards the substratum or in other ways bring the soft parts under the protection of the mantle. The foot is easily identified in most classes, but it is reduced to a narrow keel in the solenogasters and has disappeared altogether in the caudofoveates.
3. The radula: a cuticular band with teeth formed in a pocket of the ventral epithelium of the oesophagus and used in feeding. A radula is found in all the classes except the bivalves.

Ancestral adult characters shared with annelids probably include a haemal system with a heart formed by the walls of coelomic sacs and metanephridia/gonoducts originating in coelomic sacs. Pectinate gills and associated osphradial sense organs in the mantle cavity probably evolved in the larger ancestral forms. The pelagobenthic life cycle with ciliated larvae is prominent in all living classes except the cephalopods and was probably characteristic of the ancestral mollusc. The ancestral larva was probably a trochophore (see below).

The eight classes of living molluscs are clearly delimited, and the monophyly of the group Conchifera, comprising Monoplacophora, Gastropoda, Cephalopoda, Bivalvia, and Scaphopoda, is generally accepted. However, there is still disagreement about the interrelationships of Conchifera and the three remaining classes, Caudofoveata, Solenogastres, and Polyplacophora (Haszprunar *et al.* 2008).

Wingstrand (1985) found very similar radulae and associated structures in Polyplacophora and Conchifera, including hollow radula vesicles, cartilages of the odontophore, several sets of radular muscles, and a subradular sense organ. Additional shared characters were found in the structure of the gut—both groups have well-defined digestive glands and a coiled gut, whereas these are lacking in aplacophorans. This supports a monophyletic group, Testaria (Fig. 27.1). The perinotum (girdle) with spicules, and the cilia with paired roots of the usual metazoan type in the Polyplacophora, must be interpreted as plesiomorphies (Salvini-Plawen 2006; Todt *et al.* 2008b).

Two evolutionary lines can be traced from the ancestral mollusc with a ciliated foot and a mantle with a mucopolysaccharide lining with calcareous spicules.

1. Aplacophora, in which the foot lost the locomotive function and became very narrow, the lateral parts of the mantle either coming close to each other so that the foot became a ciliated keel (Solenogastres),

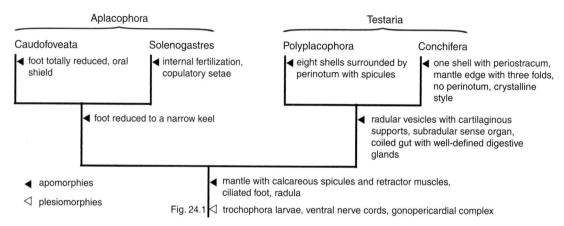

Figure 27.1. Phylogeny of the Mollusca.

or fusing along the midline so that the foot became totally reduced (Caudofoveata). Locomotion became worm-like, and the mantle retained a rather soft character with only small calcareous spicules.

2. Testaria, in which the mantle expanded laterally and secreted various types of shell plates surrounded by a perinotum with spicules. This group comprises two lineages:
   (a) Polyplacophora with eight perforate shells;
   (b) Conchifera, which lost the perinotum and specialized the mantle edge to a complex series of folds secreting a chitinous periostracum, which later became calcified on the interior side.

The mantle edge of the monoplacophorans is slightly different from that of the other conchiferans, but this is interpreted as a specialization (Schaefer and Haszprunar 1997). Conchifera is a well-founded clade, not only by the structure and formation of the shell, but also by the possession of paired statocysts, a crystalline style, and a special type of ciliary root (Haszprunar *et al.* 2008).

The microscopical anatomy was reviewed in the edited volumes by Harrison (1994, 1997). As mentioned above, the mantle can be recognized in almost all molluscs. It is covered by a layer of mucopolysaccharides, with proteins and calcareous spicules secreted by single cells in the aplacophorans (Furuhashi *et al.* 2009). The perinotum surrounding the shells of the polyplacophorans has a similar appearance, but the spi-

cules are secreted by several cells and there are numerous sensory structures, such as thin hairs and 'clappers', which are associated with a ciliated sensory cell and compound hairs with several sensory structures in a longitudinal groove. None of the hairs have a structure resembling annelid or brachiopod chaetae. The eight shells of the polyplacophorans have been interpreted as fused spicules, but this is not substantiated by direct observations, and Kniprath (1980) observed the formation of uninterrupted transverse shell plates in the larvae of two species; larvae reared at raised temperatures formed isolated calcareous granules which fused to abnormal plates. Each shell becomes secreted by a 'plate field' surrounded by other cells, which cover the plate field with flat microvilli and a cuticle, creating a crystallization chamber. The plate field grows, and the cells between the plates continue secreting a thin cuticle (periostracum) covering the shell. This development is often described as being very different to that of the conchiferans—for example, by the lack of a differentiated mantle edge, with rows of specialized cells secreting the periostracum and the periphery of the calcified shell—but it shows the same principal components, and could perhaps be interpreted as a less-specialized type of shell formation. Numerous papillae, sometimes in the shape of ocelli, penetrate the shells. Embryologically, the perinotum develops from both pre- and post-trochal cells, whereas the shells develop from the post-trochal area only (see Table 27.1). The

Table 27.1. Cell lineage of the polyplacophoran *Chaetopleura*. (Based on Henry *et al.* 2004.)

		1a		episphere (prototroch), spicules
	A	2a		secondary trochoblast, ectoderm, spicules, left ocellus, stomodaeum
		1A	3a	ectoderm, stomodaeum
AB			2A	3A	gut
		1b		episphere (prototroch)
	B	2b		ectoderm, secondary trochoblast, stomodaeum
		1B	3b	stomodaeum
			2B	3B	gut
Z		1c		episphere (apical tuft, prototroch)
	C	2c		secondary trochoblast, ectoderm, spicules, right ocellus
		1C	3c	ectoderm, spicules, shells
CD			2C	3C	gut
		1d		episphere (apical tuft, prototroch), spicules
	D	2d		ectoderm, shells
		1D	3d	secondary trochoblast, ectoderm, spicules, shells
			2D	3D	larval muscles, gut

conchiferan shells comprise a periostracum of quinone-tanned protein, sometimes with chitin, and a calcified layer with an organic matrix with β-chitin (although the presence of α-chitin is also possible) and proteins (Furuhashi *et al.* 2009). The embryonic shells are secreted by a shell gland, in which a ring of cells at the surface secretes a pellicle, the future periostracum, which becomes expanded when the shell gland spreads and begins to secrete the calcareous layer of the shell (or the two areas in the bivalves). The ring of pellicle-secreting cells follows the growth of the shell and becomes the periostracum-secreting cells in the mantle fold (Casse *et al.* 1997; Mouëza *et al.* 2006).

The foot and the retractor muscles originating at the mantle, and fanning out in the sole of the foot, can be recognized in almost all molluscs. Solenogasters have a very narrow keel-like foot with a series of foot retractors (lateroventral muscles) attached to the foot zone. Caudofoveates have lost the foot completely, by the fusion of the lateral mantle edges, but a narrow, midventral seam can be recognized in the primitive *Scutopus* (Salvini-Plawen 1972).

The radula is a band of thickened, toothed cuticle secreted by the apposed epithelia of a deep, posterior fold of the ventral side of the buccal cavity, the radular gland or sac. It consists of α-chitin and quinone-tanned proteins, and may be impregnated with iron and silicon salts. In the testarians, it can be protruded through the mouth, and pulled back and forth over the tips of a pair of radula vesicles with cartilaginous supports (Wingstrand 1985), scraping particles from the substratum. The presence of a radula in all classes, except the bivalves, indicates that the adult molluscan ancestor was a benthic deposit feeder or scraper, as a radula would be without function in a ciliary filter feeder. Cuticular thickenings in the shape of teeth or jaws occur in several protostomes—for example, rotifers and annelids—but a continuously growing band with many similar transverse rows of chitinous cuticular teeth, such as that of the radula, is not found in any other phylum.

The general epidermis of the body is a monolayered epithelium of multiciliate cells, with microvilli and a subterminal web of extracellular fibrils.

Table 27.2. Cell lineage of the gastropod *Patella*. (Based on Dictus and Damen 1997.)

					$1a^{111}$	apical organ
				$1a^{11}$	$1a^{112}$	pretrochal ectoderm (head region)
			$1a^1$		$1a^{121}$	pretrochal ectoderm (head region)
		$1a$		$1a^{12}$	$1a^{122}$	accessory trochoblasts
			$1a^2$			primary trochoblasts
	A				$2a^{11}$	secondary trochoblasts
			$2a^1$		$2a^{12}$	posttrochal ectoderm (part of foot, mantle fold and shell gland)
		$2a$		$2a^2$		posttrochal ectoderm (part of foot, mantle fold and shell gland)
AB	$1A$		$2A$	$3a$		ectomesoderm + posttrochal ectoderm
				$3A$		endoderm
Z	B					
	C	development as A (except that 3c becomes only posttrochal ectoderm)				
					$1d^{111}$	apical organ
CD			$1d^1$	$1d^{11}$	$1d^{112}$	pretrochal ectoderm (head region)
		$1d$		$1d^{12}$		accessory trochoblasts
	D		$1d^2$			primary trochoblasts
	$1D$	$2d$				ectoderm (median foot, posterior mantle and shell gland)
		$2D$	$3d$			ectomesoderm
			$3D$	$4d$		endomesodermal bands + endoderm
				$4D$		endoderm

The gut is straight in aplacophorans, and more or less coiled with paired digestive glands in the testarians; only the conchiferans have a crystalline style (Wingstrand 1985). The foregut with the radula is formed from the stomodaeum, while the midgut develops from the endoderm (Tables 27.1-3). The larvae of the gutless bivalve *Solemya* have an almost complete gut, which degenerates at metamorphosis, when nutrient uptake through symbiotic sulphur bacteria in the gills takes over (Gustafson and Reid 1988; Krueger *et al.* 1996).

The nervous system consists of a ring with ganglia surrounding the oesophagus; paired cerebral, pleural, and pedal ganglia can usually be recognized. Two pairs of prominent nerves, the ventral (pedal) and lateral (pleural) nerves, extend from the cerebral ganglia to the two last-mentioned pairs of ganglia. These nerves are generally in the shape of nerve cords in aplacophorans and polyplacophorans, but well-defined ganglia connected by nerves without cell bodies are found in the solenogaster *Genitoconia* (Salvini-Plawen 1967), and ganglia and nerves consisting of bundles of axons are the rule in the conchiferans. Transverse commissures between the pedal nerves are found in adult aplacophorans, polyplacophorans, and some gastropods, and in bivalve larvae (see below). The posterior ganglia on the lateral nerve cords of the solenogaster *Wirenia* are connected by a suprarectal commissure (Todt *et al.* 2008a). Various lines of centralization of the nervous system are seen both in gastropods and cephalopods (Moroz 2009).

Eyes are found both in larvae and adults. Larval eyes usually consist of a pigment cup and one or more receptor cells with cilia and a lens; both ciliary and

Table 27.3. Cell lineage of the gastropod *Crepidula*. (Based on Hejnol *et al.* 2007.) * The 'secondary trochoblasts' of Hejnol are here interpreted as metatroch cells.

		1a			episphere (apical ganglion, left cerebral ganglion, prototroch)
	A	2a			metatroch, stomodaeum, mantle, pedal ganglion
		1A	3a		oesophagus, ectomesoderm (body muscles)
AB			2A	4a	mesoderm
			3A	4A	endoderm
		1b			episphere (apical ganglion, prototroch)
	B	2b			food groove, mantle, anterior part of periblastoporal nerve ring
		1B	3b		stomodaeum, ectomesoderm
Z			2B	4b	mesoderm
			3B	4B	endoderm
		1c			episphere (apical ganglion, right cerebral ganglion, prototroch)
	C	2c			metatroch, mantle, nerves, larval heart, osphradium
		1C	3c		mantle and body ectoderm
CD			2C	4c	endoderm
			3C	4C	endoderm
		1d			episphere (apical ganglion)
	D	2d			body ectoderm, posterior part of periblastoporal nerve ring
		1D	3d		body ectoderm
			2D	4d	endomesoderm, gut
			3D	4D	yolk

rhabdomeric sensory cells are present in some eyes (Bartolomaeus 1992). Adult eyes vary from simple pigment-cup eyes in *Patella* to the highly complex cephalopod eyes, with a cornea, a lens, a retina with a highly organized arrangement of rhabdomeres, and ocular muscles; these eyes are innervated from the cerebral ganglia. Eyes of many different types are found on the mantle edge of bivalves, such as *Arca*, *Cardium*, and *Tridacna*, and gastropods, such as *Cerithidia*, and on the surface of the mantle, such as the esthetes of chitons. The eyes are everse or inverse, simple or compound (Nilsson 1994; Serb and Eernisse 2008), and are clearly not all homologous.

Coelomic cavities functioning as hydrostatic skeletons, used for example in burrowing, are known in many phyla, but this function is carried out by blood sinuses in molluscs (Trueman and Clarke 1988).

The circulatory system comprises a median dorsal vessel with a heart; a pair of more or less fused posterior atria is found in most classes, but two separate pairs of atria are found in some polyplacophorans, monoplacophorans, and *Nautilus*. Three to six pairs of gills are found in the monoplacophorans (see the chapter vignette), and two pairs in *Nautilus*; the polyplacophorans show many small accessory gills in the posterior part of the mantle furrow. The musculature of the heart is formed by the walls of the pericardial (coelomic) sac(s), and both muscular and non-muscular cells may have a rudimentary cilium (Bartolomaeus 1997). The peripheral part of the system comprises distinct capillaries in some organs, but large lacunae are found, for example, in the foot of burrowing bivalves (Trueman and Clarke 1988); the enormous swelling of the foot in some burrowing naticid gastropods is accomplished through intake of water into a complex sinus, which is completely isolated from the circulatory system (Bernard 1968). The blood spaces are clearly located between basal membranes, as in most invertebrates; only the cephalopods have vessels with endothelia, but these are incomplete in the capillaries; the presence of an endothelium can probably be ascribed to the high level of activity. It is sometimes stated that the cephalopods have a closed circulatory system whereas the other molluscs have open systems, but this distinction appears to be rather useless (Trueman and Clarke 1988). Surprisingly, orthologous genes regulating vascular endothelial growth have been found in developing hearts of both cephalopods and vertebrates, but this is just another example of a shared molecular signature that has been co-opted into heart organization in various lineages (Yoshida *et al.* 2010).

The excretory organs of many adult molluscs are paired metanephridia, which drain the pericardial sac. Primary urine is filtered from blood vessels/spaces to special pericardial expansions called auricles through areas with podocytes (Meyhöfer and Morse 1996; Morse and Reynolds 1996; Fahrner and Haszprunar 2002). The primary urine becomes modified during passage through the metanephridial ducts that open in the mantle cavity. The proximal part of the metanephridium is a small ciliated canal, the renopericardial canal, leading to the usually quite voluminous kidney, which is responsible for both osmoregulation and excretion. Most molluscs have one pair of nephridia (or only one; for example, in many gastropods), but *Nautilus* has two pairs of kidneys, which are not connected to the

pericardia. The monoplacophorans have three to seven pairs of nephridiopores in the mantle groove, but Schaefer and Haszprunar (1996) did not find any connection between the nephridia and the pericardium in *Laevipilina*.

The gonads are connected with the nephridia in most groups, and the gametes are spawned through the nephridiopores, but the reproductive system is, in many species, so specialized that the original structure is hard to recognize. The monoplacophoran *Laevipilina* has two pairs of large testes: the anterior pair opening through nephridia 2 and 3, and the posterior pair through nephridium 4 (Schaefer and Haszprunar 1996). Many species spawn small eggs freely in the water, but intricate egg masses are constructed in many species with internal fertilization, and brooding and vivipary occur too.

The embryology was reviewed by Nielsen (2004). Development shows much variation, ranging from free spawning of small eggs developing into planktotrophic larvae, and deposition of large eggs developing into lecithotrophic larvae, to direct development with discoidal cleavage (Fig. 27.2). The primary axis of the embryo can be recognized already in the mature oocytes, which have the apical pole facing away from the attachment of the egg in the ovary. The dorsoventral orientation and the entrance point of the sperm are apparently correlated, and experiments with eggs of *Spisula*, *Pholas*, and *Dreissena* indicate that the entrance of the sperm determines the position of the first cleavage furrow and the following bilaterality of the embryo.

Spiral cleavage can be recognized in all the classes except the cephalopods, which have very large eggs

Figure 27.2. Larval types of the molluscan classes; Monoplacophora and Cephalopoda are omitted—the development of the former is unknown and the latter has direct development. Solenogastres: *Epimenia verrucosa* (redrawn from Baba 1940); *Neomenia carinata* (redrawn from Thompson 1960). Caudofoveata: *Chaetoderma nitidulum* Lovén (an unpublished drawing by the late Dr Gunnar Gustafsson, Kristineberg Marine Biological Station, Sweden, modified on the basis of information from Nielsen *et al.* (2007). Polyplacophora: *Mopalia muscosa* (Gould) (larva reared at Friday Harbor Laboratories, WA, USA, June 1992). Gastropoda: *Crepidula fornicata* (redrawn from Werner 1955); *Lottia pelta* (Rathke) (larva reared at Friday Harbor Laboratories, WA, USA, June 1992; a later stage is a lecithotrophic veliger). Scaphopoda: *Dentalium entale* (redrawn from Lacaze-Duthiers 1859, combined with information from Wanninger and Haszprunar 2001). Bivalvia: *Barnea candida* (redrawn from Nielsen 1987); *Lyrodus pedicellatus* (Quatrefages) (pediveliger drawn after a scanning micrograph by Drs C.B. Calloway and R.D. Turner, Museum of Comparative Zoology, Harvard Univ., MA, USA); *Acila castrensis* (redrawn from Zardus and Morse 1998). The presence of compound cilia in the larvae of *Epimenia* and *Neomenia* is inferred from the descriptions.

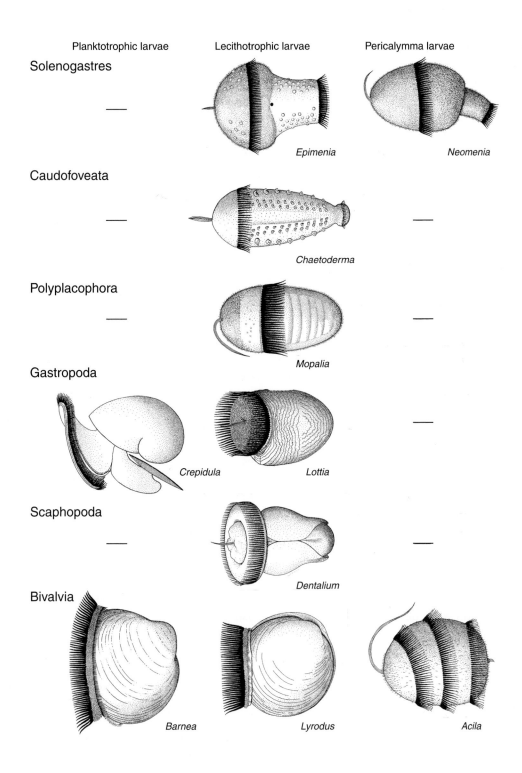

Planktotrophic larvae   Lecithotrophic larvae   Pericalymma larvae

Solenogastres

Epimenia   *Neomenia*

Caudofoveata

Chaetoderma

Polyplacophora

Mopalia

Gastropoda

Crepidula   *Lottia*

Scaphopoda

Dentalium

Bivalvia

Barnea   *Lyrodus*   *Acila*

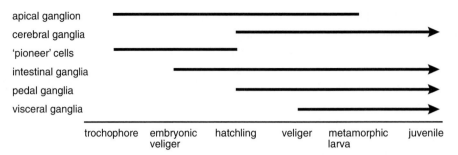

Figure 27.3. Fate of the main ganglia in developing larvae of the gastropod *Ilyanassa* revealed by immunostaining (Dickinson and Croll 2003).

and discoidal cleavage, and the holoblastic spiral cleavage type is undoubtedly ancestral in the phylum. Cleavage is equal, for example in *Haliotis*, to highly unequal with an enormous D-cell in *Busycon*. The cell lineage has been followed in a few species. The studies on the polyplacophoran *Chaetopleura* (Henry *et al.* 2004; Table 27.1) and the gastropods *Ilyanassa* (Render 1997), *Patella* (Dictus and Damen 1997; Table 27.2), and *Crepidula* (Hejnol *et al.* 2007; Table 27.3) show very similar lineages, but the studies emphasize different groups of blastomeres.

A small number of cells at the apical pole develops into a small apical ganglion with cilia forming an apical tuft (the cells $1c^{1111}$ and $1d^{1111}$ in *Dentalium*).

One or a small group of cells on each side of the apical ganglion form the cephalic plates from which the cerebral ganglia develop, with eyes and tentacles in some species. The basal parts of the apical cells come into intimate contact with the commissure between the cerebral ganglia, and the two types of ganglia are often so intimately connected that the term 'apical organ' is used. This praxis makes the interpretation of most of the older studies difficult. However, the two components have different fates, with the apical ganglion degenerating before or at metamorphosis through apoptosis, whereas the cerebral ganglia become part of the adult brain (Dickinson and Croll 2003; Leise *et al.* 2004; Gifondorwa and Leise 2006) (see Fig. 27.3).

In the several species of Polyplacophora, Gastropoda, Bivalvia, and Scaphopoda that have been studied, the prototroch, which may consist of one to three rows of cells, originates mainly from the primary trochoblasts ($1a^2$–$1d^2$). Accessory trochoblasts ($1a^{122}$–$1d^{122}$) may be added from the first quartet of micromeres, and the gaps between the four groups of primary trochoblasts may become closed anteriorly and laterally by secondary trochoblasts ($2a^{11}$–$2c^{11}$) (see Tables 27.1–3). The trochoblasts carry compound cilia, and they may be arranged in one (for example *Patella*) to three (for example *Dentalium*) rows. In *Patella*, it has been shown that some initially ciliated cells of the three rings (from accessory, primary, and secondary trochoblasts), become deciliated before the final shape of the prototroch is reached. The dorsal gap in the prototroch usually closes by fusion of the posterior tips of the prototroch. The metatroch has been observed to develop from cells of the second micromere quartet in *Crepidula* (Hejnol *et al.* 2007) (Table 27.2). The adoral ciliary zone should develop from the 2b-cell.

At the stage of 24–63 cells, one of the macromeres of the third generation stretches through the narrow blastocoel, and establishes contact with gap junctions to the central micromeres at the apical pole; this elongating cell is that of the D quadrant, and this event is probably of importance for the establishment of bilaterality, both of the prototroch and of the whole body.

As characteristic of the spiral cleavage, the main anteroposterior axis is more or less exactly through the B–D cells, but the spiral pattern shifts the plane of symmetry of the cell tiers by about 45° so that, for example, the two apical tuft cells mentioned above are from the C and D quadrants, and the ectomesodermal 3a and ectodermal 3d are situated on the left side, and 3b and 3c on the right.

Gastrulation is through invagination in species with small eggs and a coeloblastula, and through epiboly in species with large, yolky eggs and a sterroblastula. The blastopore becomes the definitive mouth in many species, where it partially closes from the posterior side but more conspicuously becomes shifted by a curving of the embryo. The blastopore may remain open while the stomodaeal invagination is formed, but there is a temporal closure at this point in most species. The anus is formed as a secondary opening from a proctodaeum. The only known exception to this general type of blastopore fate is found in *Viviparus*, in which the blastopore becomes the anus and the mouth is formed through a stomodaeal invagination.

Both endomesoderm and ectomesoderm can be recognized in the development of most species (Tables 27.1–3). The endomesoderm originates from the 4d cell, which divides into a right and a left cell. These cells usually give off two enteroblasts and then become mesoteloblasts, which produce a pair of lateral mesoderm bands (observed, for example, in the gastropod *Physa* (Wierzejski 1905) and the bivalve *Sphaerium* (Okada 1939)), but they are usually not very conspicuous. Much of the larval musculature develops from these bands, and this is also assumed for the adult heart, kidneys, and gonads, but the possibility of ectomesodermal participation cannot be excluded. Ectomesoderm is formed from the second or third micromere quartets (Hejnol *et al.* 2007), and soon becomes so intermingled with the endomesoderm that a separation becomes impossible. In *Sphaerium*, a characteristic 'cloud' of mitochondria is found in the basal part of the oocytes, and this inclusion can be followed to the 4d cell in the spiral-cell lineage, and further to the primordial germ cells (Woods 1932). Dautert (1929) reported a complete absence of endomesoderm in the embryos of *Viviparus*; his study was based on serial sections of many embryos, and the illustrations appear to support his interpretation; nevertheless, a reinvestigation seems needed.

The central nervous system develops from ectodermal areas where cells ingress and become organized in ganglia. In *Ilyanassa* (Lin and Leise 1996; Dickinson and Croll 2003), the apical sensory organ and the cerebral ganglia develop first, followed by the intestinal, pedal, and visceral ganglia, which are evident at hatching; the remaining ganglia develop during the pelagic phase (see Fig. 27.3).

The foot and the mantle, with the shell gland in the conchiferans, develop mainly from the 2d and 3d cells, but in some gastropods cells from all quadrants are involved (Table 27.2). In the polyplacophoran *Chaetopleura*, the shells develop from cells of the hyposphere, whereas the anterior part of the perinotum develops from cells of the episphere (Table 27.1). It is sometimes mentioned that the anterior plate develops from the episphere (Wanninger and Haszprunar 2002), but this is probably a misinterpretation of the anterior part of the girdle. The phylogenetic significance of this seems undecided, but the homology of the mantle of polyplacophorans and conchiferans must be re-evaluated. The embryological information indicates that the mantle of the conchiferans is homologous to the eight plate fields of the polyplacophorans only. Eight transverse rows of dorsal spicules have been indicated in the drawing of a larva of the solenogaster *Nematomenia* by Pruvot (1890) (see Fig. 27.4), but this pattern was not observed in *Wirenia* (Todt and Wanninger 2010) and appears questionable. Unpublished drawings by the late Dr Gunnar Gustafsson (Kristineberg Marine Research Station, Sweden) indicate the presence of eight transverse rows of calcareous spicules in late larvae of *Chaetoderma* (Fig. 27.4), and Nielsen *et al.* (2007) observed seven rows of papillae with a high calcium content in slightly younger larvae.

Paired protonephridia have been observed in larvae or embryos of solenogasters, polyplacophorans, gastropods, bivalves, and scaphopods. Their origin has been claimed to be either from the ectoderm, or from the mesoderm, or mixed (Raven 1966); the ultrastructure of *Lepidochitona* (Bartolomaeus 1989) supports the latter interpretation.

A ciliary apical tuft can be recognized in almost all planktonic mollusc larvae. Recent studies using transmission electron microscopy and immunocytochemistry of apical organs include a number of gastropods (e.g. Dickinson and Croll 2003; LaForge and Page 2007; Wollesen *et al.* 2007; Page and Kempf 2009),

Pedal muscles

Larvae

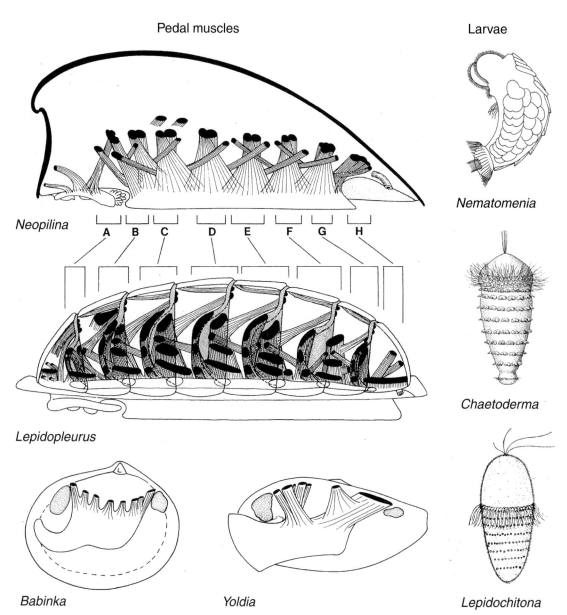

Neopilina

A B C   D   E   F G   H

Nematomenia

Chaetoderma

Lepidopleurus

Babinka

Yoldia

Lepidochitona

Figure 27.4. Indications of eight sets of iterated structures in molluscs. Pedal musculature: *Lepidopleurus* sp. and *Neopilina galatheae*: reconstructions based on serial sections (after Wingstrand 1985); *Babinka prima*: reconstruction of the pedal muscles of the mid-Ordovician fossil (after McAlester 1965); *Yoldia*: diagram of the pedal muscles (redrawn from Heath 1937). Shell plates of larvae and juveniles: newly metamorphosed larva of *Nematomenia banyulensis* (redrawn from Pruvot 1890); young larva of *Chaetoderma nitidulum* Lovén (an unpublished drawing by the late Dr Gunnar Gustafsson, Kristineberg Marine Biological Station, Sweden); larva of *Lepidopleurus asellus* (from Christiansen 1954).

bivalves (e.g. Voronezhskaya *et al.* 2008), and poly-placophorans (Wanninger 2008), with general discussion in Croll and Dickinson (2004). The apical ganglion consists of cells of characteristic types, flask-shaped cells with cilia forming the apical tuft, ampullary cells with short cilia set in a deep invagination, and parampullary cells. Some of the parampullary and non-sensory neuronal cells are serotonergic; these cells send projections to the velum and along the velar edge. The function of the apical ganglion has always been assumed to be sensory, especially chemosensory, but actual observations are few. Larvae of the opisthobranch *Phestilla* were unable to settle when certain cells of the apical ganglion had been ablated, indicating that the larvae were unable to recognize settlement cues from their coral host (Hadfield *et al.* 2000). The apical ganglion is resorbed at metamorphosis (Fig. 27.3).

Lateral groups of serotonergic cells develop a commissure, and become the cerebral ganglion (see above). In *Mytilus*, the first cells that show activity of acetylcholinesterase as a sign of neural differentiation are two pioneer sensory cells situated near the posterior pole, in the area that is sometimes called the telotroch, but that becomes the pedal ganglion with the byssus gland. These two cells send axons towards the apical pole where the apical organ develops, followed soon after by the associated cerebral ganglia (Raineri 1995).

The velum of gastropod larvae is heavily innervated and the activity of proto- and metatroch is regulated by these nerves (Braubach *et al.* 2006).

Pleural, parietal, and visceral ganglia differentiate along a paired lateral nerve cord, probably guided by the axons from the pioneer cells. connectives develop between the parietal, visceral, and pedal ganglia. In fact, the nervous system of the trochophore of *Mytilus* resembles that of a metatrochophore of an annelid. *Hox2-5* genes are expressed sequentially in the pedal, pleural, oesophageal, and branchial ganglia in early veliger larvae of *Haliotis*, but *Hox-1* is expressed only in a ring of cells around the developing shell gland (Hinman *et al.* 2003). Other Hox genes are involved in shell formation of later stages in *Gibbula* (Samadi and Steiner 2009).

Trochophora larvae with all the characteristic ciliary bands are not known in living molluscs, except perhaps in the bivalve *Pandora*, which is reported to have rudiments of a telotroch (Allen 1961). However, planktotrophic gastropod and bivalve larvae have the characteristic trochophore system of prototroch, adoral ciliary zone, and metatroch working in a downstream-collecting system with the catch-up principle (Riisgård *et al.* 2000) (Figs. 22.4–6). This ciliary system is pulled out along a pair of thin wing-like lobes, forming the velum, and the veliger larva has sometimes been interpreted as a mollusc apomorphy, but it has obviously evolved independently in gastropods and autobranch bivalves. Most lecithotrophic larvae have a locomotory prototroch, and the larvae of polyplacophorans, some solenogasters, and the caudofoveate *Chaetoderma* have a telotroch (Fig. 27.2). A study of the large gastropod genus *Conus* (Duda and Palumbi 1999) showed that the planktonic larval development is ancestral within the genus, and that direct development has evolved independently at least eight times within the genus. It appears that the ancestral larval type of the molluscs was the trochophore.

Lecithotrophic pericalymma larvae (Chapter 22; Fig. 22.7), with the prototroch zone or the prototroch plus adoral ciliary zone with the mouth expanded into a thin, ciliated sheet known as the serosa covering the hyposphere (type 1 larvae; see Fig. 27.2), are known from solenogasters and protobranch bivalves. In the solenogasters *Neomenia* and *Wirenia*, the serosa covers the episphere only partially so that the posterior end of the larva with the anus surrounded by the telotroch is exposed (Thompson 1960; Todt and Wanninger 2010) (Fig. 27.2). The bivalves *Solemya*, *Yoldia*, and *Acila* have a serosa covering the hyposphere completely, and there is no telotroch; *Yoldia* and *Acila* have three rings of cells with compound cilia, corresponding to the three rings of prototroch cells in *Dentalium* larvae. At metamorphosis, the serosa folds over anteriorly and becomes invaginated with the whole episphere in the solenogasters, whereas it becomes cast off in the bivalves. Another type of pericalymma larva is found in brooding teredinid bivalves of the genus *Lyrodus* (Fig. 22.7), where the adoral ciliary zone below the mouth is greatly expanded posteriorly so that it covers the valves almost completely (type 2 larva); this serosa retracts at a later stage, and the larvae are released as pediveligers.

The presence of repeated units of organs, such as muscles in molluscs, has been discussed by many authors (see for example Salvini-Plawen 1985). The presence of eight sets of pedal retractor muscles in polyplacophorans, where they are associated with the eight shell plates, and in monoplacophorans, which have only one shell, has been taken to indicate that the body of the ancestral mollusc had eight segments (see for example Wingstrand 1985) (Fig. 27.4). However, the eight pairs of muscles in the polyplacophoran *Mopalia* develop at metamorphosis through reorganization of a long series of equally distributed larval muscles (Wanninger and Haszprunar 2002). The presence of eight pairs of foot retractor muscles in the mid-Ordovician bivalve *Babinka* can be inferred from the well-preserved muscle scars, and the same number can, with some uncertainty, be recognized in the homologous foot and byssus retractor muscles of living bivalves, such as *Mytilus*. However, 'segmentation' of other organ systems is not well documented. Wingstrand (1985: p.43) demonstrated that serial repetition of nerve connectives, nephridiopores, gills, gonoducts, and atria (in order of decreasing numbers) correlated well with the pedal retractors in both *Neopilina* and *Vema*, but Haszprunar and Schaefer (1997) pointed out that the arrangement of these structures in monoplacophorans does not follow a strict segmental arrangement. Two pairs of gills, atria, and nephridia are found in the cephalopod *Nautilus*. The morphology of the larval nervous system of *Mytilus* (described above) could perhaps also be taken as an indication of a segmented body plan. Expression of the *engrailed* gene has been observed along the posterior edge of the growing shell plates of polyplacophorans (and at the developing spicules), and around the calcification areas in conchiferans (one in gastropods and scaphopods and a pair in bivalves) (Wanninger *et al.* 2008). This has obviously nothing to do with segmentation.

It must be concluded that the repetition of structures seen in some molluscs is not homologous with that of the annelids. It should perhaps better be described with the broader-term iteration (Scheltema and Ivanov 2002) to prevent confusion with the 'true' metameric segmentation of annelids, arthropods, and chordates, in which new segments are added from a posterior growth zone.

There is nothing to indicate that the molluscan ancestor had large coelomic cavities functioning as a hydrostatic skeleton in connection with burrowing. Many living molluscs burrow, for example, with the foot, but it is always the haemal system that functions as a hydrostatic skeleton (Trueman and Clarke 1988). All the more 'primitive' molluscs have relatively narrow coelomic cavities (pericardia), and in the classes where the embryology has been studied there are no signs of larger coelomic pouches at any stage. This could indicate that small pericardial cavities, possibly connected with gonads, were the first step in the evolution of the schizocoelic coelomates. At a later stage, burrowing habits could have favoured the enlargement of such cavities, and their specialization as the hydrostatic skeleton seen today in annelids and sipunculans.

The teloblastic growth of the 4d mesoderm, in both bivalves and gastropods, indicates that the molluscs form a monophyletic group with annelids and sipunculans. A further synapomorphy between molluscs and annelids may be the gonopericardial complex, i.e. the small pericardial coelomic sacs that contain the gonads, and that are drained by metanephridia with specialized 'nephridial' sections of the duct (Wingstrand 1985). The sipunculans have metanephridia that function as gonoducts, but they lack a haemal system (Chapter 26).

The presence of (small) coelomic cavities, originating through schizocoely in the endomesoderm mesoderm originating from the 4d cell, indicates that the molluscs are the sister group of annelids, sipunculans, and nemertines, and this now finds partial support from most of the molecular analyses (Chapter 24).

Interesting subjects for future research

1. Larval development of Solenogastres, especially the differentiation of the mesoderm and the formation of the calcareous spicules
2. Differentiation of nervous system and mesoderm in relation to the shells in Polyplacophora

3. Embryology and larval development of Monoplacophora

4. Origin and function of the pioneer cells

References

Allen, J.A. 1961. The development of *Pandora inaequivalvis* (Linné). *J. Embryol. Exp. Morphol.* **9**: 252–268.

Baba, K. 1940. The early development of a solenogastre, *Epimenia verrucosa* (Nierstrasz). *Annot. Zool. Jpn.* **19**: 107–113.

Bartolomaeus, T. 1989. Larvale Nierenorgane bei *Lepidochiton cinereus* (Polyplacophora) und *Aeolidia papillosa* (Gastropoda). *Zoomorphology* **108**: 297–307.

Bartolomaeus, T. 1992. Ultrastructure of the photoreceptor in the larvae of *Lepidochiton cinereus* (Mollusca, Polyplacophora) and *Lacuna divaricata* (Mollusca, Gastropoda). *Microfauna Mar.* **7**: 215–236.

Bartolomaeus, T. 1997. Ultrastructure of the renopericardial complex of the interstitial gastropod *Philinoglossa helgolandica* Hertling, 1932 (Mollusca, Opisthobranchia). *Zool. Anz.* **235**: 165–176.

Bernard, F.R. 1968. The aquiferous system of *Polynices lewisi* (Gastropoda, Prosobranchiata). *J. Fish. Res. Board Can.* **25**: 541–546.

Braubach, O.R., Dickinson, A.J.G., Evans, C.C.E., and Croll, R.P. 2006. Neural control of the velum in larvae of the gastropod, *Ilyanassa obsoleta*. *J. Exp. Biol.* **209**: 4676–4689.

Campbell, D.C. 2000. Molecular evidence on the evolution of the Bivalvia. In E.M. Harper (ed.): *The Evolutionary Biology of the Bivalvia*, pp. 31–46. Geological Society of London (Special publication no. 177), London.

Caron, J.B., Scheltema, A., Schander, C., and Rudkin, D. 2006. A soft-bodied mollusc with radula from the Middle Cambrian Burgess Shale. *Nature* **442**: 159–163.

Caron, J.B., Scheltema, A., Schander, C., and Rudkin, D. 2007. Reply to Butterfield on stem-group 'worms': fossil lophotrochozoans in the Burgess Shale. *BioEssays* **29**: 200–202.

Casse, N., Devauchelle, N., and Le Pennec, M. 1997. Embryonic shell formation in the scallop *Pecten maximus* (Linnaeus). *Veliger* **40**: 350–358.

Chen, H.-X., Sundberg, P., Norenburg, J.L., and Sun, S.-C. 2009. The complete mitochondrial genome of *Cephalothrix simula* (Iwata) (Nemertea: Palaeonemertini). *Gene* **442**: 8–17.

Christiansen, M.E. 1954. The life history of *Lepidopleurus asellus* (Spengler) (Placophora). *Nytt Mag. Zool.* **2**: 52–72.

Croll, R.P. and Dickinson, A.J.G. 2004. Form and function of the larval nervous system in molluscs. *Invertebr. Reprod. Dev.* **46**: 173–187.

Dautert, E. 1929. Die Bildung der Keimblätter von *Paludina vivipara*. *Zool. Jahrb., Anat.* **50**: 433–496.

Dickinson, A.J.G. and Croll, R.P. 2003. Development of the larval nervous system of the gastropod *Ilyanassa obsoleta*. *J. Comp. Neurol.* **466**: 197–218.

Dictus, W.J.A.G. and Damen, P. 1997. Cell-lineage and clonal-contribution map of the trochophore larva of *Patella vulgata* (Mollusca). *Mech. Dev.* **62**: 213–226.

Duda, T.F.J. and Palumbi, S.R. 1999. Developmental shifts and species selection in gastropods. *Proc. Natl. Acad. Sci. USA* **96**: 10272–10277.

Dunn, C.W., Hejnol, A., Matus, D.Q., *et al.* 2008. Broad phylogenomic sampling improves resolution of the animal tree of life. *Nature* **452**: 745–749.

Fahrner, A. and Haszprunar, G. 2002. Microanatomy, ultrastructure, and systematic significance of the excretory system and mantle cavity of an acochlidian gastropod (Opisthobranchia). *J. Molluscan Stud.* **68**: 87–94.

Fedonkin, M.A., Simonetta, A., and Ivantsov, A.Y. 2007. New data on *Kimberella*, the Vendian mollusc-like organism (White Sea region, Russia): palaeoecological and evolutionary implications. In P. Vickers-Rich and P. Komarower (eds): *The Rise and Fall of the Ediacaran Biota*, pp. 157–179. The Geological Society, London.

Fedonkin, M.A. and Waggoner, B.M. 1997. The Late Precambrian fossil *Kimberella* is a mollusc-like bilaterian organism. *Nature* **388**: 868–871.

Furuhashi, T., Schwarzinger, C., Miksik, I., Smrz, M., and Beran, A. 2009. Molluscan shell evolution with review of shell calcification hypothesis. *Comp. Biochem. Physiol. B* **154**: 351–371.

Gifondorwa, D.J. and Leise, E.M. 2006. Programmed cell death in the apical ganglion during larval metamorphosis of the marine mollusc *Ilyanassa obsoleta*. *Biol. Bull.* **210**: 109–120.

Giribet, G. 2008. Bivalvia. In W.F. Ponder and D.R. Lindberg (eds): *Phylogeny and Evolution of the Mollusca*, pp. 105–141. University of California Press, Berkeley.

Giribet, G. and Wheeler, W. 2002. On bivalve phylogeny: a high-level analysis of the Bivalvia (Mollusca) based on combined morphology and DNA sequence data. *Invert. Biol.* **121**: 271–324.

Gustafson, R.G. and Reid, R.G.B. 1988. Larval and post-larval morphogenesis in the gutless protobranch bivalve *Solemya reidi* (Cryptodonta: Solemyidae). *Mar. Biol.* **97**: 373–387.

Hadfield, M.G., Meleshkevitch, E.A., and Boudko, D.Y. 2000. The apical sensory organ of a gastropod veliger is a receptor for settlement cues. *Biol. Bull.* **198**: 67–76.

Harrison, F.W. (ed.) 1994. *Microscopic Anatomy of Invertebrates*, vol. 5 (*Mollusca I*). Wiley-Liss, New York.

Harrison, F.W. (ed.) 1997. *Microscopic Anatomy of Inveretbrates*, vol. 6A+B (*Mollusca II*). Wiley-Liss, New York.

Haszprunar, G. and Schaefer, K. 1997. Monoplacophora. In F.W. Harrison (ed.): *Microscopic Anatomy of Invertebrates*, vol. 6B, pp. 415–457. Wiley-Liss, New York.

Haszprunar, G., Schander, C., and Halanych, K.M. 2008. Relationships of the higher molluscan taxa. In W.F. Ponder and D.R. Lindberg (eds): *Phylogeny and Evolution of the Mollusca*, pp. 19–32. University of California Press, Berkeley.

Heath, H. 1937. The anatomy of some protobranch molluscs. *Mem. Mus. Hist. Nat. Belg.*, 2. Ser. **10**: 1–26.

Hejnol, A., Martindale, M.Q., and Henry, J.Q. 2007. High-resolution fate map of the snail *Crepidula fornicata*: the origins of ciliary bands, nervous system, and muscular elements. *Dev. Biol.* **305**: 63–76.

Hejnol, A., Obst, M., Stamatakis, A., *et al.* 2009. Assessing the root of bilaterian animals with scalable phylogenomic methods. *Proc. R. Soc. Lond. B* **276**: 4261–4270.

Henry, J.Q., Okusu, A., and Martindale, M.Q. 2004. The cell lineage of the polyplacophoran *Chaetopleura apiculata*: variation in the spiralian program and implications for molluscan evolution. *Dev. Biol.* **272**: 145–160.

Hinman, V.F., O'Brien, E.K., Richards, G.S., and Degnan, B.M. 2003. Expression of anterior *Hox* genes during larval development of the gastropod *Haliotis asinina*. *Evol. Dev.* **5**: 508–521.

Kniprath, E. 1980. Ontogenetic plate and plate field development in two chitons, *Middendorfia* and *Ischnochiton*. *Roux's Arch. Dev. Biol.* **189**: 97–106.

Krueger, D.M., Gustafson, R.G., and Cavanaugh, C.M. 1996. Vertical transmission of chemoautotrophic symbionts in the bivalve *Solemya velum* (Bivalvia: Protobranchia). *Biol. Bull.* **190**: 195–202.

Lacaze-Duthiers, H. 1859. Histoire de l'organisation et du développement du Dentale. Deuxième partie. *Ann. Sci. Nat., Zool., 4. Ser.* **7**: 171–255.

LaForge, N.L. and Page, L.R. 2007. Development of *Berthella californica* (Gastypoda: Opisthobranchia) with comparative observations on phylogenetically relevant larval characters among nudipleuran opisthobranchs. *Invert. Biol.* **126**: 318–334.

Leise, E.M., Kempf, S.C., Durham, N.R., and Gifondorwa, D.J. 2004. Induction of metamorphosis in the marine gastropod *Ilyanassa obsoleta*: 5HT, NO and programmed cell death. *Acta Biol. Hung.* **55**: 293–300.

Lemche, H. and Wingstrand, K.G. 1959. The anatomy of *Neopilina galatheae* Lemche, 1957. *Galathea Rep.* **3**: 9–71.

Lin, M.F. and Leise, E.M. 1996. NADPH-diaphorase activity changes during gangliogenesis and metamorphosis in the gastropod mollusc *Ilyanassa obsoleta*. *J. Comp. Neurol.* **374**: 194–201.

Mallatt, J., Craig, C.W., and Yoder, M.J. 2010. Nearly complete rRNA genes assembled from across the metazoan animals: Effects of more taxa, a structure-based alignment, and paired-sites evolutionary models on phylogeny reconstruction. *Mol. Phylogenet. Evol.* **55**: 1–17.

McAlester, A.L. 1965. Systematics, affinities, and life habits of *Babinka*, a transitional Ordovician lucinoid bivalve. *Palaeontology* **8**: 231–246.

Meyer, A., Todt, C., Mikkelsen, N.T., and Lieb, B. 2010. Fast evolving 18S rRNA sequences from Solenogastres (Mollusca) resist standard PCR amplification and give new insights into mollusk substitution rate heterogeneity. *BMC Evol. Biol.* **10**: 70.

Meyhöfer, E. and Morse, M.P. 1996. Characterization of the bivalve ultrafiltration system in *Mytilus edulis*, *Chlamys hastata*, and *Mercenaria mercenaria*. *Invert. Biol.* **115**: 20–29.

Moroz, L.L. 2009. On the independent origins of complex brains and neurons. *Brain Behav. Evol.* **74**: 177–190.

Morse, M.P. and Reynolds, P.D. 1996. Ultrastructure of the heart-kidney complex in smaller classes supports symplesiomorphy of molluscan coelomic characters. In J.D. Taylor (ed.): *Origin and Evolutionary Radiation of the Mollusca*, pp. 89–97. Oxford University Press, Oxford.

Mouëza, M., Gros, O., and Frenkiel, L. 2006. Embryonic development and shell differentiation in *Chione cancellata* (Bivalvia, Veneridae): an ultrastructural analysis. *Invert. Biol.* **125**: 21–33.

Nielsen, C. 1987. Structure and function of metazoan ciliary bands and their phylogenetic significance. *Acta Zool. (Stockh.)* **68**: 205–262.

Nielsen, C. 2004. Trochophora larvae: cell-lineages, ciliary bands and body regions. 1. Annelida and Mollusca. *J. Exp. Zool. (Mol. Dev. Evol.)* **302B**: 35–68.

Nielsen, C., Haszprunar, G., Ruthensteiner, B., and Wanninger, A. 2007. Early development of the aplacophoran mollusc *Chaetoderma*. *Acta Zool. (Stockh.)* **88**: 231–247.

Nilsson, D.E. 1994. Eyes as optical alarm systems in fan worms and ark clams. *Phil. Trans. R. Soc. Lond. B* **346**: 195–212.

Okada, K. 1939. The development of the primary mesoderm in *Sphaerium japonicum biwaense* Mori. *Sci. Rep. Tohoku Univ., Biol.* **14**: 25–48.

Page, L.R. and Kempf, S.C. 2009. Larval apical sensory organ in a neritimorph gastropod, an ancient gastropod lineage with feeding larvae. *Zoomorphology* **128**: 327–338.

Parkhaev, P.Y. 2008. The Early Cambrian radiation of Mollusca. In W.F. Ponder and D.R. Lindberg (eds): *Phylogeny and Evolution of the Mollusca*, pp. 33–69. University of California Press, Berkeley.

Pick, K.S., Philippe, H., Schreiber, F., *et al.* 2010. Improved phylogenomic taxon sampling noticeably affects nonbilaterian relationships. *Mol. Biol. Evol.* **27**: 1983–1987.

Pruvot, G. 1890. Sur le développement d'un Solénogastre. *C. R. Hebd. Seanc. Acad. Sci. Paris* **111**: 689–695.

Raineri, M. 1995. Is a mollusc an evolved bent metatrochophore? A histochemical investigation of neurogenesis in *Mytilus* (Mollusca: Bivalvia). *J. Mar. Biol. Assoc. UK* **75**: 571–592.

Raven, C.P. 1966. *Morphogenesis: the Analysis of Molluscan Development.* 2nd ed. Pergamon Press, Oxford.

Render, J. 1997. Cell fate maps in the *Ilyanassa obsoleta* embryo beyond the third division. *Dev. Biol.* **189**: 301–310.

Riisgård, H.U., Nielsen, C., and Larsen, P.S. 2000. Downstream collecting in ciliary suspension feeders: the catch-up principle. *Mar. Ecol. Prog. Ser.* **207**: 33–51.

Salvini-Plawen, L.v. 1967. Neue scandinavische Aplacophora (Mollusca, Aculifera). *Sarsia* **27**: 1–63.

Salvini-Plawen, L.v. 1972. Zur Morphologie und Phylogenie der Mollusken: Die Beziehungen der Caudofoveata und der Solenogastres als Aculifera, als Mollusca und als Spiralia. *Z. Wiss. Zool.* **184**: 205–394.

Salvini-Plawen, L.v. 1985. Early evolution and the primitive groups. In K.M. Wilbur (ed.): *The Mollusca*, vol. 10, pp. 59–150. Academic Press, Orlando.

Salvini-Plawen, L.v. 2006. The significance of the Placophora for molluscan plylogeny. *Venus* **65**: 1–17.

Samadi, L. and Steiner, G. 2009. Involvement of Hox genes in shell morphogenesis in the encapsulated development of a top shell gastropod (*Gibbula varia* L.). *Dev. Genes Evol.* **219**: 523–530.

Schaefer, K. and Haszprunar, G. 1996. Anatomy of *Laevipilina antarctica*, a monoplacophoran limpet (Mollusca) from Antarctic waters. *Acta Zool. (Stockh.)* **77**: 295–314.

Schaefer, K. and Haszprunar, G. 1997. Organization and fine structure of the mantle of *Laevipilina antarctica* (Mollusca, Monoplacophora). *Zool. Anz.* **236**: 13–23.

Scheltema, A. and Ivanov, D.L. 2002. An aplacophoran postlarva with iterated dorsal groups of spicules and skeletal similarities to Paleozoic fossils. *Invert. Biol.* **121**: 1–19.

Serb, J.M. and Eernisse, D.J. 2008. Charting evolution's trajectory: using molluscan eye diversity to understand parallel and convergent evolution. *Evo. Edu. Outreach* **1**: 439–447.

Sutton, M.D., Briggs, D.E.G., Siveter, D.J., and Siveter, D.J. 2001. An exceptionally preserved vermiform mollusc from the Silurian of England. *Nature* **410**: 461–463.

Sutton, M.D., Briggs, D.E.G., Siveter, D.J., and Siveter, D.J. 2004. Computer reconstructions and analysis of the vermiform mollusc *Acaenoplax hayae* from the Herefordshire Lagerstätte (Silurian, England), and implications for molluscan phylogeny. *Palaeontology* **47**: 293–318.

Thompson, T.E. 1960. The development of *Neomenia carinata* Tullberg (Mollusca Aplacophora). *Proc. R. Soc. Lond. B* **153**: 263–278.

Todt, C. and Wanninger, A. 2010. Of tests, trochs, shells, and spicules: development of the basal mollusk *Wirenia argentata* (Solenogastres) and its bearing on the evolution of trochozoan larval key features. *Front. Zool.* **7**: 6.

Todt, C., Büchinger, T., and Wanninger, A. 2008a. The nervous system of the basal mollusk *Wirenia argentea* (Solenogastres): a study employing immunocytochemical and 3D reconstruction techniques. *Mar. Biol. Res.* **4**: 290–303.

Todt, C., Okusu, A., Schander, C., and Schwabe, E. 2008b. Solenogasters, Caudofoveata, and Polyplacophora. In W.F. Ponder and D.R. Lindberg (eds): *Phylogeny and Evolution of the Mollusca*, pp. 71–96. California University Press, Berkeley.

Trueman, E.R. and Clarke, M.R. 1988. Introduction. In E.R. Trueman and M.R. Clarke (eds): *The Mollusca, vol.* 11, pp. 1–9. Academic Press, San Diego.

Vinther, J. 2009. The canal system in sclerites of Lower Cambrian *Sinosachites* (Halkieriidae: Sachitida): significance for the molluscan affinities of the sachitids. *Palaeontology* **52**: 689–712.

Vinther, J. and Nielsen, C. 2005. The Early Cambrian *Halkieria* is a mollusc. *Zool. Scr.* **34**: 81–89.

Voronezhskaya, E.E., Nezlin, L.P., Odintsova, N.A., Plummer, J.T. and Croll, R.P. 2008. Neuronal development in larval mussel *Mytilus trossulus* (Mollusca: Bivalvia). *Zoomorphology* **127**: 97–110.

Wanninger, A. 2008. Comparative lophotrochozoan neurogenesis and larval neuroanatomy: recent advances from previously neglected taxa. *Acta Biol. Hung.* **59** (Suppl.): 127–136.

Wanninger, A. and Haszprunar, G. 2001. The expression of an engrailed protein during embryonic shell formation of the tusk-shell, *Antalis entalis* (Mollusca, Scaphopoda). *Evol. Dev.* **3**: 312–321.

Wanninger, A. and Haszprunar, G. 2002. Chiton myogenesis: perspectives for the development and evolution of larval and adult muscle systems in molluscs. *J. Morphol.* **251**: 103–113.

Wanninger, A., Koop, D., Moshel-Lynch, S., and Degnan, B.M. 2008. Molluscan evolutionary development. In W.F. Ponder and D.R. Lindberg (eds): *Phylogeny and Evolution of the Mollusca*, pp. 427–445. University of California Press, Berkeley.

Werner, B. 1955. Über die Anatomie, die Entwicklung und Biologie des Veligers und der Veliconcha von *Crepidula fornicata* L. (Gastropoda, Prosobranchia). *Helgol. Wiss. Meeresunters.* **5**: 169–217.

Wierzejski, A. 1905. Embryologie von *Physa fontinalis* L. *Z. Wiss. Zool.* **83**: 502–706.

Wilson, N.G., Rouse, G.W., and Giribet, G. 2010. Assessing the molluscan hypothesis Serialia (Monoplacophora + Polyplacophora) using novel molecular data. *Mol. Phylogenet. Evol.* **54**: 187–193.

Wingstrand, K.G. 1985. On the anatomy and relationships of recent Monoplacophora. *Galathea Rep.* **16**: 7–94.

Wollesen, T., Wanninger, A., and Klussmann-Kolb, A. 2007. Neurogenesis of cephalic sensory organs of *Aplysia californica*. *Cell Tissue Res.* **330**: 361–379.

Woods, F.H. 1932. Keimbahn determinants and continuity of the germ cells in *Sphaerium striatinum* (Lam.). *J. Morphol.* **53**: 345–365.

Yoshida, M.-a., Shigeno, S., Tsuneki, K., and Furuya, H. 2010. Squid vascular endothelial growth factor receptor: a shared molecular signature in the convergent evolution of closed circulatory systems. *Evol. Dev.* **12**: 25–33.

Zardus, J.D. and Morse, M.P. 1998. Embryogenesis, morphology and ultrastructure of the pericalymma larva of *Acila castrensis* (Bivalvia: Protobranchia: Nuculoida). *Invert. Biol.* **117**: 221–244.

28

Phylum **Nemertini**

The Nemertini, or ribbon worms, is a phylum of about 1400 described species, mostly marine (benthic or pelagic), but some groups have entered freshwater and a few are found in moist, terrestrial habitats. There is no reliable fossil record. Most nemertines are cylindrical to slightly flattened, but some of the pelagic forms in particular are very flat. All species (except *Arhynchonemertes*, see below) can be recognized by the presence of a proboscis, which can be everted through a proboscis pore at the anterior end of the animal, and is contained in a coelomic cavity, the rhynchocoel, functioning as a hydrostatic skeleton during the eversion. Traditional phylogeny recognizes two classes: Anopla (Palaeonemertini + Heteronemertini) and Enopla (Hoplonemertini + Bdellonemertini). However, molecular studies show a rather different picture with the palaeonemertines as a paraphyletic stem group (Sundberg *et al.* 2001; Thollesson and Norenburg 2003). Groups with species having pilidium larvae (see below) are placed in Pilidiophora, which is the sister group of the Hoplonemertini (Enopla). The Nemertini is clearly monophyletic.

The microscopic anatomy was reviewed by Turbeville (1991, 2002). The ectoderm is pseudostratified with multiciliate cells, several glands, and sense organs. The ciliated cells have a border of branched microvilli, sometimes with a glycocalyx; there is never a cuticle of cross-arranged collagenous fibrils as that

observed in most of the other spiralian phyla, and chitinous structures have not been reported either. The ectoderm rests on a basal membrane that is traversed by myofilament-containing processes from mesodermal muscles; myoepithelial cells have not been reported. The intracellular rhabdoids and pseudocnids known from various groups are apparently not homologous with platyhelminth rhabdites (Turbeville 2006).

The gut is a straight, ciliated epithelial tube extending from the anterior part of the ventral side to the posterior end. The mouth opens into the rhynchodaeum or a shallow atrium in most of the benthic Enopla. The foregut, which may be differentiated into a buccal cavity, an oesophagus, and a stomach, is ectodermally derived like the rectum and the rhynchodaeum (see below). The midgut has various diverticula in almost all species; there are typically no intrinsic muscles associated with the gut.

The proboscis is a long, tubular, eversible, muscular structure that is used for the capture of prey, defence, and, in a few species, for locomotion; when everted, it may be lost, but regeneration is apparently very rapid. It is an invagination of the ectoderm surrounded by a muscular, coelomic sac, the rhynchocoel, which functions as a hydrostatic skeleton at eversion. There is a special musculature surrounding the proboscis coelom in some species, whereas other species

Chapter vignette: *Tubulanus sexlineatus*. (Redrawn from Kozloff 1990.)

use the body musculature both for proboscis eversion and for general peristaltic movements of the body (Senz 1995). The posterior end of the proboscis is usually attached to the posterior wall of the coelomic sac by a retractor muscle. Hoplonemertines have one or more calcified stylets in the proboscis; the function of these structures appears to be the wounding of prey so that various poisonous secretions from glands in the epithelium of the proboscis can penetrate. The stylets are formed in intracellular vacuoles and brought into position when fully formed; they become replaced after use (Stricker 1985). The proboscis has several layers of muscles, which are in fact continuations of the body wall musculature, and are capable of complex movements. Also the wall of the rhynchocoel has several muscle layers. *Arhynchonemertes* (Riser 1988) lacks any trace of a proboscis.

The brain comprises paired dorsal and ventral lobes, connected by a dorsal and a ventral commissure, surrounding the rhynchodaeum and the anterior loop of the blood vessel system, but not the alimentary canal. A pair of lateral, longitudinal nerve cords arise from the ventral lobes and extend to the posterior end where they are connected by a commissure dorsal to the anus; many dorsal and ventral, transverse commissures are present in most forms. These nerves are situated either in the connective tissue below the ectoderm or between the muscle layers; a peripheral, basiepithelial plexus is also present. A system of granule-containing cells of uncertain function is closely associated with the nervous system.

Two or more layers of muscles with different orientation are situated below the basal membrane. The muscles of some palaeonemertines and some pelagic heteronemertines are obliquely striated, while those of most heteronemertines and hoplonemertines are smooth. Striated muscles have been observed in some pilidium larvae (Norenburg and Roe 1998). In some palaeonemertines some of the muscle cells make contact with the nerves through non-contractile extensions, but normal nerve extensions are found too.

The haemal system consists of a pair of longitudinal vessels joined by anterior and posterior transverse vessels; additional longitudinal and transverse vessels are found in several groups. The vessels are lined with a continuous epithelium of mesodermal cells joined with zonulae adherentes and sometimes having a rudimentary cilium; a discontinuous layer of muscle cells is found in the surrounding extracellular matrix. This is an unusual system among the invertebrates, which as a rule have a haemal system, i.e. blood vessels surrounded by basal membrane and with no endothelium (Ruppert and Carle 1983), but a similar system is seen in hirudineans (Chapter 25). The origin of the vessels through hollowing out of narrow longitudinal bands of mesodermal cells (Turbeville 1986), i.e. schizocoely, and their epithelial character with cell junctions and cilia, show that the vessels by definition are coelomic cavities.

The excretory system consists of a pair of branched canals, with flame cells often in contact with the blood vessels (Bartolomaeus and von Döhren 2010). Some of the terminal organs are quite complicated and may resemble glands or metanephridia, but the whole system is surrounded by a basal membrane and is clearly protonephridial; there is no observation of openings between the terminal organs and the coelomic blood vessels.

The gonads are serially arranged sacs consisting of a mesodermal epithelium with germinal cells surrounded by a basal membrane; the gonoduct comprises an ectodermal invagination, but special copulatory or accessory glandular structures are generally lacking. The gametes are usually spawned freely, but several species deposit gametes in gelatinous masses, and others exhibit internal fertilization, including viviparous species (Thiel and Junoy 2006).

The embryology was reviewed by Henry and Martindale (1997a) and Nielsen (2005). The primary axis of the egg is determined in the ovary, with the blastoporal pole at the attachment point to the ovarial epithelium. The eggs are spawned and fertilized in an early phase of meiosis so the polar bodies can be followed at the apical pole inside the fertilization membrane.

The cleavage is spiral with quartets in all species studied so far. The four quadrants are of the same size, so the orientation of the embryo cannot be seen in the early stages. Deletion experiments on 8-cell stages of *Cerebratulus lacteus* (Henry 2002) indicate that the anteroposterior axis is not determined until this stage. The apical 'micromeres' of the first cleavage are often larger than the 'macromeres', and this is also the case with the 'micromeres' of the 5th and 6th cleavage for example in *Malacobdella* (Hammarsten 1918) and *Emplectonema* (Delsman 1915), but the usual terminology is used to facilitate comparisons with other spiralians. Four or more quartets of micromeres are formed, except in *Tubulanus* where Dawydoff (1928) observed only three. Hammarsten (1918) reported that *Malacobdella* forms endoderm from the 4A-D and 4a-d cells, while all the mesoderm originates from descendants of 2a-d cells (the cells 2a''''–d''''). True teloblasts, i.e. large cells budding off smaller cells in one direction, have not been reported.

Experiments with isolated blastomeres of *Cerebratulus* have shown that almost normal pilidium larvae develop from some of the isolated blastomeres of the 2-cell stage, and that only very few larvae develop from blastomeres of the 4-cell stage. However, the development of the juveniles from these larvae was not studied. Deletion experiments on 2- and 4-cell stages of the direct developing *Nemertopsis* showed much lower levels of regeneration.

Various types of 'direct' development are found in all groups, and indirect development with a planktotrophic pilidium larva is found in many pilidiophorans. The indirect development with a pilidium larva is now the best-known developmental type and it will therefore be described first (Table 28.1 and Figs. 28.1–2).

The blastula stage has a wide blastocoel, and gastrulation is through invagination that forms the gut and further a wide, deep oesophagus; the anus does not develop until at metamorphosis. The embryo becomes bell shaped, with a pair of lateral lappets. The narrow, apical cells become the apical ganglion that develops a tuft of long cilia, usually held together as one structure. A conspicuous ciliary band develops along the edge of the bell and lappets shortly before a general ciliation makes the larvae begin to rotate in the capsule and finally to hatch.

The cell lineage of species with pilidium larvae, such as *Cerebratulus lacteus* (Henry and Martindale 1998) (Table 28.1), resembles that of other spiralians (e.g. Tables 25.1 and 27.2). The descendants of the first micromere quartet form the large episphere including the upper part of the large ciliary band; the spiral shift of the micromeres is almost 45° so that the epithelium consists of symmetrical right (a and d) and left (b and c) territories (Martindale and Henry 1995). The gut develops from 3A–D and 4D, endomesoderm from 4d, and ectomesoderm (mainly larval muscles) from 3a and 3b. The large ciliary band is formed by descendants of the cells 1a–d, 2a–d and 3d; it consists of compound cilia in the fully developed, feeding larvae, at least in some species (Nielsen 1987). There is also a post-oral, transverse band of stronger ciliation (Lacalli and West 1985) that appears to function in particle retention, but the mechanism is not known (Cantell 1969). Each cell of the apical ganglion may have one or several (up to 12) cilia, and the cilia of each cell are surrounded by a ring of long microvilli connected by a mucous structure (Cantell *et al.* 1982). Two groups of small muscles extend from the apical organ to the anterolateral areas of the mouth (Wilson 1900). Nerves originating from the cells 1c–d, 2a, c–d, and 3c–d follow the basis of the ciliary band, also possibly of the postoral ciliary band, and surround the mouth at the bottom of the oesophagus. This group of cells nicely demonstrates the alternating shifts of the cleavage directions in the spiral pattern (Henry and Martindale 1998: fig. 8). The ciliary bands have associated serotonergic perikarya and basiepithelial nerves, and a few nerve cells to the gut. A FMRFamide-reactive nerve with a few perikarya runs from the anterior part of the oesophagus towards the apical organ; no nervous cells have been observed in the apical organ itself (Lacalli and West 1985; Hay-Schmidt 1990).

The development of the adult nemertine inside the pilidium larva has only been been studied in a few species (review in Maslakova 2010b), the only detailed descriptions being those of Salensky (1912: *Pilidium pyramidatum* and *P. gyrans*), Henry and Martindale (1997b: *Cerebratulus lacteus*; see Fig. 28.1), and Maslakova (2010: *Micrura alaskensis*). Soon after

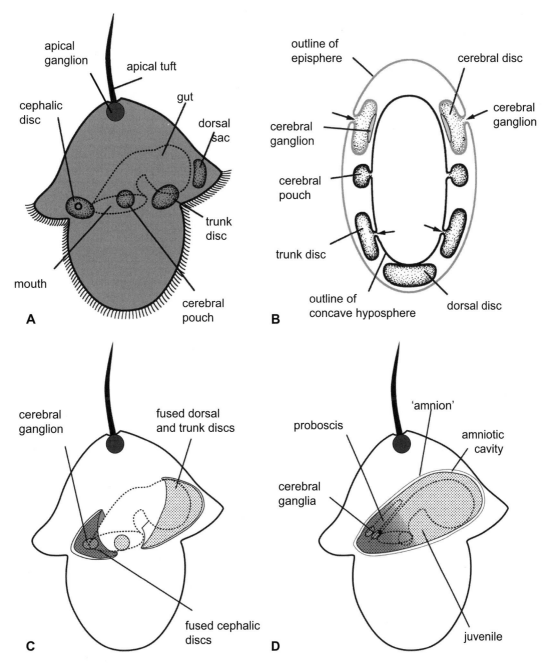

Figure 28.1. Development of the juvenile nemertine inside the pilidium of *Cerebratulus lacteus*. (Modified from Nielsen 2008.) Note that the cerebral discs develop from the episphere and the trunk discs from the hyposphere. See Colour Plate 7.

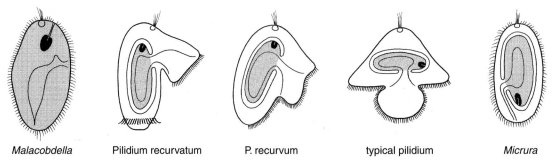

| *Malacobdella* | Pilidium recurvatum | P. recurvum | typical pilidium | *Micrura* |

Figure 28.2. Larval types of nemertines and the orientation of the juvenile inside the larval body; *Malacobdella grossa* has direct development; the larva of *Micrura akkeshiensis* is often called Iwata's larva. The 'adult' structures are shaded and the primordia of the proboscis apparatus are black. (Redrawn from Hammarsten 1918 and Jägersten 1972.)

Table 28.1. Cell lineage of *Cerebratulus lacteus*. (Based on Henry and Martindale 1998.)

Z	'AB'	A	1A	1a			apical organ, larval ectoderm, ciliary band, left cephalic disc
				2A	2a		larval ectoderm, ciliary band, oesophagus, nervous system
					3A	3a	larval muscles, oesophagus
						3A	gut
		B	1B	1b			apical organ, larval ectoderm, ciliary band, right cephalic disc
				2B	2b		larval ectoderm, ciliary band, oesophagus
					3B	3b	larval muscles, oesophagus
						3B	gut
	'CD'	C	1C	1c			apical organ, larval ectoderm, ciliary band, nervous system
				2C	2c	3c	larval ectoderm, ciliary band, nervous system, oesophagus
						3C	gut
		D	1D	1d			apical organ, larval ectoderm, ciliary band, nervous system
				2D	2d	3d	larval ectoderm, ciliary band, nervous system, oesophagus
					3D	4d	endomesoderm, gut
						4D	gut

completion of the larval gut, three pairs of ectodermal invaginations develop from the ectoderm of the larva along the equator. The cephalic sacs develop as invaginations from the episphere, whereas the trunk sacs and the cephalic pouches develop from the hyposphere. A pair of anterolateral pouches give rise to the cerebral organs. Later on, an unpaired dorsal disc develops from

the posterior part of the hyposphere, possibly through delamination. Each invagination becomes differentiated into a thick, internal cell plate, called an embryonic disc, and a very thin, exterior cell layer. The sacs expand surrounding the gut, and the embryonic discs fuse to form the ectoderm of the juvenile, while the exterior cell layers form an amnionic membrane around the juvenile. The proboscis ectoderm develops as a narrow, ventral invagination of the ectoderm at the fusion line between the cephalic discs. The cerebral ganglia develop from the ectoderm of the cephalic discs, i.e. from the episphere, lateral to the proboscis invagination, completely isolated from the apical organ of the larva; they differentiate into dorsal and ventral lobes, and the commissures around the rhynchodaeum develop; the lateral nerve cords develop as extensions from the cephalic ganglia. Mesodermal cells line the ectoderm and endoderm, resembling peritoneal epithelia, and the anterior and paired longitudinal cavities fuse and become the blood vessel coelom (Jespersen and Lützen 1987). The origin of the rhynchocoel is more uncertain. A pair of small ectodermal invaginations in front of the blastopore have been thought to give rise to the nephridia, but are now known to give rise to the juvenile foregut (Maslakova 2010a). The anus breaks through at metamorphosis. The juvenile finally breaks out of the larval body, which in several cases becomes ingested afterwards (Cantell 1966).

There are several more or less distinct types of pilidium larvae, many of which have been given names (Dawydoff 1940; Cantell 1969; Lacalli 2005) (Fig. 28.2). Only a few of these larvae have been related to known adult species (Friedrich 1979), and many larvae are known only from one developmental stage. *Pilidium incurvatum* (Cantell 1967) has a separate posterior ring of cilia, resembling a telotroch; the cilia along the edges of the anterior funnel have been described as coarse, indicating compound cilia, and the photomicrographs of sections of the posterior ciliary band resemble those of bands with compound cilia, but direct observations are lacking.

Several non-planktotrophic pilidium-like larvae have been described in species of the Pilidiophora (Henry and Martindale 1997a) (Fig. 28.2). The pelagic, lecithotrophic 'Iwata's larva' of *Micrura akkeshiensis* (Iwata 1958) resembles a pilidium, but lacks the lateral lappets and the special ciliary bands. Five ectodermal invaginations develop, corresponding to the paired cephalic and trunk sacs and the unpaired dorsal sac of the pilidium. The main axis of the juvenile is rotated so that the anterior end faces the posterior pole of the

larva. In the normal pilidium the juvenile main axis is oriented 90° from the primary axis, as usual in the protostomes, so the orientation of the juvenile inside Iwata's larva is simply a further 90° rotation. The non-planktonic 'Desor's larva' of *Lineus viridis* (Schmidt 1934) is lecithotrophic, whereas the 'Schmidt's larva' of *L. ruber* (Schmidt 1964) ingests other embryos from the same egg mass; their development resembles that of Iwata's larva, but an apical organ is not developed and the ectoderm is apparently without cilia.

Some hoplonemertines have free-swimming larvae superficially resembling cnidarian planulae, others brood the embryos in the ovaries until an advanced stage, and some species deposit the eggs in capsules from which the juveniles escape.

The development of *Paranemertes* has been described in detail (Maslakova and von Döhren 2009). Cleavage is spiral and gastrulation is by a narrow invagination. The larvae are uniformly ciliated with an apical tuft. The development of the internal organs was not studied in detail, but the stomodaeum develops isolated from the endoderm and they fuse subsequently. Early larvae are covered by about 90 large, cleavage-arrested cells with a few interspersed smaller cells; the apical organ, stomodaeum, and the invaginations to form the cerebral organs can be recognized. The smaller cells increase in number finally covering the whole surface when the benthic stage is reached. This resembles the fate of the prototrochal cells of *Carinoma*, but a cell lineage has not been established.

Other hoplonemertines have been studied by a number of authors (Friedrich 1979; Henry and Martindale 1994; Martindale and Henry 1995; Hiebert *et al.* 2010), but the descriptions show a good deal of variation between the species and are in some cases contradictory; several important details of the development are still unclear. Gastrulation shows much variation with invagination in *Prosorhochmus* (Salensky 1914), epiboly in *Oerstedia* (Iwata 1960), and ingression in *Geonemertes* (Hickman 1963). The archenteron becomes compact in most species, and a stomodaeum develops in front of the position of the closed blastopore. The compact embryos of the limnic *Prostoma* are reported to develop a new inner cavity and then to become syncytial (Reinhardt 1941). The embryos are generally ciliated with a conspicuous apical organ, but some of the terrestrial nemertines lack an apical organ and cilia completely (Hickman 1963). Studies using micro-injections of blastomeres have shown that the distribution in the larval ectoderm follows the same bilateral pattern as that observed in the episphere of *Cerebratulus* (see above). The stomodaeal mouth opening persists in *Drepanophorus* (Lebedinsky 1897), but it disappears and the oesophagus opens into the rhynchodaeum, for example in *Prostoma* (Reinhardt 1941). An ectodermal invagination between the apical organ and the stomodaeum gives rise to the proboscis; it loses the connection to the ectoderm in many species, and a special rhynchodaeum has been reported to develop from the degenerating apical organ. Hickman (1963) reported that embryos of *Geonemertes* become surrounded by a number of special, so-called ectoembryonic cells, which later on become replaced by new, ciliated epidermal cells; this resembles the development of *Paranemertes* (see above). The many different types of development of gut and proboscis (Friedrich 1979) give the impression of many lines of specialization.

'Palaeonemertine' larvae are generally planktotrophic, swallowing larger food items (Maslakova 2010b). The cell-lineage studies on *Carinoma* (Maslakova and Malakhov 1999; Maslakova *et al.* 2004a,b) (Table 28.2) appear to give a clue to the understanding of the various types of 'palaeonemertine' larvae. Cleavage is almost equal and follows the traditional spiral pattern. Forty cleavage-arrested cells characterized as trochoblasts are derived equally from the four quadrants. Unlike the trochoblasts of other spiralians, these cells show a general ciliation like that of the other blastomeres, but they can be distinguished by their special optical properties and their large size. In the young, spherical larvae, the trochoblasts cover the whole body except for narrow apical (anterior) area with an apical tuft and a blastoporal (posterior) area. The larva elongates, with strong posterior extension of the apical area and a weaker anterior extension of the blastoporal area, so that the trochoblasts shift to an oblique narrower zone and the blastopore shifts to the ventral side. During subsequent development the trochoblasts gradually diminish and finally disappear altogether. The early larvae are hollow and gastrulation is by invagination. The origin of the mesoderm was not identified.

Iwata (1960) studied the direct developing 'palaeonemertines', *Procephalothrix* and *Tubulanus*. Their blastula flattens so that the blastocoel disappears, and an invagination opposite the apical organ forms the archenteron; the outer part of the invagination is called stomodaeum in later stages. The ectodermal cells become ciliated, but their differentiation should be reinvestigated. The juvenile breaks free from the fertilization membrane. The mesodermal cells become arranged in a continuous layer

Table 28.2. Cell lineage of *Carinoma tremaphoros*. (Based on Maslakova *et al.* 2004a.)

between ectoderm and gut. The nervous system develops from two large cells at the sides of the apical organ; these cells divide and invaginate, and the invaginations detach from the ectoderm and become compact. These two cerebral ganglia become the brain and differentiate into dorsal and ventral lobes with commissures; the longitudinal nerves develop as extensions from the ventral lobes below the epidermis along the sides of the gut. The apical ganglion is not in contact with the brain, and it probably degenerates. The proboscis invagination and the anus are not formed at this stage, and their development has not been studied.

The proboscis coelom has a structure that fits the definition of a coelom, but the structure of the proboscis apparatus is not reminiscent of any structure in other spiralians and it is regarded as the most obvious apomorphy of the phylum. The absence of a proboscis in *Arhynchonemertes* (Riser 1988, 1989) could either be a specialization, i.e. that the ancestor had a proboscis, or a plesiomorphy, i.e. that the ancestral nemertine lacked a proboscis. This question seems impossible to answer without new studies. The blood-vessel system likewise fits the definition of a coelom, and its lateral position resembles the highly specialized haemal coelomic canals of the leeches, and although this is clearly a homoplasy, it demonstrates that coelomic spaces can take over the function of a blood-vessel system. The various larval types have inspired several phylogenetic speculations, and the diagrams of Jägersten (1972; see Fig. 28.2) and Maslakova (2010b) illustrate reorientations of the juvenile inside various types of pilidium and pilidium-like larvae.

Morphology, embryology, and molecular biology firmly root the nemertines in the Spiralia. The cell lineage and the fate of the trochoblasts indicate that both the pilidium larvae and the larvae of the 'direct-developing' species must be interpreted as modified trochophores. However, there is no sign of the lateral blastopore closure or the associated paired longitudinal nerve cords seen in most of the spiralians. The posterior ring of the possible compound cilia of *Pilidium recurvatum* resembles a telotroch, but the development must be followed before any conclusions

can be drawn. The cerebral ganglia develop from the episphere as in other spiralians, but the development of the peripheral nerves is in need of further studies.

I have earlier treated Nemertini and Platyhelminthes as sister groups, based on some similarities in larval morphology, but this has been the only putative homology, and molecular phylogeny does not support the idea.

Annelida, Sipuncula, Mollusca, and Nemertini have often been considered as closely related, and Zrzavý *et al.* (1998) and Peterson and Eernisse (2001) united them under the name Eutrochozoa (based on combined morphological and molecular analyses), a name previously used for the same phyla plus the brachiopods by Ghiselin (1988). Recent molecular studies generally support this view, although with various topologies of the tree. A sister group has not been identified with any certainty. Readers are referred to Chapter 24 for further discussion.

Interesting subjects for future research

1. Development of *Arhynchonemertes*
2. Organogenesis at metamorphosis of the pilidium
3. Cell lineage of the nervous system

References

Bartolomaeus, T. and von Döhren, J. 2010. Comparative morphology and evolutuon of the nephridia in Nemertea. *J. Nat. Hist.* **44**: 2255–2286.

Cantell, C.E. 1966. The devouring of the larval tissue during metamorphosis of pilidium larvae (Nemertini). *Ark. Zool.*, 2. Ser. **18**: 489–492.

Cantell, C.E. 1967. Some developmental stages of the peculiar nemertean larva pilidium recurvatum Fewkes from the Gullmar Fjord (Sweden). *Ark. Zool.*, 2. Ser. **19**: 143–147.

Cantell, C.E. 1969. Morphology, development, and biology of the pilidium larvae (Nemertini) from the Swedish west coast. *Zool. Bidr. Upps.* **38**: 61–112.

Cantell, C.E., Franzén, Å. and Sensenbaugh, T. 1982. Ultrastructure of multiciliated collar cells in the pilidium larva of *Lineus bilineatus* (Nemertini). *Zoomorphology* **101**: 1–15.

Dawydoff, C. 1928. Sur l'embryologie des protonémertes. *C. R. Hebd. Seanc. Acad. Sci. Paris* **186**: 531–533.

Dawydoff, C. 1940. Les formes larvaires de polyclades et de némertes du plancton indochinois. *Bull. Biol. Fr. Belg.* **74**: 443–496.

Delsman, H.C. 1915. Eifurchung und Gastrulation bei *Emplectonema gracile* Stimpson. *Tijdschr. Ned. Dierkd. Ver.* **14**: 68–114.

Friedrich, H. 1979. Nemertini. In F. Seidel (ed.): *Morphogenese der Tiere, Deskriptive Morphogenese, 3. Lieferung*, pp. 1–136. Gustav Fischer, Jena.

Ghiselin, M.T. 1988. The origin of molluscs in the light of molecular evidence. *Oxford Surv. Evol. Biol.* **5**: 66–99.

Hammarsten, O.D. 1918. Beitrag zur Embryonalentwicklung der *Malacobdella grossa* (Müll.). *Arb. Zootom. Inst. University Stockholm* **1**: 1–96.

Hay-Schmidt, A. 1990. Catecholamine-containing, serotonin-like and neuropeptide FMRFamide-like immunoreactive cells and processes in the nervous system of the pilidium larva (Nemertini). *Zoomorphology* **109**: 231–244.

Henry, J. and Martindale, M.Q. 1997a. Nemertines, the ribbon worms. In S.F. Gilbert and A.M. Raunio (eds): *Embryology. Constructing the Organism*, pp. 151–166. Sinauer Associates, Sunderland, MA.

Henry, J.J. 2002. Conserved mechanism of dorsoventral axis determination in equal-cleaving spiralians. *Dev. Biol.* **248**: 343–355.

Henry, J.J. and Martindale, M.Q. 1998. Conservation of the spiralian developmental program: cell lineage of the nemertean, *Cerebratulus lacteus. Dev. Biol.* **201**: 253–269.

Henry, J.Q. and Martindale, M.Q. 1994. Establishment of the dorsoventral axis in nemertean embryos: evolutionary considerations of spiralian development. *Dev. Genet.* **15**: 64–78.

Henry, J.Q. and Martindale, M.Q. 1997b. Regulation and the modification of axial properties in partial embryos of the nemertean, *Cerebratulus lacteus. Dev. Genes Evol.* **207**: 42–50.

Hickman, V.V. 1963. The occurrence in Tasmania of the land nemertine, *Geonemertes australiensis* Dendy, with some account of its distribution, habits, variations and development. *Pap. Proc. R. Soc. Tasm.* **97**: 63–75.

Hiebert, L.S., Gavelis, G., von Dassow, G. and Maslakova, S.A. 2010. Five invaginations and shedding of the larval epidermis during development of the hoplonemertean *Pantinonemertes californiensis* (Nemertea: Hoplonemertea). *J. Nat. Hist.* **44**: 2331–2347.

Iwata, F. 1958. On the development of the nemertean *Micrura akkeshiensis. Embryologia* **4**: 103–131.

Iwata, F. 1960. Studies on the comparative embryology of nemerteans with special reference to their interrelationships. *Publ. Akkeshi Mar. Biol. Stn.* **10**: 1–51.

Jespersen, Å. and Lützen, J. 1987. Ultrastructure of the nephridio-circulatory connections in *Tubulanus annulatus* (Nemertini, Anopla). *Zoomorphology* **107**: 181–189.

Jägersten, G. 1972. *Evolution of the Metazoan Life Cycle.* Academic Press, London.

Kozloff, E.N. 1990. *Invertebrates.* Saunders College Publishing, Philadelphia.

Lacalli, T.C. 2005. Diversity of form and behaviour among nemertean pilidium larvae. *Acta Zool. (Stockh.)* **86**: 267–276.

Lacalli, T.C. and West, J.E. 1985. The nervous system of a pilidium larva: evidence from electron microscope reconstructions. *Can. J. Zool.* **63**: 1909–1916.

Lebedinsky, J. 1897. Beobachtungen über die Entwicklungsgeschichte der Nemertinen. *Arch. Mikrosk. Anat.* **49**: 503–556.

Martindale, M.Q. and Henry, J.Q. 1995. Modifications of cell fate specification in equal-cleaving nemertean embryos: alternate patterns of spiralian development. *Development* **121**: 3175–3185.

Maslakova, S.A. 2010a. Development to metamorphosis of the nemertean pilidium larva. *Front. Zool.* **7**: 30.

Maslakova, S.A. 2010b. The invention of the pilidium larva in an otherwise perfectly good spiralian phylum Nemertea *Integr. Comp. Biol.* **50**: 734–743.

Maslakova, S.A. and Malakhov, V.V. 1999. A hidden larva in nemerteans of the order Hoplonemertini. *Dokl. Biol. Sci.* **366**: 314–317.

Maslakova, S.A. and von Döhren, J. 2009. Larval development with transitory epidermis in *Paranemertes peregrina* and other hoplonemerteans. *Biol. Bull.* **216**: 273–292.

Maslakova, S.A., Martindale, M.Q. and Norenburg, J.L. 2004a. Fundamental properties of the spiralian development program are displayed by the basal nemertean *Carinoma tremaphoros* (Palaeonemertea, Nemertea). *Dev. Biol.* **267**: 342–360.

Maslakova, S.A., Martindale, M.Q. and Norenburg, J.L. 2004b. Vestigial prototroch in a basal nemertean, *Carinoma tremaphoros* (Nemertea; Palaeonemertea). *Evol. Dev.* **6**: 219–226.

Nielsen, C. 1987. Structure and function of metazoan ciliary bands and their phylogenetic significance. *Acta Zool. (Stockh.)* **68**: 205–262.

Nielsen, C. 2005. Trochophora larvae: cell-lineages, ciliary bands and body regions. 2. Other groups and general discussion. *J. Exp. Zool. (Mol. Dev. Evol.)* **304B**: 401–447.

Nielsen, C. 2008. Ontogeny of the spiralian brain. In A. Minelli and G. Fusco (eds): *Evolving Pathways: Key Themes in Evolutionary Developmental Biology*, pp. 399–416. Cambridge University Press, Cambridge.

Norenburg, J.L. and Roe, P. 1998. Observations on musculature in pelagic nemerteans and on pseudostriated muscle in nemerteans. *Hydrobiologia* **365**: 109–120.

Peterson, K.J. and Eernisse, D.J. 2001. Animal phylogeny and the ancestry of bilaterians: inferences from morphology and 18S rDNA sequences. *Evol. Dev.* **3**: 170–205.

Reinhardt, H. 1941. Beiträge zur Entwicklungsgeschichte der einheimischen Süsswassernemertine *Prostoma graecense* (Böhmig). *Vierteljahrsschr. Naturforsch. Ges. Zuer.* **86**: 184–255.

Riser, N.W. 1988. *Arhynchonemertes axi* gen.n., sp.n. (Nemertini)—an insight into basic acoelomate bilaterial organology. *Fortschr. Zool.* **36**: 367–373.

Riser, N.W. 1989. Speciation and time—relationships of the nemertines to the acoelomate metazoan Bilateria. *Bull. Mar. Sci.* **45**: 531–538.

Ruppert, E.E. and Carle, K.J. 1983. Morphology of metazoan circulatory systems. *Zoomorphology* **103**: 193–208.

Salensky, W. 1912. Über die Metamorphose der Nemertinen. I. Entwicklungsgeschichte der Nemertine im Inneren des Pilidiums. *Mem. Acad. Sci. St-Petersb.*, 8. ser., Cl. Phys-math. **30(10)**: 1–74.

Salensky, W. 1914. Die Morphogenese der Nemertinen 2. Über die Entwicklungsgeschichte des *Prosorhochmus viviparus*. *Mem. Acad. Sci. St-Petersb.*, 8. ser., Cl. Phys-math. **33(2)**: 1–39.

Schmidt, G.A. 1934. Ein zweiter Entwicklungstypus von *Lineus gessneriensis* O.F. Müll. (Nemertini). *Zool. Jahrb., Anat.* **58**: 607–660.

Schmidt, G.A. 1964. Embryonic development of littoral nemertines *Lineus desori* (mihi, species nova) and *Lineus ruber* (O.F. Müller, 1774, G.A. Schmidt, 1945) in connection with ecological relation changes of mature individuals when forming the new species *Lineus ruber*. *Zool. Pol.* **14**: 75–122.

Senz, W. 1995. The 'Zentralraum': an essential character of nemertinean organisation. *Zool. Anz.* **234**: 53–62.

Stricker, S.A. 1985. The stylet apparatus of monostyliferous hoplonemerteans. *Am. Zool.* **25**: 87–97.

Sundberg, P., Turbeville, J.M. and Lindh, S. 2001. Phylogenetic relationships among higher nemertean (Nemertea) taxa inferred from 18S rDNA sequences. *Mol. Phylogenet. Evol.* **20**: 327–334.

Thiel, M. and Junoy, J. 2006. Mating behavior of nemerteans: present knowledge and future directions. *J. Nat. Hist.* **40**: 1021–1034.

Thollesson, M. and Norenburg, J.L. 2003. Ribbon worm relationships: a phylogeny of of the phylum Nemertea. *Proc. R. Soc. Lond. B* **270**: 407–415.

Turbeville, J.M. 1986. An ultrastructural analysis of coelomogenesis in the hoplonemertine *Prosorhochmus americanus* and the polychaete *Magelona* sp. *J. Morphol.* **187**: 51–60.

Turbeville, J.M. 1991. Nemertinea. In F.W. Harrison (ed.): *Microscopic Anatomy of Invertebrates*, vol. 3, pp. 285–328. Wiley-Liss, New York.

Turbeville, J.M. 2002. Progress in nemertean biology: development and phylogeny. *Integr. Comp. Biol.* **42**: 692–703.

Turbeville, J.M. 2006. Ultrastructure of the pseudocnidae of the palaeonemerteans *Cephalothrix* cf. *rufifrons* and *Carinomella lactea* and an assessment of their phylogenetic utility. *J. Nat. Hist.* **40**: 967–979.

Wilson, C.B. 1900. The habits and early development of *Cerebratulus lacteus* (Verrill). *Q. J. Microsc. Sci., N. S.* **43**: 97–198.

Zrzavý, J., Mihulka, S., Kepka, P., Bezděk, A. and Tietz, D. 1998. Phylogeny of the Metazoa based on morphological and 18S ribosomal DNA evidence. *Cladistics* **14**: 249–285.

29

Phylum **Platyhelminthes**

Platyhelminthes, or flatworms, comprise the mainly free-living, mainly aquatic 'turbellarians' and the parasitic flukes and tapeworms; about 25 000 species have been described. Many textbooks retain the economically important parasitic flukes (Digenea and Monogenea) and tapeworms (Cestoda) as classes parallel to 'Turbellaria', but this is only for practical reasons. The Acoela and Nemertodermatida have until recently been included in the Platyhelminthes, but both morphology, embryology, and molecular phylogeny have now shown that they belong to a separate clade, Acoelomorpha, that is basal to the Eubilateria (see Chapter 17). The fossil records are dubious.

The Platyhelminthes now comprises two groups, Catenulida and Rhabditophora. The rather poorly known Catenulida appears to be the sister group of the Rhabditophora, as indicated both by morphological and molecular analyses (Rieger 2001; Larsson and Jondelius 2008). The catenulids are characterized by their single, biciliate protonephridia. The Rhabditophora, characterized by the rhabdites, duo-gland adhesive systems, and paired, multiciliate protonephridia, comprises the Macrostomorpha that have aflagellate sperm, and Trepaxonemata that have biflagellate sperm (Rieger 2001). The Trepaxonemata can be divided into Polycladida that has sperm with free cilia and endolecithal eggs, and Neoophora that has internalized sperm cilia and ectolecithal eggs with nurse cells. The freshwater planarians and the parasitic flukes and tapeworms belong to Neoophora. The following discussion will concentrate on the more basal groups, i.e. the Catenulida, Macrostomida, and Polycladida, with less emphasis on the more advanced Neoophora.

The turbellarians are elongate, often dorsoventrally flattened, ciliated, with a ventral mouth leading to a gut without an anus. Their microscopic anatomy is reviewed by Bogitsch and Harrison (1991), and the structure of the epithelia by Tyler and Hooge (2004). The epidermis is generally a single layer of multiciliate cells. Catenulids have few cilia per cell that has been interpreted as a plesiomorphy (Ax 1995), but their pharyngeal cells have the usual ciliary density. A cuticle of collagenous fibrils like that observed in many other bilaterians is absent, but a more or less well-developed glycocalyx is present between microvilli in many groups. True consolidated, cuticular structures are generally absent; all the jaw-like structures in the pharynx and the male copulatory stylets are specializations of the basal membrane, and platyhelminths lack chitin completely. The peculiar epidermal scales of the ectoparasitic neoophoran *Notodactylus* have an ultrastructure resembling that of annelid chaetae, with a bundle of parallel microvilli embedded in extracellular matrix, but without chitin (Jennings *et al*. 1992). Several types of mostly unicellular glands can

Chapter vignette: The polyclad *Prostheceraeus vittatus*. (Redrawn from Lang 1884.)

be recognized, and the rhabdoid glands have often been considered characteristic of turbellarians. There are, however, various types of rhabdoids and they may not all be homologous. One special type, the lamellate rhabdite, is a unique characteristic of the group Rhabditophora. The structure and distribution of the other gland types are, as yet, incompletely known so that their phylogenetic significance cannot be evaluated. Among the platyhelminths, the adhesive duo-gland system is only found in the rhabditophorans (Tyler 1976). The epidermis rests on a basal membrane in rhabditophorans, but it is weak or lacking in catenulids. Some epidermal cells are 'insunk', i.e. the nuclear part of the cell is situated below the mesodermal layer of muscles.

Most turbellarians have a mouth that opens into a muscular, ectodermal pharynx. The pharynx is either a simple tube surrounded by muscle cells, or the epidermis is folded back around the muscles so that a more complicated, protrusible structure is formed; the uncomplicated type comprises several subtypes that are probably not homologous. The pharynx opens into the anterior end of the stomach in some forms, but a more posterior position of the mouth is characteristic of most groups.

The stomach is a simple tube of ciliated endodermal cells in catenulids and in members of several of the other groups. Lateral diverticula of many shapes characterize various groups, as reflected by names such as triclads and polyclads, and many of the more specialized groups lack the cilia. The gut is lacking completely in the cestodes and in the parasitic neoophoran family Fecampiidae. An anus is absent, and the examples of one or several openings from gut diverticula to the outside must be seen as secondary specializations (Ehlers 1985) (see below).

The nervous system generally comprises a frontal, subepidermal brain with a pair of main longitudinal nerve cords forming a complete loop, a number of smaller longitudinal nerve cords, transverse commissures, and various nerve plexuses (Halton and Maule 2004). The brain of *Macrostomum* is a ganglion with a central neuropile surrounded by perikarya, traversed by muscle fibres in various directions (Morris *et al.*

2007). The planarian *Dugesia* has large eyes and a more complex brain (Okamoto *et al.* 2005). Some of the peripheral nerve cords or nerves are basiepithelial and some subepidermal, and there is a bewildering variation in position and complexity even within the groups (Reuter *et al.* 1998; Reuter and Halton 2001; Gustafsson *et al.* 2002; Morris *et al.* 2007). Rieger *et al.* (1991: p. 34) concluded that 'the nervous system is now being viewed as showing a variable mixture of specialized and unspecialized, advanced and "primitive" components'. With this conclusion in mind it is understandable that the turbellarian nervous systems have been interpreted in many different ways (see below).

The mesoderm is a compact mass between ectoderm and endoderm, and comprises muscle cells, neoblasts, apolar mesenchymal cells, and ample extracellular material in most types. There are usually systems of longitudinal and circular muscles, but diagonal systems are found too (Hooge 2001). The myofilaments generally resemble smooth muscle, but various indications of striation have been reported in studies of the ultrastructure, and more typical cross-striated muscle cells are found in the tail of trematode cercaria (Fried and Haseb 1991).

Protonephridia are found in most forms. Catenulids have an unpaired dorsal nephridial canal with biciliate cyrtocytes, whereas the rhabditophorans usually have many flame bulbs, with many cilia arranged along a pair of lateral nephridial canals.

Hermaphroditism is the rule among the platyhelminths; the gonads are groups of cells without a special wall in catenulids, whereas the rhabditophorans have well-defined, sac-shaped gonads, in many cases with quite complicated genital organs. Fertilization is always internal and there is a copulatory structure, which in many types has hardened parts that are either intracellular or specializations of the basal membrane. The copulation is a hypodermal impregnation or simple deposition of sperm or spermatophores on the partner in several types, but sperm transfer through the female gonoducts is the rule in the more-specialized forms.

The spermatozoa are of different morphology in the main groups (Hendelberg 1986; Watson and

Rohde 1995). Catenulids and macrostomids have aciliate sperm; bristles or rods have been described in both groups, but their ultrastructure does not indicate that they are ciliary rudiments. Trepaxonemata (Polycladida and Neoophora) have highly characteristic axonemes without central tubules but with a highly characteristic electron-dense central core surrounded by a lighter middle zone and a dense outer zone, with a double-helical structure; the axonemes are internalized in most neoophorans.

The female gonads are rather uncomplicated in catenulids, macrostomids, and polyclads (collectively called 'archoophorans') that have endolecithal eggs. The neoophorans have yolk glands, and the eggs are deposited in capsules together with a number of yolk cells. The shell of the egg cases consists of quinone tanned proteins. The fertilized eggs are released through rupture of the body wall in catenulids, but the rhabditophorans have more or less complicated gonoducts.

The only catenulid larva described is the 'Luther's larva' of the limnic *Rhyncoscolex* (Reisinger 1924); it resembles the adults, but has a stronger ciliation and a different statocyst; it should probably be regarded as a specialized juvenile rather than a true larva.

The early cleavage of the macrostomid *Macrostomum* (Morris *et al.* 2004) follows the spiral pattern, but after the 8-cell stage some of the blastoporal macromeres spread over the embryo forming a 'hull' around the 'embryonic primordium'. The cells of the primordium differentiate to the various organs of the juvenile, apparently with the larger cells closest to the blastoporal pole forming the gut.

Polyclads show a normal spiral cleavage with quartets (reviewed in Nielsen 2005). A few experimental investigations show that the cleavage with quartets resembles that of other spiralians, but that the determination is less strong. Species of genera such as *Hoploplana*, *Notoplana*, *Pseudoceros*, *Imogine*, and *Maritigrella* (Rawlinson 2010) have pelagic larvae called Götte's and Müller's larvae (see below), but other species have direct development. The cell lineage appears to be similar in all species, and the following description is based on *Hoploplana inquilina* (Boyer

et al. 1998 (Table 29.1)) The first two cleavages pass through the primary egg axis, and the four resulting blastomeres are of almost equal size, although the B and D cells are sometimes slightly larger than the A and C. The third cleavage gives larger macromeres and smaller micromeres in most species, with the macromeres 1B and 1D situated in the future anteroposterior axis and the micromeres twisted clockwise, so that the micromeres 1a and 1b occupy almost bilateral positions. The micromeres form the completely ciliated episphere bordered by the band of larger cilia (see below). The 64-cell stage is highly characteristic with four quite small blastoporal 4A–D 'macromeres' and very large 4a–d cells; the 4b and 4d cells are larger than the other two. These eight cells become covered by the cells of the micromere calotte in an epibolic gastrulation, but the 4A–D and 4a–c cells disintegrate into yolk granules. Endoderm, muscles, and mesenchyme originate from descendants of 4d, and the stomodaeum from 2a, 2c and 3d. Additional ectomesoderm originates from 2b in the shape of circular body muscles in the larva. There is no sign of teloblastic growth or of coelomic cavities in the mesoderm. The endodermal cells become arranged as a small gut. The apical ganglion with the apical tuft develops at the centre of the first micromere quartet. The paired cerebral ganglia with cerebral eyes (Younossi-Hartenstein and Hartenstein 2000) develop from the episphere. The main part of the peripheral nervous system develops from the third micromere quartet. The larva develops into a Müller's larva (see below). *Pseudoceros japonicus* has unusually large macromeres and mesoblast (4d), and the micromeres develop into a completely ciliated cap, which overgrows the macromeres completely. The macromeres and the mesoblast divide and become arranged as a sac-shaped gut and a compact mesoderm. The almost spherical embryos develop lappets and become first a Götte's, and then a Müller's larva (see below). In *Prosthecereus* the 4b cell extends into the blastocoel and comes into contact with the cells of the apical pole, and a similar internalization has been observed in other spiralians, although for the 4D cell (van den Biggelaar *et al.* 1997); it is not known if this is involved in axis specification.

Two types of polyclad larvae have been named, 'Götte's larva', with four rather broad lateral lobes, and 'Müller's larva' (Fig. 29.1), with eight usually more cylindrical lobes. The larvae are completely ciliated with an apical tuft and a band of longer and denser ciliation around the body following the sides of the lobes (Fig. 29.1). This ciliary band develops along the borderline between the 1st and 2nd micromere quartets (Boyer et al. 1998) and can therefore be interpreted as a prototroch. Compound cilia have not been observed. The larvae swim with the ciliary band, which shows metachronal waves, and at least some of the larvae are planktotrophic (Ruppert 1978; Ballarin and Galleni 1987; Rawlinson 2010), but the method of particle collection has not been studied. The ciliary band of *Pseudoceros* is usually described as a ring, but the ring is broken between the two lobes at the left side, and the anterior part of the band continues across the ventral side behind the mouth to the opposite side (Lacalli 1982). The band is two to eight cells wide, and all the cells are multiciliate with the ciliary beat towards the mouth, except the median cell behind the mouth that has cilia beating away from the mouth (the suboral rejectory cell) (Ruppert 1978; Lacalli 1982, 1988). There is a system of intraepithelial nerves with mono-ciliate sensory cells in or along the band. There is no sign of an anus at any stage. The nervous system (Younossi-Hartenstein and Hartenstein 2000;

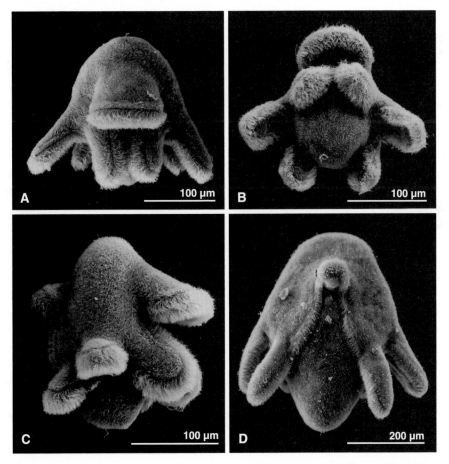

Figure 29.1. Müller's larvae (SEM). (A–C) Larvae in frontal, ventral, and lateral views (plankton, off Bamfield Marine Station, Vancouver Island, Canada, August 1988). (D) Larva in dorsal view (plankton, off Nassau, The Bahamas, September 1990).

ciliary bands on the lappets. At a very early stage, a pair of serotonergic cells differentiates near the posterior pole and sends axons to the cerebral ganglia (Rawlinson 2010). This resembles the pioneer nerve cells of other protostomes (Chapter 22).

There is a system of fine muscles under the ectodermal basement membrane with a spiral muscle around the apical pole, longitudinal muscles from the apical pole to the eight lobes, and circular muscles and a small muscle from the apical pole to the mouth (Reiter *et al.* 1996; Rawlinson 2010; Semmler and Wanninger 2010). The larval muscle system transforms gradually into the adult pattern both in indirect- and direct-developing species (Reiter *et al.* 1996; Bolañosa and Litvaitis 2009). The larvae have a pair of branched protonephridia (Ruppert 1978; Younossi-Hartenstein and Hartenstein 2000). The older larvae gradually lose the lobes and the ciliated bands, and transform into small adults. The cerebral ganglia become part of the adult brain; the fate of the ciliary-band nerves is unknown. Dawydoff (1940) described a number of polyclad larvae resembling gigantic Müller's larvae with a general ciliation, but without the prominent ciliary bands; this only shows that new investigations may reveal other types of larvae, which may add to our understanding of the development of the platyhelminths.

The most 'primitive' of the neoophorans, such as the lecithoepitheliate *Xenoprorhynchus* (Reisinger *et al.* 1974), have spiral cleavage with quartets, but the later development is influenced by the presence of the yolk cells. Eight micromeres, 2a–d and 3a–d, move to the periphery of the embryo and cover the yolk cells situated at the blastoporal pole, forming a very thin embryonic 'covering membrane'. This cell layer becomes resorbed when the yolk has been incorporated in the embryo. The 4D cell gives rise to the endoderm, and 4d divides laterally and gives rise to a pair of mesoblasts, which were stated to be mesoteloblasts; however, the illustrations show cells of equal size, and it is only stated that the cells divide—not that one large cell gives off a series of smaller cells, as in the teloblasts from both mesoderm and ectoderm of, for example, annelids (Chapter 25). Members of the Proseriata (*Minona, Monocelis*; Reisinger *et al.* 1974) show increasing volumes of yolk cells and increasingly modified embryology, and the development of the parasitic groups is highly specialized (Ehlers 1985). The parasitic forms have lecithotrophic, often ciliated larvae that shed the ciliated cells, and the following life cycles comprise stages with asexual reproduction in some groups.

The most conspicuous apomorphy of the Platyhelminthes is the lack of an anus. This was interpreted as a plesiomorphy, for example by Ax (1987, 1995), implying that they should be the sister group of the 'Eubilateria' (with a tubular gut). Most of these discussions were influenced by the older phylogenetic system that included the acoels in the 'flatworms' and this makes it difficult to disentangle many of the arguments. However, there are several characteristics that indicate that the lack of an anus is secondary. The firm rooting of the phylum in the Spiralia directly contradicts the idea that they should be 'basal' bilaterians. The Hox genes show the full series of anterior and central genes characteristic of the eubilaterians, as opposed to that of the acoelomorphs, which appear to be plesiomorphic (Fig. 21.3). The very unusual spiral cleavage pattern with the tiny 'macromeres' 4A–D, and the degeneration of both these cells and the 4a–c cells, for example in *Hoploplana* (Table 29.1), indicates that the blastopore region has been strongly modified; this indicates that the missing anus is an apomorphy.

As pointed out by Rieger *et al.* (1991), it is difficult to identify platyhelminth apomorphies, but nevertheless most morphologists now appear to agree about the monophyly of the group. The presence of spiral cleavage in several groups, the cell lineage and fate of the large ciliary band (prototroch) of the planktotrophic Götte's and Müller's larvae, and the origin of the cerebral ganglia from the episphere clearly indicate the position of the platyhelminths in the Spiralia, and the larvae as modified trochophores.

Rhabdites have sometimes been regarded as homologous with nematocysts, but the structures are quite dissimilar, and lamellate rhabdites are only found in the 'higher' turbellarians (Ehlers 1985). Reisinger (1972) and Gustafsson *et al.* (2002) interpreted the nervous system as a 'primitive, orthogonal' type with a number of longitudinal nerves connected by transverse

Table 29.1. Cell lineage of *Hoploplana inquilina*. (Based on Boyer *et al.* 1998.)

Lineage					Fate
		1a			apical tuft, ectoderm, ocellus
	A	2a			ectoderm, stomodaeum
		1A	3a		ectoderm
AB		2A	3A		yolk granules
		1b			ectoderm
	B	2b			ectoderm, muscles
		1B	3b		ectoderm
Z		2B	3B		yolk granules
	C	development as A			
		1d			ectoderm
CD	D	2d			posterior ciliary tuft, ectoderm
		1D	3d		stomodaeum
		2D	3D	4d	muscles, mesenchyme, gut
				4D	yolk granules

nerves, but, as mentioned above, one pair of the longitudinal nerves are almost always more prominent than the other pairs, and the nervous system shows so much variation that almost any interpretation can be supported by existing examples. Much more knowledge about development, structure, and function of nervous systems of many groups will be needed before a meaningful evaluation can be performed.

Platyhelminthes is recovered as a monophyletic group of spiralians in practically all molecular analyses. The catenulids are only rarely included in the analyses, but they seem always to be the sister group of the rhabditophorans (Larsson and Jondelius 2008; Paps *et al.* 2009). The topology of the Rhabditophora outlined above is usually found too, although sometimes with small variations. However, the relationship of the Platyhelminthes to other spiralian phyla is not well resolved. Many analyses place them as a sister group to the remaining spiralians, but close relationships with Gastrotricha, Gnathifera, Rotifera, and Polyzoa have been suggested too. A group called Platyzoa comprising Platyhelminthes, Gnathostomulida, Gastrotricha,

and Rotifera has been proposed by Cavalier-Smith (1998) and Giribet *et al.* (2009), but it seems impossible to point to a uniting synapomorphy. Cavalier-Smith (1998: p. 236) proposed that the flatworms are 'possibly neotenously derived from loxosomatid-like entoproct larvae'.

With no clear phylogenetic signal from morphology or molecules it appears impossible to resolve the polytomy of the Spiralia indicated in Fig. 23.4.

Interesting subjects for future research

1. Development of the nervous system in a number of types

2. Cell-lineage studies of nervous systems

References

Ax, P. 1987. *The Phylogenetic System. The Systematization of Organisms on the Basis of their Phylogenesis.* John Wiley, Chichester.

Ax, P. 1995. *Das System der Metazoa I*. Gustav Fischer, Stuttgart.

Ballarin, L. and Galleni, L. 1987. Evidence for planctonic feeding in Götte's larva of *Stylochus mediterraneus* (Turbellaria—Polycladida). *Boll. Zool.* **54**: 83–85.

Bogitsch, B.J. and Harrison, F.W. 1991. Platyhelminthes: Turbellaria. In F.W. Harrison (ed.): *Microscopic Anatomy of Invertebrates*, vol 3, pp. 7–140. Wiley-Liss, New York.

Bolañosa, D.M. and Litvaitis, M.K. 2009. Embryonic muscle development in direct and indirect developing marine flatworms (Platyhelminthes, Polycladida). *Evol. Dev.* **11**: 290–301.

Boyer, B.C., Henry, J.J. and Martindale, M.Q. 1998. The cell lineage of a polyclad turbellarian embryo reveals close similarity to coelomate spiralians. *Dev. Biol.* **204**: 111–123.

Cavalier-Smith, T. 1998. A revised six-kingdom system of life. *Biol. Rev.* **73**: 203–266.

Dawydoff, C. 1940. Les formes larvaires de polyclades et de némertes du plancton indochinois. *Bull. Biol. Fr. Belg.* **74**: 443–496.

Ehlers, U. 1985. *Das phylogenetische System der Plathelminthes*. Gustav Fischer, Stuttgart.

Fried, B. and Haseb, M.A. 1991. Platyhelminthes: Aspidogastrea, Monogenea, and Digenea. In F.W. Harrison (ed.): *Microscopic Anatomy of Invertebrates*, vol. 3, pp. 141–209. Wiley-Liss, New York.

Giribet, G., Dunn, C.W., Edgecombe, G.D., *et al.* 2009. Assembling the spiralian tree of life. In M.J. Telford and D.T.J. Littlewood (eds): *Animal Evolution. Genomes, Fossils, and Trees*, pp. 52–64. Oxford University Press, Oxford.

Gustafsson, M.K.S., Halton, D.W., Kreshchenko, N.D., *et al.* 2002. Neuropeptides in flatworms. *Peptides* **23**: 2053–2061.

Halton, D.W. and Maule, A.G. 2004. Flatworm nerve–muscle: structural and functional analysis. *Can. J. Zool.* **82**: 316–333.

Hendelberg, J. 1986. The phylogenetic significance of sperm morphology in the Platyhelminthes. *Hydrobiologia* **132**: 53–58.

Hooge, M.D. 2001. Evolution of body-wall musculature in the Platyhelminthes (Acoelomorpha, Catenulida, Rhabditophora). *J. Morphol.* **249**: 171–194.

Jennings, J.B., Cannon, L.R.G. and Hick, A.J. 1992. The nature and origin of the epidermal scales of *Notodactylus handschini*—an unusual temnocephalid turbellarian ectosymbiotic on crayfish from Northern Queensland. *Biol. Bull.* **182**: 117–128.

Lacalli, T.C. 1982. The nervous system and ciliary band of Müller's larva. *Proc. R. Soc. Lond. B* **217**: 37–58.

Lacalli, T.C. 1988. The suboral complex in the Müller's larva of *Pseudoceros canadensis* (Platyhelminthes, Polycladida). *Can. J. Zool.* **66**: 1893–1895.

Lang, A. 1884. Die Polycladen (Seeplanarien) des Golfes von Neapel. *Fauna Flora Golf. Neapel* **11**: 1–688.

Larsson, K. and Jondelius, U. 2008. Phylogeny of Catenulida and support for Platyhelminthes. *Org. Divers. Evol.* **8**: 378–387.

Morris, J., Cardona, A., De Miguel-Bonet, M.D.M. and Hartenstein, V. 2007. Neurobiology of the basal platyhelminth *Macrostomum lignano*: map and digital 3D model of the juvenile brain neuropile. *Dev. Genes Evol.* **217**: 569–584.

Morris, J., Nallur, R., Ladurner, P., *et al.* 2004. The embryonic development of the flatworm *Macrostomum* sp. *Dev. Genes Evol.* **214**: 220–239.

Nielsen, C. 2005. Trochophora larvae: cell-lineages, ciliary bands and body regions. 2. Other groups and general discussion. *J. Exp. Zool. (Mol. Dev. Evol.)* **304B**: 401–447.

Okamoto, K., Takeuchi, K. and Agata, K. 2005. Neural projections in planarian brain revealed by fluorescent dye tracing. *Zool. Sci. (Tokyo)* **22**: 535–546.

Paps, J., Baguñà, J. and Riutort, M. 2009. Bilaterian phylogeny: A broad sampling of 13 nuclear genes provides a new Lophotrochozoa phylogeny and supports a paraphyletic basal Acoelomorpha. *Mol. Biol. Evol.* **26**: 2397–2406.

Rawlinson, K.A. 2010. Embryonic and post-embryonic development of the polyclad flatworm *Maritigrella crozieri*; implications for the evolution of spiralian life history traits. *Front. Zool.* **7**: 12.

Reisinger, E. 1924. Die Gattung *Rhynchoscolex*. *Z. Morphol. Oekol. Tiere* **1**: 1–37.

Reisinger, E. 1972. Die Evolution des Orthogons der Spiralier und das Archicoelomatenproblem. *Z. Zool. Syst. Evolutionsforsch.* **10**: 1–43.

Reisinger, E., Cichocki, I., Erlach, R. and Szyskowitz, T. 1974. Ontogenetische Studien an Turbellarien: ein Beitrag zur Evolution der Dotterverarbeitung in ektolecitalen Ei. *Z. Zool. Syst. Evolutionsforsch.* **12**: 161–195.

Reiter, D., Boyer, B., Ladurner, P., *et al.* 1996. Differentiation of the body wall musculature in *Macrostomum hystricum marinum* and *Hoploplana inquilina* (Plathelminthes), as models for muscle development in lower Spiralia. *Roux's Arch. Dev. Biol.* **205**: 410–423.

Reuter, M. and Halton, D.W. 2001. Comparative neurobiology of Platyhelminthes. In D.T.J. Littlewood and R.A. Bray (eds): *Interrelationships of the Platyhelminthes (Systematics Association Special Volume 60)*, pp. 239–249. Taylor and Francis, London.

Reuter, M., Mäntyla, K. and Gustafsson, M.K.S. 1998. Organization of the orthogon—main and minor nerve cords. *Hydrobiologia* **383**: 175–182.

Rieger, R.M. 2001. Phylogenetic systematics of the Macrostomorpha. In D.T.J. Littlewood and R.A. Bray (eds): *Interrelationships of the Platyhelminthes*, pp. 28–38. Taylor & Francis, London.

Rieger, R.M., Tyler, S., Smith, J.P.S., III and Rieger, G.E. 1991. Platyhelminthes: Turbellaria. In F.W. Harrison (ed.): *Microscopic Anatomy of Invertebrates*, vol. 3, pp. 7–140. Wiley-Liss, New York.

Ruppert, E.E. 1978. A review of metamorphosis of turbellarian larvae. In F.S. Chia and M.E. Rice (eds): *Settlement and*

Metamorphosis of Marine Invertebrate Larvae, pp. 65–81. Elsevier, New York.

Semmler, H. and Wanninger, A. 2010. Myogenesis in two polyclad platyhelminths with indirect development, *Pseudoceros canadensis* and *Stylostomum sanjuana*. *Evol. Dev.* **12**: 210–221.

Tyler, S. 1976. Comparative ultrastructure of adhesive systems in the Turbellaria. *Zoomorphologie* **84**: 1–76.

Tyler, S. and Hooge, M. 2004. Comparative morphology of the body wall in flatworms (Platyhelminthes). *Can. J. Zool.* **82**: 194–210.

van den Biggelaar, J.A.M., Dictus, W.J.A.G. and van Loon, A.E. 1997. Cleavage patterns, cell-lineages and cell specification are clues to phyletic lineages in Spiralia. *Semin. Cell Dev. Biol.* **8**: 367–378.

Watson, N.A. and Rohde, K. 1995. Sperm and spermiogenesis of the 'Turbellaria' and implications for the phylogeny of the phylum Platyhelminthes. *Mem. Mus. Natl. Hist. Nat. (France)* **166**: 37–54.

Younossi-Hartenstein, A. and Hartenstein, V. 2000. The embryonic development of the polyclad flatworm *Imogine mcgrathi*. *Dev. Genes Evol.* **210**: 383–398.

30

Phylum **Gastrotricha**

Gastrotricha is a small phylum of microscopic, aquatic animals; about 500 species have been described. Traditionally, the group has been divided into the marine Macrodasyoida, with a myoepithelial pharynx with a lumen that is an inverted Y-shape in cross-section, and that has a pair of lateral pores (except in *Lepidodasys*), and the marine or limnic Chaetonotoida, with a Y-shaped pharynx lumen and no pharyngeal pores. *Neodasys* has the macrodasyoid pharynx, but is 'intermediate' between the two orders in several characters, and has been classified with both of the two orders (Todaro *et al.* 2006; Petrov *et al.* 2007). A large morphology-based analysis indicates that it could be the sister group to Macrodasyoida + Chaetonotoida (Kieneke *et al.* 2008a).

The microscopic anatomy was reviewed by Ruppert (1991) and analyzed by Kieneke *et al.* (2008a). The elongate body has a flattened ventral side with a ciliated ventral sole. The mouth is anterior and an almost cylindrical gut, consisting of a myoepithelial sucking pharynx and an intestine of cells with microvilli, leads to the ventral anus near the posterior end. The two different shapes of the pharynx naturally lead to speculations about the shape of the ancestral pharynx. Some gastrotrichs are able to regenerate both anterior and posterior ends (Manylov 1995).

The ectoderm is a monolayer of unciliated, monociliate, or multiciliate cells usually without microvilli. The whole surface, including the cilia, is covered by multiple layers of exocuticle, with each layer resembling a cell membrane and the epicuticle of nematodes, kinorhynchs, and arthropods; an inner, granular or fibrillar endocuticle, which may be thrown into complicated scales, hooks, or spines, covers the epithelial surface but not the cilia. The cuticular structures contain an extension of the epithelium in *Xenodasys*, whereas the other structures are hollow or solid. The body cuticle consists of proteinaceous compounds without chitin, but traces of chitin have been detected in the pharyngeal cuticle (Neuhaus *et al.* 1996).

The myoepithelial pharynx has radiating myomeres and a cuticle that in some species forms teeth, hooks, or more complicated, scraper-like structures. Ruppert (1982) reported that the juvenile gut of the macrodasyoid *Lepidodasys* is circular, and that both the anterior and the posterior ends of the pharynx of other genera may be circular, quadriradiate, or multiradiate, and concluded that the circular shape must be ancestral. Some of the cells bear one kinocilium, and some cells have a few microvilli that penetrate the cuticle. The tubular midgut consists of a single layer of microvillous cells surrounded by mesodermal muscles; cilia are absent except in *Xenodasys*. The chaetonotoids have a short, cuticle-lined rectum, but the other groups have a simple pore between the midgut and the epidermis.

Chapter vignette: *Turbanella cornuta*. (Redrawn from Remane 1926.)

Immunocytochemical studies have shown that the brain is dumbbell-shaped with a pair of cerebral ganglia connected by a solid dorsal and a weaker ventral commissure. There is a pair of prominent ventrolateral longitudinal nerves, which form a loop at the posterior end and a number of fine longitudinal nerves (Hochberg 2007; Rothe and Schmidt-Rhaesa 2009; Hochberg and Atherton 2011). Based on transmission electron microscopy observations of *Cephalodasys*, Wiedermann (1995) showed that only the ventral longitudinal nerves have perikarya and can be characterized as nerve cords. Gastrotrichs have poorly developed basal membranes, but the nervous system is probably basiepithelial (Wiedermann 1995). There are several types of sensory organs, chemoreceptors, mechanoreceptors, and photoreceptors, that all consist of modified monociliate cells.

The body wall has an outer layer of circular and an inner layer of longitudinal muscles, that do not form continuous sheets (Hochberg and Litvaitis 2001; Leasi and Todaro 2008). The muscles are cross-striated or obliquely striated; smooth muscles are found in *Lepidodasys*. The muscle cells have short non-contractile processes with synapses at the longitudinal nerves. This type of muscle innervation is also seen in nemertines (Chapter 28), nematodes (Chapter 46), and cephalochordates (Chapter 63).

One to several pairs of protonephridia are located laterally. Each protonephridium consists of one or a few monociliate terminal cells, a monociliate duct cell, and a sometimes monociliate pore cell situated in the epithelium (Kieneke et al. 2008a). There is no open body cavity; the only open spaces are the small lacunae around the terminal organs of the protonephridia. There is no circulatory system.

Gonad morphology was reviewed by Kieneke et al. (2009). All species are hermaphroditic, either with hermaphroditic gonads or with separate testes and ovaries. The gonads are sac-shaped and sometimes surrounded by muscle cells. Most species have complicated accessory reproductive structures associated with copulation, and it appears that all species have internal fertilization; some are viviparous. The sperm is filiform with a long, spirally coiled head and a cilium in macrodasyoids, but of variable shape in chaetonotoids (Marotta et al. 2005).

The embryology of a few species representing both orders has been studied; the macrodasyoid *Turbanella* is best known (Teuchert 1968) and will be described first (Fig. 30.1, Fig. 30.2, Table 30.1). Fertilization takes place in the ovary but the polar bodies are not given off until after spawning. The egg is ovoid and the polar bodies usually become situated at the blunt end, which has been called the animal pole, but a few eggs with the polar bodies at the more pointed end have been observed too. The movements of the cells during early developmental stages make it difficult to define the axes of the embryo, and the lack of an apical sense organ makes the use of the term 'apical' misleading (in the following description the blunt end of the egg is called the anterior pole). The first cleavage is equatorial, and the second cleavage is parallel to the longitudinal egg axis, with the two divisions being perpendicular to each other. One of the posterior cells then moves towards the anterior pole of the egg, and the resulting embryo consists of three cells at the anterior end and one at the posterior end. The embryo is now bilateral and the four cells can be named and related to the orientation of the adult: the two descendants from the anterior cell (A and B) are situated anteroventrally to the right and left, respectively; the third cell at the anterior pole (C) is anterodorsal; and the fourth cell (D) is posterior.

At the third cleavage the A, B, and C cells divide almost parallel to the longitudinal egg axis, while D divides perpendicular to the axis. The ventral descendant of the D cell slides to a midventral position, so that the embryo now consists of two dorsal cells and a ventral cell, separated by a ring of five cells. During the following cleavages the dorsal cells form two longitudinal rows of four cells each, and the ventral cell forms a longitudinal row of two cells, while the ring of cells, through two cleavages, develop into a double ring each with ten cells. The dorsal cell rows and the dorsal ring of cells become ectoderm, the cells of the ventral ring become ectoderm plus mesoderm, while the two ventral cells become endoderm. The fates of the cells are summarized in Table 30.1. The precursors of the mesodermal cells appear to form a ring

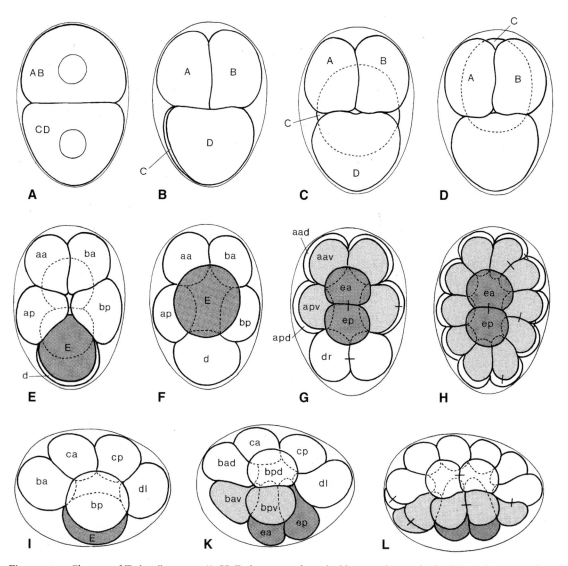

Figure 30.1. Cleavage of *Turbanella cornuta*. (A–H) Embryos seen from the blastoporal/ventral side; (I–L) embryos seen from the left side. (A) 2-cell stage. (B) 4-cell stage just after the cleavage. (C–D) 4-cell stages showing the movement of the C-cell to the anterior pole. (E) 8-cell stage with the E-cell at the posterior pole. (F and I) 8-cell stage where the E-cell has moved to the blastoporal side. (G and K) 14-cell stage. (H and L) 30-cell stage; the mesodermal cells form a ring around the two endoderm cells. Endodermal cells are dark grey and mesodermal cells light grey. (Redrawn from Teuchert 1968.)

around the coming blastoporal invagination, but it must be noted that the fate of the single cells has not been followed further on, and as the mesoderm forms a pair of lateral bands after gastrulation, it is not certain that the mesoderm surrounds the blastopore completely.

The 30-cell embryo has a small blastocoel, and during the following cell divisions the endodermal cells and the cells of the stomodaeum (pharynx) invaginate as a longitudinal furrow. The posterior parts of the blastopore lips fuse, and the endodermal cells form a compact mass of cells at the end of the stomodaeum with mesodermal cells on both sides. The blastocoel becomes obliterated, and an anus breaks through at a later stage. The brain develops from ectodermal cells

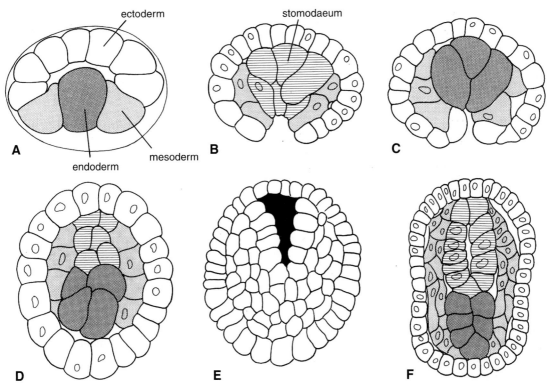

Figure 30.2. Development of *Turbanella cornuta*. (A) Cross-section of a 30-cell stage showing the infolding of the endoderm. (B–C) Cross-sections of a stage with an elongate blastopore; (B) is in the anterior region with stomodaeum and (C) in the posterior region with the endoderm. (D) Horizontal section of a similar stage. (E) Embryo in ventral view showing the elongate stomodaeal invagination (black); the endoderm is fully covered. (F) Horizontal section of a similar stage. Ectoderm, white; mesoderm, light grey; endoderm, dark grey; stomodaeum, cross-hatched. (Redrawn from Teuchert 1968.)

Table 30.1. Cell lineage of a 30-cell stage of *Turbanella cornuta* (see Fig. 30.1; based on Teuchert 1968). The first letter in each combination denotes the descendance from the first four cells A–D; the letters indicate as follows: a, anterior; p, posterior; d, dorsal; v, ventral.

	aa	aad	'secondary' ectoderm
		aav	mesendoderm
A	ap	apd	'secondary' ectoderm
		apv	mesectoderm
B		development as A	
C			'primary' ectoderm
	dr	drd	'secondary' ectoderm
d		drv	mesectoderm
	dl	dld	'secondary' ectoderm
		dlv	mesectoderm
E			endoderm

Z { AB { A, B }, CD { C, D { d }, E } }

in the pharyngeal region. Genital cells are presumed to originate from mesodermal cells.

The embryo now curves ventrally, and during the following development the endoderm differentiates as the gut and the anus opens. All other organ systems develop during the later part of the embryonic period, so that the hatching worm is a miniature adult. No traces of larval organs have been described and the development must be characterized as direct.

Two chaetonotoids have been studied, *Neogossea* (Beauchamp 1929) and *Lepidodermella* (Sacks 1955); the two authors followed the cell lineage to a stage of about 64 cells and the further development in optical sections. The two descriptions agree in all major points, but differ from that of *Turbanella* in several significant details: the primary endoderm cell should come from a descendant of the A cell and therefore be located at the anterior part of the embryo; the C and D cells should form the right and left halves of the posterior part of the embryo (compare with Fig. 30.1, A); there should be a large proctodaeum formed from the C and D cells, and the genital cells should originate from a pair of cells just lateral to the stomodaeal opening. Teuchert (1968) pointed out that a partial correspondence between the descriptions could be obtained if the anteroposterior orientation of the older descriptions was reversed. However, it appears that both Beauchamp and Sacks had rather good markers of the main axis of the embryos all along the development, whereas Teuchert (1968, pp 379–380) pointed out that the 30-cell stage has no markers that distinguish the anterior from the posterior pole, so it is also possible that Teuchert had reversed the anteroposterior axis. The differences regarding the formation of the gut are more difficult to explain, and the whole development must be studied in both orders before a meaningful discussion can be carried out.

Be that as it may, the cleavage shows the typical D-quadrant pattern, but no signs of a spiral pattern.

Two characters have had a prominent position in the discussion of the phylogenetic position of the gastrotrichs, the myoepithelial pharynx, and the mono-versus multiciliate cells of the epithelia.

Ruppert (1982) considered the myoepithelial foreguts of the gastrotrichs, nematodes, tardigrades, and ectoproct bryozoans as homologous and as a symplesiomorphy because the myoepithelial cell type is characteristic of the cnidarians. He even went so far as to indicate a sister-group relationship between nematodes and chaetonotoids on the basis of the pharynx structure. However, contractile ciliated cells are a plesiomorphy of the animals (Chapter 2), and cells with a cilium and myofibrils are a plesiomorphy of the neuralians (Chapter 12). A myoepithelial tube with a triangular lumen is simply the most efficient way of constructing a self-contained suction structure, and this has obviously evolved independently in many clades.

Several macrodasyoid genera and *Neodasys* have monociliate epithelia, whereas chaetonotoids and other macrodasyoid genera have multiciliate epithelia; the type of ciliation appears to be uniform within genera but not within families (Rieger 1976). The monociliate and some of the multiciliate cells have an accessory centriole at the base of each cilium. The monociliate condition was interpreted as primitive within the gastrotrichs, but it was admitted that it could be the result of a reduction from the multiciliate condition. It should be made clear that this question can only be resolved by comparisons with other phyla, and I have come to the conclusion (Chapter 21) that the multiciliate condition is an apomorphy of the Protostomia, and that the presence of monociliate epithelia in a few protostomian taxa, such as some gastrotrichs and annelids (*Owenia* and the gnathostomulids; Chapters 25 and 32), must represent reversals.

The multiple epicuticular layers surrounding the body, including the cilia, characterize gastrotrichs as a monophyletic group. The D-quadrant cleavage and the structure of the nervous system are typical of the protostomes and the ciliated epithelium is a spiralian character, but it is very difficult to point to characters indicating a closer relationship with any of the other spiralian phyla.

Gastrotricha is included in a number of molecular analyses but the results are ambiguous. Usually only one species is included, but with more species included they are mixed up with gnathiferans (Baguñá *et al.* 2008; Paps *et al.* 2009a). Other recent studies show sister-group relationships with gnathostomulids (Paps *et al.* 2009b), rotifers (Witek *et al.* 2009), or platyhelminths (Telford *et al.* 2005; Dunn *et al.* 2008; Hejnol *et al.* 2009). Taken together, these results indicate the existence of a Platyzoa clade (Passamaneck and Halanych 2006; Giribet *et al.* 2009), but morphological support has not been found.

Interesting subjects for future research

1. Embryology and cell lineages of both chaetonotoids and macrodasyoids

References

Baguñá, J., Martinez, P., Paps, J. and Riutort, M. 2008. Back in time: a new systematic proposal for the Bilateria. *Phil. Trans. R. Soc. Lond. B* **363**: 1481–1491.

Beauchamp, P. 1929. Le développement des Gastrotriches. *Bull. Soc. Zool. Fr.* **54**: 549–558.

Dunn, C.W., Hejnol, A., Matus, D.Q., *et al.* 2008. Broad phylogenomic sampling improves resolution of the animal tree of life. *Nature* **452**: 745–749.

Giribet, G., Dunn, C.W., Edgecombe, G.D., *et al.* 2009. Assembling the spiralian tree of life. In M.J. Telford and D.T.J. Littlewood (eds): *Animal Evolution. Genomes, Fossils, and Trees*, pp. 52–64. Oxford University Press, Oxford.

Hejnol, A., Obst, M., Stamatakis, A., *et al.* 2009. Assessing the root of bilaterian animals with scalable phylogenomic methods. *Proc. R. Soc. Lond. B* **276**: 4261–4270.

Hochberg, R. 2007. Comparative immunohistochemistry of the cerebral ganglion in Gastrotricha: an analysis of FRMFamide-like immunoreactivity in *Neodasys cirritus* (Chaetonotida), *Xenodasys riedeli* and *Turbanella hyalina* (Macrodasida). *Zoomorphology* **126**: 245–264.

Hochberg, R. and Atherton, S. 2011. A new species of *Lepidodasys* (Gastrotricha, Macrodasyida) from Panama with a description of its peptidergic nervous system using CLSM, anti-FMRFamide and anti-SCP$_B$. *Zool. Anz.* **250**: 111-122

Hochberg, R. and Litvaitis, M. 2001. The muscular system of *Dactylopodola baltica* and other macrodasyidan gastrotrichs in a functional and phylogenetic perspective. *Zool. Scr.* **30**: 325–336.

Kieneke, A., Ahlrichs, W.H. and Arbizu, P.M. 2009. Morphology and function of reproductive organs in *Neodasys chaetonotoideus* (Gastrotricha: Neodasys) with a phylogenetic assessment of the reproductive system in Gastrotricha. *Zool. Scr.* **38**: 289–311.

Kieneke, A., Riemann, O. and Ahlrichs, W.H. 2008a. Novel implications for the basal internal relationships of Gastrotricha revealed by an analysis of morphological characters. *Zool. Scr.* **37**: 429–460.

Kieneke, A., Ahlrichs, W.H., Arbizu, P.M. and Bartolomaeus, T. 2008b. Ultrastructure of protonephridia in *Xenotrichula carolinensis* syltensis and *Chaetonotus maximus* (Gastrotricha: Chaetonotida): comparative evaluation of the gastrotrich excretory organs. *Zoomorphology* **127**: 1–20.

Leasi, F. and Todaro, M.A. 2008. The muscular system of *Musellifer delamarei* (Renaud-Mornant, 1968) and other chaetonotidans with implications for the phylogeny and systematization of the Paucitubulatina (Gastrotricha). *Biol. J. Linn. Soc.* **94**: 379–398.

Manylov, O.G. 1995. Regeneration in Gastrotricha - I. Light microscopical observations on the regeneration in *Turbanella* sp. *Acta Zool. (Stockh.)* **76**: 1–6.

Marotta, R., Guidi, L., Pierboni, L., *et al.* 2005. Sperm ultrastructure of *Macrodasys caudatus* (Gastrotricha: Macrodasyida) and a sperm-based phylogenetic analysis of Gastrotricha. *Meiofauna Mar.* **14**: 9–21.

Neuhaus, B., Kristensen, R.M. and Lemburg, C. 1996. Ultrastructure of the cuticle of the Nemathelminthes and electron microscopical localization of chitin. *Verh. Dtsch. Zool. Ges.* **89(1)**: 221.

Paps, J., Baguñá, J. and Riutort, M. 2009a. Lophotrochozoa internal phylogeny: new insights from an up-to-date analysis of nuclear ribosomal genes. *Proc. R. Soc. Lond. B* **276**: 1245–1254.

Paps, J., Baguñà, J. and Riutort, M. 2009b. Bilaterian phylogeny: A broad sampling of 13 nuclear genes provides a new Lophotrochozoa phylogeny and supports a paraphyletic basal Acoelomorpha. *Mol. Biol. Evol.* **26**: 2397–2406.

Passamaneck, Y. and Halanych, K.M. 2006. Lophotrochozoan phylogeny assessed with LSU and SSU data: evidence of lophophorate polyphyly. *Mol. Phylogenet. Evol.* **40**: 20–28.

Petrov, N.B., Pegova, A.N., Manylov, O.G., *et al.* 2007. Molecular phylogeny of Gastrotricha on the basis of a comparison of the 18S rRNA genes: rejection of hypothesis of relationship between Gastrotricha and Nematoda. *Mol. Biol.* **41**: 445–452.

Remane, A. 1926. Morphologie und Verwandtschaftsbeziehungen der aberranten Gastrotrichen I. *Z. Morphol. Oekol. Tiere* **5**: 625–754.

Rieger, R.M. 1976. Monociliated epidermal cells in Gastrotricha: Significance for concepts of early metazoan evolution. *Z. Zool. Syst. Evolutionsforsch.* **14**: 198–226.

Rothe, B.H. and Schmidt-Rhaesa, A. 2009. Architecture of the nervous system in two *Dactylopodola* species (Gastrotricha, Macrodasyida). *Zoomorphology* **128**: 227–246.

Ruppert, E.E. 1982. Comparative ultrastructure of the gastrotrich pharynx and the evolution of myoepithelial foreguts in Aschelminthes. *Zoomorphology* **99**: 181–220.

Ruppert, E.E. 1991. Gastrotricha. In F.W. Harrison (ed.): *Microscopic Anatomy of Invertebrates*, vol. 4, pp. 41–109. Wiley-Liss, New York.

Sacks, M. 1955. Observations on the embryology of an aquatic gastrotrich, *Lepidodermella squamata* (Dujardin, 1841). *J. Morphol.* **96**: 473–495.

Telford, M., Wise, M.J. and Gowri-Shankar, V. 2005. Consideration of RNA secondary structure significantly improves likelihood-based estimates of phylogeny: example from the Bilateria. *Mol. Biol. Evol.* **22**: 1129–1136.

Teuchert, G. 1968. Zur Fortpflanzung und Entwicklung der Macrodasyoidea (Gastrotricha). *Z. Morph. Tiere* **63**: 343–418.

Todaro, M.A., Telford, M.J., Lockyer, A.E. and Littlewood, D.T.J. 2006. Interrelationships of the Gastrotricha and their place among the Metazoa inferred from 18SrRNA genes. *Zool. Scr.* **35**: 251–259.

Wiedermann, A. 1995. Zur Ultrastruktur des Nervensystems bei *Cephalodasys maximus* (Macrodasyoida, Gastrotricha). *Microfauna Mar.* **10**: 173–233.

Witek, A., Herlyn, H., Ebersberger, I., Welch, D.B.M. and Hankeln, T. 2009. Support for the monophyletic origin of Gnathifera from phylogenomics. *Mol. Phylogenet. Evol.* **53**: 1037–1041.

31

GNATHIFERA

The name Gnathifera was intruduced by Ahlrichs (1995; 1997) for a group comprising Gnathostomulida, Rotifera (called Syndermata), and new group 'A' (Micrognathozoa) based on striking similarities of the jaw apparatus (mastax). This group, with Micrognathozoa and Rotifera as the sister groups (Fig. 31.1), has been supported by a number of morphological analyses (Kristensen and Funch 2000; Sørensen 2003; Funch et al. 2005). Molecular analyses have placed Rotifera and Micrognathozoa as sister groups (Paps et al. 2009a; Witek et al. 2009), and Rotifera and Gnathostomulida as sister groups (Hausdorf et al. 2010), but usually mixed them with the Gastrotricha when all three groups are included (Baguñá et al. 2008; Paps et al. 2009b). Only the anasysis of Zrzavý (2003) found good support for the Gnathifera. Morphological support for a relationship with the gastrotrichs could be the presence of multiple membranes surrounding the cilia in the pharynx of

certain rotifers, and in the ectoderm of the gastrotrichs (Clément 1993).

The most conspicuous synapomorphy of the three phyla is the mastax with the complicated jaws (Fig. 31.2). The single jaw elements are built of chitinous tubes with electron-lucent canals with a central electron-dense rod. Rotifers have an unusual internal skeletal lamina, and a similar structure is found in the dorsal epithelium of *Limnognathia*, indicating a sister-group relationship.

Gnathifera is placed in the Spiralia in all morphological and molecular analyses, but a more specific position is not strongly indicated. Several of the molecular analyses indicate close relationships with the Gastroticha or the Platyhelminthes, and some authors have united Gnathifera, Gastrotricha, Cycliophora, and Platyhelminthes, in a group called Platyzoa (Giribet 2008; Giribet et al. 2009), but

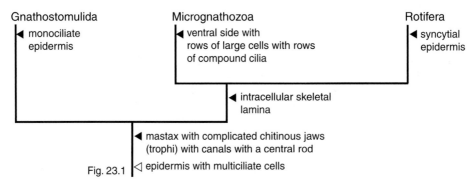

Figure 31.1. Phylogeny of the Gnathifera.

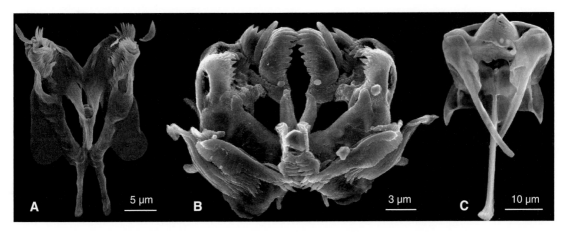

Figure 31.2. SEMs of jaws of Gnathiferan. (A) Gnathostomulida: *Gnathostomula armata*. (B) Micrognathozoa: *Limnognathia maerski*. (C) Rotifera: *Resticula nyssa*. Illustration courtesy of Dr M.V. Sørensen (University Copenhagen).

the early spiralian radiation must be characterized as unresolved.

References

Ahlrichs, W.H. 1995. *Ultrastruktur und Phylogenie von Seison nebaliae (Grube 1859) und Seison annulatus (Claus 1876)*. Dissertation, Georg-August-University, Göttingen, Cuvillier Verlag, Göttingen.

Ahlrichs, W.H. 1997. Epidermal ultrastructure of *Seison nebaliae* and *Seison annulatus*, and a comparison of epidermal structures within the Gnathifera. *Zoomorphology* **117**: 41–48.

Baguñá, J., Martinez, P., Paps, J. and Riutort, M. 2008. Back in time: a new systematic proposal for the Bilateria. *Phil. Trans. R. Soc. Lond. B* **363**: 1481–1491.

Clément, P. 1993. The phylogeny of rotifers: molecular, ultrastructural and behavioural data. *Hydrobiologia* **255/256**: 527–544.

Funch, P., Sørensen, M.V. and Obst, M. 2005. On the phylogenetic position of Rotifera – Have we come any further? *Hydrobiologia* **546**: 11–28.

Giribet, G. 2008. Assembling the lophotrochozoan (=spiralian) tree of life. *Phil. Trans. R. Soc. Lond. B.* **363**: 1513–1522.

Giribet, G., Dunn, C.W., Edgecombe, G.D., *et al.* 2009. Assembling the spiralian tree of life. In M.J. Telford and D.T.J. Littlewood (eds): *Animal Evolution. Genomes, Fossils, and Trees*, pp. 52–64. Oxford University Press, Oxford.

Hausdorf, B., Helmkampf, M., Nesnidal, M.P. and Bruchhaus, I. 2010. Phylogenetic relationships within the lophophorate lineages (Ectoprocta, Brachiopoda and Phoronida). *Mol. Phylogenet. Evol.* **55**: 1121–1127.

Kristensen, R.M. and Funch, P. 2000. Micrognathozoa: a new class with complicated jaws like those of Rotifera and Gnathostomulida. *J. Morphol.* **246**: 1–49.

Paps, J., Baguñá, J. and Riutort, M. 2009a. Lophotrochozoa internal phylogeny: new insights from an up-to-date analysis of nuclear ribosomal genes. *Proc. R. Soc. Lond. B* **276**: 1245–1254.

Paps, J., Baguñà, J. and Riutort, M. 2009b. Bilaterian phylogeny: A broad sampling of 13 nuclear genes provides a new Lophotrochozoa phylogeny and supports a paraphyletic basal Acoelomorpha. *Mol. Biol. Evol.* **26**: 2397–2406.

Sørensen, M.V. 2003. Further structures in the jaw apparatus of *Limnognathia maerski* (Micrognathozoa), with notes on the phylogeny of the Gnathifera. *J. Morphol.* **255**: 131–145.

Witek, A., Herlyn, H., Ebersberger, I., Welch, D.B.M. and Hankeln, T. 2009. Support for the monophyletic origin of Gnathifera from phylogenomics. *Mol. Phylogenet. Evol.* **53**: 1037–1041.

Zrzavý, J. 2003. Gastrotricha and metazoan phylogeny. *Zool. Scr.* **32**: 61–81.

Phylum Gnathostomulida

Gnathostomulida is a small phylum with about 100 described interstitial marine 'worms' that are mainly confined to detritus-rich sands; most species are microscopic, but a few reach sizes up to about 4 mm (Lammert 1991; Sterrer 1995). The group is most probably monophyletic, and the two orders, Filospermoida and Bursovaginoida, show differences in general body shape, reproductive organs, and sperm (Sørensen 2002; Sørensen *et al.* 2006). Most species glide on their cilia, and the gnathostomulids are unusual among the bilaterian interstitial organisms in that all epithelia are monociliate. However, some of the interstitial gastrotrichs have monociliate ectodermal cells, and this is interpreted as an apomorphy (Chapter 30). The microscopic anatomy was briefly reviewed by Lammert (1991).

The anteroventral mouth opens into a laterally compressed pharynx, with a ventral bulbus with striated mesodermal muscles and a cuticular jaw apparatus or mastax (Kristensen and Nørrevang 1977; Sterrer *et al.* 1985; Herlyn and Ehlers 1997; Sørensen *et al.* 2003). The basal parts of the jaws have a structure of electron-dense tubes surrounding an electron-lucent core with a central electron-dense rod (Sterrer *et al.* 1985; Rieger and Tyler 1995; Herlyn and Ehlers 1997). The gut consists of one layer of cells with microvilli but without cilia; there is no permanent anus, but the gut is in direct contact with the ectoderm in a small posterodorsal area

where the basal membrane is lacking, and this area may function as an anus (Knauss 1979).

The nervous system consists of a brain in front of the mouth, a small ganglion embedded in the mastax musculature, and one to three pairs of basiepithelial, longitudinal nerves (Müller and Sterrer 2004). There are several types of ciliary sense organs, one type is the 'spiral ciliary organs' that consist of one cell with a cilium spirally coiled in an interior cavity.

Longitudinal and circular striated body muscles are situated under the basement membrane; they function in body contraction but are not involved in locomotion. There is a row of separate protonephridia on each side of the body.

All species are hermaphrodites with separate testes and ovaries. The filospermoids have a simple copulatory organ, whereas some of the bursovaginoids have a more complicated copulatory organ with an intracellular stylet. The filospermoids have filiform sperm with a spirally coiled head and a long cilium, whereas the bursovaginoids have round or drop-shaped, aflagellate sperm.

Riedl (1969) described cleavage with a spiral pattern and two putative mesoblasts in *Gnathostomula*, but the following development is undescribed.

The phylogenetic position of gnathostomulids has been contentious, and relationships with almost all groups of 'worms' have been suggested (Giribet *et al.*

Chapter vignette: *Rastrognathia macrostoma*. (Based on Kristensen and Nørrevang 1977.)

2000). Ax (1987, 1995) forcefully argued for a sister-group relationship with the flatworms, especially based on the absence of an anus, which was regarded as the ancestral character within Bilateria. However, the absence of an anus is probably an apomorphy (see Chapter 21), and the gnathostomulid jaws are cuticular, whereas the jaws found in some turbellarians are formed from the basal membrane (Chapter 29). A sister-group relationship with gastrotrichs has been proposed on the basis of the monociliate epithelium (Rieger and Mainitz 1977), but the gastrotrich ectoderm and cilia are covered by a unique lamellar exocuticle (Chapter 30).

The cuticular jaws resemble those of the rotifers both in general shape and ultrastructure, and in their position on a ventral bulbus with striated muscles and an embedded ganglion. Ahlrichs (1995, 1997) proposed the name Gnathifera for a group consisting of gnathostomulids and rotifers (including acanthocephalans), and the newly discovered Micrognathozoa has been added subsequently (Kristensen and Funch 2000; Sørensen 2003; Funch *et al.* 2005). Molecular phylogeny has not resolved this complex convincingly (see Chapter 30).

Interesting subjects for future research

1. Cleavage and cell lineage

References

Ahlrichs, W.H. 1995. *Ultrastruktur und Phylogenie von Seison nebaliae (Grube 1859) und Seison annulatus (Claus 1876)*. Dissertation, Georg-August-University, Göttingen, Cuvillier Verlag, Göttingen.

Ahlrichs, W.H. 1997. Epidermal ultrastructure of *Seison nebaliae* and *Seison annulatus*, and a comparison of epidermal structures within the Gnathifera. *Zoomorphology* **117**: 41–48.

Ax, P. 1987. *The Phylogenetic System. The Systematization of Organisms on the Basis of their Phylogenesis*. John Wiley, Chichester.

Ax, P. 1995. *Das System der Metazoa I*. Gustav Fischer, Stuttgart.

Funch, P., Sørensen, M.V. and Obst, M. 2005. On the phylogenetic position of Rotifera—Have we come any further? *Hydrobiologia* **546**: 11–28.

Giribet, G., Distel, D.L., Polz, M., Sterrer, W. and Wheeler, W.C. 2000. Triploblastic relationships with emphasis on the acoelomates and the position of Gnathostomulida, Cycliophora, Plathelminthes, and Chaetognatha: a combined approach of 18S rDNA sequences and morphology. *Syst. Biol.* **49**: 539–562.

Herlyn, H. and Ehlers, U. 1997. Ultrastructure and function of the pharynx of *Gnathostomula paradoxa* (Gnathostomulida). *Zoomorphology* **117**: 135–145.

Knauss, E.B. 1979. Indication of an anal pore in Gnathostomulida. *Zool. Scr.* **8**: 181–186.

Kristensen, R.M. and Funch, P. 2000. Micrognathozoa: a new class with complicated jaws like those of Rotifera and Gnathostomulida. *J. Morphol.* **246**: 1–49.

Kristensen, R.M. and Nørrevang, A. 1977. On the fine structure of *Rastrognathia macrostoma* gen. et sp. n. placed in Rastrognathiidae fam. n. (Gnathostomulida). *Zool. Scr.* **6**: 27–41.

Lammert, V. 1991. Gnathostomulida. In F.W. Harrison (ed.): *Microscopic Anatomy of Invertebrates*, vol. 4, pp. 19–39. Wiley-Liss, New York.

Müller, M.C.M. and Sterrer, W. 2004. Musculature and nervous system of *Gnathostomula peregrina* (Gnathostomulida) shown by phalloidin labeling, immunohistochemistry, and cLSM, and their phylogenetic significance. *Zoomorphology* **123**: 169–177.

Riedl, R.J. 1969. Gnathostomulida from America. *Science* **163**: 445–462.

Rieger, R.M. and Mainitz, M. 1977. Comparative fine structure study of the body wall in Gnathostomulida and their phylogenetic position between Platyhelminthes and Aschelminthes. *Z. Zool. Syst. Evolutionsforsch.* **15**: 9–35.

Rieger, R.M. and Tyler, S. 1995. Sister-group relationships of Gnathostomulida and Rotifera-Acanthocephala. *Invert. Biol.* **114**: 186–188.

Sørensen, M.V. 2002. Phylogeny and jaw evolution in Gnathostomulida, with a cladistic analysis of the genera. *Zool. Scr.* **31**: 461–480.

Sørensen, M.V. 2003. Further structures in the jaw apparatus of *Limnognathia maerski* (Micrognathozoa), with notes on the phylogeny of the Gnathifera. *J. Morphol.* **255**: 131–145.

Sørensen, M.V., Sterrer, W. and Giribet, G. 2006. Gnathostomulid phylogeny inferred from a combined approach of four molecular loci and morphology. *Cladistics* **22**: 32–58.

Sørensen, M.V., Tyler, S., Hoge, M.D. and Funch, P. 2003. Organization of pharyngeal hard parts and musculature in *Gnathostomula armata* (Gnathostomulida: Gnathostomulidae). *Can. J. Zool.* **81**: 1463–1470.

Sterrer, W. 1995. Gnathostomulida, Kiefermäulchen. In W. Westheide and R. Rieger (eds): *Spezielle Zoologie, Teil 1: Einzeller und Wirbellose Tiere*, pp. 259–264. Gustav Fischer, Stuttgart.

Sterrer, W., Mainitz, M. and Rieger, R.M. 1985. Gnathostomulida: enigmatic as ever. In S. Conway Morris, J.D. George, R. Gibson and H.M. Platt (eds): *The Origins and Relationships of Lower Invertebrates*, pp. 181–199. Oxford University Press, Oxford.

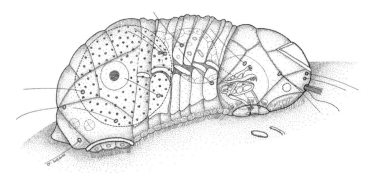

Phylum **Micrognathozoa**

This newly described phylum still comprises only one species: the microscopic *Limnognathia maerski*. It was first discovered in a cold spring at Disko Island, Greenland (Kristensen and Funch 2000), and has subsequently been reported only from the subantarctic Crozet Island (De Smet 2002).

The clumsy little worm is cylindrical with a rounded head and abdomen and a mid-body with a number of transverse wrinkles. All tissues are cellular. The dorsal and lateral epithelium shows a number of thin plates formed by one to four cells separated by thinner membranes; they show a thin intracellular skeletal lamina. The ventral surface has anterior ciliated fields, a double row of 18 large cells called ciliophores, each with four transverse rows of compound cilia, and a posterior pad of 10 smaller ciliophores.

The ventral mouth opening leads into a pharynx with a muscular mastax with complicated jaws (Fig. 31.2B). The gut is a simple tube that narrows posteriorly. There is no permanent anal opening, but a temporal opening may be formed at the narrow posterior dorsal end of the gut where the basal membrane of the ectoderm is lacking.

Two pairs of protonephridia consist of seven monociliate cells, four terminal cells, two canal cells, and one pore cell.

There is a large, slightly bilobate cerebral ganglion and a pair of lateroventral longitudinal nerves.

Only female specimens have been observed. The paired ovaries appear to consist of naked oocytes. Two types of eggs have been observed; smaller females lay smooth, sticky eggs and larger females lay highly sculptured eggs.

The presence of the unusual internal skeletal lamina strongly indicates a sister-group relationship with the rotifers, and the complicated jaws are very similar to those of both gnathostomulids and the free-living rotifers, so the position within the Gnathifera seems well established (Fig. 31.1).

The molecular analyses are discussed in Chapter 31.

References

De Smet, W.H. 2002. A new record of *Limnognathia maerski* Kristensen & Funch, 2000 (Micrognathozoa) from the subantarctic Crozet Islands, with redescription of the trophi. *J. Zool.* **258**: 381–393.

Kristensen, R.M. and Funch, P. 1995. En ny aschelminth med gnathostomulid-lignende kæber fra en kold kilde ved Isunnga. In C. Erhardt (ed.): *Arktisk Biologisk Feltkursus, Qeqertarsuaq/Godhavn 1994*, pp. 73–83. University of Copenhagen, Copenhagen.

Kristensen, R.M. and Funch, P. 2000. Micrognathozoa: a new class with complicated jaws like those of Rotifera and Gnathostomulida. *J. Morphol.* **246**: 1–49.

Chapter vignette: *Limnognathia maerski*. (From Kristensen and Funch 1995.)

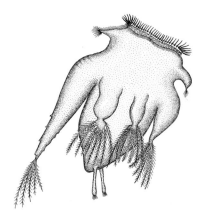

Phylum **Rotifera**

Rotifera (including Acanthocephala) consists of about 2000 described species of free-living, usually less than a millimetre-long, aquatic, mostly limnic organisms and a group of about 900 aquatic or terrestrial, completely gutless parasites (Acanthocephala), with the juveniles occurring in arthropods and the 2-mm to almost 1-metre-long adults living in the alimentary canal of vertebrates. The free-living types have direct development, whereas the acanthocephalans have complicated life cycles with more than one host.

Many of the free-living types can be recognized by the ciliary 'wheel organ' or corona that has given name to the phylum, but it is highly modified or completely absent in others. The anterior end, with the wheel organ in the free-living forms, and the proboscis of the acanthocephalans can be retracted into the main body. Four main groups are recognized: Monogononta (with parthenogenetic phases and sexual phases with small haploid males), Bdelloidea (only parthenogenetic females without meiosis), Seisonidea (with similar males and females), and Acanthocephala (dioecious, highly specialized without gut). The acanthocephalans were earlier regarded as a separate phylum, but the ultrastructure of the epidermis points to them being a sister group (or an in-group) of one of the free-living groups. The interrelationships between the four groups must be characterized as unresolved (see below), but the monophyly of the group now seems unquestioned. The name Syndermata has been used for the traditional rotifer groups plus the acanthocephalans by a number of German authors (for example Ahlrichs 1997; Ax 2001; Herlyn *et al.* 2003; Witek *et al.* 2008), but it appears completely unnecessary to introduce a new name for the group, just because the acanthocephalans have turned out to be an ingroup.

The microscopic anatomy was reviewed by Clément and Wurdak (1991) and Clément (1993) (free-living forms), and by Dunagan and Miller (1991) (acanthocephalans).

Monogononts have a single ovary; their life cycles are complicated with parthenogenetic generations of females producing diploid eggs and sexual generations of females that produce haploid eggs; non-fertilized eggs develop into haploid males and fertilized eggs become resting eggs. The males are much smaller than the females and lack the gut in most species. Several types have a wheel organ (Fig. 34.1) that is a typical protostomian downstream-collecting ciliary system with prototroch (called trochus), adoral ciliary zone, and metatroch (called cingulum), but others are strongly modified.

Bdelloids are parthenogenetic with paired ovaries. The evolution of a large clade without sexual reproduction is unique in the Metazoa (Welch and Meselson 2000). A number of karyological and ecological

Chapter vignette: *Hexarthra mira*. (Redrawn from Wesenberg-Lund 1952.)

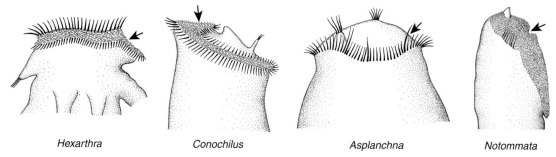

Hexarthra   Conochilus   Asplanchna   Notommata

Figure 34.1. Various types of ciliary bands in rotifers. The planktotrophic types have the trochophore type of ciliary bands (prototroch + adoral ciliary zone + metatroch). The pelagic, solitary *Hexarthra mira* has the ciliary bands of a trochophore in the unspecialized shape; the pelagic, colonial *Conochilus unicornis* has similar ciliary bands, only with the lateral parts bent to the ventral side; the carnivorous, pelagic *Asplanchna girodi* has only the prototroch; the benthic, carnivorous *Notommata pseudocerebrus* mostly creeps on the extended adoral ciliary zone, but occasionally swims with a few prominent groups of compound cilia that appear to be specialized parts of the prototroch. The arrows point at the mouth. (Redrawn from Beauchamp 1965; Nielsen 1987.)

peculiarities have been observed (Gladyshev and Meselson 2008; Welch *et al.* 2008). The integument typically forms 16 slightly thickened rings that can telescope when the animals contract. Some forms have a ciliary system, with the prototroch divided into a pair of trochal discs, an adoral ciliary zone, and a metatroch, whereas others have a field of uniform cilia, probably an extended adoral zone, around the mouth.

Seisonidea, with the only genus *Seison* with two species, are epibionts on the crustacean *Nebalia*. Males and females are similar, with the wheel organ reduced to a small ciliated field with short lateral rows of compound cilia.

Adult acanthocephalans are cylindrical or slightly flattened with a retractile, cylindrical proboscis with recurved hooks, ideal for anchoring the parasite to the intestinal wall of the host; many species have rings of smaller or larger spines or hooks on the whole body. There is no trace of an alimentary canal at any stage, and it cannot be seen directly if the proboscis represents the anterior part of the body with a reduced, terminal mouth or a dorsal attachment organ (see below). Also the dorsal-ventral orientation has been questioned. Their eggs are fertilized and develop into the acanthor stage before the eggs are shed and leave the host with the faeces. When ingested by an intermediate host the acanthor hatches in the intestine and

enters the intestinal wall; here it develops into the acanthella stage and further into the cystacanth, which is the stage capable of infecting the final host.

Many tissues of rotifers are syncytial, but the number of nuclei in most organs is nevertheless constant, and divisions do not occur after hatching; this implies that the power of regeneration is almost absent.

The body epithelium of the free-living types has a usually very thin extracellular cuticle probably consisting of glycoproteins and an intracellular skeletal lamina (sometimes referred to as an intracellular cuticle) apposed to the inner side of the apical cell membrane. This intracellular lamina may be of different thickness in various parts of the body and in different species; homogeneous in *Asplanchna*, lamellate in *Notommata*, and with a honeycomb-like structure in *Brachionus*; it has characteristic pores with drop-shaped invaginations of the cell membrane, and the general structure is identical in all the free-living groups (Ahlrichs 1997). The intracellular skeletal lamina consist of intermediate filaments of a scleroprotein of the keratine-type; chitin has not been found.

The acanthocephalan body wall consists of a syncytial ectoderm, a thick basal membrane, an outer layer of circular muscles, and an inner layer of longitudinal muscles; a rete system of tubular, anastomosing cells with lacunar canals is found on the inner side of the longitudinal muscles in *Macracanthorhynchus* and between the two muscle layers in *Oligacanthorhynchus*. The ectoderm or tegument has very few gigantic nuclei with fixed positions. The apical cell membrane shows numerous branched, tubular invaginations that penetrate an intracellular skeletal lamina consisting of a thin, outer, electron-dense layer and a thicker, somewhat less electron-dense layer.

The ectoderm of the ciliary bands found in bdelloids and monogononts consists of large cells with several nuclei and connected by various types of cell junctions, whereas the ectoderm of the main body region is a thin syncytium. All ciliated epithelial cells are multiciliate. The ectoderm of the ciliary bands has the usual surface structure with microvilli and a layer of normal, extracellular cuticle between the tips of the microvilli. Also some of the sensory organs have this type of cuticle. The buccal epithelium appears to lack a cuticle and the cilia have modified, electron-dense tips. The pharyngeal epithelium has multiple layers of double membranes that also cover the cilia. The borderline between these two epithelia marks the origin of a flattened, funnel-shaped structure called the velum that consists of two thick layers of parallel membranes lining a ring of long cilia with somewhat blown-up cell membranes. The muscular mastax carries a complicated system of cuticular jaws (trophi), which are thickened parts of a continuous membrane with more than 50% chitin in *Brachionus* (Klusemann *et al.* 1990). The whole structure is extracellular; the several reports of intracellular mastax structures are probably erroneous. The jaws have a tubular structure with basal, electron-lucent canals surrounding a cytoplasmic core (Ahlrichs 1995; Rieger and Tyler 1995). The jaws contain important systematic information (Sørensen 2002). The conspicuous hooks of larval and adult acanthocephalans are outgrowths from the connective tissue and contain chitin (Taraschewski 2000). This shows that the hooks are not homologous with the mastax.

The wheel organ shows an enormous variation (Figs. 22.4-5, 34.1). Some creeping types, such as *Dicranophorus* have a ventral, circumoral zone of single cilia used in creeping; predatory, planktonic forms, such as *Asplanchna*, have a pre-oral, almost-complete ring of compound cilia used in swimming; planktotrophic forms that may be planktonic or sessile, such as *Hexarthra*, *Conochilus*, and *Floscularia*, have an adoral zone of single cilia bordered by a pre-oral prototroch and a post-oral metatroch, with the whole ciliary system surrounding the apical field; many other variations are found, and *Acyclus* and *Cupelopagis* lack the corona in the adult stage (Beauchamp 1965). Proto- and metatroch consist of compound cilia, and the whole complex is a downstream-collecting system (Strathmann *et al.* 1972).

Particles captured by the corona are transported through the ciliated buccal tube to the mastax with the jaws (see above). The macrophagous species can protrude the jaws from the mouth and grasp algal filaments or prey. The movements of the jaws are coordinated by the mastax ganglion, which receives input from ciliated sense organs at the bottom of its lumen and from the brain. Various types of jaws are characteristic of larger systematic groups and are correlated with feeding behaviour. A partly ciliated oesophagus leads to the stomach, which is syncytial and without cilia in bdelloids and cellular with cilia in monogononts. There is a ciliated intestine opening into a short cloaca and a dorsal anus. A few genera lack the intestine, so only the protonephridia and the genital organs open into the cloaca.

The nervous system generally comprises a dorsal brain, a mastax ganglion, a pedal ganglion associated with a pair of toes ventral to the rectum, and a number of peripheral nerves with various types of cells. The brain comprises about 150 to 250 cells with species-specific numbers (Nachtwey 1925; Peters 1931). The monogononts *Notommata* and *Asplanchna* (Hochberg 2007, 2009) have rather similar brains (with about 28 identifiable serotonergic cells in *Asplanchna*); the planktonic *Asplanchna* lacks the toes and the pedal ganglion. The bdelloid *Macrotrachela* (Leasi *et al.* 2009) shows a similar general brain morphology, but with different numbers and positions of the neurons. A pair of lateroventral nerves connect the lateroposterior parts of the brain with the pedal ganglion. Photoreceptors of a number of different types, such as the phaosomes with peculiar expanded cilia, are found embedded in the brain of many species. The pedal ganglion is usually associated with the feet and the cloaca, but separate ganglia for the two regions are found in some species (Remane 1929–1933). One or a pair of dorsal antennae and a pair of lateral antennae are small sensory organs comprising one or a few primary sensory cells with a tuft of cilia. Each transverse muscle is innervated by one or two large nerve cells, which gives a superficial impression of segmentation (Zelinka 1888; Stossberg 1932).

The acanthocephalan brain comprises a low, species-specific number of cells. Nerves have been tracked to the muscles of the body wall and the proboscis, to paired genital ganglia, to a pair of sense organs at the base of the proboscis, and to a pair of sensory and glandular structures, called the apical organ, at the tip of the proboscis (Gee 1988); both structure and function of the apical organ are in need of further investigations based on a number of species before definite statements about its homology to other apical organs can be made.

Almost all the muscles of the free-living forms are narrow bands with one nucleus. They attach to the body wall through an epithelial cell with hemidesmosomes and tonofibrils. The two large retractor muscles of the corona are coupled to other muscles through gap junctions and send a cytoplasmic extension to the brain where synapses occur; other muscles are innervated by axons from the ganglia. Bdelloids have the body wall divided into a series of rings, and both the anterior and posterior end can be telescoped into the middle rings; there are one or two annular muscles in each ring and longitudinal muscles between neighbouring rings or extending over two to three rings (Zelinka 1886, 1888). There is practically no connective tissue, and collagen genes are absent in *Brachionus* (Suga *et al.* 2007). The spacious body cavity functions as a hydrostatic skeleton in protrusion of the corona.

The proboscis of the acanthocephalans has several associated sets of muscles that are involved in protrusion, eversion, and retraction (Taraschewski 2000). The proboscis region can be protruded by the muscles of the body wall and retracted by the neck-retractor muscles that surround the lemnisci and attach to the body wall. The inverted proboscis lies in a receptacle that has a single or double wall of muscles. The contraction of these muscles everts the proboscis, with the receptacle fluid functioning as the hydrostatic skeleton, and a contraction of the neck retractors squeezes fluid from the lemnisci to the wall of the proboscis that swells. A retractor muscle from the tip of the proboscis to the bottom of the receptacle inverts the proboscis, and the receptacle can be retracted further into the body by the contraction of the receptacle retractor that extends from the bottom of the receptacle to the ventral body wall. There is a spacious body cavity, which functions as a hydrostatic organ. It contains an enigmatic organ called the ligament sac(s), which develops in all types but degenerates in some forms. The ligament sac(s) and the gonads develop in the acanthella from a central mass of cells between the brain and the cloaca, and it is generally believed that a median string, called the ligament, represents endoderm. There is either a single, or a dorsal and a ventral sac that communicate anteriorly. The sacs are acellular, fibrillar structures that contain collagen (Haffner 1942). The posterior end of the (dorsal) ligament sac is connected with the uterine bell (see below).

There is a paired-protonephridial system (Riemann and Ahlrichs 2010), with one to many flame cells. Monogononts have large terminal cells with a filtering weir of longitudinal slits supported by internal pillars, the bdelloids have similar but smaller cells and lack pillars, and *Seison* has a weir with longitudinal spiral rows of pores and lack pillars. Among the acanthocephalans, only the Oligacanthorhynchidae have protonephridia. Each protonephridium is a syncytium with three nuclei situated centrally and many radiating flame bulbs with high numbers of cilia. An unpaired, ciliated excretory canal opens into the urogenital canal.

Female monogononts have an unpaired, sac-shaped germovitellarium, and the males have a single testis; both types of gonads open into the cloaca. The ovaries contain a number of oocytes, which is fixed at birth. The sperm has an elongate head with the axoneme following the nucleus in the posterior part; the tail contains the axoneme with the cell membrane expanded laterally into a longitudinal undulating membrane with a supporting structure (Melone and Ferraguti 1994, 1999). The bdelloids have paired germovitellaria. The parthenogenetic eggs become surrounded by a chitinous shell secreted by the embryo. Females of *Seison* have paired ovaria without vitellaria and the males have paired testes with a common sperm duct. The sperm superficially resembles that of the monogonents, but the cilium is free from the elongate nucleus and situated

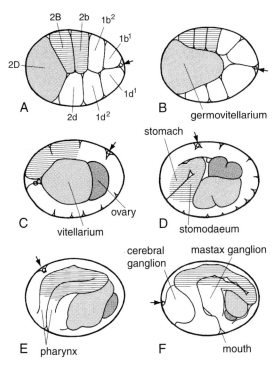

Figure 34.2. Early development of *Asplanchna girodi*; median sections with the polar bodies (apical pole) indicated by a thick arrow. (A) 16-Cell stage. (B) Internalization of the 4D cell through an epibolic gastrulation. (C) The germovitellarium is completely internalized and divided into the primordial cells of the ovary and the vitellarium; a small blastopore is formed through further gastrulation movements. (D) Gastrulation continues from the dorsal and lateral sides of the blastopore, forming the endodermal stomach, and the apical pole (indicated by the polar bodies) moves along the dorsal side. (E) Further gastrulation movements from the whole area around the blastopore give rise to the inner part of the pharynx. (F) The pharynx is now fully internalized and the mastax ganglion differentiates from its ventral side; the brain has become differentiated from the ectoderm at the apical pole. (Modified from Lechner 1966.)

in a groove on the nucleus at the whole length from the anterior basal body, and there is a row of peculiar 'dense bodies' in a double row along the elongate nucleus (Ahlrichs 1998). Acanthocephalans have their gonads suspended by the ligament strand. The testes have ducts that open into a urogenital canal, which in turn opens on the tip of a small penis at the bottom of a bursa copulatrix. The sperm resembles that of *Seison* in that the cilium is situated in a groove along the elongate nucleus, but the anterior end of the axoneme with the basal body forms a long anterior ciliary structure, so that it looks as if the sperm is swimming 'in the wrong direction' (Foata *et al.* 2004). The axoneme has zero to three central microtubuli (Carcupino and Dezfuli 1999). The male injects the sperm into the uterus, and the fertilized eggs become surrounded by an oval, resistant, chitin and keratin-containing shell with a number of layers (Peters *et al.* 1991).

Studies on monogonont development have centred on the pelagic genus *Asplanchna* (Lechner 1966), with additional observations on *Ploesoma* (Beauchamp 1956) and *Lecane* (Pray 1965). Lechner (1966) reinterpreted some of the reports on the early development (Fig. 34.2 and Table 34.1) and Nachtwey (1925) described organogenesis. The cleavage is total and unequal and the 4-cell stage has three smaller A–C blastomeres and a large D blastomere; the polar bodies are situated at the apical pole. The D cell divides unequally, and its large descendant (1D) comes to occupy the blastoporal pole, while the smaller descendant (1d) and the A–C cells form an apical ring. The 1D macromere gives off another small cell, and all the other blastomeres divide equally, with the spindles parallel to the primary axis. The embryo now consists of four rows of cells, with the large 2D cell occupying the blastoporal pole. The smaller cells divide further and slide along the macromere that becomes internalized in an epibolic gastrulation; the movements continue as an invagination, forming an archenteron, where it appears that the stomach originates either from a–c cells or exclusively from b cells, and the pharynx from all four quadrants. The 2D cell gives off two abortive micromeres and the 4D cell gives rise to the germovitellarium. The stronger gastrulation movement of the dorsal side (b cells)

Table 34.1. Cell lineage of *Asplanchna girodi*. The original notation is given in parentheses. (Modified from Lechner 1966.)

$$
Z
\begin{cases}
AB
\begin{cases}
A\ (A_3)\ \ldots\ \text{lateral ectoderm, endoderm}\\[4pt]
B\ (B_3)
\begin{cases}
1b\ (b4,2)\ \ldots\ \text{frontal ectoderm, cerebral ganglion}\\[4pt]
1B\ (b4,1)\ \ldots\ \text{endoderm (stomach, stomach glands)}
\end{cases}
\end{cases}\\[20pt]
CD
\begin{cases}
C\ (C_3)\ \ldots\ \text{lateral ectoderm, endoderm}\\[4pt]
D\ (D_3)
\begin{cases}
1d\ (d4,2)\ \ldots\ \text{dorsal and ventral ectoderm}\\[4pt]
1D\ (d4,1)
\begin{cases}
2d\ (d5,2)\ \ldots\ \text{ventral ectoderm, nephridia, bladder, uterus, cloaca}\\[4pt]
2D\ (d5,1)
\begin{cases}
3d\ (d6,2)\ \ldots\ \text{apoptosis}\\[4pt]
3D\ (6,1)
\begin{cases}
4d\ (d7,2)\ \ldots\ \text{apoptosis}\\[4pt]
4D\ (d7,1)
\begin{cases}
5d\ (d8,1)\ \ldots\ \text{vitellarium}\\[4pt]
5D\ (d8,2)
\begin{cases}
6d\ (d9,3)\ \ldots\ \text{apoptosis}\\[4pt]
6D\ (d9,4)\ \ldots\ \text{germ cells}
\end{cases}
\end{cases}
\end{cases}
\end{cases}
\end{cases}
\end{cases}
\end{cases}
$$

moves the apical pole with the polar bodies towards the blastopore, so that the cells of the D quadrant finally cover almost the whole dorsal and ventral side. The small ectodermal cells of the apical region multiply and differentiate into the cerebral ganglion, which finally sinks in and becomes overgrown by the surrounding ectoderm. The mastax ganglion differentiates from the epithelium of the posterior (ventral) side of the pharynx, and the caudal ganglion differentiates from the ectoderm behind the blastopore/mouth. A small caudal appendix, perhaps with a pair of rudimentary toes (Car 1899), develops at an early stage but disappears in the adult *Asplanchna*. Protonephridia, bladder, oviduct, and cloaca develop from the 2d cell. The origin of the ciliary bands and mesoderm is poorly known; muscles of the body wall have been reported to differentiate from ectodermal cells (Nachtwey 1925), but this should be studied with modern methods.

The eggs are highly determined already before the polar bodies are given off, and the powers of regulation are very limited (Lechner 1966).

The cleavage pattern shows no sign of a spiral arrangement of the blastomeres, but as far as the cell lineage is known, the cleavage is clearly of the D-quadrant type (Fig. 21.1). The cerebral ganglion develops from cells near the apical pole, and the lack

of a larval stage may have caused a loss of a ciliated apical ganglion. The ciliary bands have the very same structure and function as those of larvae and adults of spiralians, such as annelids and molluscs (Fig. 22.4-5), but the cell lineage has not been studied.

Bdelloid embryology is poorly known; there is no meiosis. Zelinka (1891) studied the development of *Callidina* (now *Mniobia*). The embryos become curved and the report is difficult to follow in detail, but the development appears to resemble that of the monogononts.

The development of *Seison* has not been studied.

Acanthocephalan embryology has been studied by a number of authors (Schmidt 1985). The polar bodies are situated at one pole of the ellipsoidal egg and mark the future anterior end. The first two cleavages result in an embryo with one anterior (B), two median (A and C), and one posterior (D) cell; the blastomeres are usually of equal size, but the posterior cell is larger than the others in a few species. The embryo becomes syncytial at a stage of 4–36 cells according to the species. Meyer (1928, 1932–1933, 1936, 1938) elegantly followed the cell lineage (or rather the nuclear lineage) of *Macracanthorhynchus* (Fig. 34.3), and reported a cleavage with a primary axis slightly oblique to the longitudinal axis of the egg; the A and C cells of the

4-cell stage are in contact along the whole primary axis, and the spindles of the following cleavages are almost parallel. After the 34-cell stage, the cells begin to fuse, soon forming one large syncytium, and the cleavage pattern and the movements of the nuclei become difficult to follow. At the stage of 163 nuclei, small inner nuclei of the ganglion and of the muscula-

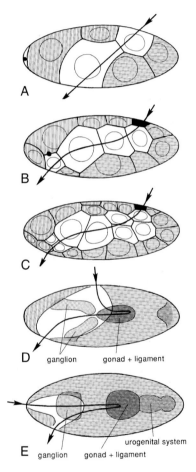

Figure 34.3. Embryology of *Macracanthorhynchus hirudinaceus*; embryos seen from the left side. The three first stages are cellular and the two latest syncytial. (A) 8-cell stage. (B) 17-cell stage. (C) 34-cell stage, the last stage with a regular cell pattern. (D) Early stage of internalization of the condensed nuclei of the inner organs and of the movement of the apical pole. (E) Stage with fully organized primordia of the inner organs and with the apical pole at the anterior end. A quadrant, white; B quadrant, vertically hatched; C quadrant, black; D quadrant, horizontally hatched; inner primordia are shown by shading. (Redrawn from Meyer 1928, 1938.)

ture can be recognized. Soon after, the proboscis forms from an anterior invagination, and the urogenital system forms from posterior ingression of cells. The origin of the gonads plus the ligament seems uncertain. Concomitantly, the external areas of the embryo make differential growth, so that the D quadrant extends dorsally all the way to the anterior pole, this results in a strongly bent egg axis, resembling that observed in *Asplanchna* (Fig. 34.2). A ring of spines or hooks with associated myofibrils develop in the anterior end, and the acanthor larva is ready for hatching. After entering the first host, the acanthor loses the hooks, their associated muscles degenerate, and the early acanthella stage is reached. The various organ systems differentiate from the groups of nuclei seen already in the acanthor stage (Hamann 1891; Meyer 1932–1933, 1938), but the details of organogenesis have not been studied. Most tissues remain syncytial, but the nervous system and the muscles become cellular. The lemnisci develop as a pair of long, syncytial protrusions from the ectoderm around the proboscis invagination (Hamann 1891); in *Macracanthorhynchus*, a ring of 12 very large nuclei migrate into the early, cytoplasmic protrusions (Meyer 1938). The proboscis apparatus is at first enclosed by the syncytial ectoderm, but an opening is formed, and the larva is now in the cystacanth stage, which has almost the adult morphology and is ready for infection of the final host.

The monophyly of a group comprising the free-living rotifers and the acanthocephalans is supported by the presence of the unique epidermis with the intracellular skeletal lamina, and molecular studies almost unanimously support monophyly (Giribet *et al.* 2004; García-Varela and Nadler 2006; Sørensen and Giribet 2006; Witek *et al.* 2008). However, almost every possible phylogenetic hypothesis for interrelationships of the four groups has been proposed. They are probably all monophyletic, and a number of molecular studies favour a sister-group relationship between Bdelloidea and Acanthocephala (Giribet *et al.* 2004; García-Varela and Nadler 2006). I have chosen to retain the name Rotifera for the whole group, and to treat the four subgroups separately pending additional information.

Conway Morris and Crompton (1982) found so many similarities between the Burgess Shale priapulan *Ancalagon* and the living acanthocephalans that they considered priapulans and acanthocephalans as sister groups. The overall resemblance between the two groups is considerable, but the intracellular nature of both the 'cuticle' and the spines on the 'proboscis' in acanthocephalans is in strong contrast to the true cuticular structure of these organs in priapulans, which demonstrates that the resemblance is completely superficial.

Already Hatschek (1878, 1891) stressed the similarities of the ciliary bands of rotifers and trochophora larvae of annelids and molluscs and proposed that the common ancestor of these groups had a larva of this type. Lang (1888) proposed that the rotifers are neotenic, i.e. sexually mature trochophores. However, the idea of ancestral trochophore-like ciliary bands in rotifers fell into disregard when Beauchamp (1907, 1909) published his comparative studies on the ciliary bands of several rotifers. His conclusions were that the types with the trochophore-type ciliation have evolved several times from an ancestral type with a circumoral ciliary field used in creeping, and that the rotifer ciliation could be derived from the general ciliation of a flatworm via the ventral ciliation of the gastrotrichs. Jägersten (1972) hesitantly supported the old idea that the rotifers have the trochophore ciliation and that this is an 'original larval feature', and this was also favoured by Clément (1993).

I believe that the rotiferan wheel organ, with proto- and metatroch of compound cilia bordering an adoral zone of single cilia and functioning as a downstream-collecting system, is homologous with the similar bands of the trochophores of annelids, molluscs, and entoprocts. The various other types of wheel organs can be interpreted as adaptations to other feeding types, and the parasitic acanthocephalans are highly derived. The trochophore is definitely a larval form (Chapter 22) and the rotifers must therefore be interpreted as neotenic—not as neotenic annelids, but as neotenic descendants of the protostomian ancestor, gastroneuron. The planktotrophic rotifers must therefore represent the ancestral type, which have become temporarily or permanently attached; sessile forms have planktonic juvenile stages, and changes between pelagic and sessile habits may have taken place several times; the macrophagous types, which may be pelagic or creeping, have reduced ciliary bands and must be regarded as specialized.

A close relationship with the micrognathozoans and gnathostomulids is indicated both by the general structure of the mastax jaws (Fig. 31.2) and by the ultrastructure of the jaws that consists of parallel cuticular tubules with a dense core.

The molecular analyses are discussed further in Chapter 31.

Interesting subjects for future research

1. Cell lineage of a species with prototroch—traits of spiral cleavage
2. Hox genes of species with and without anus
3. Embryology of *Seison* and bdelloids

References

Ahlrichs, W.H. 1995. *Ultrastruktur und Phylogenie von Seison nebaliae (Grube 1859) und Seison annulatus (Claus 1876)*. Dissertation, Georg-August-University, Göttingen, Cuvillier Verlag, Göttingen.

Ahlrichs, W.H. 1997. Epidermal ultrastructure of *Seison nebaliae* and *Seison annulatus*, and a comparison of epidermal structures within the Gnathifera. *Zoomorphology* **117**: 41–48.

Ahlrichs, W.H. 1998. Spermatogenesis and ultrastructure of the spermatozoa of *Seison nebaliae* (Syndermata). *Zoomorphology* **118**: 255–261.

Ax, P. 2001. *Das System der Metazoa III*. Spektrum, Heidelberg.

Beauchamp, P. 1907. Morphologie et variations de l'apparail rotateur dans la série des Rotifères. *Arch. Zool. Exp. Gen.*, 4. Ser. **6**: 1–29.

Beauchamp, P. 1909. Recherches sur les Rotifères: les formations tégumentaires et l'appareil digestif. *Arch. Zool. Exp. Gen.*, 4. Ser. **10**: 1–410.

Beauchamp, P. 1956. Le développement de *Ploesoma hudsoni* (Imhof) et l'origine des feuillets chez les Rotifères. *Bull. Soc. Zool. Fr.* **81**: 374–383.

Beauchamp, P. 1965. Classe des Rotifères. *Traité de Zoologie*, vol. 4(3), pp. 1225–1379. Masson, Paris.

Car, L. 1899. Die embryonale Entwicklung von *Asplanchna brightwellii. Biol. Zentbl.* **19**: 59–74.

Carcupino, M. and Dezfuli, B.S. 1999. Acanthocephala. In K.G. Adiyodi and R.G. Adiyodi (eds): *Reproductive Biology of Invertebrates, vol 9A*, pp. 229–241. Wiley, Chichester.

Clément, P. 1993. The phylogeny of rotifers: molecular, ultrastructural and behavioural data. *Hydrobiologia* **255/256**: 527–544.

Clément, P. and Wurdak, E. 1991. Rotifera. In F.W. Harrison (ed.): *Microscopic Anatomy of Invertebrates*, vol. 4, pp. 219–297. Wiley-Liss, New York.

Conway Morris, S. and Crompton, D.W.T. 1982. The origins and evolution of the Acanthocephala. *Biol. Rev.* **57**: 85–115.

Dunagan, T.T. and Miller, D.M. 1991. Acanthocephala. In F.W. Harrison (ed.): *Microscopic Anatomy of Invertebrates*, vol. 4, pp. 299–332. Wiley-Liss, New York.

Foata, J., Dezfuli, B.S., Pinelli, B. and Marchand, B. 2004. Ultrastructure of spermiogenesis and spermatozoon of *Leptorhynchoides plagicephalus* (Acanthocephala, Palaeacanthocephala), a parasite of the sturgeon *Acipenser naccarii* (Osteichthyes, Acipenseriformes). *Parasitol. Res.* **93**: 56–63.

García-Varela, M. and Nadler, S.A. 2006. Phylogenetic relationships among Syndermata inferred from nuclear and mitochondrial gene sequences. *Mol. Phylogenet. Evol.* **40**: 61–72.

Gee, R.J. 1988. A morphological study of the nervous system of the praesoma of *Octospinifer malicentus* (Acanthocephala: Noechinorhynchidae). *J. Morphol.* **196**: 23–31.

Giribet, G., Sørensen, M.V., Funch, P., Kristensen, R.M. and Sterrer, W. 2004. Investigations into the phylogenetic position of Micrognathozoa using four molecular loci. *Cladistics* **20**: 1–13.

Gladyshev, E. and Meselson, M. 2008. Extreme resistance of bdelloid rotifers to ionizing radiation. *Proc. Natl. Acad. Sci. USA* **105**: 5139–5144

Haffner, K. 1942. Untersuchungen über das Urogenitalsystem der Acanthocephalen. I–III. *Z. Morphol. Oekol. Tiere* **38**: 251–333.

Hamann, O. 1891. Monographie der Acanthocephalen. *Jena. Z. Naturw.* **25**: 113–231.

Hatschek, B. 1878. Studien über Entwicklungsgeschichte der Anneliden. *Arb. Zool. Inst. University Wien.* **1**: 277–404.

Hatschek, B. 1891. *Lehrbuch der Zoologie, 3. Lieferung* (pp 305–432). Gustav Fischer, Jena.

Herlyn, H., Piskurek, O., Schmitz, J., Ehlers, U. and Zischler, H. 2003. The syndermatan phylogeny and the evolution of acanthocephalan endoparasitism as inferred from 18S rDNA sequences. *Mol. Phylogenet. Evol.* **26**: 155–164.

Hochberg, R. 2007. Topology of the nervous system of *Notommata copeus* (Rotifera: Monogononta) revealed with anti-FRMFamide, -SCPb, and -serotonin (5-HT) immunohistochemistry. *Invert. Biol.* **126**: 247–256.

Hochberg, R. 2009. Three-dimensional reconstruction and neural map of the serotonergic brain of *Asplanchna brightwellii* (Rotifera, Monogononta). *J. Morphol.* **270**: 430–441.

Jägersten, G. 1972. *Evolution of the Metazoan Life Cycle*. Academic Press, London.

Klusemann, J., Kleinow, W. and Peters, W. 1990. The hard parts (trophi) of the rotifer mastax do contain chitin: evidence from studies on *Brachionus plicatilis. Histochemistry* **94**: 277–283.

Lang, A. 1888. *Lehrbuch der vergleichenden Anatomie der wirbellosen Tiere.* Gustav Fischer, Jena.

Leasi, F., Pennati, R. and Ricci, C. 2009. First description of the serotonergic nervous system in a bdelloid rotifer: *Macrotrachela quadricornifera* Milne 1886 (Philodinidae). *Zool. Anz.* **248**: 47–55.

Lechner, M. 1966. Untersuchungen zur Embryonalentwicklung des Rädertieres *Asplanchna girodi* de Guerne. *Wilhelm Roux' Arch. Entwicklungmech. Org.* **157**: 117–173.

Melone, G. and Ferraguti, M. 1994. The spermatozoon of *Brachionus plicatilis* (Rotifera, Monogononta) with some notes on sperm ultrastructure in Rotifera. *Acta Zool. (Stockh.)* **75**: 81–88.

Melone, G. and Ferraguti, M. 1999. Rotifera. In K.G. Adiyodi and R.G. Adiyodi (eds): *Reproductive Biology of Invertebrates*, vol. 9A, pp. 157–169. Wiley, Chichester.

Meyer, A. 1928. Die Furchung nebst Eibildung, Reifung und Befruchtung des *Gigantorhynchus gigas. Zool. Jahrb., Anat.* **50**: 117–218.

Meyer, A. 1932–1933. Acanthocephala. *Bronn's Klassen und Ordnungen des Tierreichs*, 4. Band, 2. Abt., 2. Buch, Akademische Verlagsgesellschaft, Leipzig.

Meyer, A. 1936. Die plasmodiale Entwicklung und Formbildung des Riesenkratzers (*Macracanthorhynchus hirudinaceus*). I. Teil. *Zool. Jahrb., Anat.* **62**: 111–172.

Meyer, A. 1938. Die plasmodiale Entwicklung und Formbildung des Riesenkratzers (*Macracanthorhynchus hirudinaceus* (Pallas)). III. Teil. *Zool. Jahrb., Anat.* **64**: 131–197.

Nachtwey, R. 1925. Untersuchungen über die Keimbahn, Organogenese und Anatomie von *Asplanchna priodonta* Gosse. *Z. Wiss. Zool.* **126**: 239–492.

Nielsen, C. 1987. Structure and function of metazoan ciliary bands and their phylogenetic significance. *Acta Zool. (Stockh.)* **68**: 205–262.

Peters, F. 1931. Untersuchungen über Anatomie und Zellkonstanz von *Synchaeta* (*S. grimpei* Remane, *S. baltica* Ehrenb., *S. tavina* Hood und *S. triophthalma* Lauterborn). *Z. Wiss. Zool.* **139**: 1–119.

Peters, W., Taraschewski, H. and Latka, I. 1991. Comparative investigations of the morphology and chemical composition of the eggshells of Acanthocephala. *Macracanthorhynchus hirudinaceus* (Archiacanthocephala). *Parasitol. Res.* **77**: 542–549.

Pray, F.A. 1965. Studies on the early development of the rotifer *Monostyla cornuta* Müller. *Trans. Am. Microsc. Soc.* **84**: 210–216.

Remane, A. 1929–1933. Rotifera. *Bronn's Klassen und Ordningen des Tierreichs*, 4. Band, 2. Abteilung, 1. Buch, pp. 1–576. Akademische Verlagsgesellschaft, Leipzig.

Rieger, R.M. and Tyler, S. 1995. Sister-group relationships of Gnathostomulida and Rotifera-Acanthocephala. *Invert. Biol.* **114**: 186–188.

Riemann, O. and Ahlrichs, W.H. 2010. The evolution of the protonephridial terminal organs across Rotifera with particular emphasis on *Dicranophorus forcipatus*, *Encentrum mucronatum* and *Erignatha clastopis* (Rotifera: Dicranophoridae). *Acta Zool. (Stockh.)* **91**: 199–211.

Schmidt, G.D. 1985. Development and life cycles. In D.W.T. Crompton and B.B. Nickol (eds): *Biology of the Acanthocephala*, pp. 273–305. Cambridge University Press, Cambridge.

Sørensen, M.V. 2002. On the evolution and morphology of the rotiferan trophi, with a cladistic analysis of Rotifera. *J. Zool. Syst. Evol. Res.* **40**: 129–154.

Sørensen, M.V. and Giribet, G. 2006. A modern approach to rotiferan phylogeny: combining morphological and molecular data. *Mol. Phylogenet. Evol.* **40**: 585–608.

Stossberg, K. 1932. Zur Morphologie der Rädertiergattungen *Euchlanis*, *Brachionus* und *Rhinoglaena*. *Z. Wiss. Zool.* **142**: 313–424.

Strathmann, R.R., Jahn, T.L. and Fonseca, J.R. 1972. Suspension feeding by marine invertebrate larvae: Clearance of particles by ciliated bands of a rotifer, pluteus, and trochophore. *Biol. Bull.* **142**: 505–519.

Suga, K., Welch, D.M., Tanaka, Y., Sakakura, Y. and Hagiwara, A. 2007. Analysis of expressed sequence tags of the cyclically parthenogenetic rotifer *Brachionus plicatilis*. *PLoS ONE* **2(8)**: e671.

Taraschewski, H. 2000. Host-parasite interactions in Acanthocephala: a morphological approach. *Adv. Parasitol.* **46**: 1–179.

Welch, D.M. and Meselson, M. 2000. Evidence for the evolution of bdelloid rotifers without sexual reproduction or genetic exchange. *Science* **288**: 1211–1215.

Welch, D.B.M., Welch, J.L.M. and Meselson, M. 2008. Evidence for degenerate tetraploidy in bdelloid rotifers. *Proc. Natl. Acad. Sci. USA* **105**: 5145–5149.

Wesenberg-Lund, C. 1952. *De danske søers og dammes dyriske plankton*. Munksgaard, Copenhagen.

Witek, A., Herlyn, H., Meyer, A., *et al.* 2008. EST based phylogenomics of Syndermata questions monophyly of Eurotatoria. *BMC Evol. Biol.* **8**: 345.

Zelinka, C. 1886. Studien über Räderthiere. I. Über die Symbiose und Anatomie von Rotatorien aus dem Genus *Callidina*. *Z. Wiss. Zool.* **44**: 396–506.

Zelinka, C. 1888. Studien über Räderthiere. II. Der Raumparasitismus und die Anatomie von *Discopus synaptae* n.g., nov.sp. *Z. Wiss. Zool.* **47**: 141–246.

Zelinka, C. 1891. Studien über Räderthiere. III. Zur Entwicklungsgeschichte der Räderthiere nebst Bemerkungen über ihre Anatomie und Biologie. *Z. Wiss. Zool.* **53**: 1–159.

35

POLYZOA (BRYOZOA s.l.)

The nomenclature of the group here called Polyzoa, i.e. Entoprocta + Cycliophora + Bryozoa, has a confusing history. Thompson (1830) recognized the group here called Bryozoa (i.e. Ectoprocta), and gave it the name Polyzoa. The following year Ehrenberg (1831) gave the name Bryozoa to the same group. Over time, Bryozoa has gained preference. The entoproct *Pedicellina* was simply placed in the Bryozoa by Gervais (1837). However, Nitsche (1869) noted important differences between *Pedicellina* and *Loxosoma* and the other bryozoans and divided the Bryozoa into Entoprocta and Ectoprocta. Hatschek (1891) raised the two groups to phylum rank; he stressed the similarity between the entoproct larvae and rotifers, and placed the entoprocts next to the rotifers, whereas the ectoprocts were united with phoronids and brachiopods in the group Tentaculata. Hyman (1959) introduced the name Lophophorata for the same group, and her name is now in common use, although most of the molecular phylogenies do not support such a clade (see below).

The group Lophophorata was defined as archimeric (or trimeric) animals with a ciliated lophophore and a U-shaped gut. This definition fits the pterobranchs equally well, but this is usually ignored. An archimeric body should consist of three regions: prosome, mesosome, and metasome, each with a paired or unpaired coelomic compartment: protocoel, mesocoel, and metacoel. A lophophore is defined as a mesosomal extension with ciliated tentacles containing mesocoelomic canals. However, coelomic sacs are not observed in bryozoan embryology, and the adult polypides comprise a large body cavity in the cystid, connected through a wide opening with a smaller, ring-shaped tentacle coelom around the mouth, with extensions into the tentacles. The phylactolaemates have a lip at the posterior side of the mouth with a median extension from the body coelom. Phoronids and brachiopods (Brachiozoans; Chapters 54–56) and pterobranchs (Chapter 60) show strong indications of archimery, with ciliated tentacles on a mesosome and large metacoelomic cavities with metanephridia that also function as gonoducts. Metanephridia are not found in bryozoans and their male and female gametes or embryos are shed through different openings. Other important differences between bryozoans and brachiozoans include: brachiozoans have only monociliate cells, whereas multiciliate cells are found in all bryozoan epithelia. The embryology of the two groups show no specific similarities. Bryozoans have a well-defined ganglion at the posterior (anal) side of the oesophagus; it bears no resemblance to the more diffuse nervous system of the brachiozoans. I see no characters that should unite these phyla. I have earlier argued for a reunification of Ectoprocta (Bryozoa) and Entoprocta in the supraphyletic group Bryozoa (Polyzoa), placed within the Spiralia, inspired by studies of entoproct ontogeny (Nielsen 1971, 1985, 1987). The newly described phylum Cycliophora has been placed as sister group of the Entoprocta (Funch and Kristensen 1995).

There is no unequivocal set of names for the groups. Entoprocta and Ectoprocta are unambiguous, but Bryozoa is by far the most-used name for Ectoprocta (and a few authors use Kamptozoa for the Entoprocta). Also the name of the supraphyletic group is problematic. Cavalier-Smith (1998) suggested to use Thompson's name Polyzoa, and this was followed by Hejnol *et al.* (2009) and Hejnol (2010). I have chosen to use the names Polyzoa = Entoprocta + Cycliophora + Bryozoa (i.e. Ectoprocta), because this seems to retain most names in their familiar use (Fig. 35.1).

It is difficult to point to good synapomorphies of the three phyla of the Polyzoa. Some entoprocts have a settling process resembling that of some of the bryozoans (Fig. 35.2; compare also Figs. 36.1 and 38.2). Entoprocts and *Symbion* have the typical protostomian downstream-collecting ciliary system, with the ciliary ring around the mouth of *Symbion* probably being homologous with prototroch + metatroch in other spiralians; the ciliary feeding structures and methods of the bryozoans are completely different. Only one synapomorphy of Entoprocta and Cycliophora has been pointed out, the presence of unusual, mushroom-shaped extensions of the basal membrane into the ectodermal cells (Sørensen *et al.* 2000).

However, recent molecular studies lend strong support to the phylogeny chosen here. Entoprocta and Bryozoa were found to be sister groups in several studies in which Cycliophora was not included (Helmkampf *et al.* 2008; Bleidorn *et al.* 2009; Witek *et al.* 2009; Hausdorf *et al.* 2010). The topology preferred here was recovered in all the studies that include all three groups (Baguñá *et al.* 2008; Hejnol *et al.* 2009; Paps *et al.* 2009a; Mallatt *et al.* 2010), and the sister-group relationship of Cycliophora and Entoprocta was found in a study including a large number of entoprocts (Fuchs *et al.* 2010).

The 'Lophophorata' has been recovered only in few molecular analyses (Dordel *et al.* 2010). One of the analyses of complete mitochondrial genomes (Jang and Hwang 2009) shows a monophyletic Lophotrochozoa (including Entoprocta), but both this and other mitogenomic analyses show several bizarre topologies. Conversely, the Brachiozoa (Phoronida + Brachiopoda, see Chapter 53) is found separated from the Polyzoa in very many recent analyses, and the Brachiozoa is often placed as sister group of the Nemertini or Mollusca (Dunn *et al.* 2008; Hejnol *et al.* 2009; Paps *et al.* 2009a; Paps *et al.* 2009b; Hausdorf *et al.* 2010). I hope that the name Lophophorata will disappear from the zoological vocabulary.

It seems certain that Polyzoa belongs to the Spiralia. The entoprocts have spiral cleavage and their larvae are trochophores, with the characteristic downstream-collecting ciliary bands. Both cycliophorans and bryozoans appear highly specialized with unique larval types and life cycles. At present it seems impossible to point to a sister-group relationship with any of the other spiralian groups (Fig. 23.1). Molecular phylogenies show no consistent picture either.

References

Baguñá, J., Martinez, P., Paps, J. and Riutort, M. 2008. Back in time: a new systematic proposal for the Bilateria. *Phil. Trans. R. Soc. Lond. B* **363**: 1481–1491.

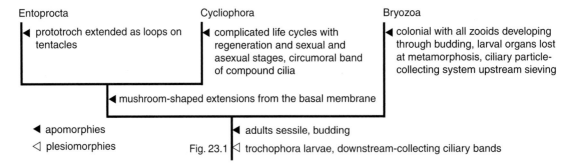

Figure 35.1. Phylogeny of the Polyzoa.

Pedicellina

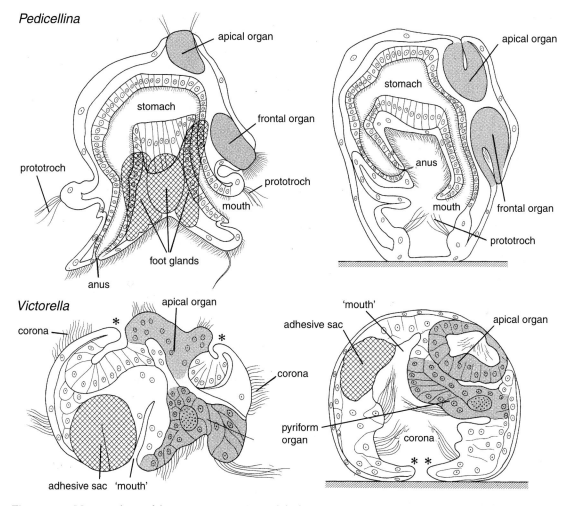

Figure 35.2. Metamorphosis of the entoproct *Pedicellina* and the bryozoan *Victorella*. (From Nielsen 1971.)

Bleidorn, C., Podsiadlowski, L., Zhong, M., *et al.* 2009. On the phylogenetic position of Myzostomida: can 77 genes get it wrong? *BMC Evol. Biol.* **9**: 150.

Cavalier-Smith, T. 1998. A revised six-kingdom system of life. *Biol. Rev.* **73**: 203–266.

Dordel, J., Fisse, F., Purschke, G. and Struck, T. 2010. Phylogenetic position of Sipuncula derived from multi-gene and phylogenomic data and its implication for the evolution of segmentation. *J. Zool. Syst. Evol. Res.* **48**: 197–207.

Dunn, C.W., Hejnol, A., Matus, D.Q., *et al.* 2008. Broad phylogenomic sampling improves resolution of the animal tree of life. *Nature* **452**: 745–749.

Ehrenberg, C.G. 1831. *Symbolae Physicae, seu Icones et Descriptiones Animalium Evertebratorum sepositis Insectis, quae ex itinere per African Borealium et Asiam Occidentalem.* Berolini.

Fuchs, J., Iseto, T., Hirose, M., Sundberg, P. and Obst, M. 2010. The first internal molecular phylogeny of the animal phylum Entoprocta (Kamptozoa). *Mol. Phylogenet. Evol.* **56**: 370–379.

Funch, P. and Kristensen, R.M. 1995. Cycliophora is a new phylum with affinities to Entoprocta and Ectoprocta. *Nature* **378**: 711–714.

Gervais, M. 1837. Recherches sur les polypes d'eau douce des genres *Plumatella, Cristatella* et *Paludicella. Ann. Sci. Nat., Zool., 2. Ser.* **7**: 74–93.

Hatschek, B. 1891. *Lehrbuch der Zoologie, 3. Lieferung* (pp 305–432). Gustav Fischer, Jena.

Hausdorf, B., Helmkampf, M., Nesnidal, M.P. and Bruchhaus, I. 2010. Phylogenetic relationships within the lophophorate lineages (Ectoprocta, Brachiopoda and Phoronida). *Mol. Phylogenet. Evol.* **55**: 1121–1127.

Hejnol, A. 2010. A twist in time—The evolution of spiral cleavage in the light of animal phylogeny. *Integr. Comp. Biol.* **50**: 695–706.

Hejnol, A., Obst, M., Stamatakis, A., *et al.* 2009. Assessing the root of bilaterian animals with scalable phylogenomic methods. *Proc. R. Soc. Lond. B* **276**: 4261–4270.

Helmkampf, M., Bruchhaus, I. and Hausdorf, B. 2008. Phylogenomic analyses of lophophorates (brachiopods, phoronids and bryozoans) confirm the Lophotrochozoa concept. *Proc. R. Soc. Lond. B* **275**: 1927–1933.

Hyman, L.H. 1959. *The Invertebrates, vol. 5. Smaller Coelomate Groups.* McGraw-Hill, New York.

Jang, K.H. and Hwang, U.W. 2009. Complete mitochondrial genome of *Bugula neritina* (Bryozoa, Gymnolaemata, Cheilostomata): phylogenetic position of Bryozoa and phylogeny of lophophorates within the Lophotrochozoa. *BMC Genomics* **10**: 167.

Mallatt, J., Craig, C.W. and Yoder, M.J. 2010. Nearly complete rRNA genes assembled from across the metazoan animals: Effects of more taxa, a structure-based alignment, and paired-sites evolutionary models on phylogeny reconstruction. *Mol. Phylogenet. Evol.* **55**: 1–17.

Nielsen, C. 1971. Entoproct life-cycles and the entoproct/ectoproct relationship. *Ophelia* **9**: 209–341.

Nielsen, C. 1985. Animal phylogeny in the light of the trochaea theory. *Biol. J. Linn. Soc.* **25**: 243–299.

Nielsen, C. 1987. Structure and function of metazoan ciliary bands and their phylogenetic significance. *Acta Zool. (Stockh.)* **68**: 205–262.

Nitsche, H. 1869. Beiträge zur Kentniss der Bryozoen I–II. *Z. Wiss. Zool.* **20**: 1–36.

Paps, J., Baguñá, J. and Riutort, M. 2009a. Lophotrochozoa internal phylogeny: new insights from an up-to-date analysis of nuclear ribosomal genes. *Proc. R. Soc. Lond. B* **276**: 1245–1254.

Paps, J., Baguñà, J. and Riutort, M. 2009b. Bilaterian phylogeny: A broad sampling of 13 nuclear genes provides a new Lophotrochozoa phylogeny and supports a paraphyletic basal Acoelomorpha. *Mol. Biol. Evol.* **26**: 2397–2406.

Sørensen, M.V., Funch, P., Willerslev, E., Hansen, A.J. and Olesen, J. 2000. On the phylogeny of the Metazoa in the light of Cycliophora and Micrognathozoa. *Zool. Anz.* **239**: 297–318.

Thompson, J.V. 1830. On Polyzoa, a new animal discovered as an inhabitant of some Zoophites—with a description of the newly instituted genera of *Pedicellaria* and *Vesicularia*, and their species. *Zoological Researches, and Illustrations; or Natural History of Nondescript or Imperfectly Known Animals, in a Series of Memoires.* Memoire 5, pp. 89–102. King & Ridings, Cork.

Witek, A., Herlyn, H., Ebersberger, I., Welch, D.B.M. and Hankeln, T. 2009. Support for the monophyletic origin of Gnathifera from phylogenomics. *Mol. Phylogenet. Evol.* **53**: 1037–1041.

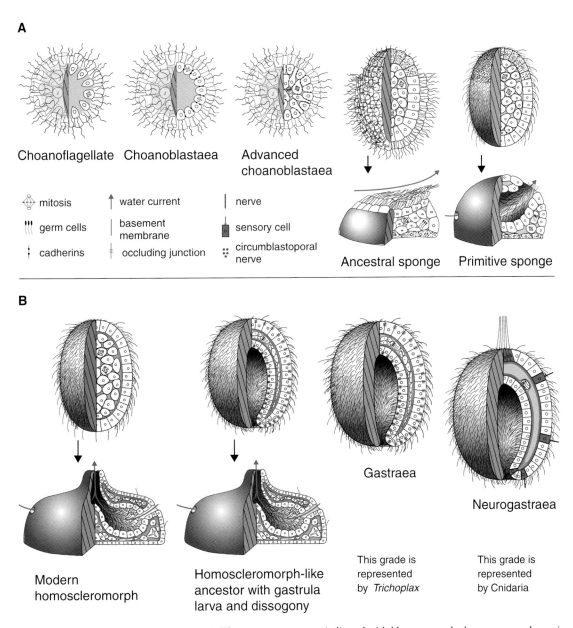

A

Choanoflagellate Choanoblastaea Advanced choanoblastaea

⟨⟩ mitosis   ↑ water current | nerve

ⵝ germ cells | basement membrane   ▮ sensory cell

⁞ cadherins   ǂ occluding junction   ⁛ circumblastoporal nerve

Ancestral sponge   Primitive sponge

B

Gastraea

Neurogastraea

Modern homoscleromorph

Homoscleromorph-like ancestor with gastrula larva and dissogony

This grade is represented by *Trichoplax*

This grade is represented by Cnidaria

Plate 1. The evolutionary scenario (see text). The water currents are indicated with blue arrows, the basement membrane is red, and the nervous cells green. The muscle cells of the gastraea have been omitted for clarity. See Figure 4.3, page 18.

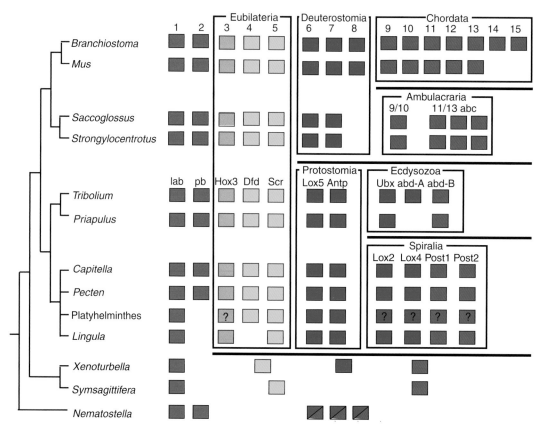

Plate 2. Hox genes of a cnidarian and selected representatives of the major bilaterian groups, demonstrating the existence of series of 'Hox signatures' (see the text). The Eubilaterian Hox signature consists of a group 3 gene and the two anteriormost central genes. Genes belonging to the same orthology group have the same colour: anterior genes, red; group three, orange; central group, green; and posterior group, blue. Duplications are not shown. Diagram based on: *Branchiostoma*: Holland *et al.* (2008), *Mus*: Carroll *et al.* (2005), *Saccoglossus*: Aronowicz and Lowe (2006), *Strongylocentrotus*: Cameron *et al.* (2006), *Tribolium*: Shippy *et al.* (2008), *Priapulus*: de Rosa *et al.* (1999), *Capitella*: Fröbius *et al.* (2008), *Pecten*: Canapa *et al.* (2005), Platyhelminthes: Olson (2008); Badets and Verneau (2009), *Lingula*: de Rosa *et al.* (1999), *Xenoturbella*: Fritsch *et al.* (2008), *Symsagittifera*: Moreno *et al.* (2009), *Nematostella*: Ryan *et al.* (2007). The platyhelminth sequences are rather divergent so the genes have been inserted with question marks. See Figure 21.3, page 83.

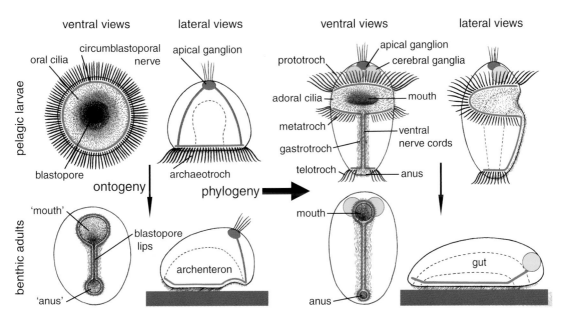

Plate 3. The trochaea theory. Evolution of the pelago-benthic protostomian ancestor gastroneuron (right) from a radially symmetric, holopelagic trochaea (upper left), through an advanced, bilaterally symmetric pelago-benthic trochaea (left). Short thin lines, separate cilia; long thick lines, compound cilia; red, apical organ; yellow, cerebral ganglia; green, circumblastoporal nerve ring. (See Nielsen and Nørrevang 1985; Nielsen 2010). See Figure 22.8, page 97.

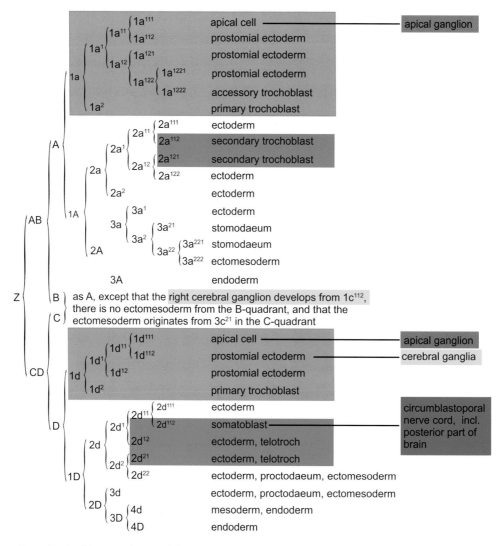

Plate 4. Generalized cell lineage of an annelid. (Based on *Podarke* and *Arenicola*; Child 1900; Treadwell 1901). See Figure 22.9, page 98.

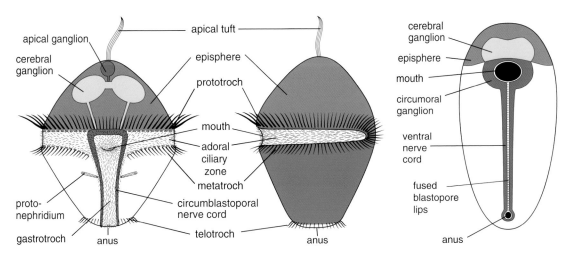

Plate 5. Larval regions of epidermis and nervous systems of a larva and an adult of the ancestral spiralian, gastroneuron, with a trochophora larva and a benthic adult. The two left diagrams show the trochophore from the ventral and the left side, and the right diagram shows the adult from the ventral side. The right and left diagrams show the development of the nervous system, and the middle diagram shows the embryological origin of the main epithelial areas. The blue areas in the middle diagram indicate the secondary trochoblasts originating from the second micromere quartet, from the cells $2a^2$–$2d^2$, and the whole area covered by the descendents of the 2d cell, which have spread enormously to cover almost the whole post-trochal region. See Figure 22.10, page 99.

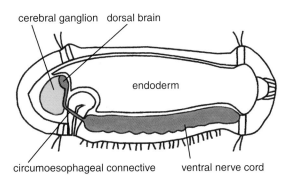

Plate 6. Central nervous system of a stage-6 larva of *Capitella telata* in left view. The left cerebral ganglion (yellow) is derived from 1c cells, and the circumblastoporal nerves (green, posterodorsal brain, circumoesophageal connectives, and ventral nerve cord) are derived from 2d cells. (Based on Meyer and Seaver 2010). See Figure 22.11, page 99.

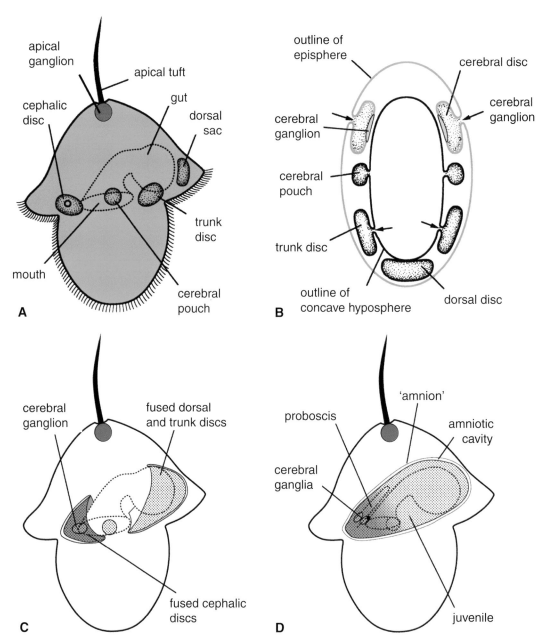

Plate 7. Development of the juvenile nemertine inside the pilidium of Cerebratulus lacteus. (Modified from Nielsen 2008.) Note that the cerebral discs develop from the episphere and the trunk discs from the hyposphere. See Figure 28.1, page 159.

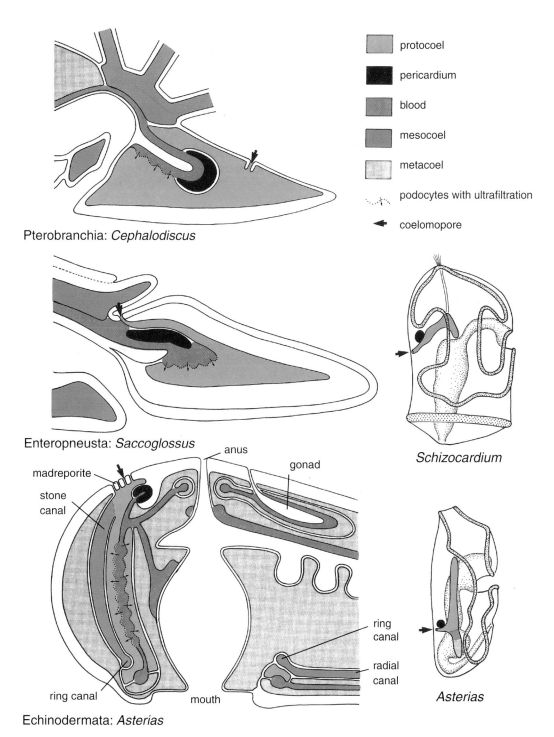

protocoel
pericardium
blood
mesocoel
metacoel
podocytes with ultrafiltration
coelomopore

Pterobranchia: *Cephalodiscus*

Enteropneusta: *Saccoglossus*

Schizocardium

anus
gonad
madreporite
stone canal
ring canal
radial canal
ring canal
mouth

Asterias

Echinodermata: *Asterias*

Plate 8. Morphology of the anterior coelomic cavities in larvae and adults of ambulacrarians with emphasis on the transformations of parts of the larval coelomic sacs into the adult's axial complex: The larvae are drawn on the basis of photographs in Ruppert and Balser (1986), while the adults are diagrammatic median sections with some of the structures shifted slightly to get the necessary details into the plane of the drawing. Pterobranchs: adult *Cephalodiscus gracilis* (based on Lester 1985 and Dilly *et al.* 1986); enteropneusts: larva of *Schizocardium brasiliense*, adult of *Saccoglossus kowalevskii* (based on Balser and Ruppert 1990); echinoderms: larva of *Asterias forbesi*, adult of *Asterias* (based on Nichols 1962 and Ruppert and Balser 1986). Mesocoel, blue; renal chamber (protocoel), green; pericardium, black; blood, red. The lines with dots indicate a coelomic layer consisting of podocytes. The small arrows indicate the direction of the presumed or proven ultrafiltration of primary urine. See Figure 57.1, page 317.

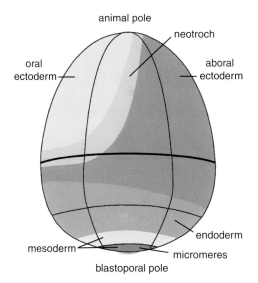

Plate 9. Fate map of an embryo of *Strongylocentrotus purpuratus* at a stage of about 400 cells. The position of the cleavage furrow between the apical (animal) and blastoporal (vegetal) cells of the 8-cell stage is indicated by the thick horizontal line. Note that the neotroch does not follow that line (compare with Fig. 22.10). (Adapted with permission from Davidson *et al.* 1998). See Figure 58.2, page 326.

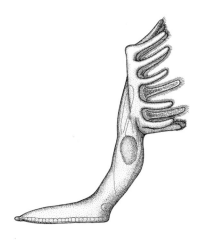

36

Phylum **Entoprocta**

Entoprocta is a small phylum comprising about 200 described species; the fossil record consists of Upper Jurassic, bioimmured colonies of a species of *Barentsia* very similar to the recent *B. matsushimana* (Todd and Taylor 1992). All species are benthic with trochophora-type larvae. The family Loxosomatidae (order Solitaria) comprises solitary, usually commensal species, whereas the other three families (order Coloniales) are colonial. The monophyly of the group and the sister-group relationship of the two orders is supported by molecular analyses (Fuchs *et al.* 2010). Most species are marine; only a few enter the brackish zone and one is limnic.

Each individual or zooid consists of a more or less globular body with a horseshoe of ciliated tentacles surrounding the depressed ventral side of the body, called the atrium, and a shorter or longer cylindrical stalk. The microscopic anatomy is reviewed in Nielsen and Jespersen (1997). There is no sign of a secondary body cavity, but spaces between ectodermal, endodermal, and mesodermal elements form a narrow primary body cavity. The fluid of this cavity can be moved between body and stalk in the pedicellinids and barentsiids, by the action of a small organ consisting of a stack of star-shaped cells with contractile peripheral rays (Emschermann 1969).

Most areas of the ectoderm are covered by a cuticle of crossing fibrils between branching microvilli with swollen tips, but there is only a thin glycocalyx without fibres at the ciliated faces of the tentacles. The basal membrane shows peculiar mushroom-shaped extensions into the ectoderm. The stiff portions of the stalks of the barentsiids, and the thick cuticle around the characteristic resting bodies of some barentsiid species, have an outer layer of fibrils with small knobs resembling microvillar tips, and an inner layer with chitin; this may suggest that the surface layer has been secreted by microvilli that have then retracted or degenerated. The ciliated cells of both ectodermal and endodermal epithelia, as well as in the protonephridia, are all multiciliate.

The tentacles (Nielsen and Rostgaard 1976; Emschermann 1982) are almost cylindrical with a frontal row of cells (i.e. at the side of the tentacle facing the atrium), with separate cilia beating towards the base of the tentacle, where they join a similar band, the adoral ciliary zone, that beats along the tentacle bases towards the mouth. A row of cells along the lateral sides of the tentacles carries a band of long compound lateral cilia that create a water current between the tentacles toward the atrial space. The lateral bands of compound cilia on the abfrontal pair of tentacles bend and follow the adoral ciliary zone around the atrium. The opposed lateral bands of the tentacles function as a downstream-collecting system working on the 'catch-up principle' (Riisgård *et al.* 2000). The frontal cilia transport the strained particles to the adoral ciliary zone that transports them to the mouth. A laterofrontal row of cells with separate cilia lies between the two types of bands; their function is unknown. The laterofrontal ciliary

Chapter vignette: *Loxosomella elegans*. (Redrawn from Nielsen 1964.)

cells are myoepithelial, and a row of mesodermal muscle cells is found along the lateral ciliary cells.

There is a U-shaped gut with the mouth centrally in the adoral ciliary zone and the anus situated near the aboral opening in the horseshoe of tentacles. Most of the cells of the gut are ciliated; the cilia surrounding the opening from the pharynx to the stomach are of a peculiar shape with 'swollen' cell membranes.

The nervous system comprises a dumbbell-shaped ganglion situated ventrally at the bottom of the atrium (Nielsen and Jespersen 1997; Fuchs *et al.* 2006). Fine nerves lead to sensory organs on the abfrontal side of the tentacles and, in some species, to lateral sense organs on the body. The lateral ciliated cells of the tentacles are innervated, and it is probable that this innervation controls the beat of the lateral cilia. A pair of lateral nerves is situated in the stalk of *Pedicellina*, and the longitudinal muscles send cytoplasmic strands to the nerves where synapses are formed. Other parts of the peripheral nervous system, including the nerves probably connecting the zooids in the colonies, are poorly known. Hilton (1923) described a nervous network with numerous sensory pits staining with methylene blue in *Barentsia*, but the nature of these cells is uncertain.

A pair of protonephridia, each consisting of a few cells and with a common nephridiopore, are situated at the bottom of the atrium. The freshwater species *Urnatella gracilis* has several protonephridia on branched ducts in the body and several protonephridia in each joint of the stalk; all the protonephridia develop from ectodermal cells (Emschermann 1965a).

The gonads are simple, sac-shaped structures opening at the bottom of the atrium. Many of the solitary forms appear to be protandric; the colonial *Barentsia* comprises species that have male and female colonies, species with male and female zooids in the same colony, and species with simultaneously hermaphroditic zooids (Wasson 1997). Fertilization has not been observed directly, but Marcus (1939) found spermatozoa in eggs in the ovaries of *Pedicellina*. The fertilized eggs usually become enveloped in a secretion from glands in the thickened, unpaired part of the gonoduct, and this secretion is pulled out into a string that attaches the egg to a special area of the atrial epithelium. When the egg membrane bursts open and

the larva starts feeding, a ring around the apical organ remains, and the larvae are tethered to the atrium for a period. *Loxosomella vivipara* has very small eggs that are taken up in a narrow invagination from the atrial epithelium where a placenta develops, nourishing the embryo through the apical area; the embryo increases enormously in size, and the fully grown larva has a large internal bud (Nielsen 1971). *Urnatella gracilis* is viviparous (Emschermann 1965b).

Cleavage is spiral with quartets (Marcus 1939: *Pedicellina*; Malakhov 1990: *Barentsia*). The cleavage follows the normal spiral pattern, and Marcus (1939) observed a 56-cell stage, with the cells $1a^{111}$–$1d^{111}$ forming the apical rosette, a ring of trochoblasts consisting of the daughter cells of primary ($1a^2$–$1d^2$) and accessory ($1a^{122}$–$1d^{122}$) trochoblasts, a 4d mesoblast, and the endoderm cells 4a–c, 5a–d and 5A–D (Fig. 23.3). In *Loxosomella vivipara*, the spiral pattern is lost after the 8-cell stage (Nielsen 1971).

Almost all entoproct larvae (review in Nielsen 1971) have the typical protostomian downstream-collecting ciliary system with a large prototroch of compound cilia, a smaller metatroch of compound cilia, and an adoral ciliary zone with separate cilia (Figs. 22.4, 22.5). The prototroch comprises two rows of cells in *Barentsia discreta* (Malakhov 1990). The cilia are used in filter-feeding both in the older larvae still in the maternal atrium and after liberation. The gastrotroch is a small ventral band in some species, for example *Loxosomella elegans*, but most species have a conspicuous expansion of the ventral area, resembling the molluscan foot, for example *L. harmeri*, and most of the colonial forms. The nervous system (Nielsen 1971; Wanninger *et al.* 2007) comprises a large apical organ with eight flask-shaped serotonergic cells, a pair of thin nerves to a paired frontal ganglion, then further to a ring nerve at the base of the large prototroch cells (Fig. 23.4), and to paired nerves along the foot (in *L. murmanica*). Some larvae have a pair of lateral sense organs innervated from the frontal ganglion. The apical organ of *Loxosoma pectinaricola* consists of a circle of multiciliate cells, a number of monociliate cells, myoepithelial cells, and vacuolated cells (Sensenbaugh 1987). Most larvae have a large, retractable, ciliated frontal organ in contact with the frontal ganglion, but the organ is very small and lacks cilia in others (Nielsen 1971). Many species of *Loxosomella* have a ring of gland cells around the organ (see below). Larvae of most loxosomatids have a pair of eyes in the frontal organ. Each eye consists of a cup-shaped pigment cell, a lens cell, and a photoreceptor cell with a bundle of cilia oriented perpendicular to the incoming light. The cilia have a normal $(9 \times 2) + 2$ axoneme, but the dynein arms are lacking (Woollacott and Eakin 1973). Many of the loxosomatid larvae and all the pedicellinid and barentsiid larvae have a large ciliated foot with a transverse row of long compound cilia at the anterior end; the larvae of the colonial species have three pairs of large glands with different types of secretion granules in the foot. There is a pair of protonephridia, each consisting of three cells, and opening into the groove surrounding the foot.

Most larvae have a free period of only a few hours before they settle, but some of the loxosomatid larvae apparently stay in the plankton for weeks; their development is mostly unknown.

Solitary and colonial entoprocts show considerable differences in the life cycles (Nielsen 1971) (Fig. 36.1), but the variations within both types make it possible to interpret the types as variations over one common theme.

When the larvae with a foot are ready to settle, they creep on the foot and test the substratum with the frontal organ. Some species of *Loxosomella* have a very straightforward metamorphosis. The larva settles on the frontal organ, which apparently becomes glued to the substratum by a secretion from the gland cells around the ciliated sensory cells, with the hyposphere retracted and the muscles along the prototroch constricted so that all the larval organs are enclosed. During the following metamorphosis the apical and frontal organs disintegrate and the gut rotates slightly. The larval ciliary bands disintegrate, but similar bands develop from neighbouring areas on tentacle buds at the frontal side of the closed atrium. A new ganglion forms apparently from an invagination of the ventral epithelium, but a contribution from the nerves of the larval foot cannot be excluded. The atrium reopens exposing the short adult tentacles. The larval protonephridia may be retained, but this has not been studied.

Other species of *Loxosomella* and probably all species of *Loxosoma* have precocious budding from areas

of the episphere, corresponding to the laterofrontal budding zones of the adult, and the larval body disintegrates after having given off the buds. The budding points are situated at the bottom of ectodermal invaginations, and the buds appear to be internal; these larvae become disrupted when the buds are liberated.

When the larvae of pedicellinids and barentsiids settle, the large glands in the foot apparently give off their secretion, and the larva then retracts the hyposphere and contracts the prototroch muscle so that the contracted larva becomes glued to the substratum with the ring-shaped zone above the constricted prototroch. The frontal organ is obviously only acting as a sensory organ testing potential settling spots. The retracted apical and frontal organs disintegrate and the gut rotates about 180° with the mouth in front; the degeneration of the larval ciliary bands, the development of the adult bands, the formation of a new brain, and the reopening of the atrium takes place as in the loxosomatids.

Adult loxosomatids form buds from laterofrontal areas of the body, and the buds detach after having reached the shape of a small adult, while the buds in the colonial species are formed at the base of the stalk (*Loxokalypus*, see Emschermann 1972), or from the growing tips of stolons (pedicellinids and barentsiids). The buds develop from small thickened ectodermal areas, which develop an atrium and a gut from an invagination of the body wall. Thick-walled resting buds are formed from the stolons of some barentsiids; at germination they develop new zooids through the same budding process.

The entoprocts are clearly a monophyletic group belonging to the Spiralia. Their larvae are typical or modified trochophores, but the sessile adults do not resemble any other spiralian type. The ciliated foot of many entoproct larvae is a specialization of the gastrotroch corresponding to the creeping area of the ancestral protostome. It resembles the molluscan foot, and the pattern of crossing ventral muscle fibres resembles that of some molluscs. This, together with some similarities, for example in the larval nervous systems, has been interpreted as synapomorphies of a group called Lacunifera (Ax 1999; Haszprunar and Wanninger 2008). However, most of these characters are open to interpretation, and the sister-group relationship is not

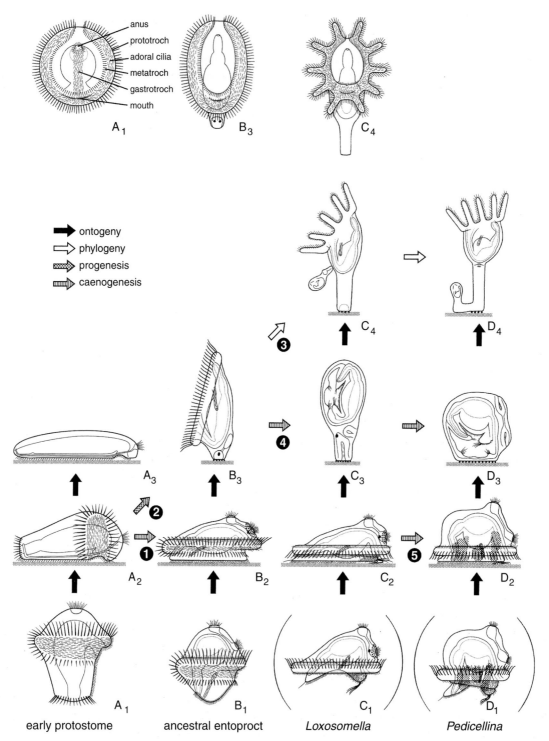

Figure 36.1. The evolution of living entoproct types from a hypothetic ancestor. (A) The early ancestor had a trochophora larva and a benthic adult creeping on the enlarged gastrotroch. (B) The ancestral entoproct had an older larva with the gastro-troch extended as a creeping sole and a frontal sense organ surrounded by gland cells (1); at settling, the larva cemented the frontal organ to the substratum by secretion from the gland cells; the adult had the same general structure as the larva (2). (C) *Loxosomella* evolved from this ancestor probably in two steps: first, the prototroch became extended onto a horseshoe of tentacles (3); and second, the metamorphosis involved a temporary closure of the atrium by constriction of the ring of cells at the apical side of the prototroch (4). (D) *Pedicellina* probably evolved through a specialization of the settling mechanism (5), which comprised the evolution of a set of attachment glands in the larval foot and the attachment of the settling larva to the substratum by the area above the contracted prototroch. (Modified from Nielsen 1971.)

seen in the molecular analyses. The foot is lacking in some larvae of *Loxosomella* and *Loxosoma*, which then resemble polychaete trochophores, but this may be a secondary reduction; the following development and metamorphosis of these larvae are unknown. The central nervous system of the adults deviates from the protostomian pattern in lacking the cerebral ganglia and the longitudinal ventral nervous system originating from the fused blastopore lips, but strongly concentrated, quite aberrant central nervous systems are found in many sessile forms, such as bryozoans.

Structurally, adult entoprocts resemble a larva that has settled and developed tentacles with loops from the prototroch, i.e. a larval stage that has become sexually mature. If the ancestral protostome had a creeping adult stage that was a deposit feeder, then the entoprocts must be interpreted as a group evolved through progenesis. Relationships with other spiralian phyla are not obvious, but as pointed out in Chapter 35, similarities in settling mode and metamorphosis of the colonial entoprocts resembles one type of ectoproct settling and metamorphosis (Fig. 35.1), and a number of molecular analyses now indicate a sister-group relationship with Cycliophora, with the ectoprocts as their sister group (see discussion in Chapter 35).

Interesting subjects for future research

1. The nervous system of stalk and stolon in the colonial species
2. Origin of the adult ganglion at metamorphosis

References

Ax, P. 1999. *Das System der Metazoa II.* Gustav Fischer, Stuttgart.

Emschermann, P. 1965a. Das Protonephridiensystem von *Urnatella gracilis* Leidy (Kamptozoa). Bau, Entwicklung und Funktion. *Z. Morphol. Oekol. Tiere* **55**: 859–914.

Emschermann, P. 1965b. Über die sexuelle Fortpflanzung und die Larve von *Urnatella gracilis* Leidy (Kamptozoa). *Z. Morphol. Oekol. Tiere* **55**: 100–114.

Emschermann, P. 1969. Ein Kreislauforgan bei Kamptozoen. *Z. Zellforsch.* **97**: 576–607.

Emschermann, P. 1972. *Loxokalypus socialis* gen. et sp. nov. (Kamptozoa, Loxokalypodidae fam. nov.), ein neuer Kamptozoentyp aus dem nördlichen Pazifischen Ozean. Ein Vorslag zur Neufassung der Kamptozoensystematik. *Mar. Biol.* **12**: 237–254.

Emschermann, P. 1982. Les Kamptozoaires. État actuel de nos connaissances sur leur anatomie, leur développement, leur biologie et leur position phylogénétique. *Bull. Soc. Zool. Fr.* **107**: 317–344.

Fuchs, J., Bright, M., Funch, P. and Wanninger, A. 2006. Immunocytochemistry of the neuromuscular systems of *Loxosomella vivipara* and *L. parguerensis* (Entoprocta: Loxosomatidae). *J. Morphol.* **267**: 866–883.

Fuchs, J., Iseto, T., Hirose, M., Sundberg, P. and Obst, M. 2010. The first internal molecular phylogeny of the animal phylum Entoprocta (Kamptozoa). *Mol. Phylogenet. Evol.* **56**: 370–379.

Haszprunar, G. and Wanninger, A. 2008. On the fine structure of the creeping larva of *Loxosomella murmanica*: additional evidence for a clade of Kamptozoa and Mollusca. *Acta Zool. (Stockh.)* **89**: 137–148.

Hilton, W.A. 1923. A study of the movements of entoproctan bryozoans. *Trans. Am. Microsc. Soc.* **42**: 135–143.

Malakhov, V.V. 1990. Description of the development of *Ascopodaria discreta* (Coloniales, Barentsiidae) and discussion of the Kamptozoa status in the animal kingdom. *Zool. Zh.* **69**: 20–30.

Marcus, E. 1939. Briozoários marinhos brasileiros III. *Bol. Fac. Filos. Cienc. Let. Univ S. Paulo, Zool.* **3**: 111–354.

Nielsen, C. 1964. Studies on Danish Entoprocta. *Ophelia* **1**: 1–76.

Nielsen, C. 1971. Entoproct life-cycles and the entoproct/ectoproct relationship. *Ophelia* **9**: 209–341.

Nielsen, C. and Jespersen, Å. 1997. Entoprocta. In F.W. Harrison (ed.): *Microscopic Anatomy of Invertebrates*, vol. 13, pp. 13–43. Wiley-Liss, New York.

Nielsen, C. and Rostgaard, J. 1976. Structure and function of an entoproct tentacle with discussion of ciliary feeding types. *Ophelia* **15**: 115–140.

Riisgård, H.U., Nielsen, C. and Larsen, P.S. 2000. Downstream collecting in ciliary suspension feeders: the catch-up principle. *Mar. Ecol. Prog. Ser.* **207**: 33–51.

Sensenbaugh, T. 1987. Ultrastructural observations on the larva of *Loxosoma pectinaricola* Franzén (Entoprocta, Loxosomatidae). *Acta Zool. (Stockh.)* **68**: 135–145.

Todd, J.A. and Taylor, P.D. 1992. The first fossil entoproct. *Naturwissenschaften* **79**: 311–314.

Wanninger, A., Fuchs, J. and Haszprunar, G. 2007. Anatomy of the serotonergic nervous system of an entoproct creeping-type larva and its phylogenetic implications. *Invert. Biol.* **126**: 268–278.

Wasson, K. 1997. Sexual modes in the colonial kamptozoan genus *Barentsia*. *Biol. Bull.* **193**: 163–170.

Woollacott, R.M. and Eakin, R.M. 1973. Ultrastructure of a potential photoreceptor organ in the larva of an entoproct. *J. Ultrastruct. Res.* **43**: 412–425.

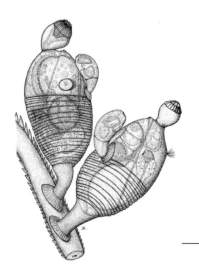

37

Phylum Cycliophora

This phylum of microscopic organisms, which occur on the mouth parts of lobsters, comprises only the genus *Symbion* with two described species, although the presence of more species has been revealed by molecular analyses (Obst *et al.* 2005; Baker and Giribet 2007). Both the structure of the various stages and the unusually complicated life cycle show that *Symbion* must be treated as a separate phylum.

The life cycle shows both sexual and asexual cycles (Fig. 37.1). The large feeding stages show peculiar cyclic degeneration and regeneration of the gut and the buccal funnel, and both sexual and asexual stages show development of internal 'embryos' that escape as non-feeding 'larvae' that settle on the host and give rise to new 'larvae' or to feeding stages (Kristensen 2002). All the 'larval' stages lack a gut.

In the asexual cycle, a 'Pandora larva' develops inside a feeding individual. The larva contains a group of cells that differentiate into a gut and buccal funnel; these structures become functional after settling, making a new feeding stage.

The sexual cycle begins with feeding stages that, after replacement of buccal funnel and gut, appear to follow either a female or a male line. Individuals of the female line develop small 'female larvae', which must be interpreted as progenetic. Individuals of the male line develop 'Prometheus larvae', which settle on an individual of the female line, encyst, and develop tiny dwarf males. Fully differentiated sperm, fertilization, and embryology have not been studied. The 'female larvae' settle on the same host and encyst, and their fertilized egg develops into a 'chordoid larva', which absorbs all the tissues of the female and escapes as the dispersal stage. It settles on a new host, encysts, and develops into the feeding stage.

The microscopic anatomy of a number of stages was described by Funch and Kristensen (1997), and a number of subsequent papers have dealt with details of single stages.

The adult feeding stages are about 350 µm long, with a goblet-shaped cuticle with the foot attached to the host. The distal part of the body is a buccal funnel with a circular mouth opening surrounded by a ring of large cells with compound cilia interspersed with smaller myoepithelial cells, which form outer and inner sphincter muscles. The ciliated cells carry several compound cilia each consisting of one row of cilia. The ring of compound cilia functions as a downstream-collecting system using the 'catch-up principle' (Riisgård *et al.* 2000). The buccal cavity is covered by compound cilia, whereas the multiciliate cells of oesophagus-stomach-rectum have separate cilia. The

Chapter vignette: Two female *Symbion pandora* with attached prometheus larvae. (Courtesy of Dr R.M. Kristensen, University Copenhagen; see also Funch and Kristensen 1997.)

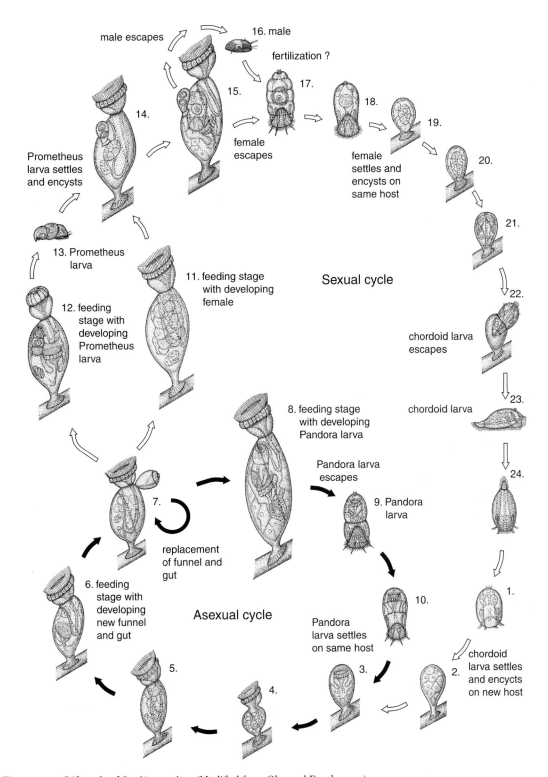

male escapes

16. male

fertilization ?

15.

17.

14.

18.

19.

Prometheus
larva settles
and encysts

female
escapes

20.

female
settles and
encysts on
same host

21.

13. Prometheus
larva

11. feeding stage
with developing
female

Sexual cycle

22.

12. feeding
stage with
developing
Prometheus
larva

chordoid larva
escapes

23.

chordoid larva

8. feeding stage
with developing
Pandora larva

24.

Pandora larva
escapes

7.

9. Pandora
larva

replacement
of funnel and
gut

1.

6. feeding
stage with
developing
new funnel
and gut

Asexual cycle

10.

Pandora
larva settles
on same host

2.

chordoid
larva settles
and encycts
on new host

5.

3.

4.

Figure 37.1. Life cycle of *Symbion pandora*. (Modified from Obst and Funch 2003.)

anus is situated just behind the base of the buccal funnel. A new buccal funnel and gut develop from a mass of undifferentiated cells in the basal part of the body, and the old structure is cast off; this process may be repeated. The cuticle consists of a thin trilaminar epicuticle and a thicker procuticle with or without fibrils. The basal membrane shows peculiar mushroom-shaped extensions into the epithelial cells. The muscle system comprises a few longitudinal muscles, which can make the funnel 'nod', and a series of weak anal-sphincter muscles (Neves *et al.* 2009a, 2010c). There is a small ganglion at the dorsal wall of the buccal cavity and a ganglion partially encircling the oesophagus.

The attached Prometheus larva (Obst and Funch 2003) (called male or dwarf male in previous papers) is ovoid, with a flattened, ciliated anteroventral field. There are many longitudinal and some oblique muscles, but only very weak circular muscles (Neves *et al.* 2009a, 2010c). The nervous system comprises ring-shaped anterior nerves with paired cerebral ganglia and paired lateral nerves (Neves *et al.* 2010a).

The dwarf male (Obst and Funch 2003; Neves *et al.* 2010d) is ovoid with an anteroventral ciliary field consisting of five rows of cells and an oval anterodorsal ciliated area. There is a complicated system of muscles with longitudinal, crossing oblique, and semicircular muscles (Neves *et al.* 2009a, 2010c). The nervous system resembles a generalized protostomian system, with a pair of large 'cerebral' ganglia connected with a pair of ventral nerves (Neves *et al.* 2010a). There is a small unpaired testis and a short penis (Neves *et al.* 2009b). It is uncertain whether the males fertilize the females still inside the feeding stage or after release.

The Pandora larva is elongate with a cuticle and an anteroventral ciliated area (Neves *et al.* 2009a, 2010a). A buccal funnel develops inside the larva. The muscular system comprises both longitudinal, oblique, and semicircular muscles (Neves *et al.* 2009a). The nervous system comprises a cerebral ganglion with three serotonergic perikarya and paired lateral nerves (Neves *et al.* 2010a).

The sexually produced chordoid larva (Funch 1996) has a characteristic, chorda-like row of muscle cells along the ventral midline. The ventral side shows a cili-ated area that is narrow at the posterior end and expands almost to the full width anteriorly, where it is bordered by two rings of compound cilia. There is a small dorsal ciliated organ that is probably not an apical organ. There are a few longitudinal, a series of weak semicircular, and a few oblique muscles (Wanninger 2005; Neves *et al.* 2009a, 2010c). The nervous system resembles a generalized protostomian system with a pair of 'cerebral' ganglia connected with a pair of ventral nerves (Funch 1996; Wanninger 2005; Neves *et al.* 2010b). A pair of protonephridia is situated mid-ventrally.

Based on the morphology, especially the 'mushroom-shaped' extensions of the basal membrane, *Symbion* was originally placed close to entoprocts and ectoprocts (Funch and Kristensen 1995; Sørensen *et al.* 2000), and the resemblance between the chordoid larva and a trochophore has been pointed out by Funch (1996). The cyclic degeneration of gut and buccal funnel resembles the cyclic renewal of the polypides of the bryozoans (Chapter 38). The ring of compound cilia can be interpreted as the circumoral part of an archaeotroch (Fig. 22.8), which has not been expanded into the lateral loops with prototroch and metatroch, and this is in agreement with its function. The few molecular analyses that include *Symbion*, generally places it together with entoprocts and bryozoans, with the entoprocts as the sister group (Hejnol *et al.* 2009; Paps *et al.* 2009; Fuchs *et al.* 2010); an earlier study placed it with the rotifers (Winnepenninckx *et al.* 1998) (see discussion in Chapter 35).

References

Baker, J.M. and Giribet, G. 2007. A molecular phylogenetic approach to the phylum Cycliophora provides further evidence for cryptic speciation in *Symbion americanus*. *Zool. Scr.* **36**: 353–359.

Fuchs, J., Iseto, T., Hirose, M., Sundberg, P. and Obst, M. 2010. The first internal molecular phylogeny of the animal phylum Entoprocta (Kamptozoa). *Mol. Phylogenet. Evol.* **56**: 370–379.

Funch, P. 1996. The chordoid larva of *Symbion pandora* (Cycliophora) is a modified trochophore. *J. Morphol.* **230**: 231–263.

Funch, P. and Kristensen, R.M. 1995. Cycliophora is a new phylum with affinities to Entoprocta and Ectoprocta. *Nature* **378**: 711–714.

Funch, P. and Kristensen, R.M. 1997. Cycliophora. In F.W. Harrison (ed.): *Microscopic Anatomy of Invertebrates*, vol. 13, pp. 409–474. Wiley-Liss, New York.

Hejnol, A., Obst, M., Stamatakis, A., *et al.* 2009. Assessing the root of bilaterian animals with scalable phylogenomic methods. *Proc. R. Soc. Lond. B* **276**: 4261–4270.

Kristensen, R.M. 2002. An introduction to Loricifera, Cycliophora, and Micrognathozoa. *Integr. Comp. Biol.* **42**: 641–651.

Neves, R.C., Kristensen, R.M. and Wanninger, A. 2009a. Three-dimensional reconstruction of the musculature of various life cycle stages of the cycliophoran *Symbion americanus. J. Morphol.* **270**: 257–270.

Neves, R.C., Kristensen, R.M. and Wanninger, A. 2010a. Serotonin immunoreactivity in the nervous system of the Pandora larva, the Prometheus larva, and the dwarf male of *Symbion americanus* (Cycliophora). *Zool. Anz.* **249**: 1–12.

Neves, R.C., da Cunha, M.R., Kristensen, R.M. and Wanninger, A. 2010b. Expression of synapsin and co-localization with serotonin and RFamide-like immunoreactivity in the nervous system of the chordoid larva of *Symbion pandora* (Cycliophora). *Invert. Biol.* **129**: 17–26.

Neves, R.C., Cunha, M.R., Funch, P., Kristensen, R.M. and Wanninger, A. 2010c. Comparative myoanatomy of cycliophoran life cycle stages. *J. Morphol.* **271**: 596–611.

Neves, R.C., da Cunha, M.R., Funch, P., Wanninger, A. and Kristensen, R.M. 2010d. External morphology of the cycliophoran dwarf male: a comparative study of *Symbion pandora* and *S. americanus. Helgol. Mar. Res.* **64**: 257–267.

Neves, R.C., Sørensen, K.J.K., Kristensen, R.M. and Wanninger, A. 2009b. Cycliophoran dwarf males break the rule: high complexity with low cell numbers. *Biol. Bull.* **217**: 2–5.

Obst, M. and Funch, P. 2003. Dwarf male of *Symbion pandora* (Cycliophora). *J. Morphol.* **255**: 261–278.

Obst, M., Funch, P. and Giribet, G. 2005. Hidden diversity and host specificity in cycliophorans: a phylogenetic analysis along the North Atlantic and Mediterranean Sea. *Mol. Ecol.* **14**: 4427–4440.

Paps, J., Baguñá, J. and Riutort, M. 2009. Lophotrochozoa internal phylogeny: new insights from an up-to-date analysis of nuclear ribosomal genes. *Proc. R. Soc. Lond. B* **276**: 1245–1254.

Riisgård, H.U., Nielsen, C. and Larsen, P.S. 2000. Downstream collecting in ciliary suspension feeders: the catch-up principle. *Mar. Ecol. Prog. Ser.* **207**: 33–51.

Sørensen, M.V., Funch, P., Willerslev, E., Hansen, A.J. and Olesen, J. 2000. On the phylogeny of the Metazoa in the light of Cycliophora and Micrognathozoa. *Zool. Anz.* **239**: 297–318.

Wanninger, A. 2005. Immunocytochemistry of the nervous system and the musculature of the cordoid larva of *Symbion pandora* (Cycliophora). *J. Morphol.* **265**: 237–243.

Winnepenninckx, B.M.H., Backeljau, T. and Kristensen, R.M. 1998. Relations of the new phylum Cycliophora. *Nature* **393**: 636–637.

Phylum **Bryozoa (Ectoprocta)**

Bryozoans, or moss animals, is a quite isolated phylum of sessile, colonial, aquatic organisms; about 6000 living species are known, and there is an extensive fossil record going back to the Early Ordovician (Xia *et al.* 2007). Two classes are recognized, Gymnolaemata, that are marine, brackish, or limnic, with small zooids with a circular tentacle crown, and Phylactolaemata, that are limnic, with larger zooids, and in most species have a horseshoe-shaped tentacle crown. This classification is followed by all modern morphologists and finds support from some molecular studies (for example Fuchs *et al.* 2009; Hausdorf *et al.* 2010), although quite bizarre results have been published (Tsyganov-Bodounov *et al.* 2009). The living gymnolaemates are usually classified in Eurystomata, that have a 'normal' embryology with planktotrophic or lecithotrophic larvae, and Cyclostomata (the only living order of the group Stenolaemata), that have specialized gonozooids and polyembryony. The living Eurystomata comprises the calcified Cheilostomata and the uncalcified Ctenostomata, but it appears that the calcified lineage originated from uncalcified 'ctenostomes' in the Upper Jurassic (Taylor 1990). Palaeozoic stenolaemates lacked gonozooids, so they may have had the 'normal' type of embryology (Taylor and Larwood 1990), and some of the very early forms, such as *Corynotrypa*,

resemble the Living uncalcified *Arachnidium*. Together, this indicates that also the stenolaemates evolved from uncalcified, ctenostome-like ancestors (Taylor 1985; Todd 2000).

A unique character of the phylum is that all zooids develop by budding, either from the body wall of the metamorphosed larva (or a germinating statoblast), or from the body wall of a zooid, or a stolon. This means that it is impossible to relate the orientation of the zooids to the larval organization, and that the terms 'dorsal' and 'ventral' cannot be applied to the zooids. Budding begins as an invagination of ectoderm and mesoderm of the body wall, forming a small sac at the inner side of the body wall; the sac becomes bilobed, with the outer cavity differentiating into the vestibule with a tentacle sheath, tentacles, and pharynx, whereas the inner portion differentiates into stomach, intestine, and rectum (review in Nielsen 1971). Some types show polymorphism, with zooids specialized for defence, cleaning, reproduction, anchoring, or other functions. Most of these special zooids lack feeding structures and are nourished by the neighbouring zooids.

The microscopic anatomy is reviewed by Mukai *et al.* (1997). All ciliated epithelia consist of multiciliate cells. A generalized feeding zooid has a box- or tube-shaped, mostly rather-stiff body wall called the cystid

Chapter vignette: A branch of the ctenostome *Farrella repens*. (Redrawn after Marcus 1926a.)

and a moveable polypide consisting of the gut and a ring of ciliated tentacles around the mouth. The ectoderm of the polypide has a thin cuticle between branched microvilli. The ectoderm of the cystid secretes a cuticle but lacks microvilli; the ctenostomes and phylactolaemates have a sometimes quite thick cuticle with proteins and chitin, whereas the cheilostomes and cyclostomes have calcified areas, where the cuticle consists of an outer, organic periostracum and an inner, calcified layer with an organic matrix; the calcified wall alone is usually called the zooecium.

The tentacle crown (often called the lophophore, but this is misleading, see below) can be retracted into the cystid that closes either through constriction or with a small operculum. Retraction is caused by strong retractor muscles extending from the basal part of the cystid to the thickened basement membrane at the base of the tentacle crown. Protrusion is caused by contraction of various muscles of the cystid wall, with the body cavity acting as a hydrostatic skeleton. In ctenostomes and phylactolaemates, the muscles constrict the whole cystid; in cheilostomes, muscles retract special, non-calcified parts of the cystid wall; and in cyclostomes, only the detached peritoneum of the cystid wall, called the membranous sac, is contracted. The first zooid in a colony develops from the metamorphosed larva, either through rearrangement and differentiation of larval tissues or blastemas, or through a budding process like that of all the following polypides. The body cavity remains continuous in phylactolaemate colonies, where only incomplete walls separate the zooids. The developing gymnolaemate zooids form almost complete septa between neighbouring zooids, but the ectoderm remains continuous through openings in pore plates that are plugged by special rosette cells. The cyclostome colonies have a continuous primary body cavity, but the zooids have individual coelomic cavities bordered by the membranous sac.

The tentacle crown surrounds the ciliated mouth. Most of the phylactolaemates have the lateral sides of the tentacle crown extended posteriorly (in the direction of the anus), so that the tentacle crown becomes horseshoe-shaped; they also have a lip (epistome)

originating from the body wall at the posterior side of the mouth. The tentacles show characteristic patterns of ciliated cells, thickened basement membrane, and peritoneal cells with frontal and abfrontal muscles (Nielsen 2002; Riisgård et al. 2004). Each tentacle has two rows of multiciliate ectodermal cells along the lateral sides. The frontal side of the tentacles has one row of cells in gymnolaemates, whereas the phylactolaemates have a wide band of cells with interspersed sensory cells; the cyclostomes have a row of unciliated, apparently secretory frontal cells. One or more basepithelial nerves run beneath the frontal cells. A row of monociliate laterofrontal sensory cells with a basiepithelial nerve borders the frontal cells. The lateral ciliary bands create a water current towards the centre of the tentacle crown and out between the tentacles, and food particles are strained from this current. The cyclostome tentacle is the least complicated; its long, stiff laterofrontal cilia function as a mechanical filter, and when a particle is caught the deflection of the cilia triggers a flicking movement of the tentacle towards the centre of the tentacle crown, where the water current carries the particle to the mouth (Nielsen and Riisgård 1998). The gymnolaemate tentacle shows the same filtering mechanism (Riisgård and Manríquez 1997), but this rather passive type of particle capture is in almost all species complicated by various types of tentacle movements that result in capture of particles of different characters (Winston 1978). Particle capture by reversal of the ciliary beat has been proposed by Strathmann (1982), but direct observation has not been possible. The phylactolaemates have a wide band of frontal ciliated cells flanked by laterofrontal cells with a cilium, and the particle collection generally resembles that of the gymnolaemates (Riisgård et al. 2010). However, *Plumatella* apparently lacks muscles in the tentacles and tentacle flicks have not been observed (Riisgård et al. 2004).

The gut is a U-shaped tube with a number of ciliated regions; it develops from the ectoderm during budding, so it is not possible to make direct statements about ecto- and endodermal regions. The pharynx of gymnolaemates and cyclostomes is triradial and consists of myoepithelial cells each with a large vacuole;

the contraction of the radial, cross-striated myofilaments shortens and widens the cells thereby expanding the pharynx, so that food particles collected in front of the mouth become swallowed (Gordon 1975; Nielsen and Riisgård 1998). The phylactolaemate pharynx is non-muscular, and the epistome is apparently controlling the ingestion of captured particles (Brien 1953). Some gymnolaemates have a gizzard at the entrance to the stomach. Each tooth is secreted by an epithelial cell with microvilli; the teeth have a honeycomb-structure with cylindrical canals, with the microvilli extending into the basal part of the canals. The presence of a cuticle indicates that the whole pharynx is ectodermal. The basal point of the stomach is attached to the cystid by a tissue strand called the funiculus. It is simple, comprising only muscle cells and peritoneal cells (with testes) in the cyclostomes. In the phylactolaemates it additionally contains an extension of the ectoderm that secretes the cuticle of the resting bodies called statoblasts. The eurystomes have a more complicated system of additional, hollow, branching mesodermal strands, which attach to the cystid wall in connection with the interzooidal pores. This system apparently transports substances within a zooid, and the polarized rosette cells of the pores have been shown to transport organic molecules across the pores (Lutaud 1982). Carle and Ruppert (1983) interpreted the funiculi and funicular systems of all bryozoans as homologues of the haemal systems of brachiopods and phoronids, based on their structure, i.e. blastocoelic cavities surrounded by basement membrane, and their function, i.e. transport. However, there is nothing to indicate that the small, unbranched funiculi of phylactolaemates and cyclostomes function as circulatory organs, and the ground plan of the gymnolaemate funicular system bears no resemblance to the haemal systems of phoronids and brachiopods. If this homology is followed further, it leads to the conclusion that all blastocoels are homologous, and the information about homology of specialized haemal systems is lost.

Each polypide has a ganglion at the posterior (anal) side of the mouth with lateral nerves or extensions following the tentacle bases around the mouth. The peripheral nervous system of the individual zooid is delicate, and most of the studies are based on vital staining supplemented by a few electron microscopic studies. A few nerves connect the ganglion with a fine nerve net that connects the zooids. The connection is through intermediate cells in the rosettes of the pore plates in the gymnolaemates, and Thorpe (1982) observed electrical impulses across colonies of *Flustrellidra*. The interzooidal connections are more uncomplicated in the phylactolaemates that lack walls between the zooids; interzooidal connections are not known in cyclostomes.

The colonies have species-specific growth patterns, and the growth areas vary from wide zones along the edge of the colonies, as in phylactolaemates and *Membranipora*, to narrow points at the tips of stolons, as in stolonate ctenostomes, or to certain points of the cystid, as in *Electra* and *Crisia*. Ectoderm and mesoderm can be difficult to distinguish in these areas (Borg 1926; Brien and Huysmans 1938), but the two layers become distinct a short distance from the growth zone. The peritoneum surrounds a spacious coelom, which shows some important differences between the classes.

Gymnolaemate zooids each have a well-delimited coelom, with the cystid wall consisting of ectoderm, muscles, and peritoneum. There is a ring-shaped coelom around the mouth, with extensions into the tentacles and a posterior opening to the main cystid coelom. Cyclostomes have very unusual body cavities (Nielsen and Pedersen 1979), with the peritoneum of the cystid with its basement membrane and a series of very thin, annular muscles detached from the ectoderm to form the membranous sac. The coelomic cavity of each polypide is completely separated from that of the neighbouring zooids, whereas the spacious pseudocoel surrounding the membranous sac is in open connection with that of the neighbouring zooids through communication pores, or via the common extrazooidal cavity at the surface of the colonies. The tentacle coelom is narrow and continuous with the body coelom through a pore above the ganglion.

Phylactolaemates have a somewhat more complicated coelom in the polypide, but the descriptions of

Brien (1960) and Gruhl et al. (2009) settle much of the uncertainty found in the older literature. There is a coelomic canal in the tentacle basis with branches extending into the tentacles. A pair of strongly ciliated canals along the posterior (anal) side of the buccal cavity and the ganglion connects the posterior part of the tentacle coelom to the main coelom. In *Cristatella*, the median part of the tentacle coelom canal (at the upper part of the ciliated canals) is expanded into a small posterior bladder, in which amoebocytes with excretory products accumulate. The amoebocytes may become expelled from the bladder, but there is no permanent excretory pore. A narrow canal from the main coelom extends between the two ciliated canals to a more spacious cavity in the lip. The cilia of the peritoneum create a circulation of the coelomic fluid. It is clear that there is one, rather complicated coelomic cavity in the polypide, and that archimery and metanephridia are not present.

Special excretory organs are not present, but waste products may accumulate in the cells of the gut; the whole polypide degenerates periodically, and a new polypide forms by budding from the cystid wall.

Bryozoan colonies are hermaphroditic, usually with hermaphroditic zooids, but male and female zooids are found in a number of cheilostomes and in all cyclostomes. The testes are usually situated on the funiculus and the ovaries on the cystid wall. The ripe gametes float in the coelom, and the spermatozoa are liberated through a small, transitory pore at the tentacle tips in the gymnolaemates (Silén 1966; 1972). Gymnolaemate sperm shows considerable variation, but both head and midpiece are always elongate (Franzén 1956). The spermatozoa are usually spawned separately, but, for example, *Membranipora* spawns 32 or 64 spermatozoa in spermatozeugmata that enter the 'female' through the ciliated funnel (Temkin 1994; Temkin and Bortolami 2004). Phylactolaemate sperm is highly characteristic, with a short head, a mid-piece with spirally coiled mitochondria, and a rather short and thick tail (Lützen et al. 2009).

The mode of fertilization has been much discussed, and self-fertilization has been suggested in species where the zooids are simultaneous hermaphrodites.

However, observations and experiments with the cheilostome *Celleporella* indicated that outcrossing is prevalent (Hoare et al. 1999; Hughes et al. 2009). Observations on polymorphic DNA in phylactolaemates have given evidence for outcrossing in this group (Jones et al. 1994).

Gymnolaemate reproduction, development, and metamorphosis have been studied by a number of authors (extensive review in Reed 1991), but there are still considerable gaps in our knowledge. The eggs are fertilized in the ovaries, but the fertilization membrane is not formed until after spawning. The eggs are usually shed through a median supraneural pore at the posterior (anal) side of the tentacle crown. The supraneural pore is simple in most species, but in the spawning period of free spawners it is extended into a ciliated funnel, the intertentacular organ, situated between the two posterior tentacles. *Epistomia* is viviparous with a single tiny egg nourished in the maternal cystid after the degeneration of the polypide.

The free spawners have planktotrophic, shelled cyphonautes larvae (Nielsen and Worsaae 2010), but the majority of species brood the embryos in one of a bewildering variety of ways, and their larvae are lecithotrophic or placentally nourished, and are usually without shells. Simple retention of the fertilized eggs attached to the tentacle sheath is seen, for example, in the ctenostome *Triticella*, and brooding in the retracted tentacle sheath of zooids with partially degenerated polypide is known, for example, in *Bowerbankia*. Brooding in special ovicells formed by the zooid distal to the maternal zooid is found in many cheilostomes (Ostrovsky et al. 2009). The eggs of the brooding species are usually quite large and the development lecithotrophic, but a few species, such as *Bugula neritina*, deposit the very small egg in an ovicell closed by an extension of the maternal epithelium that becomes a placenta nourishing the developing embryo through the epithelium of the developing adhesive sac.

The regulative powers of the embryos are almost unstudied, but Zimmer (1973) isolated the blastomeres of 2-cell stages of *Membranipora isabelleana* and found complete regulation, although it was not stated how far the development was followed.

The small, freely spawned eggs cleave equally so that a radially symmetrical 8-cell stage with two tiers of four cells is formed, but the planes of the following cleavage are lateral to the primary, apical-blastoporal axis, so that the 16-cell stage is transversely elongate consisting of two apical and two blastoporal rows of four cells each. The polar bodies remain visible near the apical pole until gastrulation. In *Membranipora*, the following cleavages result in a spherical coeloblastula, and four or eight cells at the blastoporal pole move into the blastocoel forming the endoderm in a solid gastrula (Gruhl 2010). At this stage, one ectodermal cell just above the equator of the embryo becomes internalized and differentiates as mesoderm. The following development shows that this side of the embryo becomes anterior. The apical cells differentiate into an apical organ, and the endoderm develops a cavity connected to an ectodermal invagination, the vestibule. The mesoderm spreads and differentiates into the anterior and posterior neuromuscular strands. The endoderm stretches and becomes differentiated into a funnel-shaped pharynx and the more spacious stomach. The embryos develop cilia protruding through the egg membrane on the apical cells and on a ring of large coronal cells (which may be interpreted as prototrochal cells, see below). Finally the egg membrane breaks along the corona, exposing the vestibule with the ciliated ridge. The young cyphonautes has a pair of delicate lateral shells.

Cyphonautes larvae are planktotrophic and spend weeks in the plankton before settling. The cyphonautes larvae of the cheilostomes *Electra* (Kupelwieser 1905) and *Membranipora* (Stricker *et al.* 1988a, b), and the ctenostome *Hislopia* (Nielsen and Worsaae 2010) have been studied in detail. The apical organ consists of concentric rings of unciliated and monociliate cells, and myoepithelial cells. A median nerve flanked by a pair of lateral muscles connect the apical organ and the pyriform organ, which is a strongly ciliated cleft of thick ventral epithelium bordered by the anterior loop of the corona; it contains both gland cells and nerve cells and is obviously active in the explorative phase of the settling. The prototroch is situated along the lower edge of the shells and consists of two rows of large,

multiciliate cells, with a row of cells with a single row of cilia on the apical side. The coronal cilia are not organized as compound cilia; they show metachronal waves, and the effective stroke makes the larva swim with the apical organ in front, but the stroke can be reversed for short periods (Nielsen 1987). The spacious vestibule is divided into an anterior-inhalant and a posterior-exhalant chamber by a U-shaped ciliated ridge. The ridge of the large larvae of *Electra* and *Membranipora* bears three rows of ciliated cells: a row of multiciliate lateral cells creating the current through the gap between the ridges, a row of biciliate laterofrontal cells, and a row of multiciliate frontal cells (Nielsen 2002). Young but feeding larvae of *Membranipora* and the small larvae of *Hislopia* lack the frontal row of cilia. Strathmann and McEdward (1986) observed that the laterofrontal cilia are stiff and function as a mechanical filter, and Strathmann (2006) suggested that the laterofrontal cilia may push captured particles towards the frontal cilia. The frontal cilia and cilia in the upper part of the vestibule carry the food particles to the mouth. The median epithelium of the exhalant chamber forms a large, complicated, glandular invagination, the adhesive sac. The gut consists of a ciliated oesophagus and stomach, and an unciliated rectum. Species with non-planktotrophic development also show a biradial cleavage pattern (Fig. 38.1). All species studied so far appear to have the same cleavage pattern, but only few species have been followed in more detail, and it has only been possible to follow the cell lineage to the 64-cell stage. The cleavages do not show the alternating oblique cleavage furrows of the spiral cleavage, but if the notation of the spiral cleavage is used (Table 38.1), it turns that the corona cells that give rise to the large ciliary band are descendants of the cells $1q^{12}$ and $1q^{22}$, which is practically identical to the pattern seen in typical spiralians, where the primary prototroch cells derive from the cells $1q^2$ and secondary prototroch cells from $1q^{12}$ (Chapter 23). The corona can therefore perhaps be interpreted as a prototroch. The eight coronal cells divide twice, and the 32 cells become the ring of very large corona cells. This number of prototroch cells has been observed in many gymnolaemate larvae, but

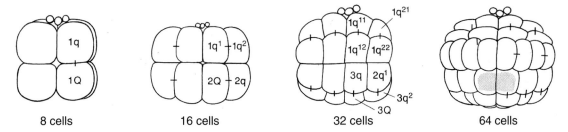

Figure 38.1. Embryology of *Bugula flabellata*. All embryos are in lateral view. The 4q cells inside the blastocoel of the 64-cell stage are indicated by shading. (Based on Corrêa 1948.)

Table 38.1. Cell lineage of one quadrant of *Bugula flabellata* (based on Corrêa 1948); the four quadrants show biradial symmetry so the individual quadrants cannot be identified and the letter q/Q is therefore used. The first notation for each blastomere is given in accordance with the usual spiralian pattern with '1' indicating the cells closest to the apical pole along the meridian of the embryo; the numbers in parenthesis are the notation given by Corrêa.

4 cells	8 cells	16 cells	32 cells	64 cells	
				$1q^{111}$ (2)	apical plate
			$1q^{11}$ (2)	$1q^{112}$ (10)	
		$1q^1$		$1q^{121}$ (18)	corona
			$1q^{12}$ (10)	$1q^{122}$ (22)	
	$1q$			$1q^{211}$ (6)	
			$1q^{21}$ (6)	$1q^{212}$ (14)	
		$1q^2$		$1q^{221}$ (26)	corona
			$1q^{22}$ (14)	$1q^{222}$ (30)	
Q				$2q^{11}$ (38)	
			$2q^1$ (22)	$2q^{12}$ (54)	
		$2q$		$2q^{21}$ (42)	
			$2q^2$ (26)	$2q^{22}$ (58)	
	$1Q$			$3q^1$ (34)	
			$3q$ (18)	$3q^2$ (50)	
		$2Q$		$4q$ (46)	mesoderm+endoderm
			$3Q$ (30)	$4Q$ (62)	

higher numbers are observed in other species. The four cells at the blastoporal pole of the 32-cell stage divide horizontally so that four cells become situated inside the blastula; these cells migrate towards the periphery of the blastocoel and can be recognized in embryos of many species (Corrêa 1948; d'Hondt 1983). The following cell divisions are more difficult to follow, and it is unclear whether more cells enter the blastocoel. The labelling of the eight lower cells is therefore tentative. The following development has not been studied in detail.

The non-planktotrophic larvae are of several types (review in Reed 1991). Both shelled larvae with a non-functioning but almost complete gut and cyphonautes-like larvae without shells are known. Various types of a more spherical or elongate larvae, so-called coronate larvae, with the prototroch expanded to cover the surface of the larva, except the apical plate and a smaller or larger area at the opening of the adhesive sac, are found in many genera. The well-described larva of *Bugula neritina* has a large apical organ with a number of cell types, covering a thick blastema of undifferentiated cells, and similar, although less conspicuous structures have been reported in most other gymnolaemate larvae. The nervous and muscular systems of the lecithotrophic larvae are generally less complex than those of the cyphonautes. Many of the coronate larvae have pigment spots, which are presumably photosensory, associated with the corona; some spots consist of a number of pigment cells surrounding a depression with a putative sensory cell (which may have pigment too), with a bundle of almost unmodified cilia, whereas others consist of only one pigmented cell; non-pigmented cells may be photosensitive too.

Settling behaviour and the first phases of the metamorphosis are similar in all the gymnolaemates, but the later stages show considerable variation (Zimmer and Woollacott 1977; see Fig. 38.2). The larvae creep on the substratum exploring with the pyriform organ, and when a suitable spot has been found, the adhesive sac everts and gives off its secretion. Settling cyphonautes larvae expand the adhesive sac over the substratum and contract muscles between the adhesive sac and the apical organ, so that the adductor muscles of the shells rupture and the larva changes from a laterally compressed to a dorsoventrally compressed shape. The pal-lial epithelium releases the shells and extends over the upper side of the metamorphosing larva. The lower periphery of the pallial epithelium fuses with the periphery of the adhesive sac, enclosing the prototroch and all the larval organs that degenerate; the apical organ becomes invaginated. The newly metamorphosed larva, the primary disc, is completely covered by ectoderm, but its internal morphology is rather chaotic, consisting of a ring-shaped portion of the former exterior (at first surrounded by the infolded prototroch) and various partially degenerating larval organs. Two opposite lines can be followed from the cyphonautes type (Fig. 38.2): the *Bugula* type, characterized by an extreme expansion of the adhesive sac that finally covers the upper side of the primary disc; and the *Bowerbankia* type that goes to the opposite extreme by withdrawing the adhesive sac and expanding the pallial epithelium along the lower side of the primary disc (Fig. 35.2).

The polypide of the ancestrula develops from a blastema on the underside of the apical organ, around the adhesive sac, or possibly from an infracoronal ring (Reed 1991). The cystid wall expands in various patterns, and new polypides develop from the cystid wall through small invaginations of the body wall (review in Nielsen 1971). The invagination becomes the vestibule, but there is some uncertainty about the origin of the various parts of the gut. Most accounts describe a posterior pouch from the vestibule differentiating into rectum-intestine-stomach, and an anterior pouch that becomes the oesophagus; a secondary opening is then formed between oesophagus and stomach. The ganglion develops from an ectodermal invagination at the posterior side of the area of the mouth.

Figure 38.2. Types of metamorphosis in bryozoans emphasizing variations in the origin of the cystid epithelium. Eurystomata: *Electra pilosa*, free-swimming and just-formed primary disc; the cystid epithelium originates from the adhesive sac and pallial epithelium (mostly underlying the shells). (Based on Nielsen 1971.) *Bugula neritina*, free-swimming larva, newly settled larva, and just formed primary disc; the cystid epithelium originates from the adhesive sac. (Based on Reed and Woollacott 1982, 1983.) *Bowerbankia gracilis*, free-swimming larva, just settled larva with everted adhesive sac, and young primary disc; the cystid epithelium originates from the pallial epithelium. (Based on Reed and Cloney 1982a,b.) Cyclostomata: *Crisia eburnea*, free-swimming larva, settling larva, and just formed primary disc; the cystid epithelium originates from the adhesive sac and the cuticle-lined epithelium of the apical invagination, i.e. a pallial epithelium. (Based on Nielsen 1970.) Phylactolaemata: *Plumatella fungosa*, free-swimming stage with precociously developed polypides at the bottom of the apical invagination, free-swimming stage with everted polypides, and fully metamorphosed stage; the cystid epithelium originates from the epithelium of the apical invagination, i.e. a pallial epithelium. (Based on Brien 1953.)

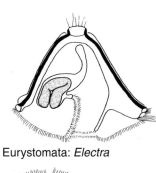

pallial epithelium
adhesive sac

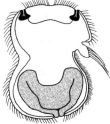

Eurystomata: *Electra*

Eurystomata: *Bugula*

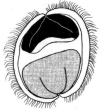
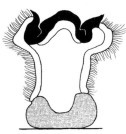
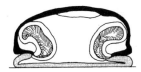

Eurystomata: *Bowerbankia*

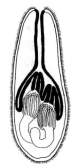
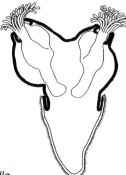
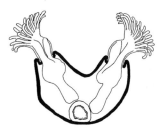

Cyclostomata: *Crisia*

Phylactolaemata: *Plumatella*

Cyclostomes have a highly specialized type of reproduction with polyembryony (Harmer 1893). Only a few zooids in special positions in the colonies are female, and when their single egg has been fertilized, the polypide degenerates and the zooid becomes a large gonozooid, nourished by the neighbouring zooids. The embryo becomes irregular and gives off secondary and even tertiary embryos. The embryos finally differentiate into almost spherical larvae without any trace of apical organ, pyriform organ, or gut. The ectoderm forms a large invagination of the apical side and secretes a cuticle, and a corresponding invagination of the opposite side becomes a large adhesive sac. The whole outer part of the ectoderm is ciliated. The liberated larvae (Nielsen 1970) swim for a few hours and settle by everting the adhesive sac onto the substratum; the apical invagination everts at the same time, and the ciliated ectoderm becomes internalized as a ring-shaped cavity with all the cilia, and degenerates (Fig. 38.2). The first polypide develops from a layer of cells below the ectoderm of the apical area. The origin and differentiation of the mesoderm including the membranous sac have not been studied.

The polyembryony and lack of an apical organ make it difficult to ascertain the axes of the cyclostome larva, but the cuticle is probably secreted by the epithelial area corresponding to the pallial area that secretes the shells in the cyphonautes larva, and the adhesive organs of the larvae are probably homologous. The ciliated epithelium is probably homologous with the corona, although it consists of many small cells instead of one ring of large cells. The primary disc is thus covered by pallial and adhesive sac epithelia, like the primary disc of *Electra* (Fig. 38.2).

Subsequently, buds are formed through a process resembling that of the gymnolaemates, but the origin of the various cell layers has not been studied.

The phylactolaemates are viviparous, and fertilization takes place in the ovary. Species of most of the eight genera have been studied, but the most detailed information comes from *Plumatella* (Brien 1953; review in Reed 1991). The small fertilized egg enters an embryo sac, which is an invagination of the body wall. The early development shows no definite patterns of cleavage or germ-layer formation, but an elongate, two-layered embryo without any indication of a gut has been observed in all species investigated. The embryos become nourished by placental structures situated either in an equatorial zone, for example in *Plumatella*, or an 'apical' zone (see below), for example in *Fredericella*. One or more polypide buds develop from invaginations of the body wall above the annular placenta, or lateral to the apical placenta, and the polypides become fully developed before the 'larva' is liberated. There is one polypide in *Fredericella*, usually two in *Plumatella*, and from four to several in *Cristatella* and *Pectinatella*. The 'larva' hatches from the maternal zooid and swims with the pole opposite the invagination with polypides in front. Marcus (1926b) noted a concentration of nerve cells at this 'anterior' pole, but there is no well-defined epithelial thickening as in other spiralian apical organs. Franzén and Sensenbaugh (1983) studied the ultrastructure of the anterior pole and found a number of different cell types, including ciliated cells, sensory cells, nerve cells, and gland cells, but not a single cell type showing specific similarities, with cells of the apical organ of the gymnolaemate larvae could be pointed out. On the other hand, it was stated that the glandular cells resemble the cells of the adhesive sac of the gymnolaemate larvae and that their secretion is released at settling. My conclusion is that the anterior pole of the phylactolaemate larva is homologous with the adhesive-sac epithelium of the gymnolaemate larvae, and that the larva therefore swims 'backwards' with the apical end trailing (Nielsen 1971) (Fig. 38.2). This is in no way contradicted in the beat direction of the cilia of the phylactolaemate larvae because the gymnolaemate larvae are known to be able to reverse the beat of the coronal cilia (Reed and Cloney 1982b; Nielsen 1987). This orientation makes the position of the precociously formed polypide buds agree with the position of the twin buds in *Membranipora*, which develop from an area of the pallial epithelium. The free-swimming colonies settle after a short pelagic phase, and the whole ciliated epithelium, representing the extended prototroch and the adhesive sac, becomes invaginated.

The polypides of the 'larva' are normal polypide buds, as are the buds from the germinating statoblasts, and their development from the two-layered body wall has been studied in a number of species (Brien 1953). The development of the polypides resembles that of the gymnolaemates. The shells of the statoblasts are formed by ectodermal cells in the funiculus that secrete chitinous shells around a mass of cells rich in stored nutrients.

The monophyly of the Bryozoa is almost unquestioned, although some earlier molecular phylogenies failed to show this. The structure and development both of the colonies and the individual polypides show unique characters, the most conspicuous being the division of the body into polypide and cystid, and the loss of larval structures with subsequent development of all zooids through budding from the body wall of the settled larva, or precociously in the free-swimming larval stage.

The relationships with entoprocts, cycliophorans, and brachiozoans are discussed in Chapter 35.

Interesting subjects for future research

1. Early embryology of phylactolaemates
2. Polypide development in ancestrulae of a number of groups
3. Budding of cyclostomes
4. Colonial nervous system of cyclostomes

References

Borg, F. 1926. Studies on Recent cyclostomatous Bryozoa. *Zool. Bidr. Upps.* **10**: 181–507.

Brien, P. 1953. Étude sur les Phylactolémates. *Ann. Soc. R. Zool. Belg.* **84**: 301–444.

Brien, P. 1960. Classe des Bryozoaires. *Traité de Zoologie*, vol. 5(2), pp. 1054–1335. Masson, Paris.

Brien, P. and Huysmans, G. 1938. La croissance et le bourgeonnement du stolon chez les Stolonifera (*Bowerbankia* (Farre)). *Ann. Soc. R. Zool. Belg.* **68**: 13–40.

Carle, K.J. and Ruppert, E.E. 1983. Comparative ultrastructure of the bryozoan funiculus: a blood vessel homologue. *Z. Zool. Syst. Evolutionsforsch.* **21**: 181–193.

Corrêa, D.D. 1948. A embriologia de *Bugula flabellata* (J.V. Thompson) (Bryozoa Ectoprocta). *Bol. Fac. Filos. Cienc. Let. Univ S. Paulo, Zool.* **13**: 7–71.

d'Hondt, J.L. 1983. Sur l'évolution des quatre macromères du pôle végétatif chez les embryons de Bryozoaires Eurystomes. *Cah. Biol. Mar.* **24**: 177–185.

Franzén, Å. 1956. On spermiogenesis, morphology of the spermatozoon, and biology of fertilization among invertebrates. *Zool. Bidr. Upps.* **31**: 355–480.

Franzén, Å. and Sensenbaugh, T. 1983. Fine structure of the apical plate of the freshwater bryozoan *Plumatella fungosa* (Pallas) (Bryozoa: Phylactolaemata). *Zoomorphology* **102**: 87–98.

Fuchs, J., Obst, M. and Sundberg, P. 2009. The first comprehensive molecular phylogeny of Bryozoa (Ectoprocta) based on combined analyses of nuclear and mitochondrial genes. *Mol. Phylogenet. Evol.* **52**: 225–233.

Gordon, D.P. 1975. Ultrastructure and function of the gut of a marine bryozoan. *Cah. Biol. Mar.* **16**: 367–382.

Gruhl, A. 2010. Ultrastructure of mesoderm formation and development in *Membranipora membranacea* (Bryozoa: Gymnolaemata). *Zoomorphology* **129**: 45–60.

Gruhl, A., Wegener, I. and Bartolomaeus, T. 2009. Ultrastructure of the body cavities in Phylactolaemata (Bryozoa). *J. Morphol.* **270**: 306–318.

Harmer, S.F. 1893. On the occurrence of embryonic fission in cyclostomatous Polyzoa. *Q. J. Microsc. Sci., N. S.* **34**: 199–241.

Hausdorf, B., Helmkampf, M., Nesnidal, M.P. and Bruchhaus, I. 2010. Phylogenetic relationships within the lophophorate lineages (Ectoprocta, Brachiopoda and Phoronida). *Mol. Phylogenet. Evol.* **55**: 1121–1127.

Hoare, K., Hughes, R.N. and Goldson, A.J. 1999. Molecular genetic evidence for the prevalence of outcrossing in the hermaphroditic brooding bryozoan *Celleporella hyalina*. *Mar. Ecol. Prog. Ser.* **188**: 73–79.

Hughes, R.N., Wright, P.J., Carvalho, G.R. and Hutchingson, W.F. 2009. Patterns of self compatibility, inbreeding depression, outcrossing, and sex allocation in a marine bryozoan suggest the predominating influence of sperm competition. *Biol. J. Linn. Soc.* **98**: 519–531.

Jones, C.S., Okamura, B. and Noble, L.R. 1994. Parent and larval RAPD fingerprints reveal outcrossing in freshwater bryozoans. *Mol. Ecol.* **3**: 193–199.

Kupelwieser, H. 1905. Untersuchungen über den feineren Bau und die Metamorphose des Cyphonautes. *Zoologica (Stuttg.)* **47**: 1–50.

Lutaud, G. 1982. Étude morphologique et ultrastructurale du funicule lacunaire chez le Bryozoaire Chilostome *Electra pilosa* (Linné). *Cah. Biol. Mar.* **23**: 71–81.

Lützen, J., Jespersen, Å. and Nielsen, C. 2009. Ultrastructure of spermiogenesis in *Cristatella mucedo* Cuvier (Bryozoa: Phylactolaemata: Cristatellidae). *Zoomorphology* **128**: 275–283.

Marcus, E. 1926a. Beobachtungen und Versuche an lebenden Meeresbryozoen. *Zool. Jahrb., Syst.* **52**: 1–102.

Marcus, E. 1926b. Beobachtungen und Versuche an lebenden Süsswasserbryozoen. *Zool. Jahrb., Syst.* **52**: 279–350.

Mukai, H., Terakado, K. and Reed, C.G. 1997. Bryozoa. In F.W. Harrison (ed.): *Microscopic Anatomy of Invertebrates*, vol. 13, pp. 45–206. Wiley-Liss, New York.

Nielsen, C. 1970. On metamorphosis and ancestrula formation in cyclostomatous bryozoans. *Ophelia* **7**: 217–256.

Nielsen, C. 1971. Entoproct life-cycles and the entoproct/ectoproct relationship. *Ophelia* **9**: 209–341.

Nielsen, C. 1987. Structure and function of metazoan ciliary bands and their phylogenetic significance. *Acta Zool. (Stockh.)* **68**: 205–262.

Nielsen, C. 2002. Ciliary filter-feeding structures in adult and larval gymnolaemate bryozoans. *Invert. Biol.* **121**: 255–261.

Nielsen, C. and Pedersen, K.J. 1979. Cystid structure and protrusion of the polypide in *Crisia* (Bryozoa, Cyclostomata). *Acta Zool. (Stockh.)* **60**: 65–88.

Nielsen, C. and Riisgård, H.U. 1998. Tentacle structure and filter-feeding in *Crisia eburnea* and other cyclostomatous bryozoans, with a review of upstream-collecting mechanisms. *Mar. Ecol. Prog. Ser.* **168**: 163–186.

Nielsen, C. and Worsaae, K. 2010. Structure and occurrence of cyphonautes larvae (Bryozoa, Ectoprocta). *J. Morphol.* **271**: 1094–1109.

Ostrovsky, A.N., Nielsen, C., Vávra, N. and Yagunova, E.B. 2009. Diversity of brood chambers in calloporid bryozoans (Gymnolaemata, Cheilostomata): comparative anatomy and evolutionary trends. *Zoomorphology* **128**: 13–35.

Reed, C.G. 1991. Bryozoa. In A.C. Giese, J.S. Pearse and V.B. Pearse (eds): *Reproduction of Marine Invertebrates*, vol. 6, pp. 85–245. Boxwood Press, Pacific Grove, CA.

Reed, C.G. and Cloney, R.A. 1982a. The larval morphology of the marine bryozoan *Bowerbankia gracilis* (Ctenostomata: Vesicularioidea). *Zoomorphology* **100**: 23–54.

Reed, C.G. and Cloney, R.A. 1982b. The settlement and metamorphosis of the marine bryozoan *Bowerbankia gracilis* (Ctenostomata: Vesicularioidea). *Zoomorphology* **100**: 103–132.

Reed, C.G. and Woollacott, R.M. 1982. Mechanisms of rapid morphogenetic movements in the metamorphosis of the bryozoan *Bugula neritina* (Cheilostomata, Cellularioidea): I. Attachment to the substratum. *J. Morphol.* **172**: 335–348.

Reed, C.G. and Woollacott, R.M. 1983. Mechanisms of rapid morphogenetic movements in the metamorphosis of the bryozoan *Bugula neritina* (Cheilostomata, Cellularioidea): II. The role of dynamic assemblages of microfilaments in the pallial epithelium. *J. Morphol.* **177**: 127–143.

Riisgård, H.U. and Manríquez, P. 1997. Filter-feeding in fifteen marine ectoprocts (Bryozoa): particle capture and water pumping. *Mar. Ecol. Prog. Ser.* **154**: 223–239.

Riisgård, H.U., Nielsen, K.K., Fuchs, J., *et al.* 2004. Ciliary feeding structures and particle capture mechanism in the freshwater bryozoan *Plumatella repens* (Phylactolaemata). *Invert. Biol.* **123**: 156–167.

Riisgård, H.U., Okamura, B. and Funch, P. 2010. Particle capture in ciliary filter-feeding gymnolaemate and phylac-

tolaemate bryozoans - a comparative study. *Acta Zool. (Stockh.)* **91**: 416–425.

Silén, L. 1966. On the fertilization problem in gymnolaematous Bryozoa. *Ophelia* **3**: 113–140.

Silén, L. 1972. Fertilization in Bryozoa. *Ophelia* **10**: 27–34.

Strathmann, R.R. 1982. Cinefilms of particle capture by an induced local change of beat of lateral cilia of a bryozoan. *J. Exp. Mar. Biol. Ecol.* **62**: 225–236.

Strathmann, R.R. 2006. Versatile ciliary behaviour in capture of particles by the bryozoan cyphonautes larva. *Acta Zool. (Stockh.)* **87**: 83–89.

Strathmann, R.R. and McEdward, L.R. 1986. Cyphonautes' ciliary sieve brakes a biological rule of inference. *Biol. Bull.* **171**: 754–760.

Stricker, S.A., Reed, C.G. and Zimmer, R.L. 1988a. The cyphonautes larva of the marine bryozoan *Membranipora membranacea*. I. General morphology, body wall, and gut. *Can. J. Zool.* **66**: 368–383.

Stricker, S.A., Reed, C.G. and Zimmer, R.L. 1988b. The cyphonautes larva of the marine bryozoan *Membranipora membranacea*. II. Internal sac, musculature, and pyriform organ. *Can. J. Zool.* **66**: 384–398.

Taylor, P.D. 1985. Carboniferous and Permian species of the cyclostome bryozoan *Corynotrypa* Bassler, 1911 and their clonal propagation. *Bull. Br. Mus. (Nat. Hist.), Geol.* **38**: 359–372.

Taylor, P.D. 1990. Bioimmured ctenostomes from the Jurassic and the origin of the cheilostome Bryozoa. *Palaeontology* **33**: 19–34.

Taylor, P.D. and Larwood, G.P. 1990. Major evolutionary radiations in the Bryozoa. In P.D. Taylor and G.P. Larwood (eds): *Major Evolutionary Radiations (The Systematic Association Special volume no. 42)*, pp. 209–233. Oxford University Press, Oxford.

Temkin, M.H. 1994. Gamete spawning and fertilization in the gymnolaemate bryozoan *Membranipora membranacea*. *Biol. Bull.* **187**: 143–155.

Temkin, M.H. and Bortolami, S.B. 2004. Waveform dynamics of spermatozeugmata during the transfer from paternal to maternal individuals of *Membranipora membranacea*. *Biol. Bull.* **206**: 35–45.

Thorpe, J.P. 1982. Bryozoa. In G.A.B. Shelton (ed.): *Electrical Conduction and Behaviour in 'Simple' Invertebrates*, pp. 393–439. Clarendon Press, Oxford.

Todd, J.A. 2000. The central role of ctenostomes in bryozoan phylogeny. In A. Herrera Cubilla and J.B.C. Jackson (eds): *Proceedings of the 11th International Bryozoology Association Conference*, pp. 104–135. Smithsonian Tropical Research Institute, Balboa, Panama.

Tsyganov-Bodounov, A., Hayward, P.J., Porter, J.S. and Skibinski, D.O.F. 2009. Bayesian phylogenetics of Bryozoa. *Mol. Phylogenet. Evol.* **52**: 904–910.

Winston, J.E. 1978. Polypide morphology and feeding behavior in marine ectoprocts. *Bull. Mar. Sci.* **28**: 1–31.

Xia, F.-S., Zhang, S.-G. and Wang, Z.-Z. 2007. The oldest bryozoans: new evidence from the Late Tremadocian

(Early Ordovician) of East Yangtse Gorges in China. *J. Paleont.* **81**: 1308–1326.

Zimmer, R.L. 1973. Morphological and developmental affinities of the lophophorates. In G.P. Larwood (ed.): *Living and Fossil Bryozoa*, pp. 593–599. Academic Press, London.

Zimmer, R.L. and Woollacott, R.M. 1977. Metamorphosis, ancestrulae, and coloniality in bryozoan life cycles. In R.M. Woollacott and R.L. Zimmer (eds): *Biology of Bryozoans*, pp. 91–142. Academic Press, New York.

39

BRACHIOZOA

Phoronids and brachiopods are traditionally regarded as closely related because of many anatomical and embryological characters. The common name, Brachiozoa, for the two phyla was introduced by Cavalier-Smith (1998), and the alternative name Phoronozoa by Zrzavý *et al.* (1998). It seems that Brachiozoa is the more commonly used name, so I have chosen to use it here.

Surprisingly, some molecular studies show phoronids as an ingroup of the brachiopods (Cohen *et al.* 1998; Cohen 2000; Cohen and Weydmann 2005; Santagata and Cohen 2009), but as embryology shows that the ventral side is short in brachiopods and long in the phoronids (Fig. 39.1), this appears very unlikely, and it is not shown in other molecular analyses (Paps *et al.* 2009; Hausdorf *et al.* 2010). The ancestral brachiozoan was probably vagile (or perhaps tube building) with a straight anterior-posterior axis. The ontogeny of the phoronids shows that the gut becomes U-shaped by an enormous elongation of the ventral side, while the ontogeny of *Novocrania* and *Lingula* shows the U-shape of the brachiopods to be the result of an elongation of the dorsal side, and the two valves as both being dorsal. It appears natural to interpret these two life cycles as derived independently from the life cycle of a common ancestor having a straight gut and a lophophore with ciliated tentacles (Nielsen 1991). The planktotrophic larvae of the two groups show striking similarities (Fig. 39.2).

The Bryozoa was united for a century with the two phyla in the Tentaculata (Hatschek 1888) or Lophophorata (Hyman 1959), but, as discussed in Chapter 35, there are no convincing synapomorphies between brachiozoans and bryozoans (which are now placed with entoprocts and cycliophorans in the Polyzoa, Chapter 35). Molecular phylogenetic analyses almost never show Brachiozoa and Bryozoa as sister groups. I have earlier (Nielsen 2001) regarded phoronids and brachiopods as deuterostomes closely related to the pterobranchs, based mainly on similarities in the 'lophophores', and on the similarity of mesoderm development in *Novocrania* and ambulacrarians, but new information and a renewed scrutiny of the old literature have made me change my mind.

The new information about structure and function of the ciliary filter-feeding bands of the tentacles of bryozoans, phoronids, and brachiopods (Chapters 38, 40, 41) shows hitherto unknown similarities, with laterofrontal cilia functioning as a sieve, although the cells of the bryozoan tentacles are multiciliate and those of the brachiozoans monociliate. The ciliated tentacles on the arms of pterobranchs lack laterofrontal cilia (Chapter 61) and their particle capture method is probably the same as that of the larvae of echinoderms and enteropneusts (Chapter 57). The similarities in the coelomic compartments in the tentacle region of brachiozoans and pterobranchs are rather striking, but as the bryozoans unquestionably belong to the Protostomia, and the pterobranchs to the Deuterostomia, convergent evolution must necessarily have been involved.

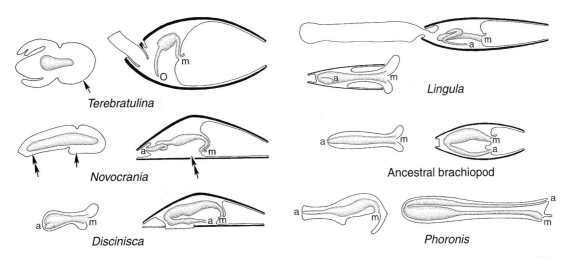

Terebratulina

Novocrania

Discinisca

Lingula

Ancestral brachiopod

Phoronis

Figure 39.1. Larvae and adults of the four main types of brachiopods, a hypothetical brachiopod ancestor, and a phoronid. The guts are shaded, with 'm' indicating the mouth and 'a' the anus (the circle in *Terebratulina* indicates the end of the intestine); in *Terebratulina* and *Novocrania* the single arrows indicate the position of the stomodaeum, and the double arrows indicate the position of the closed blastopore. (From Nielsen 1991.)

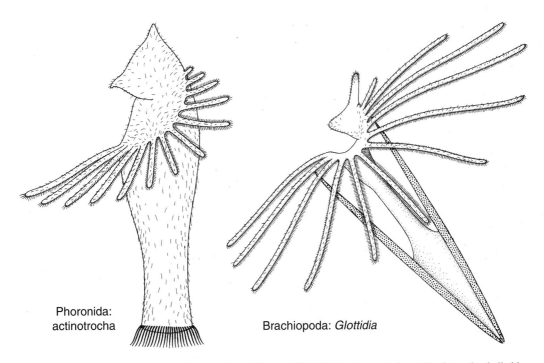

Phoronida: actinotrocha

Brachiopoda: *Glottidia*

Figure 39.2. Planktotrophic brachiozoan larvae. Phoronida: an unidentified actinotrocha larva; Brachiopoda: shelled larva of *Glottidia* sp. (From Nielsen 1985.)

Another strong character indicating the position of brachiozoans as spiralians is the presence of the spiralian Hox signature in a brachiopod (Fig. 21.3).

A spiralian clade of Brachiozoa and Nemertini, called Kryptrochozoa, has been recovered in a number of recent molecular studies (Dunn *et al.* 2008; Helmkampf *et al.* 2008; Hejnol *et al.* 2009; Hausdorf *et al.* 2010; Pick *et al.* 2010), but the morphological implications have generally not been considered. Stereotyped cleavage patterns can of course be lost, as seen in clades of molluscs and annelids. In the extreme case of the cyclostome bryozoans (Chapter 38) no trace of an ancestral cleavage pattern has been found, and the larvae are of a generalized lecithotrophic metazoan type. If this happened in the ancestral brachiozoan, all the developmental information used to characterize protostomes/spiralians was lost. The idea of interpreting the cilia on the edge of the actinotroch pre-oral hood or the edge of the apical lobe of articulate brachiopod larvae as modified prototrochs (Lacalli 1990; Hejnol 2010) lacks any morphological support. Neither a cell lineage nor structure function of these bands bear any resemblance to a prototroch. The cilia at the edge of the apical lobe of early actinotrochs and articulate brachiopod larvae are longer than those of the other parts of the body, but not 'larger' as stated by Hejnol (2010), and they are not engaged in filter feeding. Multiciliarity is another spiralian characteristic, but even this character can be lost, as shown, for example, by the monociliate cells in the prototroch of the annelid *Owenia* (Chapter 25). The other of the most important morphological characteristics of the Protostomia/Spiralia, the morphology of the nervous system with paired cerebral ganglia and a paired or fused ventral nerve cord, is not seen in the brachiozoans. There is no trace of cerebral ganglia or of a ventral nerve cord along fused blastopore lips. The nervous systems of the early larvae resemble those of ambulacrarian larvae (Hay-Schmidt 2000).

The brachiopod valves have a structure resembling that of mollusc valves, and the general structure of all calcified exoskeletons is rather similar (Lowenstam and Weiner 1989); this just shows that a calcareous exoskeleton in an organic matrix can easily be secreted by any ectodermal epithelium. The stereom endoskeleton of the lophophore of some terebratulids has no counterparts in protostomes or cnidarians, but resembles the structure of the echionoderm endoskeleton (Chapter 58).

The mouth of the *Novocrania* larva develops in the anterior part of the metamorphosing larva as in the ambulacrarians, but blastopore fate is a notoriously weak phylogenetic character, and a blastopore becoming the adult anus is found in certain annelids and molluscs (Chapters 25, 27).

The protonephridia of the phoronids is probably a protostomian synapomorphy, because the cyrtopodocytes of the cephalochordates are clearly not protonephridia (Chapter 63).

The flask-shaped cells found in the apical organ of the *Novocrania* larva resemble cells in other spiralian apical ganglia (Altenburger and Wanninger 2010).

The sum of the above-mentioned characters has led me to interpret the Brachiozoa as a spiralian clade, although its relationships with the other groups seems uncertain. The schizocoely and the chaetae indicate annelid relationships, but new information is needed before any further conclusions can be drawn.

References

Altenburger, A. and Wanninger, A. 2010. Neuromuscular development in *Novocrania anomala*: evidence for the presence of serotonin and spiralian-like apical organ in lecithotrophic brachiopod larvae. *Evol. Dev.* **12**: 16–24.

Cavalier-Smith, T. 1998. A revised six-kingdom system of life. *Biol. Rev.* **73**: 203–266.

Cohen, B. and Weydmann, A. 2005. Molecular evidence that phoronids are a subtaxon of brachiopods (Brachiopoda: Phoronata) and that genetic divergence of metazoan phyla began long before the early Cambrian. *Org. Divers. Evol.* **5**: 253–273.

Cohen, B.L. 2000. Monophyly of brachiopods and phoronids: reconciliation of molecular evidence with Linnaean classification (the subphylum Phoroniformea nov.). *Proc. R. Soc. Lond. B* **267**: 225–231.

Cohen, B.L., Gawthrop, A. and Cavalier-Smith, T. 1998. Molecular phylogeny of brachiopods and phoronids based on nuclear-encoded small subunit ribosomal RNA gene sequences. *Phil. Trans. R. Soc. Lond. B* **353**: 2040–2060.

Dunn, C.W., Hejnol, A., Matus, D.Q., *et al.* 2008. Broad phylogenomic sampling improves resolution of the animal tree of life. *Nature* **452**: 745–749.

Hatschek, B. 1888. *Lehrbuch der Zoologie, 1. Lieferung* (pp. 1–144). Gustav Fischer, Jena.

Hausdorf, B., Helmkampf, M., Nesnidal, M.P. and Bruchhaus, I. 2010. Phylogenetic relationships within the lophophorate lineages (Ectoprocta, Brachiopoda and Phoronida). *Mol. Phylogenet. Evol.* **55**: 1121–1127.

Hay-Schmidt, A. 2000. The evolution of the serotonergic nervous system. *Proc. R. Soc. Lond. B* **267**: 1071–1079.

Hejnol, A. 2010. A twist in time—the evolution of spiral cleavage in the light of animal phylogeny. *Integr. Comp. Biol.* **50**: 695–706.

Hejnol, A., Obst, M., Stamatakis, A., *et al.* 2009. Assessing the root of bilaterian animals with scalable phylogenomic methods. *Proc. R. Soc. Lond. B* **276**: 4261–4270.

Helmkampf, M., Bruchhaus, I. and Hausdorf, B. 2008. Phylogenomic analyses of lophophorates (brachiopods, phoronids and bryozoans) confirm the Lophotrochozoa concept. *Proc. R. Soc. Lond. B* **275**: 1927–1933.

Hyman, L.H. 1959. *The Invertebrates, Vol. 5: Smaller Coelomate Groups.* McGraw-Hill, New York.

Lacalli, T.C. 1990. Structure and organization of the nervous system in the actinotroch larva of *Phoronis vancouverensis. Phil. Trans. R. Soc. Lond. B* **327**: 655–685.

Lowenstam, H.A. and Weiner, S. 1989. *On Biomineralization.* Oxford University Press, New York.

Nielsen, C. 1985. Animal phylogeny in the light of the trochaea theory. *Biol. J. Linn. Soc.* **25**: 243–299.

Nielsen, C. 1991. The development of the brachiopod *Crania (Neocrania) anomala* (O. F. Müller) and its phylogenetic significance. *Acta Zool.* (*Stockh.*) **72**: 7–28.

Nielsen, C. 2001. *Animal Evolution: Interrelationships of the Living Phyla,* 2nd ed. Oxford University Press, Oxford.

Paps, J., Baguñà, J. and Riutort, M. 2009. Bilaterian phylogeny: A broad sampling of 13 nuclear genes provides a new Lophotrochozoa phylogeny and supports a paraphyletic basal Acoelomorpha. *Mol. Biol. Evol.* **26**: 2397–2406.

Pick, K.S., Philippe, H., Schreiber, F., *et al.* 2010. Improved phylogenomic taxon sampling noticeably affects nonbilaterian relationships. *Mol. Biol. Evol.* **27**: 1983–1987.

Santagata, S. and Cohen, B.L. 2009. Phoronid phylogenetics (Brachiopoda: Phoronata): evidence from morphological cladistics, small and large subunit rDNA sequences, and mitochondrial *cox1. Zool. J. Linn. Soc.* **157**: 34–50.

Zrzavý, J., Mihulka, S., Kepka, P., Bezděk, A. and Tietz, D. 1998. Phylogeny of the Metazoa based on morphological and 18S ribosomal DNA evidence. *Cladistics* **14**: 249–285.

40

Phylum **Phoronida**

Phoronida is one of the smallest animal phyla, with only about 12 species in two genera, *Phoronis* and *Phoronopsis*. There is no reliable fossil record, although the Cretaceous ichnofossil *Talpina* and some older burrows in calcareous material have been interpreted as phoronid burrows (Voigt 1975). All species are marine, and are benthic, building chitinous tubes covered by mud or sand, or bored into calcareous material. The general life form is solitary, but several species occur in smaller or larger masses, and *P. ovalis* may form lateral buds from the body so that small, temporary colonies arise (du Bois-Reymond Marcus 1949). Regenerative powers are considerable: an autotomized tentacle crown is easily regenerated, and autotomized tentacle crowns of *P. ovalis* appear to regenerate completely (Silén 1955; Marsden 1957).

The microscopic anatomy was reviewed by Herrmann (1997). The adult phoronid has a cylindrical body with a lophophore carrying cylindrical, ciliated tentacles around the mouth. The smallest species, *P. ovalis*, has an almost circular tentacle crown, whereas the somewhat larger species have lophophores, with the lateral parts pulled out into an arm on each side forming a simple horseshoe with a double row of tentacles, and the largest species have spirally coiled lophophore arms (Abele *et al.* 1983; Temereva and Malakhov 2009b). The gut is U-shaped with the anus situated near the mouth, and the ontogeny shows that the short area between mouth and anus is the dorsal side. The epithelia of all three germ layers are monolayered and their ciliated cells are all monociliate. The interpretation of the body regions and their coelomic cavities is a contentious subject. The body of both larvae and adults appears to be archimeric with prosome, mesosome, and metasome, but the terms 'pre-oral hood', 'lophophore region' (with the tentacle coelom), and 'trunk' are used in the following to prevent misunderstandings. There is no chitin in the epidermis.

The small pre-oral hood is a round or crescentic flap on the dorsal side of the mouth in the small species, with extensions along the lophophore in the larger species. Its mesoderm has been interpreted differently, but there appear to be differences between small and large species. The large *P. harmeri* has a spacious cavity in the hood (Temereva and Malakhov 2001). The medium-sized *P. muelleri* has a hood filled with a central extracellular matrix (Bartolomaeus 2001). In the small *P. ovalis* the hood has an extension of the epithelium of the tentacle coelom, but with no separating septum and no cavity (Gruhl *et al.* 2005).

The short lophophore region carries ten to several hundred tentacles. All the cells are ciliated, and the cilia form frontal, laterofrontal, and lateral bands, and a general abfrontal ciliation (Temereva and Malakhov 2009b). The most conspicuous ciliary band is the lateral band that creates an abfrontally directed (away

Chapter vignette: *Phoronis hippocrepia*. (Redrawn from Emig 1982.)

from the centre) water current between the tentacles. One or two rows of laterofrontal sensory cells, with rather stiff cilia, border the frontal band and these cilia appear to function as a sieve in particle capture. The deformation of the cilia by a particle either elicits a tentacle flick that transports the particle to the current towards the tentacle basis, or makes the cilia push the particle to the frontal ciliary band that then transports it to the ciliated groove between the tentacle bases and the hood, and then to the mouth (Riisgård 2002; Temereva and Malakhov 2010). A crescentic coelomic cavity at the frontal side of the oesophagus, with lateral lobes along the lophophore basis, sends a small channel into each tentacle. The ciliated coelomic epithelium contains muscle cells and shows a longitudinal invagination forming a blind-ending blood vessel.

The elongate trunk has a long smooth anterior region and a slightly swollen 'end bulb' that anchors the animal in the tube. There is a large coelom, which is separated from the lophophore coelom by a conspicuous transverse septum, often called diaphragm, at the base of the lophophore. The peritoneum forms a median and paired lateral mesenteries that suspend the gut.

The nervous system is intraepithelial (basiepithelial) with a concentration in the short dorsal side between mouth and anus, overlying the dorsal side of the lophophore coelom, and a nerve ring at the lophophore base (Silén 1954b; Fernández et al. 1996; Santagata 2002; Temereva and Malakhov 2009a). Bundles of axons are found along the laterofrontal cells of the tentacles, and median bundles are found on both the frontal and the abfrontal side.

The traditional view of the haemal system describes well-defined vessels formed between the basal membranes of coelomic epithelium and endoderm, or surrounded by the coelomic epithelium, as in the tentacles (see above). A horseshoe-shaped vessel runs at the base of the lophophore and two or three longitudinal vessels run along the gut; two of the vessels extend to the 'posterior' end of the body, where a system of capillaries and lacunae surround the gut; the median vessel is contractile and pumps the blood anteriorly into the lophophore vessel. However, Temereva and Malakhov (2001, 2004a,b) describe a more complicated haemal

system in the large *P. harmeri*, with an afferent and an efferent lophophoral vessel, and most vessels with an endothelium. Circulating haemoglobin aids oxygen transport to the body, which is surrounded by a tube or calcareous matter, and even makes it possible for the animals to endure in habitats where most of the tubes are surrounded by anoxic sediment (Vandergon and Colacino 1991). Podocytes have been observed in many of the blood vessels of the trunk region of *P. muelleri*, and they are probably the site of the formation of primary urine.

A pair of large metanephridia, with large funnels formed by ciliated epithelio-muscular cells, are situated in the trunk coelom. The gonads are formed from the peritoneum at the stomach part of the gut, and the metanephridia function as gonoducts. The sperm, which has a long cylindrical head lying parallel to the cilium, becomes enclosed in elaborate spermatophores that are shed and float in the water (Zimmer 1991). The spermatophores become caught by the tentacles of another specimen (or even engulfed); the mass of sperm becomes amoeboid and lyses the body wall to enter the trunk coelom (and the septum if caught by a tentacle) (Zimmer 1991). This type of internal fertilization has been observed in a number of species; only *P. ovalis* lacks spermatophoral glands and may not produce spermatophores (Zimmer 1997). The polar bodies are given off after the eggs have been shed; they have been reported to move back into the egg and become resorbed in some species (Herrmann 1986), or to become internalized at a later stage (Santagata 2004).

Development has been studied in a number of species (Zimmer 1991). Silén (1954a) showed that *P. ovalis* with the largest eggs has direct development, whereas other species with comparatively small eggs develop through the well-known actinotrocha larva. Some species shed the eggs free in the water but others retain the embryos in the lophophore until a stage of about four tentacles. Both isolated blastoporal halves of 8- and 16-cell stages, and early cleavage to blastula stages divided along the median plane, develop into normal actinotrochs (Freeman 1991).

The two first cleavages take place along the primary axis of the egg (Freeman 1991). Cleavage is total

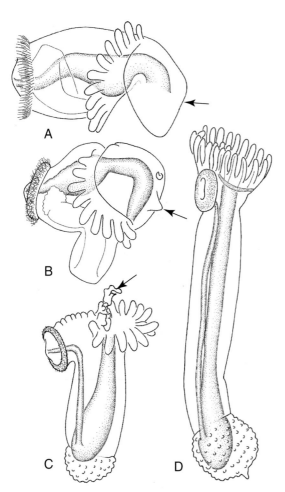

Figure 40.1. Metamorphosis of *Phoronopsis harmeri*; the position of the apical organ is indicated by an arrowhead. (A) A full-grown actinotrocha larva. (B) Beginning of metamorphosis, the trunk sac is halfway everted. (C) Eversion of the trunk sac completed, the pre-oral hood is disintegrating and the dorsal side is strongly shortened. (D) Metamorphosis completed, the pre-oral hood has been cast off and the perianal ciliary ring will soon become discarded too. (Based on Zimmer 1964, 1991.)

and usually equal. Traces of a spiral pattern of early blastomeres have been reported by a few authors, but none of the other characteristics of spiral cleavage have been reported and it is now generally agreed that the cleavage is radial. The 16-cell stage often consists of two tiers of eight cells each. Experiments with vital staining of blastomeres have shown that the orientation of the first cleavage shows considerable variation, but the first cleavage is transverse in about 70% of the embryos (Freeman 1991). Marking of individual blastomeres of stages up to 16 cells showed no consistent developmental pattern (Freeman and Martindale 2002). A coeloblastula is formed, and gastrulation is by invagination. The gastrula elongates so that the apical pole becomes situated in the anterior end of the embryo, where an apical organ differentiates. The blastopore elongates and constricts from the posterior end; it closes almost completely in some species, but its anterior end remains open as the larval mouth, for example in *P. ijimai (P. vancouverensis)* (Zimmer 1980; Malakhov and Temereva 2000; Freeman and Martindale 2002). At a later stage the anus breaks through at the posterior pole of the larva, possibly at the posterior end of the constricted blastopore. The region in front of the mouth becomes extended into a flat hood overhanging the mouth. Its margin shows a more prominent ciliation than that of the other parts of the body; it is probably locomotory (Lacalli 1990).

Origin and differentiation of the mesoderm have been interpreted variously, and some variation between the species cannot be excluded. The European *P. muelleri* has been studied by many authors (for example by Herrmann 1986; Bartolomaeus 2001), the US West Coast *P. vancouverensis (P. ijimai)* (for example by Zimmer 1980; Freeman 1991; Malakhov and Temereva 2000; Freeman and Martindale 2002), and the N. Pacific *Phoronopsis harmeri (Phoronopsis viridis)* (for example by Temereva and Malakhov 2006; 2007). Zimmer (1978) studied a number of species.

It appears that all authors agree that the first mesenchymal cells originate through ingression from the periphery of the invaginating archenteron. These cells proliferate and become arranged as a thin epithelium in the anterior part of the embryo, in the region of the future hood and tentacle region. In *P. vancouverensis* and *P. muelleri* the anterior part of this mesoderm differentiates into a coelomic sac filling the hood, which later on collapses. In *P. ijimai*, the hood coelom is first reduced to a small sac beneath the apical organ, but this sac finally collapses too. In *P. harmeri*, the small coelomic sac is apparently retained. The tentacle coelom originates from the posterior part of the mesenchymal cells, which differentiate into the coelomic walls. The mesoderm of the trunk develops slightly later but is poorly documented. Mesenchymal cells, possibly originated through ingression from the gut, are observed around the posterior part of the gut in *P. vancouverensis*, and these cells differentiate into the unpaired trunk coelom (Zimmer 1980). Temereva and Malakhov (2006) reported that the trunk coelom in *P. harmeri* develops as a dorsal pocket from the posterior part of the gut.

The blastulae are uniformly ciliated, but when the first tentacles develop, a band of longer and more closely set cilia can be recognized along the lateral faces of the tentacles. This band becomes the lateral ciliary band, which is the sole locomotory and feeding organ in the early stages. The first pair of tentacle buds develops on either side of the ventral midline behind the mouth, and additional tentacles develop laterally along curved lines almost reaching each other dorsally behind the apical organ. The ciliary bands of the tentacles are identical to those of the adults (see above) (Strathmann and Bone 1997; Riisgård 2002). The hood reacts to particles caught on the tentacles by lifting the neighbouring part, thereby sucking the particle towards the mouth (Strathmann and Bone 1997). At a later stage, a peri-anal ring of large compound cilia, likewise on monociliate cells (Nielsen 1987), develops and becomes the main locomotory organ of the larva. The larval tentacles of *P. vancouverensis* contain narrow extensions of the lophophoral coelom, and it appears that the adult tentacles develop directly from the larval ones (Zimmer 1964). In other species, such as *P. psammophila*, small abfrontal knobs at the bases of the larval tentacles become the adult tentacles with the coelomic canals, whereas the distal parts of the tentacles, which lack the coelomic canals, are shed after metamorphosis (Herrmann 1979). Advanced larvae of *P. muelleri* develop a row of adult tentacles just behind the larval tentacles (Silén 1954a), and the larval tentacles are shed together with their common bases (Herrmann 1980).

The apical organ with the ciliary tuft can be followed to metamorphosis (Santagata 2002). An additional sensory organ with cilia in front of the apical organ is found in advanced larvae of, for example, *P.*

muelleri and *P. architecta* (Zimmer 1978, 1991); this organ contains bipolar sensory cells (Santagata 2002) and is protruded when the larva is testing the substratum ready for settling (Silén 1954a). A U-shaped apical ganglion sends nerves to the edge of the hood and to the developing tentacles (Hay-Schmidt 1989; Lacalli 1990; Freeman and Martindale 2002).

A long, tubular invagination of the ventral body wall, the trunk ('metasomal') sac, develops at about the mid stage of the larval life, and, in larvae that are about to be ready for metamorphosis, it occupies much of the space around the gut (see Fig. 40.1).

A pair of protonephridia, each with multiple solenocytes, develops from a median or paired ectodermal invagination(s) just anterior to the anus in the early larvae; the common part of the nephridial duct soon disappears, and the nephridiopores of older larvae are situated below the larval tentacles. In the advanced larvae the solenocytes form clusters in the tentacle region where they drain the blastocoel (Hay-Schmidt 1987; Bartolomaeus 1989).

Metamorphosis is rapid and dramatic (Fig. 40.1). The trunk sac everts, pulling the gut into a U-shape, and contraction of the dorsal larval muscles brings the mouth and anus close to each other; this establishes a new main body axis perpendicular to the larval main axis. The major part of the hood with the apical organ (and the accessory sensory organ when present) is cast off or ingested together with the larval tentacles in species where these do not become the adult tentacles; the large ciliary ring around the anus is either resorbed or cast off (Herrmann 1997).

The juvenile has the general shape of the adult, and the lophophoral and trunk coeloms are taken over almost without modifications. The hood coelom is partially or completely lost (Zimmer 1997). The dorsal central nervous concentration develops in the ectoderm without any connection with the apical region. The nerves of the lophophore are retained. The larval nephridia undergo a major reorganization at metamorphosis (Bartolomaeus 1989). The solenocytes and the inner parts of the ducts break off and become phagocytosed by other duct cells, and the ducts now end blindly. At a later stage, areas of the tentacle coelom differentiate into a pair of ciliated funnels that gain connection with the ducts to form metanephridia.

The actinotrocha larva has been central in many phylogenetic discussions. Hatschek (1891) interpreted it as a trochophore, but later, more-detailed studies have shown that the two larval types are very different. Studies of innervation of larval ciliary bands led Lacalli (1990) to interpret the ciliation along the edge of the pre-oral hood as a modified prototroch. This view is also expressed in some of the papers on molecular phylogeny, which show the phoronids within the Lophotrochozoa/Trochozoa. Hejnol (2010) directly stated that the 'actinotroch larva would have to be interpreted as a derived trochophore that evolved beyond recognition'. However, it must be stressed that this finds no support from morphology. The edge of the hood does not develop from trochoblasts of a spiral-cleaving embryo, the cells are monociliate in contrast to the multiciliate prototroch cells with compound cilia, and the cilia are not engaged in particle collection.

The phylogenetic relationships are discussed in Chapter 39.

Interesting subjects for future research

1. TEM studies of the morphology and ontogeny of the hood in several species

2. Development and function of the nervous system of one species through all stages from the youngest actinotroch to the adult

3. Hox genes

References

Abele, L.G., Gilmour, T. and Gilchrist, S. 1983. Size and shape in the phylum Phoronida. *J. Zool.* **200**: 317–323.

Bartolomaeus, T. 1989. Ultrastructure and relationship between protonephridia and metanephridia in *Phoronis muelleri* (Phoronida). *Zoomorphology* **109**: 113–122.

Bartolomaeus, T. 2001. Ultrastructure and formation of the body cavity lining in *Phoronis muelleri* (Phoronida, Lophophorata). *Zoomorphology* **120**: 135–148.

du Bois-Reymond Marcus, E. 1949. *Phoronis ovalis* from Brazil. *Bol. Fac. Filos. Cienc. Let. Univ S. Paulo, Zool.* **14**: 157–166.

Emig, C.C. 1982. Phoronida. In S.P. Parker (ed.): *Synopsis and Classification of Living Organisms*, vol. 2, pp. 741. McGraw-Hill, New Yourk.

Fernández, I., Pardos, F., Benito, J. and Roldán, C. 1996. Ultrastructural observations on the phoronid nervous system. *J. Morphol.* **230**: 265–281.

Freeman, G. 1991. The bases for and timing of regional specification during larval development of *Phoronis*. *Dev. Biol.* **147**: 157–173.

Freeman, G. and Martindale, M.Q. 2002. The origin of mesoderm in phoronids. *Dev. Biol.* **252**: 301–311.

Gruhl, A., Grobe, P. and Bartolomaeus, T. 2005. Fine structure of the epistome in *Phoronis ovalis*: significance for the coelomic organization in Phoronida. *Invert. Biol.* **124**: 332–343.

Hatschek, B. 1891. *Lehrbuch der Zoologie, 3. Lieferung* (pp 305–432). Gustav Fischer, Jena.

Hay-Schmidt, A. 1987. The ultrastructure of the protonephridium of the actinotroch larva (Phoronida). *Acta Zool. (Stockh.)* **68**: 35–47.

Hay-Schmidt, A. 1989. The nervous system of the actinotroch larva of *Phoronis muelleri* (Phoronida). *Zoomorphology* **108**: 333–351.

Hejnol, A. 2010. A twist in time—The evolution of spiral cleavage in the light of animal phylogeny. *Integr. Comp. Biol.* **50**: 695–706.

Herrmann, K. 1979. Larvalentwicklung und Metamorphose von *Phoronis psammophila* (Phoronida, Tentaculata). *Helgol. Wiss. Meeresunters.* **32**: 550–581.

Herrmann, K. 1980. Die archimere Gliederung bei *Phoronis mülleri* (Tentaculata). *Zool. Jahrb., Anat.* **103**: 234–249.

Herrmann, K. 1986. Die Ontogenese von *Phoronis mülleri* (Tentaculata) unter besonderer Berücksichtigung der Mesodermdifferenzierung und Phylogenese des Coeloms. *Zool. Jahrb., Anat.* **114**: 441–463.

Herrmann, K. 1997. Phoronida. In F.W. Harrison (ed.): *Microscopic Anatomy of Invertebrates*, vol. 13, pp. 207–235. Wiley-Liss, New York.

Lacalli, T.C. 1990. Structure and organization of the nervous system in the actinotroch larva of *Phoronis vancouverensis*. *Phil. Trans. R. Soc. Lond. B* **327**: 655–685.

Malakhov, V.V. and Temereva, E.N. 2000. Embryonic development of the phoronid *Phoronis ijimai*. *Russ. J. Mar. Biol.* **26**: 412–421.

Marsden, J.R. 1957. Regeneration in *Phoronis vancouverensis*. *J. Morphol.* **101**: 307–323.

Nielsen, C. 1987. Structure and function of metazoan ciliary bands and their phylogenetic significance. *Acta Zool. (Stockh.)* **68**: 205–262.

Riisgård, H.U. 2002. Methods of ciliary filter feeding in adult *Phoronis muelleri* (phylum Phoronida) and its free-swimming actinotroch larva. *Mar. Biol.* **141**: 75–87.

Santagata, S. 2002. Structure and metamorphic remodeling of the larval nervous system and musculature of *Phoronis pallida* (Phoronida). *Evol. Dev.* **4**: 28–42.

Santagata, S. 2004. Larval development of *Phoronis pallida* (Phoronida): implications for morphological convergence and divergence among larval body plans. *J. Morphol.* **259**: 347–358.

Silén, L. 1954a. Developmental biology of Phoronidea of the Gullmar Fiord area (West coast of Sweden). *Acta Zool. (Stockh.)* **35**: 215–257.

Silén, L. 1954b. On the nervous system of *Phoronis. Ark. Zool.*, 2. Ser. **6**: 1–40.

Silén, L. 1955. Autotomized tentacle crowns as propagative bodies in *Phoronis. Acta Zool. (Stockh.)* **36**: 159–165.

Strathmann, R.R. and Bone, Q. 1997. Ciliary feeding assisted by suction from the muscular oral hood of phoronid larvae. *Biol. Bull.* **193**: 153–162.

Temereva, E.N. and Malakhov, V.V. 2001. The morphology of the phoronid *Phoronopsis harmeri. Russ. J. Mar. Biol.* **27**: 21–30.

Temereva, E.N. and Malakhov, V.V. 2004a. Ultrastructure of the blood system in the phoronid *Phoronopsis harmeri* Pixell, 1912: 1. Capillaries. *Russ. J. Mar. Biol.* **30**: 28–36.

Temereva, E.N. and Malakhov, V.V. 2004b. Ultrastructure of the circulatory system of the phoronid *Phoronopsis harmeri* Pixell, 1912. 2. Main vessels. *Russ. J. Mar. Biol.* **30**: 101–112.

Temereva, E.N. and Malakhov, V.V. 2006. Development of excretory organs in *Phoronopsis harmeri* (Phoronida): from protonephridium to metanephridium. *Entomol. Rev.* **86, suppl.** 2: S201–S209.

Temereva, E.N. and Malakhov, V.V. 2007. Embryogenesis and larval development of *Phoronopsis harmeri* Pixell, 1912 (Phoronida): dual origin of the coelomic mesoderm. *Invertebr. Reprod. Dev.* **50**: 57–66.

Temereva, E.N. and Malakhov, V.V. 2009a. Microscopic anatomy and ultrastructure of the nervous system of *Phoronopsis harmeri* Pixell, 1912 (Lophophorata: Phoronida). *Russ. J. Mar. Biol.* **35**: 388–404.

Temereva, E.N. and Malakhov, V.V. 2009b. On the organization of the lophophore in phoronids (Lophophorata: Phoronida). *Russ. J. Mar. Biol.* **35**: 479–489.

Temereva, E.N. and Malakhov, V.V. 2010. Filter feeding mechanism in the phoronid *Phoronopsis harmeri* (Phoronida, Lophophorata). *Russ. J. Mar. Biol.* **36**: 109–116.

Vandergon, T.L. and Colacino, J.M. 1991. Hemoglobin function in the lophophorate *Phoronis architecta* (Phoronida). *Physiol. Zool.* **64**: 1561–1577.

Voigt, E. 1975. Tunnelbaue rezenter und fossiler Phoronidea. *Palaeontol. Z.* **49**: 135–167.

Zimmer, R.L. 1964. *Reproductive Biology and Development of Phoronida*. Ph.D., University of Washington.

Zimmer, R.L. 1978. The comparative structure of the preoral hood coelom in Phoronida and the fate of this cavity during and after metamorphosis. In F.S. Chia and M.E. Rice (eds): *Settlement and Metamorphosis of Marine Invertebrate Larvae*, pp. 23–40. Elsevier, New York.

Zimmer, R.L. 1980. Mesoderm proliferation and formation of the protocoel and metacoel in early embryos of *Phoronis vancouverensis* (Phoronida). *Zool. Jahrb., Anat.* **103**: 219–233.

Zimmer, R.L. 1991. Phoronida. In A.C. Giese, J.S. Pearse and V.B. Pearse (eds): *Reproduction of Marine Invertebrates*, vol. 6, pp. 1–45. Boxwood Press, Pacific Grove, CA.

Zimmer, R.L. 1997. Phoronids, brachiopods, and bryozoans, the lophophorates. In S.F. Gilbert and A.M. Raunio (eds): *Embryology. Constructing the Organism*, pp. 279–305. Sinauer Associates, Sunderland.

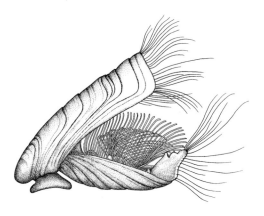

41

Phylum **Brachiopoda**

Brachiopods, or lamp-shells, are a highly characteristic group of benthic, marine organisms. The two valves, usually called dorsal and ventral (but see discussion below) make both living and extinct brachiopods immediately recognizable. Nearly 400 living and over 12 000 fossil species have been recognized, with the fossil record going back to the earliest Cambrian (Williams and Carlson 2007). With such an extensive fossil record, which comprises many extinct major groups, the phylogeny is largely built on the fossils. The living forms were earlier arranged in two main groups, Inarticulata (with Lingulacea, Discinacea, and Craniacea) and Articulata, but newer studies have concluded that lingulaceans and discinaceans form a monophyletic group, Linguliformea. The interrelationships of the three living groups, Linguliformea, Craniiformea, and Rhynchonelliformea, along with the extinct lineages remains uncertain (Carlson 2007). Some studies of molecular phylogeny have shown phoronids as an ingroup of the brachiopods, with the phoronids as the sister group of the Craniiformea plus Linguliformea (Cohen *et al.* 1998; Cohen 2000; Cohen and Weydmann 2005; Santagata and Cohen 2009), but this finds no support from morphology (see Chapter 39). Molecular studies almost unanimously show Brachiopoda as a monophyletic group; only the study of Passamaneck and Halanych (2006) indicates polyphyly.

The anatomy of brachiopods has been reviewed a couple of times recently (James 1997; Williams 1997; Williams *et al.* 1997; Lüter 2007), but the three 'inarticulate' groups are still poorly known, the only comprehensive study being that of Blochmann (1892–1900) on *Lingula, Discinisca,* and *Crania (Novocrania)*. All epithelia are monolayered, and the many types of ciliated cells are all monociliate, both in ectoderm, endoderm, and peritoneum. Compound cilia have not been reported.

The body of the adult brachiopods has an upper and lower mantle fold, which secrete calcareous or chitinophosphatic valves. The mantle folds with the valves enclose the lophophore and the main body with gut, gonads, and excretory organs. One valve, usually called the dorsal valve, has the gut and the lophophore attached to it, and will here be called the brachial valve, whereas the other valve that carries the stalk, or pedicle, or is cemented to the substratum, will be called the pedicle valve. The linguliforms have chitinophosphatic shells; the linguloids are anchored in the bottom of a burrow by a stalk that protrudes between the posterior edges of the valves, whereas the discinoids are attached to a hard substratum by a short stalk that protrudes through a slit in the pedicle valve. The craniiforms have calcareous valves and are cemented to a hard substratum by the pedicle valve. The rhynchonelliforms, by far the most diverse of the

Chapter vignette: The rhynchonelliform brachiopod *Pumilus antiquatus*. (Redrawn from Atkins 1958.)

living groups, have calcareous shells and are almost all attached by a stalk that protrudes through a hole in the pedicle valve; the umbo of the pedicle valve, with the hole for the stalk, is usually curved towards the brachial valve so that the brachial valve faces the substratum (see the chapter vignette). The outer layer of the valves, the periostracum, is mainly proteinaceous and contains β-chitin in linguliforms and *Novocrania*, whereas chitin is absent from the periostracum of rhynchonelliforms. The periostracum is secreted by a narrow band of cells at the inner side of the mantle edge. The mineralized shell material is secreted by the outer surface of the mantle epithelium; it is mainly composed of calcium carbonate in rhynchonelliforms and craniiforms, and of calcium phosphate in linguliforms. Craniiforms and certain rhynchonelliforms have characteristic small extensions of the mantle epithelium into channels in the calcified shells. The extensions (unfortunately called caeca) are branched and do not reach the periostracum in the craniiforms. The more stout, simple extensions of the rhynchonelliforms have a distal 'brush border' of microvilli, which each extend through a narrow canal in the shell to the periostracum in the young stage; in the later stages the microvilli retract and additional layers of periostracum separate the epithelial extension from the outer periostracum. The function of these structures is unknown. The mantle edges carry chitinous chaetae (setae), each formed by one ectodermal chaetoblast with additional material added from surrounding follicle cells.

The rhynchonelliforms have a gut with a blind-ending intestine at the ventral side of the stomach; the inarticulate groups have a complete gut, with the anus situated in the right side of the mantle cavity in linguliforms and in the posterior midline in craniiforms.

The lophophore of newly metamorphosed specimens is shaped as an almost closed horseshoe; it is situated just behind the mouth with the arms extending anterodorsally, almost meeting in the midline some distance in front of the mouth. In the larger species it becomes more complicated with the two arms coiled or wound into various complicated shapes. The cylindrical tentacles are arranged in a single row in the juveniles, but as additional tentacles become

added at the tips of the two rows they become arranged alternatingly in two parallel rows, frontal (inner, adlabial) tentacles closest to the mouth/food groove and abfrontal (outer, ablabial) tentacles. A narrow upper lip (brachial lip, epistome) borders the anterior side of the mouth and follows the base of the tentacle row, so that a ciliated food groove is formed between the tentacle bases and the lip as an extension of the lateral corners of the mouth. The frontal and abfrontal tentacles show the same ciliary bands, but with somewhat different positions on the two types of tentacle. The frontal surfaces have a narrow longitudinal band of frontal cilia, which beat towards the base of the tentacles where the bands unite with the ciliation of the food groove leading to the mouth. A row of laterofrontal cilia borders the frontal band. A lateral ciliary band is found on each side of the tentacle; this band is situated on thickened epithelial ridges on the laterofrontal sides of the outer tentacles and on the lateroabfrontal sides of the inner tentacles, so that the cilia of adjacent tentacles can bridge the gaps between the tentacles. Other cells are mainly unciliated, but *Lingula* has cilia on most tentacle cells.

The lateral ciliary bands create the water currents of the lophophore, and the frontal cilia transport captured particles to the food groove and to the mouth. Particle collection has not been observed in adult brachiopods, but the mechanism is supposedly the same as in the larvae (see below).

The peritoneum of the tentacles and of the lophophoral arms comprises myoepithelial cells, with smooth or striated myofilaments that control the various flexions of the tentacles and the more restricted movements of the arms, whereas the thick basement membrane with collagenous fibrils is responsible for the straight relaxed posture. The arms of the lophophore are supported by a quite complicated connective tissue, consisting of a hyaline matrix with scattered cells. A calcareous skeleton, consisting of more or less fused spicules, is secreted by cells of the connective tissue of the lophophores, and sometimes also of the mantle folds of several terebratulid rhynchonelliforms. The spicules are secreted as single crystals in special scleroblasts, and the spicules may later fuse to form a stiff endoskeleton supporting the lophophore base and short, almost tubular 'joints' in the tentacles.

The structure of lophophore and tentacles resembles that of the phoronids, and the embryology of *Novocrania* (see below) indicates that the coelomic canal along the tentacle bases (the small brachial canal), with channels into the tentacles, is homologous with the tentacle coelom of the phoronids. The other large coelomic cavity in the lophophore, the large brachial channel, has by some authors been interpreted as a protocoel (see for example Pross 1980), but its ontogenetic development is unknown, and the apparent absence of an anterior coelomic cavity in the embryos of linguliforms and rhynchonelliforms makes this interpretation entirely conjectural.

The main body cavity is spacious and sends extensions into the mantle folds and into the stalk of linguliforms. It is completely separated from the tentacle coelom in *Novocrania*, but the septum is incomplete in other forms. The peritoneum is monolayered.

The nervous systems are not well known; the only detailed study of the rhynchonelliforms is that of Bemmelen (1883), mainly based on *Gryphus*; the inarticulates were studied by Blochmann (1892–1900). The overall pattern of the nervous systems appears similar in the four groups; the largest nervous concentration is a suboesophageal ganglion that sends nerves to the mantle folds, lophophore (with small nerves into each tentacle), adductor muscles, and stalk. The far less conspicuous supraoesophageal ganglion is transversely elongate and continues laterally into the main nerve of the lophophore. Small nerves connect the two systems.

A haemal system surrounded by basal membranes occurs in all brachiopods, but is poorly known. A contractile vessel is found in the dorsal mesentery of rhynchonelliforms, and *Novocrania* is said to have several such vessels, whereas the haemal system should be poorly developed in *Discinisca*. At least some of the peripheral circulation appears to be through larger haemal spaces. Each tentacle has a small vessel in the shape of a fold of the frontal side of the peritoneum, with an inner basal lamina and an outer layer of peritoneal cells with myofibrils.

A pair of large metanephridia, each with a large ciliated funnel, open in the trunk coelom (two pairs in certain rhynchonelliforms); they also function as gonoducts. The gonads are formed from the peritoneum and extend into the mantle canals in most species. The sperm is of the 'primitive' type.

Most species are free spawners, but most rhynchonelliforms retain the embryos in the lophophore for a period. Rhynchonelliforms and craniiforms have lecithotrophic larvae, whereas the linguliforms have planktotrophic stages that are almost free-swimming juveniles (Fig. 41.1). The embryology has been reviewed by Long and Stricker (1991), Nielsen (2005), and Lüter (2007).

The descriptions of the differentiation of gut and mesoderm of rhynchonellids can be difficult to interpret, but the descriptions of *Terebratulina* (Conklin 1902) and *Terebratalia* (Long 1964; Long and Stricker 1991) together give a rather clear picture. Other important studies of development comprise Freeman (1993, 2003: *Hemithiris* and *Terebratalia*) and Lüter (2000a: *Notosaria* and *Calloria*). The plane of the first cleavage bears no fixed relation to the median plane of the larva, and there is some variation in the early cleavage pattern, for example with the 8-cell stage consisting of one tier of eight cells or two tiers of four cells. A coeloblastula develops and gastrulation begins at the blastoporal pole. Apical halves of horizontally bisected embryos of this stage form only apical lobe ectoderm, whereas the blastoporal halves develop into rather normal larvae (Freeman 1993, 2003). In *Terebratulina* and *Terebratalia* the blastocoel is soon obliterated, and the rather wide archenteron becomes divided by a U-shaped fold of its dorsal lining, so that an antero-median gut becomes separated from a posterior and lateral coelomic cavity still connected to the exterior through the longitudinally elongate blastopore, which closes from the posterior end. An apical organ with a tuft of long cilia develops at the apical pole. In *Notosaria* and *Calloria*, the development of the mesoderm was not followed, but the early three-lobe stadium showed a compact mesoderm forming a rather thick layer along the posterior and lateral sides of the endoderm, and a thin layer at the anterior end

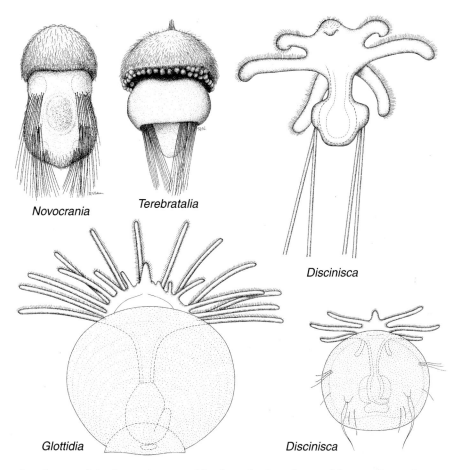

Novocrania

Terebratalia

Discinisca

Glottidia

Discinisca

Figure 41.1. Larval types of the four main types of brachiopods. Craniiformea: full-grown larva of *Novocrania anomala*. Rhynchonelliformea: full-grown larva of *Terebratalia transversa*. Linguliformea Discinacea: young and full-grown larvae of *Discinisca* sp. (not to scale, the distal parts of the larval chaetae have been omitted). Linguliformea Lingulacea: full-grown larva of *Glottidia* sp. (not to scale). (Modified from Nielsen 1991.)

of the gut. Coelomic cavities, as indicated by the presence of stages of ciliogenesis, develop through schizocoely, and lateral mesodermal cavities become divided into anterior and posterior coelomic sacs. In all species, the anterior part of the larva forms a swelling with a ring of longer, locomotory cilia at the posterior side. An annular, later on skirt-shaped thickening, containing extensions of the posterior coelomic sacs and developing two pairs of chaetal bundles, forms around the equator of the larva, behind the closing blastopore. The larval chaetae are accompanied by a sensory cell, with a cilium surrounded by a ring of

microvilli (Lüter 2000b). The full-grown, lecithotrophic larva swims with the chaetae held together in the posterior direction (Fig. 41.1), but when disturbed the larvae can contract the muscles of the skirt so that the chaetae form a defensive belt (Pennington *et al.* 1999). At settling, the skirt-shaped fold and the chaetal bundles fold anteriorly covering the anterior lobe, and the larva attaches with the posterior pole, where the stalk develops. The outer faces of the reflexed folds begin to secrete the two valves (Stricker and Reed 1985a,b). The tentacles develop from an area near the anterior part of the closed blastopore. The larval chaetae are shed and the adult chaetae develop from the mantle edges.

The development of *Novocrania* (Nielsen 1991; Freeman 2000) (Fig. 41.2) resembles that of the rhynchonelliforms in many respects. The cleavage is total and radial and the following stages are a coeloblastula and an invagination gastrula. Nielsen (1991) observed that the apical/anterior and lateral parts of the early archenteron showed different cytological morphology and the two parts could be traced during the following development. The gastrula elongates and becomes bilateral, with the blastopore situated ventrally at the posterior end. The anterior cells of the early archenteron become situated anterodorsally and become the endodermal gut, whereas the more peripheral cells become situated posteroventrally and become the mesoderm, which subsequently slides forwards laterally as a thin layer between ectoderm and endoderm. The mesoderm of each side appears to fold up so that four coelomic sacs are finally found along each side of the larva. The anterior sac forms a thin cap around the anterior part of the gut, and the three more posterior sacs have thick, well-defined walls. The blastopore remains as a ventral opening from the posterior part of the archenteron, which becomes the fourth pair of mesodermal sacs when the blastopore finally closes. Surprisingly, Freeman (2000) reported a different origin and differentiation of the mesoderm in embryos from the same locality as those studied by Nielsen (1991). He reported that the mesoderm should originate through ingression of cells along the lateral sides of the archenteron. The observations are unfortunately poorly documented, and the reported origin of the mesoderm does not resemble that reported for the rhynchonelliforms (see above). The full-grown, planktonic larva has a pair of ectodermal thickenings with bundles of chaetae at the dorsal side of each of the three posterior coelomic sacs (Fig. 41.2). The chaetae resemble those of the adults and are accompanied by small cilia; they are held close to the body in the undisturbed larva, but can be spread out when the larva contracts after disturbance. The competent larvae show four flask-shaped, serotonergic apical cells and a pair of lateral neurites; the flask cells are lost at metamorphosis. Immunostainings of juveniles showed an apical concentration, a thin ring along the periphery of the posterior lobe, and two more prominent anterior commissures, possibly representing the supra- and suboesophageal nerve concentrations (Altenburger and Wanninger 2010). At settling, the larva curls up through the contraction of a pair of muscles from the first pair of coelomic sacs to the posterior end of the larva, just behind the area where the blastopore closed (see also Altenburger and Wanninger 2010). A secretion from epithelial cells at the posterior end attaches the larva to the substratum, and the posterior pair of chaetae is usually shed. The brachial valve becomes secreted from a special area of the dorsal ectoderm in the region of the second and third pair of coelomic sacs; this area expands strongly at the periphery, and the whole organism soon becomes covered by the valve. The periphery of the attachment area also expands and a thin pedicle valve becomes secreted; the pedicle valve is thus secreted by an area behind the closed blastopore and can therefore not be described as ventral. The unpaired anterior coelomic sac apparently disappears, while the first and second pair become the tentacle coelom and the large body cavity, respectively. The fate of the fourth pair of coelomic sacs has not been followed. The adult mouth breaks through at the anteroventral side of the metamorphosed larva, and the anus apparently develops from a proctodaeal invagination of the posterior ectoderm between the valves, i.e. behind the attachment area and, accordingly, quite some distance from the closed blastopore.

The early development of the linguloids *Lingula* (Yatsu 1902) and *Glottidia* (Freeman 1995) shows the first two cleavages along the primary axis, with the first one in the median plane; the third cleavage is horizontal, and a blastula and an invagination gastrula follow. Very surprisingly, the apical pole with the polar bodies ends up at the middle of the brachial valve (Freeman 1995), but an apical ganglion develops in the median tentacle (see below). The blastopore elongates and closes from behind, and the definitive mouth is believed to develop from its anterior part; the anus develops at a later stage. The embryo flattens along

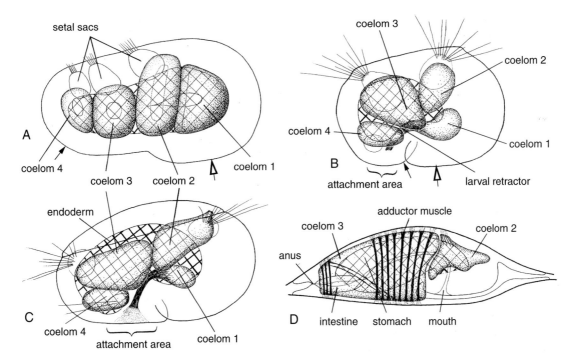

Figure 41.2. Four stages of the development of *Novocrania anomala*. Reconstructions based on complete serial sections. (A) Larva with short chaetae and fully formed coelomic sacs, the longitudinal muscles from the first coelomic sac are not developed yet. (B) Contracted, full-grown larva, the contracted muscles from the first coelomic sac are conspicuous. (C) Young bottom-stage, the first stage of the brachial valve is indicated above the two remaining bundles of chaetae, the attachment of the second coelomic sac to the valve can be seen. (D) Juvenile with mouth, anus, the two shells with adductor muscles, and the short extensions of the second coelomic sacs to the first three pairs of tentacles can be seen. The black arrows point at the position of the closed blastopore and the white arrows to the position of the adult mouth. (Modified from Nielsen 1991.)

the primary axis and the anterior part develops a ventral, bean-shaped thickening curved around a stomodaeum. Its edges become ciliated, a median protrusion becomes the median tentacle, and pair of lateral protrusions becomes the first pair of tentacles; additional tentacles develop laterally to the median tentacle. The dorsal side secretes a circular chitinous valve, which during growth bends and becomes divided into the brachial and pedicle valves. Mesoderm proliferates from the anterior and lateral parts of the invaginating gut; coelomic cavities develop through schizocoely, but the details are unknown. The median tentacle contains muscle cells and sensory cells, with a cilium surrounded by microvilli; a narrow coelomic cavity with a few cilia was observed by Hay-Schmidt (1992), but a coelomic cavity was not found by Lüter (1996).

The free-swimming larvae are covered by the two valves, and have a short median tentacle and a horseshoe of ciliated tentacles used both in swimming and feeding (Fig. 41.1). The larval nervous system was studied by Hay-Schmidt (1992). The median tentacle contains an apical organ with 6–8 serotonergic perikarya and axons extending along the mouth, with branches into the tentacles, and ending in a small ventral ganglion in the older larvae. Nerve cells containing FMRFamide follow a similar pattern, with additional perikarya at the basis of each tentacle and the median tentacle. The tentacles apparently have the same structure as those of the adults, and the lophophore continues its filtering function after the settling. Each tentacle has a lateral band of cilia that create an abfrontally directed current, used both in swimming and particle collection, and a frontal band bordered by a row of laterofrontal cilia. It appears that the laterofrontal cilia act as a sieve, but the mechanism transferring the particles to the frontal ciliary band is still unknown (Strathmann 2005). A small stalk develops from the posterior part of the pedicle-valve tissue and lies curved up between the valves ready for stretching out at metamorphosis.

The early development of the linguliform *Discinisca* was studied by Freeman (1999), who found that the apical pole of the oocytes in the ovary becomes the apical pole with the polar bodies after fertilization of the egg. The first two cleavages go through the apical-blastoporal axis, and the first one is median. The subsequent development includes a 16-cell stage, with an apical and a blastoporal layer of 2×4 cells each, a 32-cell stage of two 4×4 cell layers, and a 64-cell stage of four 4×4 cell layers; thereafter the cells rearrange into a hollow, one-layered, ciliated blastula. The blastoporal pole invaginates, and the embryo begins to elongate with the apical pole with a ciliary tuft moving towards the anterior end. The blastoporal pole becomes elongate and mesodermal cells separate from the endoderm at the area of the blastopore, which soon begins to close from behind; it remains open at the anterior end, which becomes the adult mouth, whereas a new anus breaks through posteroventrally. The anterior end of the larva swells and the posterior half becomes more cylindrical, with a small group of long larval setae developing on each side (Fig. 41.1). The anterior lobe becomes triangular in outline, with the apical organ situated on the mediodorsal tip, which becomes the median tentacle. The later development was studied by Chuang (1977), who observed development of the two initially circular shells, shedding of the long larval chaetae, development of curved larval chaetae, and finally of straight adult chaetae. New tentacles become added laterally to the median tentacle, the stalk develops posteriorly, and the posterior side of the pedicle valve becomes concave, as a first stage in the development of the slit through which the stalk protrudes in the adult. Freeman (1999) bisected cleavage stages and blastulae of *Discinisca*, and the abilities of the various regions to regenerate resemble those reported in the classical studies of regeneration in sea-urchin embryos (Chapter 58).

The observed variation makes it very difficult to infer the ancestral embryology of the brachiopods. The first cleavage is either median or unspecified through the primary axis, never oblique and forming characteristic spiralian quadrants (Chapter 21, Fig. 21.1). An analysis of early brachiopod fossils indicates that the ancestral brachiopod had a planktotrophic larva (Popov *et al.* 2010).

I have chosen to regard phoronids and brachiopods as sister groups; see discussion in Chapter 39.

Interesting subjects for future research

1. Development and structure of nervous systems in all groups

2. Origin and differentiation of the coeloms, especially the origin of the large arm sinus

3. Development of the shells in discinids

References

Altenburger, A. and Wanninger, A. 2010. Neuromuscular development in *Novocrania anomala*: evidence for the presence of serotonin and spiralian-like apical organ in lecithotrophic brachiopod larvae. *Evol. Dev.* **12**: 16–24.

Atkins, D. 1958. A new species and genus of Kraussinidae (Brachiopoda) with a note on feeding. *Proc. Zool. Soc. Lond.* **131**: 559–581.

Bemmelen, J.F. 1883. Untersuchungen über den anatomischen und histologischen Bau der Brachiopoda Testicardinia. *Jena. Z. Naturw.* **16**: 88–161.

Blochmann, F. 1892–1900. *Untersuchungen über den Bau der Brachiopoden I–II.* Gustav Fischer, Jena.

Carlson, S.J. 2007. Recent research on brachiopod evolution. In R.C. Moore and P.A. Selden (eds): *Treatise on Invertebrate Paleontology, part H (revised)*, vol.6, pp. 2878–2900. The Geological Society of America, Boulder, CO.

Chuang, S.H. 1977. Larval development in *Discinisca* (Inarticulate brachiopod). *Am. Zool.* **17**: 39–53.

Cohen, B. and Weydmann, A. 2005. Molecular evidence that phoronids are a subtaxon of brachiopods (Brachiopoda: Phoronata) and that genetic divergence of metazoan phyla began long before the early Cambrian. *Org. Divers. Evol.* **5**: 253–273.

Cohen, B.L. 2000. Monophyly of brachiopods and phoronids: reconciliation of molecular evidence with Linnaean classification (the subphylum Phoroniformea nov.). *Proc. R. Soc. Lond. B* **267**: 225–231.

Cohen, B.L., Gawthrop, A. and Cavalier-Smith, T. 1998. Molecular phylogeny of brachiopods and phoronids based on nuclear-encoded small subunit ribosomal RNA gene sequences. *Phil. Trans. R. Soc. Lond. B* **353**: 2040–2060.

Conklin, E.G. 1902. The embryology of a brachiopod, *Terebratulina septentrionalis* Couthouy. *Proc. Am. Philos. Soc.* **41**: 41–76.

Freeman, G. 1993. Regional specification during embryogenesis in the articulate brachiopod *Terebratalia. Dev. Biol.* **160**: 196–213.

Freeman, G. 1995. Regional specification during embryogenesis in the inarticulate brachiopod *Glottidia. Dev. Biol.* **172**: 15–36.

Freeman, G. 1999. Regional specification during embryogenesis in the inarticulate brachiopod *Discinisca. Dev. Biol.* **209**: 321–339.

Freeman, G. 2000. Regional specification during embryogenesis in the craniiform brachiopod *Crania anomala. Dev. Biol.* **227**: 219–238.

Freeman, G. 2003. Regional specification during embryogenesis in Rhynchonelliform brachiopods. *Dev. Biol.* **261**: 268–287.

Hay-Schmidt, A. 1992. Ultrastructure and immunocytochemistry of the nervous system of the larvae of *Lingula anatina* and *Glottidia* sp. (Brachiopoda). *Zoomorphology* **112**: 189–205.

James, M.A. 1997. Brachiopoda: Internal anatomy, embryology, and development. In F.W. Harrison (ed.): *Microscopic Anatomy of Invertebrates*, vol. 13, pp. 297–407. Wiley-Liss, New York.

Long, J.A. 1964 *The embryology of three species representing three superfamilies of articulate brachiopods.* Ph.D. Thesis, University of Washington.

Long, J.A. and Stricker, S.A. 1991. Brachiopoda. In A.C. Giese, J.S. Pearse and V.B. Pearse (eds): *Reproduction of Marine Invertebrates*, vol. 6, pp. 47–84. Blackwell Scientific Publs., Boston/Boxwood Press, Pacific Grove.

Lüter, C. 1996. The median tentacle of the larva of *Lingula anatina* (Brachiopda) from Queensland, Australia. *Aust. J. Zool.* **44**: 355–366.

Lüter, C. 2000a. The origin of the coelom in Brachiopoda and its phylogenetic significance. *Zoomorphology* **120**: 15–28.

Lüter, C. 2000b. Ultrastructure of larval and adult setae of Brachiopoda. *Zool. Anz.* **239**: 75–90.

Lüter, C. 2007. Anatomy. In R.C. Moore and P.A. Selden (eds): *Treatise on Invertebrate Paleontology, part H (revised)*, vol.6, pp. 2321–2355. The Geological Society of America, Boulder, CO.

Nielsen, C. 1991. The development of the brachiopod *Crania (Neocrania) anomala* (O. F. Müller) and its phylogenetic significance. *Acta Zool. (Stockh.)* **72**: 7–28.

Nielsen, C. 2005. Trochophora larvae: cell-lineages, ciliary bands and body regions. 2. Other groups and general discussion. *J. Exp. Zool. (Mol. Dev. Evol.)* **304B**: 401–447.

Passamaneck, Y. and Halanych, K.M. 2006. Lophotrochozoan phylogeny assessed with LSU and SSU data: evidence of lophophorate polyphyly. *Mol. Phylogenet. Evol.* **40**: 20–28.

Pennington, J.T., Tamburri, M.N. and Barry, J.P. 1999. Development, temperature tolerance, and settlement preference of embryos and larvae of the articulate brachiopod *Laqueus californianus. Biol. Bull.* **196**: 245–256.

Popov, L.E., Bassett, M.G., Holmer, L.E., Skovsted, C.B. and Zukov, M.A. 2010. Earliest ontogeny of Early Palaeozoic Craniiformea: implications for brachiopod phylogeny. *Lethaia* **40**: 85–96.

Pross, A. 1980. Untersuchungen zur Gliederung von *Lingula anatina* (Brachiopoda).—Archimerie bei Brachiopoden. *Zool. Jahrb., Anat.* **103**: 250–263.

Santagata, S. and Cohen, B.L. 2009. Phoronid phylogenetics (Brachiopoda: Phoronata): evidence from morphological cladistics, small and large subunit rDNA sequences, and mitochondrial *cox1. Zool. J. Linn. Soc.* **157**: 34–50.

Strathmann, R.R. 2005. Ciliary sieving and active ciliary response in capture of particles by suspension-feeding brachiopod larvae. *Acta Zool. (Stockh.)* **86**: 41–54.

Stricker, S.A. and Reed, C.G. 1985a. The ontogeny of shell secretion in *Terebratalia transversa* (Brachiopoda, Articulata) I. Development of the mantle. *J. Morphol.* **183**: 233–250.

Stricker, S.A. and Reed, C.G. 1985b. The ontogeny of shell secretion in *Terebratalia transversa* (Brachiopoda, Articulata) II. Formation of the protegulum and juvenile shell. *J. Morphol.* **183**: 251–271.

Williams, A. 1997. Brachiopoda: Introduction and integumentary system. In F.W. Harrison (ed.): *Microscopic Anatomy of Invertebrates*, vol. 13, pp. 237–296. Wiley-Liss, New York.

Williams, A. and Carlson, S.J. 2007. Affinities of brachiopods and trends in their evolution. In R.C. Moore and P.A. Selden (eds): *Treatise on Invertebrate Paleontology, Part H (revised)*, vol. 6, pp. 2822–2877. The Geological Society of America, Boulder, CO.

Williams, A., James, M.A., Emig, C.C., Mackay, S. and Rhodes, M.C. 1997. Anatomy. In R.C. Moore (ed.): *Treatise on Invertebrate Paleontology, part H (revised)*, vol.1, pp. 7–188. Geological Society of America, Lawrence, KS.

Yatsu, N. 1902. On the development of *Lingula anatina. J. Coll. Sci. Imp. University Tokyo* **17**: 1–112.

42

ECDYSOZOA

The 'defining' apomorphy of this clade is the moulting of the cuticle (ecdysis), although complete or partial moulting have been observed in a few annelids (Chapter 25). Ecdysis is apparently governed by the hormone 20-hydroxyecdysone in arthropods (Chapter 44), but other hormones are involved in the ecdysis of nematodes (Chapter 49). Whether this is associated with the different chemical composition of the cuticle in the two groups, chitin in arthropods, and collagen in nematodes, remains to be seen. The hormone 20-hydroxyecdysone is found in most of the lower invertebrates and protostomes, but it is not certain that it is endogenous in all groups. In some species it is a deterrent against predators (Lafont and Koolman 2009). The moulting is intimately connected with the lack of a ciliated epithelium, and this combination is clearly an apomorphy, because both the sister group Spiralia and the ancestral metazoan groups are ciliated. See also discussion in Chapter 23.

Other morphological apomorphies are difficult to point out, but anti-horseradish peroxidase immunoreactivity has been reported from the central nervous systems of arthropods, onychophorans, nematodes, nematomorphs, and priapulans, but not from any of the spiralians, deuterostomes, or cnidarians studied (Haase *et al.* 2001).

Monophyly of the Ecdysozoa is shown in almost all newer molecular phylogenetic analyses (for example Dunn *et al.* 2008; Hejnol *et al.* 2009; Pick *et al.* 2010).

Two clades are recognized, Panarthropoda with segments with limbs, and a brain with paired ganglia, and Cycloneuralia without limbs and with a collar-shaped brain (Fig. 42.1). The two groups are morphologically well delimited, but the position of the tardigrades has been unstable both in morphological and molecular analyses (see discussion for example in Edgecombe 2009). They are placed in the Panarthropoda in morphological analyses, such as Brusca and Brusca (2003), and molecular analyses, such as Dunn *et al.* (2008) and Rota-Stabelli *et al.* (2010), but in the Cycloneuralia in a few morphological analyses, such as Kristensen (2003), and molecular analyses, such as Roeding *et al.* (2007) and Hejnol *et al.* (2009). The analyses of Dunn *et al.* (2008) showed that the position of the tardigrades is sensitive to taxon sampling, and the analyses of Rota-Stabelli *et al.* (2010) indicate that the trees with tardigrades as cycloneuralians are strongly influenced by the long branches of both tardigrades and many of the nematodes. New studies on the embryology of the tardigrades strongly support the arthropod relationships (Chapter 46). Thus, the intuitive interpretation of the tardigrades as arthropods now finds support from both morphology and molecular phylogeny.

Ecdysozoa has a rich fossil record because chitinous cuticles fossilize easily; some strata even show fossils with astonishingly well-preserved three-dimensional details, for example the famous 'Orsten' fauna (Maas *et al.* 2006) and Early Cambrian shales from Australia (Harvey and Butterfield 2008). Many arthropod and stem-arthropod fossils provide much information about the origin and radiation of the group

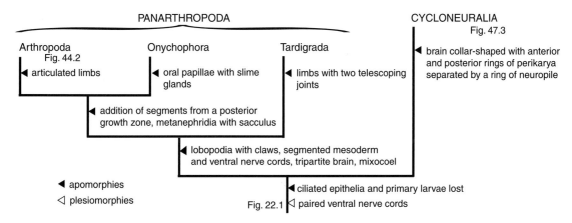

Figure 42.1. Phylogeny of the Ecdysozoa and the Panarthropoda.

(Edgecombe 2010), whereas many of the cycloneuralian fossils are difficult to interpret. The anomalocaridids show both a mouth, with radially arranged sclerites resembling a scalidophoran mouth, and a segmented body with appendages (Daley *et al.* 2009). They have been interpreted as stem-group arthropods, but the structure of the mouth could indicate a position as stem-group ecdysozoans.

The palaeoscolecids have been interpreted as stem-group ecdysozoans, but are probably stem-group priapulids (Harvey *et al.* 2010).

References

Brusca, R.C. and Brusca, G.J. 2003. *Invertebrates*, 2nd ed. Sinauer, Sunderland, MA.

Daley, A.C., Budd, G.E., Caron, J.B., Edgecombe, G.D. and Collins, D. 2009. The Burgess Shale anomalocaridid *Hurdia* and its significance for early arthropod evolution. *Science* **323**: 1597–1600.

Dunn, C.W., Hejnol, A., Matus, D.Q., *et al.* 2008. Broad phylogenomic sampling improves resolution of the animal tree of life. *Nature* **452**: 745–749.

Edgecombe, G.D. 2009. Palaeontological and molecular evidence linking arthropods, onychophorans, and other Ecdysozoa. *Evo. Edu. Outreach* **2**: 178–190.

Edgecombe, G.D. 2010. Arthropod phylogeny: An overview from the perspectives of morphology, molecular data and the fossil record. *Arthropod Struct. Dev.* **39**: 74–87.

Haase, A., Stern, M., Wächter, K. and Bicker, G. 2001. A tissue-specific marker of Ecdysozoa. *Dev. Genes Evol.* **211**: 428–433.

Harvey, T.H.P. and Butterfield, N.J. 2008. Sophisticated particle-feeding in a large Early Cambrian crustacean. *Nature* **452**: 868–871.

Harvey, T.H.P., Dong, X. and Donoghue, P.C.J. 2010. Are palaeoscolecids ancestral ecdysozoans? *Evol. Dev.* **12**: 177–200.

Hejnol, A., Obst, M., Stamatakis, A., *et al.* 2009. Assessing the root of bilaterian animals with scalable phylogenomic methods. *Proc. R. Soc. Lond. B* **276**: 4261–4270.

Kristensen, R.M. 2003. Comparative morphology: do the ultrastructural investigations of Loricifera and Tardigrada support the clade Ecdysozoa? In A. Legakis, S. Sfenthourakis, R. Polymeni and M. Thessalou-Legaki (eds): *The New Panorama of Animal Evolution*, pp. 467–477. Pensoft, Sofia.

Lafont, R. and Koolman, J. 2009. Diversity of ecdysteroids in animal species. In G. Smagghe (ed.): *Ecdysone: Structures and Functions*, pp. 47–71. Springer, Dordrecht.

Maas, A., Braun, A., Dong, X.-P., *et al.* 2006. The 'Orsten'—More than a Cambrian Konservat-Lagerstätte yielding exceptional preservation. *Palaeoworld* **15**: 266–282.

Pick, K.S., Philippe, H., Schreiber, F., *et al.* 2010. Improved phylogenomic taxon sampling noticeably affects nonbilaterian relationships. *Mol. Biol. Evol.* **27**: 1983–1987.

Roeding, F., Hagner-Holler, S., Ruhberg, H., *et al.* 2007. EST sequencing of Onychophora and phylogenomic analysis of Metazoa. *Mol. Phylogenet. Evol.* **45**: 942–951.

Rota-Stabelli, O., Campbell, L., Brinkmann, H., *et al.* 2010. A congruent solution to arthropod phylogeny: phylogenomics, microRNAs and morphology support monophyletic Mandibulata. *Proc. R. Soc. Lond. B* **278**: 298–306.

43

PANARTHROPODA

The group Panarthropoda, comprising the living Arthropoda, Onychophora, and Tardigrada (Fig. 42.1), is now almost unanimously regarded as monophyletic (Edgecombe 2010; Rota-Stabelli *et al.* 2010). They are segmented ecdysozoans with paired appendages and a cuticle containing α-chitin and protein, but lacking collagen (the type of chitin in the tardigrade cuticle has not been identified). The appendages show nerves from the paired ventral nerve cords, usually from segmental ganglia with commissures, intrinsic muscles, and one or more terminal claws; there are no chaetae (as those of the annelids). The food is typically manipulated with modified anterior limbs. There are essential similarities between amino-acid compositions in the non-sclerotized cuticles of crustaceans, insects, merostomes, and onychophorans (Hackman and Goldberg 1976). *Engrailed* is expressed in segmental stripes in embryos of all three phyla, but only the arthropods show clear parasegments (Chapter 44); it is expressed in the mesoderm before ectodermal segmentation becomes visible in tardigrades and onychophorans (Gabriel and Goldstein 2007; Eriksson *et al.* 2009).

This is a group in which the fossils play an important role in the phylogenetic discussions. The rich Middle-Cambrian faunas of the Burgess Shale and Chengjiang show variations over the 'panarthropod theme' that are unknown today. Two main groups of panarthropod fossils can be recognized: 'lobopods', which are forms with unjointed appendages and that must be interpreted as basal panarthropods, and stem-group arthropods with jointed appendages. Already

the Lower Cambrian, Maotianshan-Shale *Fuxianhuia* had articulated limbs with an exopod, i.e. a biramous limb (Waloszek *et al.* 2005), and was definitely part of the arthropod stem group. Many lobopods are known from the Cambrian deposits. Some of them are probably onychophoran stem-group species, such as the famous *Hallucigenia* (Hou and Bergström 1995). The Upper Cambrian Orsten fossil *Orstenotubulus* is most probably a marine stem-group onychophoran (Maas *et al.* 2007). The tardigrade ancestor would probably be classified as a lobopod, and the panarthropod lineage probably evolved from a lobopod, which developed articulations of the appendages.

Homology of the various head segments has been a matter of much controversy, but Scholtz and Edgecombe (2006) and Eriksson *et al.* (2003) interpreted the onychophoran 'primary antennae' as belonging to the protocerebral/ocular brain, whereas the chelicerae/antennules of the chelicerates/mandibulates belong to the deutocerebral segment. The protocerebral brain could well be homologous with the cerebral ganglia of the spiralians (see Chapter 22). The homologies of the tardigrade brain are still discussed (Chapter 46).

A series of additional synapomorphies can be recognized (Weygoldt 1986; Paulus 2007): coelomic sacs develop during ontogeny, but their walls disintegrate partially later on, so that the body cavity is a combination of the blastocoel and the coelomic cavities, i.e. a mixocoel or haemocoel. Some of the 'crustaceans' develop a 'naupliar mesoderm' with unclear segmentation, and add segments from a posterior-growth zone (Chapter 44). There is a dorsal heart (except in the minute tardigrades) that is a tubular structure consisting of circular muscles and having a pair of ostia per segment. These are situated at the borders between segments, as expected from their ontogenetic origin from the mediodorsal walls of the coelomic sacs. The blood enters the heart through the ostia from the haemocoel, is pumped anteriorly, and leaves the heart through its anterior end. Protonephridia are lacking, but modified metanephridia are found in most segments in the onychophorans and in a few segments of most arthropods. These metanephridia develop from small groups of mesodermal cells and become differentiated into a sacculus, with basal membrane and podocytes, and a duct, which modifies the primary urine from the sacculus. The onychophorans have a ciliated funnel as the beginning of the duct, but cilia are lacking in the arthropods. The tardigrades lack a haemal system and metanephridia, probably as a function of their small size (see Chapter 46).

The morphology and life cycle of the panarthropod ancestor is difficult to infer. The embryology of tardigrades and onychophorans give no clue, and the nauplius larva is probably an arthropod character.

The sister-group relationship between onychophorans and arthropods is found in almost all recent analyses, based on mitochondrial genomes (Braband *et al.* 2010), expressed sequence tag data (Dunn *et al.* 2008), and phylogenomics (Hejnol *et al.* 2009; Rota-Stabelli *et al.* 2010), although the position of the tardigrades is unstable.

References

Braband, A., Cameron, S.L., Podsiadlowski, L., Daniels, S.R. and Mayer, G. 2010. The mitochondrial genome of the onychophoran *Opisthopatus cinctipes* (Peripatopsidae) reflects the ancestral mitochondrial gene arrangement of Panarthropoda and Ecdysozoa. *Mol. Phylogenet. Evol.* **57**: 285–292.

Dunn, C.W., Hejnol, A., Matus, D.Q., *et al.* 2008. Broad phylogenomic sampling improves resolution of the animal tree of life. *Nature* **452**: 745–749.

Edgecombe, G.D. 2010. Arthropod phylogeny: An overview from the perspectives of morphology, molecular data and the fossil record. *Arthropod Struct. Dev.* **39**: 74–87.

Eriksson, B.J., Tait, N.N. and Budd, G.E. 2003. Head development in the onychophoran *Euperipatoides kanangrensis* with particular reference to the central nervous system. *J. Morphol.* **255**: 1–23.

Eriksson, B.J., Tait, N.N., Budd, G.E. and Akam, M. 2009. The involvement of engrailed and wingless during segmentation in the onychophoran *Euperipatoides kanangrensis* (Peripatopsidae: Onychophora) (Reid 1996). *Dev. Genes Evol.* **219**: 249–264.

Gabriel, W.N. and Goldstein, B. 2007. Segmental expression of Pax3/7 and engrailed homologs in tardigrade development. *Dev. Genes Evol.* **217**: 421–433.

Hackman, R.H. and Goldberg, M. 1976. Comparative chemistry of arthropod cuticular proteins. *Comp. Biochem. Physiol.* **55B**: 201–206.

Hejnol, A., Obst, M., Stamatakis, A., *et al.* 2009. Assessing the root of bilaterian animals with scalable phylogenomic methods. *Proc. R. Soc. Lond. B* **276**: 4261–4270.

Hou, X. and Bergström, J. 1995. Cambrian lobopodians—ancestors of extant onychophorans? *Zool. J. Linn. Soc.* **114**: 3–19.

Maas, A., Mayer, G., Kristensen, R.M. and Waloszek, D. 2007. A Cambrian micro-lobopodian and the evolution of arthropod locomotion and reproduction. *Chin. Sci. Bull.* **52**: 3385–3392.

Paulus, H. 2007. Arthropoda, Gliederfüsser. In W. Westheide and R. Rieger (eds): *Spezielle Zoologie, Teil 1: Einzeller und Wirbellose Tiere*, 2nd ed., pp. 438–462. Elsevier, München.

Rota-Stabelli, O., Campbell, L., Brinkmann, H., *et al.* 2010. A congruent solution to arthropod phylogeny: phylogenomics, microRNAs and morphology support monophyletic Mandibulata. *Proc. R. Soc. Lond. B* **278**: 298–306.

Scholtz, G. and Edgecombe, G.D. 2006. The evolution of arthropod heads: reconciling morphological, developmental and palaeontological evidence. *Dev. Genes Evol.* **216**: 395–415.

Waloszek, D., Chen, J., Maas, A. and Wang, X. 2005. Early Cambrian arthropods—new insights into arthropod head and structural evolution. *Arthropod Struct. Dev.* **34**: 189–205.

Weygoldt, P. 1986. Arthropod interrelationships—the phylogenetic-systematic approach. *Z. Zool. Syst. Evolutionsforsch.* **24**: 19–35.

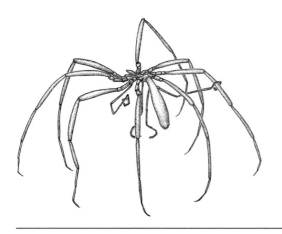

44

Phylum **Arthropoda**

Arthropoda is one of the largest animal phyla; the insects alone are now believed to comprise several million living species, while the other arthropods number more than 100 000. They inhabit almost every conceivable habitat, from the deep sea to the deserts, and some of the parasitic forms are so modified that their arthropod nature can only be recognized in certain developmental stages. There is a substantial fossil record both of living and extinct groups reaching back to the Early Cambrian, and the fossils are very important for our understanding of the evolution of the phylum (see below). The monophyly of the group was questioned by the Manton school (Manton 1972,1977; Anderson 1973), but recent morphological, palaeontological, and molecular investigations agree on monophyly (Wills *et al.* 1998; Scholtz and Edgecombe 2005; Dunn *et al.* 2008; Hejnol *et al.* 2009; Edgecombe 2010; Regier *et al.* 2010; Rota-Stabelli *et al.* 2010).

Conspicuous arthropod apomorphies (Edgecombe 2010) include: an articulated chitinous exoskeleton with thick sclerites corresponding to the segments, separated by rings with a thin, flexible cuticle. Each segment may carry a pair of articulated legs with intrinsic musculature (Shear 1999). A number of anterior segments are fused to form a cephalon, which in most groups carries eyes and two or more pairs of limbs. Compound eyes have new elements being added from a proliferation zone (Harzsch and Hafner 2006). Free-

living larvae developing from small eggs are found in pycnogonids (pantopods), and crustaceans and their early stages are rather similar, with eyes and three pairs of appendages (Fig. 44.1). The pycnogonid larva has a pair of cheliceres (chelifores) and two uniramous appendages, whereas the crustacean nauplius larva has a pair of antennules and two pairs of biramous appendages. A (meta)nauplius has been reported from the Lower Cambrian (Zhang *et al.* 2010). New studies of gene expression support the homology of these three pairs of appendages (Manuel *et al.* 2006), and the homology of the various appendages in the main arthropod groups is now well-established (Table 44.1). In adults, the body is divided into regions, tagmata, that consist of fused sclerites with different types of appendages. Many forms have one or two pairs of metanephridia that lack cilia. The rubber-like protein resilin has been found in scorpions, crustaceans, and insects (Govindarajan and Rajulu 1974; Burrows 2009), and may be an arthropod apomorphy. These characters are discussed further below.

The more traditional phylogenetic scheme of living arthropods recognized two or three subphyla, with a number of subgroups: Pycnogonida (of uncertain position), Chelicerata (Xiphosura + Arachnida), and Mandibulata (Crustacea + Tracheata (= Hexapoda + Myriapoda)). However, newer morphological and molecular studies have indicated that Hexapoda is an

Chapter vignette: The pycnogonid *Colossendeis scotti*. (Redrawn from Brusca and Brusca 1990.)

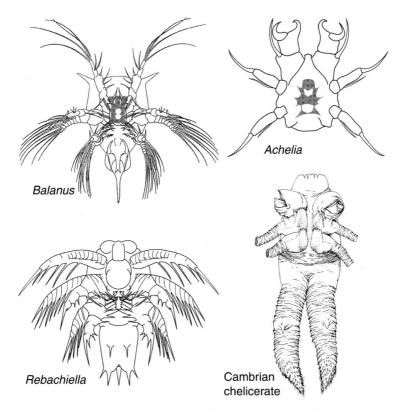

Figure 44.1. Early larval stages of living and fossil arthropods (the nervous systems of the living forms are indicated in grey). Nauplius of the crustacean *Semibalanus balanoides*. (Redrawn from Sanders 1963 and Walley 1969.) Protonymphon of the pycnogonid *Achelia echinata* (Redrawn from Meisenheimer 1902.) Second larval stage of the Cambrian crustacean *Rebachiella* (Redrawn from Walossek 1993.) Early larva of an Upper Cambrian chelicerate, possibly a pycnogonid (Redrawn from Müller and Walossek 1986.) Note the gnathobases on the two posterior limb bases.

Table 44.1. Somites and their ganglia and eyes, or appendages in Chelicerata (Pycnogonida, Xiphosura, and Arachnida) and Mandibulata (Myriapoda, 'Crustacea', and Hexapoda). (Based on Carroll *et al.* 2001; Harzsch *et al.* 2005; Janssen and Damen 2006; Manuel *et al.* 2006.) With few exceptions, the interpretation agrees with the expression domains of the Hox genes.

Segment	Ganglion	Pycnogonida	Xiphosura	Arachnida	Myriapoda	'Crustacea'	Hexapoda
0	protocerebrum	eyes	eyes	eyes	eyes	eyes	eyes
1	deutocerebrum	chelifores	cheliceres	cheliceres	antennae	antennules	antennae
2	tritocerebrum	pedipalps	pedipalps	pedipalps	no limbs	antennae	no limbs
3	1st ventral g.	ovigers	1st legs	1st legs	mandibles	mandibles	mandibles
4	2nd ventral g.	1st legs	2nd legs	2nd legs	maxillae	1 maxillae	maxillae
5	3rd ventral g.	2nd legs	3rd legs	3rd legs	no limbs	2 maxillae	labium

ingroup of the Crustacea, which creates a name problem. It would be logical just to include the Hexapoda in Crustacea, but this would undoubtedly create much confusion, and the alternative names Tetraconata and Pancrustacea have been proposed for Hexapoda plus the paraphyletic Crustacea. I have chosen to use the name Pancrustacea for the clade, and to use 'crustaceans' for the non-hexapod groups (in parallel to 'polychaetes' and 'invertebrates'). Further, it appears that the Myriapoda is the sister group of the Pancrustacea.

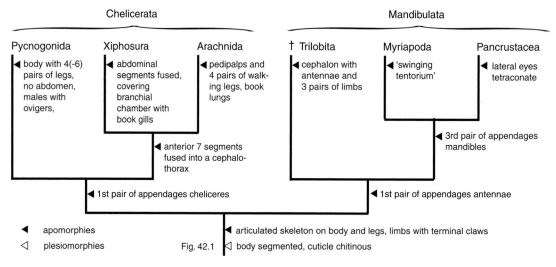

Figure 44.2. Arthropod phylogeny. The 'swinging tentorium' (see Koch 2003). See also Table 44.1.

The position of the Pycnogonida has been a matter of controversy, but the latest studies strengthen a traditional view that they are the sister group of the Euchelicerata (Xiphosura + Arachnida), so a consensus about a phylogeny as that in Fig. 44.2, seems to be developing. This is discussed further below.

The parasitic Pentastomida (Storch 1993) was earlier interpreted as a 'proarthropod' group, but studies on morphology and molecules indicate a position within the Arthropoda as the sister group of the crustacean group Branchiura (Lavrov *et al.* 2003; Møller *et al.* 2008; Regier *et al.* 2010). Waloszek *et al.* (2006) described a Late Cambrian marine fossil that definitely resembles a juvenile, parasitic pentastomid. The adults of living pentastomids are parasites of terrestrial vertebrates, but the host of the fossil species has not been identified. The dating of this fossil threw doubt over the systematic position of the pentastomids within the Arthropoda, but new information about molecular divergence times have removed the problem (Sanders and Lee 2010). The embryology of pentastomids is poorly known, and the regions of the central nervous system are difficult to homologize with those of the other arthropods. Embryos show four pairs of coelomic pouches and seven pairs of ganglionic cell groups, the four anterior ganglia

being connected with the coelomic pouches. The three anterior pairs of ganglia fuse to form a 'brain', and the three posterior ones fuse too (Böckeler 1984). These observations indicate the arthropod character of the group, but give no certain hints about its more precise position. The sperm of *Raillietiella* and the branchiuran *Argulus* are very similar and of a highly specialized type (Wingstrand 1972; Storch and Jamieson 1992), and both this and observations on embryology and structure and moulting of the cuticle (Riley *et al.* 1978) indicate that the group clusters with the crustaceans.

Numerous studies of arthropod and especially mandibulate phylogeny have focussed on limb morphology, with emphasis on the occurrence of uniramous versus biramous limbs. Chelicerates, myriapods, and hexapods have uniramous legs (with the possible exception of Xiphosura), so the biramous condition could be a crustacean apomorphy, but most authors have favoured the biramous leg as the ancestral type, and derived the uniramous types through simplifications. The evolution of uniramous legs in malacostracans from an ancestral biramous condition is supported by the fact that several of the 'higher crustaceans', such as brachyurans, have biramous legs in larval or juvenile stages and uniramous legs as adults. However, new studies of astonishingly well-preserved Cambrian fossils, including a number of extinct stem-group arthropods, have more or less settled the discussion, because a number of the Cambrian stem arthropods had biramous limbs (Hou and Bergström 1997; Waloszek et al. 2005; Edgecombe 2010; however, see also Wolff and Scholtz 2008). Waloszek (1993) and Haug et al. (2010) proposed that the proximal endite during the evolution of the stem crustaceans became extended all around the limb, forming the coxa of at least the second antenna and the mandible. This development is possibly seen during ontogenesis of living crustaceans, such as *Lightiella* (Sanders and Hessler 1964). The coxa should thus be an apomorphy of the 'higher' crustaceans.

In the Mandibulata (also called Antennata), the first pair of appendages, the antennae/antennules, are not involved in feeding, but all the following appendages may at some stage be engaged in creating food currents, or collecting and handling particles. A number of the more basal crustaceans, such as cephalocarids, branchiopods, maxillopods, and the primitive malacostracan group Phyllocarida, have biramous limbs, with endopodites used in filter feeding and bases or coxae with gnathobases used in handling of food particles (Walossek 1993). Antennae, mandibles, and maxillae of the larval stages are, in many cases, of this type too, and the ancestral mandibulate was most probably a swimming, filter-feeding organism with a series of similar, biramous limbs, transporting food particles to the mouth along the food groove between the coxae to a posteriorly directed mouth partially covered by a labrum. This fits well with Lower Cambrian stem-group arthropods, such as *Canadaspis* (Hou and Bergström 1997).

Myriapoda and Hexapoda show many specializations to terrestrial life, and they have often been united under the name Tracheata or Atelocerata, with emphasis on a number of morphological similarities (Klass and Kristensen 2001; Bitsch and Bitsch 2004). However, this relationship is not shown in any of the molecular analyses, and it is contradicted by numerous studies on nervous systems and eyes (see for example Fanenbruck et al. 2004; Strausfeld 2009).

Boore et al. (1998) found a mitochondrial gene translocation in all the investigated pancrustaceans relative to all chelicerates, myriapods, onychophorans, and tardigrades, and it now seems unquestionable that the Hexapoda is an ingroup of the Crustacea. No molecular studies indicate a sister-group relationship between Crustacea and Hexapoda, but the position of the Hexapoda within the Crustacea is unresolved (Edgecombe 2010). However, it should be mentioned that a number of neuroanatomical characters, including the optic chiasmata, strongly support the malacostracan sister-group relationship of the insects (Harzsch 2004; Strausfeld 2009).

The monophyly of the Myriapoda is supported by morphological analyses, Hox gene data, a unique microRNA family, and RNA sequence analyses (Janssen et al. 2010; Regier et al. 2010; Shear and Edgecombe 2010). Their sister-group relationship with the Pancrustacea likewise appears well founded.

The Lower Cambrian to Upper Permian trilobites form a well-defined, monophyletic group (Fortey 2001). Their first appendage is an antenna that indicates that they belong to the Mandibulata, most probably as the sister group of the remaining mandibulates (Scholtz and Edgecombe 2006). A logical nomenclature would be to use the terms 'Mandibulata' for Myriapoda + Pancrustacea, and 'Antennata' for Trilobita + Mandibulata, but the more traditional scheme has been chosen here.

Chelicerates are generally carnivores that manipulate food with the cheliceres and transport it to the anteriorly directed mouth. The Xiphosura (Merostomata) and Arachnida seem unquestionably to be sister groups (Giribet et al. 2005; Regier et al. 2008; Edgecombe 2010), but the position of the Pycnogonida has been more controversial. The first larval stage of the pycnogonids has a pair of eyes, a pair of cheliceres, and two pairs of walking legs. New morphological studies of the development of a number of pycnogonids show that the cheliceres are innervated from the deutocerebrum as in the euchelicerates (Brenneis et al. 2008), and this interpretation is in full agreement with studies on the expression of Hox genes (Manuel et al. 2006). A Cambrian larval form has very similar appendages (see Fig. 44.2). The sister-group relationship between the pycnogonids and the euchelicerates is further supported by new studies of nuclear protein-coding sequences (Regier *et al.* 2010).

It thus seems well-documented, that the living chelicerates and mandibulates are both monophyletic, and that the stem lineages of both groups were present already in the Lower Cambrian. The diverse fossil faunas from Chengjiang (Hou *et al.* 2004) and the Burgess Shale (Briggs *et al.* 1994) comprise many forms that appear to be stem-group arthropods of considerable interest for the understanding of arthropod origin and evolution.

The microscopic anatomy was reviewed in the volumes edited by Harrison (1992a,b; 1993; 1998; 1999).

The ectoderm is simple, cuboid, and covered with a cuticle consisting of α-chitin with non-collagenous protein; locomotory cilia are completely absent. The cuticle is often heavily sclerotized and/or calcified in the stiff plates or sclerites, while the articulation membranes connecting the sclerites are thin and flexible. The special protein resilin is probably an arthropod apomorphy. The cuticle can only expand slightly, so growth is restricted to moultings, where the old cuticle breaks open along preformed sutures, and a thin, folded, unsclerotized cuticle preformed below the old cuticle becomes stretched out.

Moulting is controlled by ecdysones, especially 20-hydroxyecdysone. The process is best known in malacostracans and insects, and there are several unexpected differences between the organs involved in secreting the ecdysteroids and in their ways of control (Brusca and Brusca 2003). In malacostracans, X-organs in the eye stalks secrete a moult-inhibiting hormone that is stored in the sinus gland and released via the blood to the Y-organs, located either at the base of the antennae or near the mouthparts. When the nervous system is stimulated to induce ecdysis, the hormone production of the X-organs is inhibited, and the Y-organs begin to produce ecdysone and moulting is initiated. In insects, moulting is induced by secretion of ecdysone from the thoracic glands, situated in the ventral side of the prothoracic segment; an inhibitory system is apparently not involved. Copepods apparently

lack Y organs and the ecdysteroids may originate directly in the ectoderm (Hopkins 2009). In the myriapod *Lithobius* ecdysteroids have been found in the salivary glands and are believed to regulate ecdysis (Seifert and Bidmon 1988). In chelicerates a special gland has not been observed, and the physiology of moulting is poorly known. 20-hydroxyecdysone stimulates moulting in pycnogonids (Bückmann and Tomaschko 1992). These differences in function and the widespread occurrence of the ecdysial hormones (Lafont and Koolman 2009), especially related to reproduction, make it difficult to use these hormones in phylogenetic argumentation at this point.

Respiration is through the thin areas of the body wall in small, aquatic crustaceans, but special gill structures are developed in the larger forms. Terrestrial groups, such as some arachnids, myriapods, insects, and oniscoid isopods, have developed tracheae with variously arranged stigmata, but these organs appear to have evolved independently several times, as indicated by their scattered occurrence in the groups and their variable position.

The pharynx-oesophagus-stomach, or foregut, is a stomodaeum with cuticular lining. The endodermal midgut is a shorter or longer intestine, with a pair of digestive midgut glands (remipedes have serially arranged diverticula) lacking motile cilia, but biciliate cells, with cilia lacking central tubules and dynein-arms, have been reported from the midgut of a branchiopod (Rieder and Schlecht 1978). The rectum or hindgut is a proctodaeum lined with a cuticle.

The central nervous system of the arthropods is of the usual protostomian type, with a circumoesophageal brain and a pair of ventral longitudinal cords, sometimes with one or three thin longitudinal nerves between the cords (Harzsch 2004). It has generally been believed that the ancestral arthropod had a ladder-like central nervous system, with only the eye-bearing protocerebrum situated anterior to the mouth. However, it appears that the deutocerebrum is perioral in all arthropods (Fanenbruck *et al.* 2004; Harzsch 2004), which indicates that the ancestral arthropod brain consisted of a pre-oral protocerebrum, probably homologous to the spiralian cerebral ganglia (Figs.

22.9–11), and a peri-oral ganglion homologous of the anterior part of the periblastoporal nerve of the protostomian ancestor (Chapter 22). During evolution, additional segments have moved anteriorly and fused with the protocerebrum to form a larger brain, and this is seen during ontogeny of many forms (Scholtz and Edgecombe 2006). In living forms, the protocerebrum innervates the various types of eyes (Table 44.1), and the deutocerebrum innervates the antennules/antennae or cheliceres; the two anterior parts are fused into an anterodorsal brain. The tritocerebrum innervates the antennae in the crustaceans and the pedipalps in the chelicerates. In mandibulates, it is situated lateral to or behind the oesophagus on the connectives to the first ventral ganglion in nauplius larvae (Vilpoux *et al.* 2006; Lacalli 2009) and in adults of 'primitive' forms, such as *Hutchinsoniella* (Elofsson and Hessler 1990), *Derocheilocaris* (Baccari and Renaud-Mornant 1974), and *Triops* (Henry 1948), but it has moved forwards and fused with the deutocerebrum in the more advanced types. The more primitive, elongate arthropods have a chain of paired ventral ganglia (Fig. 22.3), but forms with a short body have the ganglia fused into a single mass.

Arthropod eyes have photoreceptor cells of the rhabdomeric type, innervated from the protocerebrum. Their development, as well as that of the optic lobes of the brain, are controlled by eye genes, such as *sine oculis* (Blanco *et al.* 2010). Two main types are usually recognized, median and lateral eyes.

Median eyes are usually called ocelli; they show much variation from small pigment-cup ocelli with few cells in nauplius eyes (Vaissière 1961; Elofsson 2006), to large eyes with a common lens and high image resolution in spiders and dragonflies (Land and Barth 1985; Berry *et al.* 2007). Paulus (1979) proposed that the ancestral arthropod had four pairs of medial photoreceptors as, for example, in living spiders, which have then differentiated in the various groups, for example the four ocelli on a dorsal tubercle in pycnogonids, the three ocelli in the nauplius eye of the crustaceans, and the ventral and dorsal frontal organs of all crustaceans. The myriapods lack median eyes.

Lateral eyes are usually quite complicated, consisting of several units of characteristic structure called ommatidia.

Pancrustaceans have facetted (compound) eyes consisting of a usually hexagonal array of ommatidia that each comprise two corneagenous cells, four crystalline cone cells (which has given the alternative name Tetraconata to the group), eight long retinular cells, plus various types of pigment cells (Paulus 2000). The central core of the retinular cells is the rhabdome, which is a stack of microvilli arranged in characteristic orthogonal layers (Paulus 1979). Addition of new ommatidia is of the 'morphogenetic front type' (Harzsch and Hafner 2006), where cells of the developing eye begin differentiation along a line, or furrow across the imaginal eye disc. In *Drosophila*, a number of cells along the front become designated as ommatidial progenitor cells, and these cells finally differentiate as the R8-photoreceptor cell. Neighbouring cells become recruited into the developing ommatidium in a fixed sequence, and when all 17 cells of each fully differentiated ommatidium are in place, the superfluous cells die (Ready 1989; Baker 2001; Harzsch and Hafner 2006). This may of course be a highly specialized process characteristic of the holometabolous insects, but highly similar ommatidial units with 18 cells have been observed in the eyes of *Homarus* (Hafner and Tokarski 2001).

Myriapods have various types of lateral eyes, with more scattered ommatidia or with ommatidia closely apposed resembling facetted eyes (Paulus 1979; Harzsch and Hafner 2006; Müller and Meyer-Rochow 2006; Harzsch *et al.* 2007). Diplopods have 'facetted' eyes, consisting of ommatidia with a high number of cells of the different types. New units are added to the eyes from the median side of the eye, as opposed to the 'morphogenic front type' of the Pancrustacea (Harzsch and Hafner 2006). Scutigeromorph chilopods have lateral eyes with hundreds of ommatidia, each with a crystalline cone of four cells, resembling that of the pancrustacean ommatidium (Müller *et al.* 2003). The pleurostigmophoran chilopod *Scolopendra* has ocelli of unique structure, with a stack of radiating photoreceptor cells (Müller *et al.* 2003; Müller and Meyer-Rochow 2006). The myriapod eyes have been interpreted in various ways, either as facetted eyes that have become dispersed or vice versa. Harzsch *et al.* (2007) concluded that diplopod, chilopod, and xiphosuran (see below) eye growth probably represents the ancestral arthropod type of eye formation.

The xiphosurans have very large facetted eyes, with ommatidia consisting of more than 300 cells (Battelle 2006). New ommatidia are added along the anterior edge of the eyes as in the diplopods.

Other sense organs are sensilla, i.e. one or a few primary receptor cells, usually with a strongly modified cilium surrounded by three cell types, tormogen, trichogen, and thecogen cells, the two last-mentioned cells secreting a cuticular structure (see for example Shanbhag *et al.* 1999). All the cells of a sensillum develop from one epithelial cell through a fixed sequence of differential cell divisions (Hartenstein and Posakony 1989). Both the mechanoreceptive sensory hairs and the chemoreceptive sensilla have very thin cuticular areas or small pores through the cuticle.

The mesoderm forms small segmental, coelomic cavities in early developmental stages, but later divides into a number of different structures (see below). The main part of the mesoderm loses the epithelial character and splits up into muscles and other organs, so that the coelomic cavities fuse with extracoelomic spaces forming a mixocoel or haemocoel. The musculature is striated, and the locomotory muscles are usually attached to the exoskeleton of neighbouring sclerites, often on long, internal extensions called apodemes. Typically, locomotion is thus not of the hydrostatic-skeleton type. The appendages have intrinsic musculature, i.e. muscles between the joints.

Modified metanephridia without cilia each originating in a small coelomic compartment, the sacculus, are found in xiphosurans, arachnids, and crustaceans, which may have antennal glands (at the base of the antennae) or maxillary glands (at the base of the second maxillae); many groups show antennal glands in the larval stage and maxillary glands in the adult stage, but some retain the antennal glands as adults; a few types, such as mysids, have both types. The sacculi

consist of podocytes resting on a basal membrane and resemble those of the nephridia of onychophorans (Chapter 45). Cephalocarids have the usual pair of maxillary glands, and in addition a small group of podocytes (without a duct) at the base of the antenna and the eight thoracic limbs, suggesting modified nephridia (Hessler and Elofsson 1995). Similar structures have been found in copepods and syncarids (Hosfeld and Schminke 1997). The formation of primary urine through ultrafiltration from the haemocoel to the sacculus, and its modification during the passage through the duct, are well documented. *Limulus* has a pair of coxal glands that develop from the ventral somites of the six prosomal segments, with the nephridial funnel and duct originating from segment number five (Patten and Hazen 1900). The arachnids have one or two pairs of coxal glands that develop from tubular outgrowths of the splanchnic mesoderm (Moritz 1959). Hexapods have a pair of labial glands that are excretory in collembolans (Feustel 1958). In terrestrial forms, the excretion is usually taken over more or less completely by Malpighian tubules, which arise from the posterior part of the endodermal midgut in arachnids and from the hindgut (proctodaeum) in insects, indicating separate origins.

The heart is a dorsal tube with a pair of ostia in each segment. The haemocoelic fluid enters through the ostia and is pumped anteriorly by the contractions of the circular muscles that constitute the wall of the heart.

The gonads are derived from the coelomic pouches, but their cavity is normally formed secondarily; they occupy various regions of the trunk, but they are situated in the head region in the cirripedes. There is one pair of gonoducts with gonopores located on different segments in different groups. The germ cells develop from the mesoderm, but the lineage of these cells has not been followed in many groups. In the cirripede *Semibalanus* (Fig. 44.3) the germ cells develop from descendants of the D cell, in the amphipod *Parhyale*, the g cell of the 8-cell stage is the primordial germ cell (Gerberding *et al.* 2002), and in the shrimp *Sicyonia*, the germ line can be followed too (Table 44.3).

Most arthropods have copulation and aflagellate sperm. Among the chelicerates, pycnogonids, xipho-

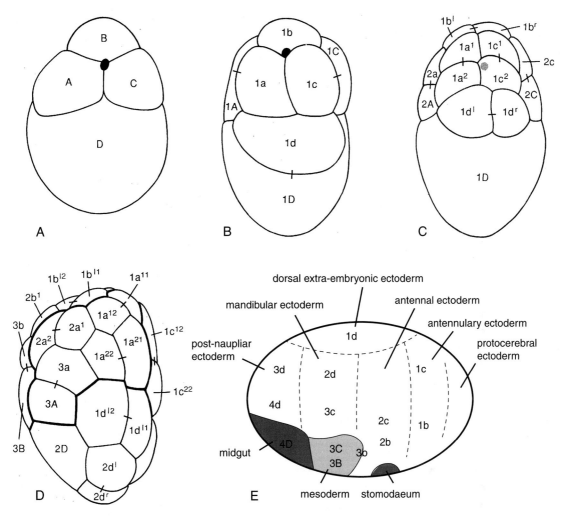

Figure 44.3. Early cirripede development. (A–D) The cleavage of *Semibalanus balanoides* (redrawn from Delsman 1917) with notation changed to resemble the spiral cleavage notation. (A) 4-cell stage in dorsal view. (B) 8-cell stage in dorsal view. (C) 15-cell stage in dorsal view. (D) 31-cell stage seen from the left side. (E) Fate map of a cirripede. (Redrawn from Anderson 1969.)

surans, and some scorpions have flagellate sperm (van Deurs 1974; Michalik and Mercati 2009). Among the mandibulates, a few myriapods have flagellate sperm, which is transferred by means of spermatophores (Dallai and Afzelius 2000), and a few groups of the aquatic crustaceans, viz. Phyllopoda, Cirripedia, Mystacocarida, and Branchiura have flagellate sperm (Ruppert *et al.* 2004).

Xiphosurans have a pseudocopulation, where the female deposits the eggs in a small cavity in the sand,

and the males shed the sperm over the eggs (Shuster 1950). In barnacles, the hermaphrodite stretches its long penis into the mantle cavity of a neighbouring specimen, where already deposited eggs are fertilized. All other arthropods have copulation in many cases with spermatophores.

Arthropod embryology shows tremendous variation (Anderson 1973; Scholtz 1997).

Among the chelicerates, the pycnogonid males carry the eggs. The embryology is rather poorly known

Table 44.2. Cell lineage of *Semibalanus balanoides*. (After Delsman 1917.) The notation is that of the spiral cleavage followed by Delsman's original notation. l, left; r, right.

```
                                  ┌ 1a=a^4.2                                    ectoderm
                   ┌ A=a^3  ┤                     ┌ 2a =a^5.1                   ectoderm
                   │        └ 1A=a^4.1 ┤          │          ┌ 3a=a^6.4         ectoderm
            ┌ AB ┤                     └ 2A =a^5.2 ┤         │        ┌ 4a=a^7.6 ectoderm
            │      │                               └ 3A=a^6.3 ┤       └ 4A=a^7.5 mesoderm
            │      └ B=b^3
  Z ┤               C=c^3 ┤ development as A
            │      ┌ C=c^3
            └ CD ┤        ┌ 1d=d^4.2                          ectoderm + mesoderm
                   └ D=d^3 ┤          ┌ 2d =d^5.2             mesoderm
                           └ 1D=d^4.1 ┤          ┌ 2D^l=d^6.2 ┤ 2D^l1=d^7.4 mesoderm
                                      └ 2D =d^5.1 ┤           └ 2D^l2=d^7.3 endoderm
                                                  └ 2D^r=d^6.1 ┤ 2D^r1=d^7.2 mesoderm
                                                               └ 2D^r2=d^7.1 endoderm
```

and the cell lineage has not been studied. Species with rather small eggs and only little yolk, such as *Achelia*, *Phoxichilidium*, and *Nymphon*, have total cleavage (Meisenheimer 1902; Dogiel 1913; Sanchez 1959; Ungerer and Scholtz 2009), whereas species with large, yolk-rich eggs, such as *Chaetonymphon* and *Callipallene*, have initial total cleavages but go through a blastoderm stage (Dogiel 1913; Sanchez 1959; Winter 1980). Early stages may resemble spiral-cleaving embryos, but the cleavage pattern is not spiral. Blastula stages are compact or have a narrow blastocoel. Gastrulation is through epiboly, and there seems to be one cell that elongates into the blastula towards the opposite pole, and this resembles the formation of the D quadrant in mollusc embryology (see Chapter 27). This elongate cell should become the endoderm, and surrounding cells should ingress and become mesoderm. In *Callipallene* the ventral ganglia develop in connection with paired ectodermal pockets (ventral organs) situated behind the stomodaeum.

Xiphosuran development is very poorly known. The large yolk-rich eggs show superficial cleavage; a thin blastoderm is formed, and a germ band resem-

bling that of spiders, crustaceans, and insects develop (Iwanoff 1933). Mesoderm migrates laterally from a median ventral groove and becomes organized in paired coelomic sacs. The heart with lateral ostia develops from the mediodorsal walls of the coelomic sacs (Kishinouye 1891).

All arachnids have direct development with yolk-rich eggs and blastoderm formation (Anderson 1973; Gilbert 1997). Well-defined coelomic sacs are seen in embryos of several groups, for example scorpions, where the vestiges of metanephridial ducts are found in prosomal segments three, four, and six; the coelomic sacs and ducts of segment five become the coxal glands, and the ducts of the second opisthosomal segment becomes the gonoducts (Brauer 1895; Dawydoff 1928). The origin of the heart from the mediodorsal walls of the coelomic sacs are well documented, for example in spiders (Anderson 1973).

In the mandibulate line, the almost exclusively terrestrial Myriapoda (Gilbert 1997) and Hexapoda (Schwalm 1997) generally have large, yolk-rich eggs with discoidal or superficial cleavage and formation of a blastoderm. This is regarded as highly specialized and will not be discussed here.

The mainly aquatic crustaceans show a wide variation in larval types, but representatives of almost all major groups go through the characteristic nauplius stage with paired antennules, antennae, and mandibles (Fig. 44.1) (Brusca and Brusca 2003). However, the embryological development leading to the nauplius stage shows all intermediate types, from small eggs with total cleavage to large eggs with blastoderm formation. Several species in most major groups have small eggs and total cleavage, which presumably represents the ancestral mode of development, and the following discussion will centre around these species.

Cirripedes with small eggs and nauplius larvae exhibit many supposedly ancestral characters. Both barnacles, acorn barnacles, and rhizocephalans have been studied (Biegelow 1902; Delsman 1917; Anderson 1969; Scholtz et al. 2009). All the studied species have total cleavage, either equal or unequal, and certain 'spiralian' traits can be recognized in early stages, so it is possible to 'translate' Delsman's (1917) cell-lineage diagram of Semibalanus to a spiral-like notation (Table 44.2 and Fig. 44.3). The large D cell gives rise to all the endoderm, whereas the mesoderm comes collectively from all four quadrants. Exact orientation of the embryos is difficult because the polar bodies apparently move, and an apical organ is missing. The blastopore closes completely, and mouth and anus are formed at a later stage from stomodaeal and proctodaeal invaginations. None of the two new openings are parts of the blastopore, but the general pattern seems to be that the stomodaeum forms in front of the blastopore, while the proctodaeum forms in the region of the closed blastopore, and this is often observed in other arthropods too (Manton 1949). The mesoderm proliferates laterally along the endodermal cells of the midgut, and fills the ectodermal evaginations that become the appendages of the nauplius larva (Delsman 1917; Anderson 1969). The mesodermal cells differentiate into the muscles of the body and appendages without passing a stage resembling coelomic sacs, with the exception of the small coelomic sacs from which the antennal organs develop. A small caudal papilla at the posterior tip of the embryo contains a compact mass of mesodermal cells, and a pair of these cells grow larger than the others; these cells each divide twice and become a group of four teloblasts on each side, giving rise to the mesoderm of the following segments after the hatching of the larva (Anderson 1969).

The ganglia develop from ectodermal thickenings that eventually separate from the epithelium. The protocerebral and deutocerebral ganglia arise from pre-oral areas, while the tritocerebral ganglia and the ganglia of the following segments develop from postoral areas and have postoral commissures (Walley

Table 44.3. Cell lineage of *Sicyonia ingentis*. (After Hertzler 2002.) d, dorsal; c, ventral.

1969). The first larval stage of *Semibalanus* is a plank-totrophic nauplius, a stage studied by several authors, for example Walley (1969). The larva has three pairs of appendages: uniramous antennules, and biramous antennae and mandibles with well-developed gnatho-bases (Fig. 44.1); all three pairs of appendages are used in swimming, and the food particles are handled by the gnathobases and transported to the mouth, which is covered by a labrum. There is a pair of small anten-nal glands, but these excretory organs degenerate dur-ing the last nauplius stages and their function is taken over by the maxillary glands from the cypris stage. The nervous system shows the three ganglia associ-ated with the three segments (Semmler *et al.* 2008).

In crustaceans with nauplius larvae, the following development is through a series of moults, with addi-tion of posterior segments with characteristic append-ages: two pairs of maxillae belonging to the cephalon, and further limbs that are all locomotory. The meso-derm of the body behind the three naupliar segments develops from the paired groups of mesoteloblasts. Two rows of compact cell groups are given off anteri-orly, and a coelomic cavity develops in each cell mass in many groups (Anderson 1973). In each segment, the mesoderm spreads around the gut towards the dorsal side (Weygoldt 1958). The mesodermal cell groups split up, and the coelomic cavities fuse with the pri-mary body cavity forming a mixocoel or haemocoel. Some cell groups become muscles between the seg-ments and in the limbs, others become arranged around a longitudinal haemocoelic cavity, forming the heart with a pair of ostia in each segment, and others form the pericardial septum, connective tissue, and segmental organs. Coelomic cavities are found only in nephridia and gonads.

Two malacostracans with total cleavage have been studied in much detail, the shrimp *Sicyonia* and the amphipod *Parhyale*. They represent two main types of development, *Sicyonia* with a freely spawned egg and pelagic development with a non-feeding nauplius stage, and *Parhyale* with brooding, development of a blastoderm and direct development.

The cell lineage of *Sicyonia* (Table 44.3) from the egg to the nauplius was documented by Hertzler and

Clark (1992) and Hertzler (2002). The first cleavage is almost equal, with a cell designated AB in contact with the polar bodies. The second cleavage shows oblique spindles and the 4-cell stage is tetrahedral, with the B cell in contact with the polar body. The following cleavages lead to a 32-cell coeloblastula, in which two cells divide slower than the others and initiate gastrula-tion. The resulting four cells are the mesendodermal cells and become situated at the bottom of the arch-enteron. The two dorsal cells (Ya and Yp, see Table 44.3) become the yolk endoderm, whereas the ventral cells become the endoderm and the primordial germ cell + the primordial mesoteloblast, respectively. The following gastrulation movement invaginates cells that differentiate into naupliar mesoderm. The blastopore closes and the naupliar mesoderm spreads along the inner side of the developing limb buds. The primordial mesoteloblast divides three times to form the eight cells of the mesoteloblast band.

Parhyale shows total, slightly unequal cleavage, but following the first cleavages, the development becomes a typical germ-band development and is direct (Gerberding *et al.* 2002; Browne *et al.* 2005). The 4-cell stage has a clear bilateral orientation with left, anterior, right, and posterior cells. The polar bodies indicate the apical pole. The egg and the embryos are elongate, and there are two mirror-image cleavage patterns, resembling the dextral and sinistral spiralian patterns. The third cleavage is unequal, with large blastomeres at the apical pole and small blast-omeres at the blastoporal pole. The left (El), posterior (Ep), and right (Er) macromeres give rise to ecto-derm, whereas the posterior macromere gives rise to mesoderm. The left (ml) and right (mr) micromeres give rise to mesoderm, the anterior micromere gives rise to endoderm, and the posterior micromere gives rise to the germ cells. After several rounds of cleav-ages, the very small blastomeres become arranged in a characteristic pattern, which has been recognized in a number of species (see below). The Er and El prog-eny form lateral areas separated by a midline of cells from the Ep cell. The mesodermal cells give off mes-odermal areas in the head region and become organ-ized as four large mesoblasts on each side. Similar development has been observed in *Orchestia* (Scholtz and Wolff 2002).

A superficial cleavage with oblique spindles has been observed in the cladoceran *Holopedium* (Baldass 1937; Anderson 1973). If the spiralian cleavage nomenclature is used, the 2D cell, situated at the blastopore, becomes the germ cells, the 2d cell becomes endoderm, and the mesoderm comes from the 3A-C cells. The 62-cell stage is totally cellular, but from the 124-cell stage onwards a blastoderm separates from a central mass of yolk; gastrulation is epibolic.

The development of the segments has been studied both through direct observations of cell lineage and gene expression (Dohle 1970; Dohle and Scholtz 1988, 1997; Dohle *et al.* 2004). The cell lineage in several malacostracans showed that ectoteloblasts and mesoteloblasts could be distinguished from an early stage. A row of ectoteloblasts on each side of the blastopore gives rise to the ectoderm of the thoracic and more posterior regions, with each row of cells given off anteriorly giving rise to one parasegment, i.e. a zone of three rows of cells of the posterior part of one morphological segment, and the anterior cell row of the following one. The ancestral number of ectoteloblasts is believed to be 19, but variations occur. Two mesoteloblast mother cells were recognized on each side at the posterior end of the germ disc, and these cells divide in a complicated pattern to give rise to four mesoteloblasts from the thoracic region. Four mesoteloblasts on each side are found in all malacostracans investigated so far (Dohle *et al.* 2004). The antennal nephridia differentiate from undifferentiated mesodermal cells in *Astacus* (Khodabandeh *et al.* 2005).

The limb buds develop from the parasegments; at the stage of four cell rows in a parasegment the limb buds develop from cells from the posterior two cell rows of a segment and the anteriormost row of cells in the following segment (Dohle *et al.* 2004). This is in contrast to the parapodia of annelids, which develop from separate segments (Chapter 25) (Prud'homme *et al.* 2003; Prpic 2008; Dray *et al.* 2010).

The 'acron' with the protocerebrum has been homologized with the episphere (with the cerebral

ganglia) of the annelids, and this has been taken to support the 'Articulata' hypothesis (which proposes that annelids and arthropods are sister groups) (Scholtz and Edgecombe 2006), but it may just as well be interpreted as the ancestral protostomian body regionation, with the cerebral ganglion from the episphere and the circumblastoporal ventral nerve cord (Fig. 22.11). Two pairs of pioneer cells resembling those of annelids and molluscs (Chapters 25 and 27) have been found in *Artemia* and *Gonodactylus* (Fischer and Scholtz 2010).

The role of *engrailed* in segmentation has been studied extensively in all major arthropod groups (Carroll *et al.* 2005), especially in insects, such as *Drosophila* and *Tribolium*, malacostracans, such as *Gammarus* and *Homarus*, and arachnids, such as *Cupiennius*. The gene is at first expressed in the head or thorax region in transverse rows of ventral ectodermal nuclei situated at the anterior margin of each parasegment (in malacostracans; Scholtz 1997). At the stages when each parasegment consists of two and four rows of cells, *engrailed* is expressed in the anterior row of cells, and the border between the adult segments will develop just behind this row of cells of the four-row stage. Together with *even-skipped*, *engrailed* demonstrates the segmentation and the orientation of the segments. The limbs develop from the parasegments, so that the *engrailed* expression is located at the posterior side of the limb (Dray *et al.* 2010).

Newer studies on morphology and molecules are unanimous about the monophyly of the Arthropoda, and the phylogeny pictured in Fig. 44.2 likewise finds general support (see for example Regier *et al.* 2010). The segmented chitinous cuticle, with sclerites separated by thinner arthrodial membranes and articulated limbs, are conspicuous arthropod apomorphies. The chitinous, moulted cuticle with α-chitin but without collagen, the morphology and ontogeny of nephridia with a sacculus with podocytes, and ontogeny and morphology of the longitudinal heart with segmental ostia indicate the sister-group relationships with the onychophorans. This is further supported by the expression of *engrailed* in the posterior side of the limbs in both groups (Prpic 2008; Dray *et al.* 2010), although morphological indications of parasegments have not been reported from onychophorans. The tardigrades are microscopic and apparently highly specialized, and all three phyla appear to have evolved from a Precambrian group of 'lobopodians' (Chapter 42).

Interesting subjects for future research

1. Embryology/cell lineage of pycnogonids with small eggs

References

Anderson, D.T. 1969. On the embryology of the cirripede crustaceans *Tetraclita rosea* (Krauss), *Tetraclita purpurascens* (Wood), *Chthamalus antennatus* (Darwin) and *Chamaesipho columna* (Spengler) and some considerations of crustacean phylogenetic relationships. *Phil. Trans. R. Soc. Lond. B* **256**: 183–235.

Anderson, D.T. 1973. *Embryology and Phylogeny of Annelids and Arthropods (International Series of Monographs in pure and applied Biology, Zoology 50)*. Pergamon Press, Oxford.

Baccari, S. and Renaud-Mornant, J. 1974. Étude du système nerveux de *Derocheilocaris remanei* Delamare et Chappuis 1951 (Crustacea, Mystacocarida). *Cah. Biol. Mar.* **15**: 589–604.

Baker, N.E. 2001. Cell proliferation, survival, and death in the *Drosophila* eye. *Semin. Cell Dev. Biol.* **12**: 499–507.

Baldass, F. 1937. Entwicklung von *Holopedium gibberum*. *Zool. Jahrb., Anat.* **63**: 399–454.

Battelle, B.-A. 2006. The eyes of *Limulus polyphemus* (Xiphosura, Chelicerata) and their afferent and efferent projections. *Arthropod Struct. Dev.* **35**: 261–274.

Berry, R.P., Stange, G. and Warrant, E.J. 2007. Form vision in the insect dorsal ocelli: An anatomical. *Vision Res.* **47**: 1394–1409.

Biegelow, M.A. 1902. The early development of *Lepas*. A study of cell-lineage and germ-layers. *Bull. Mus. Comp. Zool. Harv Coll.* **40**: 61–144.

Bitsch, C. and Bitsch, J. 2004. Phylogenetic relationships of basal hexapods among the mandibulate arthropods: a cladistic analysis based on comparative morphological characters. *Zool. Scr.* **33**: 511–550.

Blanco, J., Pauli, T., Seimiya, M., Udolph, G. and Gehring, W.J. 2010. Genetic interactions of *eyes absent*, *twin of eyeless* and *orthodenticle* regulate sine oculis expression during ocellar development in *Drosophila*. *Dev. Biol.* **344**: 1088–1099.

Boore, J.L., Lavrov, D.V. and Brown, W.M. 1998. Gene translocation links insects and crustaceans. *Nature* **392**: 667–668.

Brauer, A. 1895. Beiträge zur Kenntnis der Entwicklungsgeschichte des Skorpions. II. *Z. Wiss. Zool.* **59**: 351–435.

Brenneis, G., Ungerer, P. and Scholtz, G. 2008. The chelifores of sea spiders (Arthropoda, Pycnogonidda) are the appendages of the deutocerebral segment. *Evol. Dev.* **10**: 717–724.

Briggs, D.E.G., Erwin, D.H. and Collier, F.J. 1994. *The Fossils of the Burgess Shale*. Smithsonian Institution Press, Washington.

Browne, W.E., Price, A.L., Gerberding, M. and Patel, M.N. 2005. Stages of embryonic development in the amphipod crustacean, *Parhyale hawaiensis*. *Genesis* **42**: 124–149.

Brusca, R.C. and Brusca, G.J. 1990. *Invertebrates*. Sinauer Associates, Sunderland, MA.

Brusca, R.C. and Brusca, G.J. 2003. *Invertebrates, 2nd ed.* Sinauer, Sunderland, MA.

Burrows, M. 2009. A single muscle moves a crustacean limb joint rhythmically by acting against a spring containing resilin. *BMC Biology* **7**: 27.

Bückmann, D. and Tomaschko, K.-H. 1992. 20-hydroxyecdysone stimulates molting in pycnogonid larvae (Arthropoda, Pantopoda). *Gen. Com. Endocrinol.* **88**: 261–266.

Böckeler, W. 1984. Embryogenese und ZNS-Differenzierung bei *Reighardia sternae*. Licht- und elektronenmikroskopische Untersuchungen zur Tagmosis und systematischen Stellung der Pentastomiden. *Zool. Jahrb., Anat.* **111**: 297–342.

Carroll, S.B., Grenier, J.K. and Weatherbee, S.D. 2001. *From DNA to Diversity. Molecular Genetics and the Evolution of Animal Design*. Blackwell Science, Malden, MA.

Carroll, S.B., Grenier, J.K. and Weatherbee, S.D. 2005. *From DNA to Diversity. Molecular Genetics and the Evolution of Animal Design, 2nd ed.* Blackwell Publishing, Malden, MA.

Dallai, R. and Afzelius, B.A. 2000. Spermatozoa of the 'primitive type' in *Scutigerella* (Myriapoda, Symphyla). *Tissue Cell* **32**: 1–8.

Dawydoff, C. 1928. *Traité d'Embryologie Comparée des Invertébrés*. Masson, Paris.

Delsman, H.C. 1917. Die Embryonalentwicklung von *Balanus balanoides* Linn. *Tijdschr. Ned. Dierkd. Ver.* **15**: 419–520.

Dogiel, V. 1913. Embryologische Studien an Pantopoden. *Z. Wiss. Zool.* **107**: 575–741.

Dohle, W. 1970. Die Bildung und Differenzierung des postnauplialen Keimstreifes von *Diastylis rathkei* (Crustacea, Cumacea) I. Die Bildung der Teloblasten ind ihrer Derivate. *Z. Morph. Tiere* **67**: 307–392.

Dohle, W. and Scholtz, G. 1988. Clonal analysis of the crustacean segment: the discordance between genealogical and segmental borders. *Development* **104** (**Suppl.**): 147–169.

Dohle, W. and Scholtz, G. 1997. How far does cell lineage influence cell fate specification in crustacean embryos? *Semin. Cell Dev. Biol.* **8**: 379–390.

Dohle, W., Gerberding, M., Hejnol, A and Scholtz, G. 2004. Cell lineage, segment differentiation, and gene expression in crustaceans. In G. Scholtz (ed.): *Evolutionary Developmental Biology of Crustacea (Crustacean Issues 15)*, pp. 95–133. Balkema, Lisse.

Dray, N., Tessmar-Raible, K., Le Gouar, M., *et al.* 2010. Hedgehog signaling regulates segment formation in the annelid *Platynereis*. *Science* **329**: 339–342.

Dunn, C.W., Hejnol, A., Matus, D.Q., *et al.* 2008. Broad phylogenomic sampling improves resolution of the animal tree of life. *Nature* **452**: 745–749.

Edgecombe, G.D. 2010. Arthropod phylogeny: An overview from the perspectives of morphology, molecular data and the fossil record. *Arthropod Struct. Dev.* **39**: 74–87.

Elofsson, R. 2006. The frontal eyes of crustaceans. *Arthropod Struct. Dev.* **35**: 275–291.

Elofsson, R. and Hessler, R.R. 1990. Central nervous system of *Hutchinsoniella macracantha* (Cephalocarida). *J. Crustac. Biol.* **10**: 423–439.

Fanenbruck, M., Harzsch, S. and Wägele, W. 2004. The brain of the Remipedia (Crustacea) and an alternative hypothesis on their phylogenetic relationships. *Proc. Natl. Acad. Sci. USA* **101**: 3868–3873.

Feustel, H. 1958. Untersuchungen über die Exkretion bei Collembolen. *Z. Wiss. Zool.* **161**: 209–238.

Fischer, A.H.L. and Scholtz, G. 2010. Axogenesis in the stomatopod crustacean *Gonodactylus falcatus* (Malacostraca). *Invert. Biol.* **129**: 59–76.

Fortey, R.A. 2001. Trilobite systematics: the last 75 years. *J. Paleont.* **75**: 1141–1151.

Gerberding, M., Browne, W.E. and Tatel, N.H. 2002. Cell lineage analysis of the amphiopod crustacean *Parhyale hawaiensis* reveals an early restriction of cell fates. *Development* **129**: 5789–5801.

Gilbert, S.F. 1997. Arthropods: The crustaceans, spiders, and myriapods. In S.F. Gilbert and A.M. Raunio (eds): *Embryology. Constructing the Organism*, pp. 237–257. Sinauer, Sunderland, MA.

Giribet, G., Richter, S., Edgecombe, G.D. and Wheeler, W.C. 2005. The position of crustaceans within Arthropoda—Evidence from nine molecular loci and morphology. *Crustac. Issues* **16**: 307–352.

Govindarajan, S. and Rajulu, G.S. 1974. Presence of resilin in a scorpion *Palamnaeus swammerdami* and its role in the food-capturing and sound-producing mechanisms. *Experientia* **30**: 909–910.

Hafner, G.S. and Tokarski, T.R. 2001. Retinal development in the lobster *Homarus americanus*. Comparison with compound eyes of insects and other crustaceans. *Cell Tissue Res.* **305**: 147–158.

Harrison, F.W. (ed.) 1992a. *Microscopic Anatomy of Invertebrates*, vol. 9 *(Crustacea)*. Wiley-Liss, New York.

Harrison, F.W. (ed.) 1992b. *Microscopic Anatomy of Invertebrates*, vol. 10 *(Decapod Crustacea)*. Wiley-Liss, New York.

Harrison, F.W. (ed.) 1993. *Microscopic Anatomy of Invertebrates*, vol. 12 *(Onychophora, Chilopoda, and Lesser Protostomes)*. Wiley-Liss, New York.

Harrison, F.W. (ed.) 1998. *Microscopic Anatomy of Invertebrates*, vol. 11A–C *(Insecta)*. Wiley-Liss, New York.

Harrison, F.W. (ed.) 1999. *Microscopic Anatomy of Invertebrates*, vol. 8A–C *(Chelicerate Arthropoda)*. Wiley-Liss, New York.

Hartenstein, V. and Posakony, J.W. 1989. Development of adult sensilla on the wing and notum of *Drosophila melanogaster*. *Development* **107**: 389–405.

Harzsch, S. 2004. The tritocerebrum of Euarthropoda: a 'non-drosophilo centric' perspective. *Evol. Dev.* **6**: 303–309.

Harzsch, S. and Hafner, G. 2006. Evolution of eye development in arthropods: phylogenetic aspects. *Arthropod Struct. Dev.* **35**: 319–340.

Harzsch, S., Melzer, R.R. and Müller, C.H.G. 2007. Mechanisms of eye development and evolution of the arthropod visual system: The lateral eyes of myriapoda are not modified insect ommatidia. *Org. Divers. Evol.* **7**: 20–32.

Harzsch, S., Wildt, M., Battelle, B. and Waloszek, D. 2005. Immunohistochemical localization of neurotransmitters in the nervous system of larval *Limulus polyphemus* (Chelicerata, Xiphosura): evidence for a conserved protocerebral architecture in Euarthropoda. *Arthropod Struct. Dev.* **34**: 327–342.

Haug, J.T., Waloszek, D., Haug, C. and Maas, A. 2010. Early Cambrian arthropods—new insights into arthropod head and structural evolution. *Arthropod Struct. Dev.* **39**: 154–173.

Hejnol, A., Obst, M., Stamatakis, A., *et al.* 2009. Assessing the root of bilaterian animals with scalable phylogenomic methods. *Proc. R. Soc. Lond. B* **276**: 4261–4270.

Henry, L.M. 1948. The nervous system and the segmentation of the head in the Annulata. *Microentomology* **13**: 1–26.

Hertzler, P.L. 2002. Development of the mesendoderm in the dendrobranchiate shrimp *Sicyonia ingentis*. *Arthropod Struct. Dev.* **31**: 33–49.

Hertzler, P.L. and Clark, W.H., Jr 1992. Cleavage and gastrulation in the shrimp *Sicyonia ingentis*: invagination is accompanied by oriented cell division. *Development* **116**: 127–140.

Hessler, R.R. and Elofsson, R. 1995. Segmental podocytic excretory glands in the thorax of *Hutchinsoniella macracantha* (Cephalocarida). *J. Crustac. Biol.* **15**: 61–69.

Hopkins, P.M. 2009. Crustacean ecdysteroids and their receptors. In G. Smagghe (ed.): *Ecdysone: Structures and Functions*, pp. 73–97. Springer, USA.

Hosfeld, B. and Schminke, H.K. 1997. Discovery of segmental extranephridial podocytes in Harpacticoida (Copepoda) and Bathynellacea (Syncarida). *J. Crustac. Biol.* **17**: 13–20.

Hou, X.-G., Aldridge, R.J., Bergström, J., *et al.* 2004. *The Cambrian Fossils of Chengjiang, China*. Blackwell, Malden, MA.

Hou, X. and Bergström, J. 1997. Arthropods from the Lower Cambrian Chengjiang fauna, southwest China. *Fossils Strata* **45**: 1–116.

Iwanoff, P.P. 1933. Die embryonale Entwicklung von *Limulus moluccanus*. *Zool. Jahrb., Anat.* **56**: 163–348.

Janssen, R. and Damen, W.G.M. 2006. The ten *Hox* genes of the millipede *Glomeris marginata*. *Dev. Genes Evol.* **216**: 451–465.

Janssen, R., Eriksson, B.J., Budd, G., Akam, M. and Prpic, N.M. 2010. Gene expression patterns in an onychophoran reveal that regionalization predates limb segmentation in pan-arthropods. *Evol. Dev.* **12**: 363–372.

Khodabandeh, S., Charmantier, G., Blasco, C., Grousset, E. and Charmantier-Daures, M. 2005. Ontogeny of the antennal glands in the crayfish *Astacus leptodactylus* (Crustacea, Decapoda): anatomical and cell differentiation. *Cell Tissue Res.* **319**: 153–165.

Kishinouye, K. 1891. On the develooment of *Limulus longispinus*. *J. Coll. Sci. Imp. University Tokyo* **5**: 53–100.

Klass, K.-D. and Kristensen, N.P. 2001. The ground plan and affinities of hexapods: recent progress and open problems. *Ann. Soc. Entomol. Fr., N.S.* **37**: 265–298.

Koch, M. 2003. Monophyly of the Myriapoda ? Reliability of current arguments. *Afr. Invertebr.* **44**: 137–153.

Lacalli, T.C. 2009. Serial EM analysis of a copepod larval nervous system: Naupliar eye, optic circuitry, and prospects for full CNS reconstruction. *Arthropod Struct. Dev.* **38**: 361–375.

Lafont, R. and Koolman, J. 2009. Diversity of ecdysteroids in animal species. In G. Smagghe (ed.): *Ecdysone: Structures and Functions*, pp. 47–71. Springer, Dordrecht.

Land, M.F. and Barth, F.G. 1985. The morphology and optics of spider eyes. *Neurobiology of Arachnida*, pp. 53–78. Springer-Verlag, Berlin.

Lavrov, D.V., Brown, W.M. and Boore, J.L. 2003. Phylogenetic position of the Pentastomida and (pan)crustacean relationships. *Proc. R. Soc. Lond. B* **271**: 537–544.

Manton, S.F. 1949. Studies on the Onychophora VII. The early embryonic stages of *Peripatopsis*, and some general considerations concerning the morphology and phylogeny of the Arthropoda. *Phil. Trans. R. Soc. Lond. B* **233**: 483–580.

Manton, S.F. 1977. *The Arthropoda. Habits, Functional Morphology, and Evolution*. Oxford University Press, Oxford.

Manton, S.M. 1972. The evolution of arthropodan locomotory mechanisms. Part 10. Locomotory habits, morphology and evolution of the hexapod classes. *Zool. J. Linn. Soc.* **51**: 203–400.

Manuel, M., Jager, M., Murienne, J., Clabaut, C. and Le Guyader, H. 2006. Hox genes in sea spiders (Pycnogonida) and the homology of arthropod head segments. *Dev. Genes Evol.* **216**: 481–491.

Meisenheimer, J. 1902. Beiträge zur Entwicklungsgeschichte der Pantopoden. I. Die Entwicklung von *Ammothea echinata* Hodge bis zur Ausbildung der Larvenform. *Z. Wiss. Zool.* **72**: 191–248.

Michalik, P. and Mercati, D. 2009. First investigation of the spermatozoa of a species of the superfamily Scorpionoidea (*Opistophthalmus penrithorum*, Scorpionidae) with a revision of the evolutionary and phylogenetic implications of sperm structures in scorpions (Chelicerata, Scorpiones). *J. Zool. Syst. Evol. Res.* **48**: 89–101.

Møller, O.S., Olesen, J., Avenant-Oldewage, A., Thomsen, P.F. and Glenner, H. 2008. First maxillae suction discs in Branchiura (Crustacea): Development and evolution in light of the first molecular phylogeny of Branchiura, Pentastomida, and other 'Maxillopoda'. *Arthropod Struct. Dev.* **37**: 333–246.

Moritz, M. 1959. Zur Embryonalentwicklung der Phalangiidae (Opiliones, Palpatores) II. Die Anlage unt Entwicklung der Coxaldrüse bei *Phalangium opilio* L. *Zool. Jahrb., Anat.* **77**: 229–240.

Müller, C.H.G. and Meyer-Rochow, V.B. 2006. Fine structural organization of the lateral ocelli in two species of Scolopendra (Chilopoda: Pleurostigmophora): an evolutionary evaluation. *Zoomorphology* **125**: 13–26.

Müller, C.H.G., Rosenberg, J., Richter, S. and Meyer-Rochow, V.B. 2003. The compound eye of *Scutigera coleoptrata* (Linnaeus, 1758) (Chilopoda: Notostigmophora): an ultrastructural reinvestigation that adds support to the Mandibulata concept. *Zoomorphology* **122**: 191–209.

Müller, K.J. and Walossek, D. 1986. Arthropod larvae from the Upper Cambrian of Sweden. *Trans. R. Soc. Edinb., Earth Sci.* **77**: 157–179.

Patten, W. and Hazen, A.P. 1900. The development of the coxal gland, branchial cartilages, and genital ducts of *Limulus polyphemus*. *J. Morphol.* **16**: 459–502.

Paulus, H.F. 1979. Eye structure and the monophyly af the Arthropoda. In A.P. Gupta (ed.): *Arthropod Phylogeny*, pp. 299–383. Van Norstrand Reinhold Co., New York.

Paulus, H.F. 2000. Phylogeny of the Myriapoda—Crustacea—Insecta: a new attempt using photoreceptor structure. *J. Zool. Syst. Evol. Res.* **38**: 189–208.

Prpic, N.-M. 2008. Parasegmental appendage allocation in annelids and arthropods and the homology of parapodia and arthropodia. *Front. Zool.* **5**: 17.

Prud'homme, B., de Rosa, R., Arendt, D., *et al.* 2003. Arthropod-like expression patterns of *engrailed* and *wingless* in the annelid *Platynereis dumerilii* suggest a role in segment formation. *Curr. Biol.* **13**: 1876–1881.

Ready, D.F. 1989. A multifaceted approach to neural development. *Trends Neurosci.* **12**: 102–110.

Regier, J.C., Schultz, J.W., Ganley, A.R.D., *et al.* 2008. Resolving arthropod phylogeny: exploring phylogenetic signal within 41 kb of protein-coding nuclear gene sequence. *Syst. Biol.* **57**: 920–938.

Regier, J.C., Schultz, J.W., Zwick, A., *et al.* 2010. Arthropod relationships revealed by phylogenomic analysis of nuclear protein-coding sequences. *Nature* **463**: 1079–1083.

Rieder, N. and Schlecht, F. 1978. Erster Nachweis von freien Cilien im Mitteldarm von Arthropoden. *Z. Naturforsch.* **33c**: 598–599.

Riley, J., Banaja, A.A. and James, J.L. 1978. The phylogenetic relationships of the Pentastomida: the case for their inclusion within the Crustacea. *Int. J. Parasitol.* **8**: 245–254.

Rota-Stabelli, O., Campbell, L., Brinkmann, H., *et al.* 2010. A congruent solution to arthropod phylogeny: phylogenomics, microRNAs and morphology support monophyletic Mandibulata. *Proc. R. Soc. Lond. B* **278**: 298–306.

Ruppert, E.E., Fox, R.S. and Barnes, R.D. 2004. *Invertebrate Zoology: A Functional Evolutionary Approach (7th ed. of R.D. Barnes' Invertebrate Zoology)*. Brooks/Cole, Belmont, CA.

Sanchez, S. 1959. Le développement des Pycnogonides et leurs affinités avec les Arachnides. *Arch. Zool. Exp. Gen.* **98**: 1–101.

Sanders, H.L. 1963. The Cephalocarida. Functional morphology, larval development, comparative external anatomy. *Mem. Conn. Acad. Arts Sci.* **15**: 1–80.

Sanders, H.L. and Hessler, R.R. 1964. The larval development of *Lightiella incisa* Gooding (Cephalocarida). *Crustaceana* **7**: 81–97.

Sanders, K.L. and Lee, M.S.Y. 2010. Arthropod molecular divergence times and the Cambrian origin of pentastomids. *Syst. Biodivers.* **8**: 63–74.

Scholtz, G. 1997. Cleavage pattern, germ band formation and head segmentation: the ground pattern of the Euarthropoda. In R.A. Fortey (ed.): *Arthropod Relationships*, pp. 317–332. Chapman & Hall, London.

Scholtz, G. and Edgecombe, G.D. 2005. Heads, Hox and the phylogenetic position of trilobites. *Crustac. Issues* **16**: 139–165.

Scholtz, G. and Edgecombe, G.D. 2006. The evolution of arthropod heads: reconciling morphological, developmental and palaeontological evidence. *Dev. Genes Evol.* **216**: 395–415.

Scholtz, G. and Wolff, C. 2002. Cleavage, gastrulation, and germ disc formation of the amphipod *Orchestia cavimana* (Crustacea, Malacostraca, Peracarida). *Contrib. Zool.* **71**: 9–28.

Scholtz, G., Ponomarenko, E. and Wolff, C. 2009. Cirripede cleavage patterns and the origin of the Rhizocephala (Crustacea: Thecostraca). *Arthropod Syst. Phyl.* **67**: 219–228.

Schwalm, F.E. 1997. Arthropods: The Insects. In S.F. Gilbert and A.M. Raunio (eds): *Embryology. Constructing the Organism*, pp. 259–278. Sinauer Associates, Sunderland, MA.

Seifert, G. and Bidmon, H.J. 1988. Immunohistochemical evidence for ecdysteroid-like material in the putative molting glands of *Lithobius forficatus* (Chilopoda). *Cell Tissue Res.* **253**: 263–266.

Semmler, H., Wanninger, A., Høeg, J.T. and Scholtz, G. 2008. Immunocytochemical studies on the naupliar nervous system of *Balanus improvisus* (Crustacea, Cirripedia, Thecostraca). *Arthropod Struct. Dev.* **37**: 383–395.

Shanbhag, S.R., Müller, B. and Steinbrecht, R.A. 1999. Atlas of olfactory organs of *Drosophila melanogaster* 1. Types, external organization, innervation and distribution of olfactory sensilla. *Int. J. Insect Morphol. Embryol.* **28**: 377–397.

Shear, W.A. 1999. Introduction to Arthropoda and Cheliceriformes. In F.W. Harrison (ed.): *Microscopic Anatomy of Invertebrates*, vol. 8A, pp. 1–19. Wiley-Liss, New York.

Shear, W.A. and Edgecombe, G.D. 2010. The geological record and phylogeny of the Myriapoda. *Arthropod Struct. Dev.* **39**: 174–190.

Shuster, C.N., Jr 1950. Observations on the natural history of the American horseshoe crab, *Limulus polyphemus*. *Woods Hole Oceanogr. Inst. Contr.* **564**: 18–23.

Storch, V. 1993. Pentastomida. In F.W. Harrison (ed.): *Microscopic Anatomy of Invertebrates*, vol. 13, pp. 115–142. Wiley-Liss, New York.

Storch, V. and Jamieson, B.G.M. 1992. Further spermatological evidence for including the Pentastomida (Tongue worms) in the Crustacea. *Int. J. Parasitol.* **22**: 95–108.

Strausfeld, N.J. 2009. Brain organization and the origin of insects: an assessment. *Proc. R. Soc. Lond. B* **276**: 1929–1937.

Ungerer, P. and Scholtz, G. 2009. Cleavage and gastrulation in *Pycnogonum littorale* (Arthropoda, Pycnogonidae): morphological support for the Ecdysozoa? *Zoomorphology* **128**: 263–274.

Vaissière, R. 1961. Morphologie et histologie comparées des yeux des Crustacés Copépodes. *Arch. Zool. Exp. Gen.* **100**: 1–125.

van Deurs, B. 1974. Spermatology of some Pycnogonida (Arthropoda), with special reference to a microtubule-nuclear envelope complex. *Acta Zool. (Stockh.)* **55**: 151–162.

Vilpoux, K., Sandeman, R. and Harzsch, S. 2006. Early embryonic development of the central nervous system in the Australian crayfish and the marbled crayfish (Marmorkrebs). *Dev. Genes Evol.* **216**: 209–223.

Walley, L.J. 1969. Studies on the larval structure and metamorphosis of *Balanus balanoides* (L.). *Phil. Trans. R. Soc. Lond. B* **256**: 237–280.

Walossek, D. 1993. The Upper Cambrian *Rehbachiella* and the phylogeny of Branchiopoda and Crustacea. *Fossils Strata* **32**: 1–202.

Waloszek, D., Chen, J., Maas, A. and Wang, X. 2005. Early Cambrian arthropods—new insights into arthropod head and structural evolution. *Arthropod Struct. Dev.* **34**: 189–205.

Waloszek, D., Repetski, J.E. and Maas, A. 2006. A new Late Cambrian pentastomid and a review of the relationships of this parasitic group. *Trans. R. Soc. Edinb., Earth Sci.* **96**: 163–176.

Weygoldt, P. 1958. Die Embryonalentwicklung des Amphipoden *Gammarus pulex pulex* (L.). *Zool. Jahrb., Anat.* **77**: 51–110.

Wills, M.A., Briggs, D.E.G., Fortey, R.A., Wilkinson, M. and Sneath, P.H.A. 1998. An arthropod phylogeny based on fossil and recent taxa. In G.D. Edgecombe (ed.): *Arthropod Fossils and Phylogeny*, pp. 33–105. Columbia University Press, New York.

Wingstrand, K.G. 1972. Comparative spermatology of a pentastomid, *Raillietiella hemidactyli*, and a branchiuran crustacean, *Argulus foliaceus*, with a discussion of pentastomid relationships. *Biol. Skr.—K. Dan. Vidensk. Selsk.* **19**: 1–72.

Winter, G. 1980. Beiträge zur Morphologie und Embryologie des vorderen Körperabschnitts (Cephalosoma) der Pantopoda Gerstaecker, 1863. I. Entstehung und Struktur des Zentralnervensystems. *Z. Zool. Syst. Evolutionsforsch.* **18**: 27–61.

Wolff, C. and Scholtz, G. 2008. The clonal composition of biramous and uniramous arthropod limbs. *Proc. R. Soc. Lond. B* **275**: 1023–1028.

Zhang, X.-G., Maas, A., Haug, J.T., Siveter, D.J. and Waloszek, D. 2010. A eucrustacean metanauplius from the Lower Cambrian. *Curr. Biol.* **20**: 1075–1079.

45

Phylum Onychophora

Onychophora is a small phylum of terrestrial animals mainly found in humid tropical and southern temperate regions; only about 200 species have been described, representing two families, Peripatidae and Peripatopsidae. Lower and Middle Cambrian lobopodians, such as *Ostenotubulus* (Maas *et al.* 2007), appear to represent the marine stem group of the living onychophorans (see also Chapter 43), and the earliest reasonably certain fossils are from Cretaceous amber (Grimaldi *et al.* 2002).

The body is cylindrical, with a pair of long anterior antennae, paired chitinous jaws inside the mouth, a pair of oral (or slime) papillae with slime glands, and a number of segments each with a pair of lobopods, i.e. sac-shaped, unarticulated legs with internal muscles, each terminating in two claws. The cuticle and the nervous system do not show any segmentation, but excretory organs/gonads are organized segmentally. The microscopic anatomy was reviewed by Storch and Ruhberg (1993).

The epidermis is covered with scaled papillae, many of which are sensillae (see below). The cuticle is velvety with micro-annulations (Budd 2001) and consists of α-chitin and proteins. There is a thin epicuticle, and a thicker exo- and endocuticle. Cuticular pore canals are lacking. The exocuticle may be tanned, for example in the sclerotized claws and jaws. The cuticle is shed (ecdysis) at regular intervals throughout life,

and the new cuticle is secreted underneath the old one by the ectodermal cells, which develop microvilli, but these are subsequently withdrawn; the whole process is very similar to that of the arthropods (Robson 1964). Ecdysteroids have been found in various tissues, but their function remains unknown (Hoffmann 1997). Various types of larger, multicellular glands are present. Crural glands are ectodermal invaginations of the underside of the legs. Both the very large slime glands opening on the oral papillae and the accessory genital glands found in varying numbers in the males are modified crural glands.

Each segment has a high number of small tracheal spiracles that are the openings of the respiratory organs. The spiracles, which may be situated at the bottom of a wide ectodermal invagination, the atrium, are the openings for numerous tubular tracheae extending to the organs they supply.

The musculature comprises ring muscles, strong ventral, lateral, and dorsal, non-segmented longitudinal muscles, and muscles associated with the legs. The whole body, including the limbs, functions as parts of a hydrostatic unit (Manton 1977). The muscles are of an unusual type, perhaps representing a step in the evolution of striated muscles.

The mouth harbours a pair of lateral jaws, which represent the claws of modified legs. The oral papillae contain the openings of the large slime glands. The

Chapter vignette: *Macroperipatus geayi*. (Redrawn from Pearse *et al.* 1987.)

corresponding nephridia are modified to salivary glands each with a small sacculus. The gut comprises a muscular pharynx and a thin-walled oesophagus, both with a cuticle, a wide, cylindrical gut with absorptive cells and secretory cells that secrete a peritrophic membrane surrounding the gut content, and a rectum with a thin cuticle.

The central nervous system consists of a bilobed brain above the pharynx, a pair of connectives around the pharynx, and a pair of ventral, longitudinal nerve cords (Eriksson and Budd 2001). Many small annular bundles of axons and a few longitudinal bundles emerge from the ventral nerve cords (Mayer and Harzsch 2007; Mayer and Whitington 2009). The brain is rather compact, and neither the brain nor nerve cords show ganglionic organization. The organization of the brain and of the central/arcuate bodies shows similarities with that of the chelicerates, but differences to that of mandibulates; this could be interpreted as the onychophoran-chelicerate organization being plesiomorphic (Strausfeld et al. 2006; Homberg 2008). A pair of tiny eyes are situated at the base of the antennae, and the two structures are innervated from the same part of the brain. The eyes (Mayer 2006b) develop from an epithelial invagination and consist of a domed lens covered by a cornea and a circular, almost-flat basal retina, with sensory and pigment cells. The cornea consists of a thin cuticle-covered ectoderm and a thin inner layer that peripherally is continuous with the retina. The sensory cells are of the rhabdomeric type, with a long, thick distal process with numerous orderly arranged microvilli at the periphery and a small cilium situated among the microvilli. The innervation indicates that the eyes are homologous of the median ocelli of the arthropods (Mayer 2006b).

The skin bears several types of sensory organs; some are apparently chemoreceptors, whereas others obviously are mechanoreceptors. They all contain various types of bipolar sensory cells, with more or less strongly modified cilia often surrounded by microvilli.

The spacious body cavity, the haemocoel, develops through confluence of coelomic sacs and the primary body cavity, i.e. it is a mixocoel (Mayer 2006a). There is a dorsal, longitudinal heart that is open at both ends and has a pair of lateral ostia in each segment; the lumen of the heart is surrounded by a basal membrane with muscle cells (Nylund et al. 1988); this is the normal structure of the heart of a bilaterian, but the openings between the primary blood space surrounded by a basal membrane and the coelomic pericardial cavity is an panarthropod apomorphy. The haemolymph enters the heart through the ostia, is pumped anteriorly, and leaves the heart through the anterior opening.

The excretory organs (Mayer 2006a) are paired, segmental metanephridia with a small coelomic vesicle (sacculus), a ciliated funnel, and a duct that opens through a small ventral pore near the base of the legs. The wall of the sacculus consists of podocytes resting on an outer basal membrane, obviously the site of the ultrafiltration of primary urine from the haemocoel. The multiciliate cells of the funnel and the various cell types of the duct show many structures associated with absorption and secretion. Also the cells of the mid-gut have excretory functions. The salivary glands, which open into the foregut through a common ventral duct, are modified nephridia with a small sacculus with podocytes.

The gonads develop through fusion of dorsal series of coelomic compartments, and the gonoducts are modified nephridia. The spermatozoa have a long head with a helically coiled nuclear surface and a long tail consisting of a cilium with accessory tubules (Marotta and Ruhberg 2004). The sperm is transferred in a spermatophore that in some species becomes deposited in the genital opening of the young female, while the males of other species attach the spermatophore to the females back in a random area, and the spermatozoa wander through the epidermis to the haemocoel to reach the ovary.

A few species deposit large (up to 2 mm in diameter), yolk-rich eggs with chitinous shells, but many more species are either ovoviviparous with medium-sized eggs, or viviparous with small eggs (down to 40 μm in diameter) nourished by secretions from the uterus. The ovoviviparous type of development is regarded as ancestral and the viviparous type with or without a placenta as the more specialized (Reid 1996), although the embryology of the species with a placenta is seemingly the more primitive, with total cleavage (see below). The oviparous and the ovoviviparous species have superficial cleavage resembling that of the insects, but their early development is incompletely described (Sheldon 1889; Anderson 1973).

The development of the viviparous species shows a wide variation (Manton 1949; Anderson 1966). Species with rather small eggs and without a placenta show superficial cleavage with formation of a blastoderm. Species with very small eggs form a placenta and have total, almost equal cleavage that leads to the formation of a coeloblastula or a sterroblastula (Kennel 1885; Sclater 1888). The placental structures are enlarged anterodorsal areas of the embryo, which in the least developed types have the shape of a dorsal sac of

ectoderm surrounding the yolk cells. This dorsal structure becomes increasingly larger with decreasing amounts of yolk, and includes mesoderm and endoderm; in the most specialized types there is an 'amnion' and a dorsal 'umbilical cord' (Kennel 1885; Anderson and Manton 1972). The development of species with very small eggs, such as *Peripatopsis capensis* and *P. balfouri* (Sedgwick 1885, 1886, 1887, 1888; Manton 1949), is well described and perhaps the most easy to follow (Fig. 45.1). The cleavage leads to the formation of a plate of cells on one side of the elongate embryo; somewhat later stages become spherical, and show one half of the embryo covered by a layer of smaller cells with small amounts of yolk covering a mass of larger, yolky cells; the smaller cells become the ectoderm and

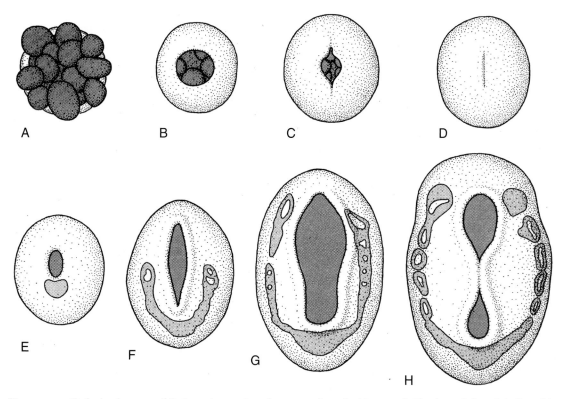

Figure 45.1. Early development of *Peripatopsis capensis*; embryos seen from the blastoporal side; the endoderm is indicated in dark shading and the mesoderm, which is seen through transparency, in light shading. (A) 'Blastula'. (B) 'Gastrula'. (C) Lateral blastopore closure. (D) Blastopore lips in contact. (E) Blastopore reopens; the mesoderm is seen behind the blastopore. (F) Blastopore elongates; the mesoderm extends laterally along the blastopore. (G) The mesoderm has reached the anterior part of the embryo and coelomic cavities are developing. (H) The blastopore closes laterally, leaving mouth and anus; coelomic sacs separate from the mesodermal bands. (Redrawn from Manton 1949.)

the larger ones the endoderm. A gastrulation takes place, resulting in a gastrula with a compressed blastopore, but a wide opening, called mouth-anus or stomo-proctodaeum, soon opens in the same area. The mouth-anus divides by lateral fusion of the mouth-anus lips (Fig. 22.2). Mesoderm is now being produced from a small area at the posterior side of the anus, and compact mesodermal bands grow forwards along the sides of the mouth-anus; teloblasts have not been reported. Schizocoelic spaces develop in these bands, which divide to form a row of coelomic sacs on each side. Coelomic sacs of the antennal, jaw, oral papilla, and trunk segments can clearly be distinguished (Anderson 1973). Bartolomaeus and Ruhberg (1999) described the ultrastructure of the coelomic sacs in embryos of *Epiperipatus biolleyi*, where they found that the wall apposed to the gut was a thin layer of epithelial cells, whereas the outer wall was thicker and consisted of more undifferentiated cells; a rudimentary cilium with an accessory centriole was found on many cells. Other species show modifications to this pattern, probably induced by larger amounts of yolk (Manton 1949).

Nerve cells proliferate from the ectoderm, and there are permanent or transitory connections between ectoderm and nerve cords, so-called ventral organs.

Each coelomic sac divides into dorsal and ventral compartments, but the fates of the two sacs have been interpreted somewhat differently by the various authors (Kennel 1885; Sedgwick 1887; Evans 1901; Mayer *et al.* 2004; Mayer 2006a). Only new investigations can show whether this is owing to differences between the species or to misinterpretations. The ventral sac apparently differentiates into the nephridium with a thin-walled sacculus and a thick-walled ciliated funnel, which becomes connected to the exterior through a short ectodermal duct. In *Epiperipatus*, the nephridia differentiate from the thick lateral wall of the coelomic sacs (Mayer 2006a). The ciliated funnel develops first, in open connection with the coelom. The podocytes of the sacculus become added soon after. The fate of the dorsal sac is less well known. Kennel (1885: *Epiperipatus edwardsii* and *E. torquatus*)

was of the opinion that the walls of the sacs disintegrate, so that the coelomic and pseudocoelomic spaces become confluent, whereas Sedgwick (1887: *Peripatopsis capensis*) and Evans (1901: *Eoperipatus weldoni*) reported that the sacs collapse and that the cells become incorporated into the wall of the heart.

Expression of an *engrailed*-class gene has been seen in an early stage of mesoderm segment development in *Acanthokara* (Wedeen *et al.* 1997), and Eriksson *et al.* (2009) found both this gene and *wingless* arranged in stripes along the mesodermal segment borders. *Engrailed* is first expressed just before the somite boundaries become visible, but becomes expressed in stripes in later stages. *Wingless* becomes expressed after somite formation. Both genes become expressed in ectodermal stripes in later stages, but the lack of overt external segment borders makes interpretation difficult. However, *engrailed* is expressed in the posterior side of the limbs, as in the arthropods (Dray *et al.* 2010).

Gene expression indicates that the onychophoran limbs are regionalized along the axis, and that this represents the stage before the true segmentation seen in the arthropods (Janssen *et al.* 2010).

Several apomorphies characterize the onychophorans as a monophyletic group: the second pair of limbs transformed into a pair of knobs carrying very large oral hooks or jaws. The third pair of limbs is transformed into oral papillae with slime glands, which are modified crural glands, and salivary glands, which are modified metanephridia. The ventral nerve cords are widely separated and have several commissures in each segment. A large number of stigmata with tracheal bundles can be found on each segment. Recent onychophorans show several specializations that indicate that the group originated in a terrestrial environment. The tracheae and slime glands cannot function in water. The taxon must be seen as a specialized, terrestrial offshoot from the diverse Cambrian fauna of early aquatic panarthropods (lobopodes).

The development of the eyes shows how the rhabdomeric type characteristic of the panarthropods may have evolved from the ciliary type. The innervation of eyes and antennae indicate that they belong to the same segment, which may well be homologous with the protocerebrum of the arthropods (Chapter 41) and with the pre-oral cerebral ganglia of the annelids (Chapter 25).

The phylogenetic position of the Onychophora as sister group of the Arthropoda is supported by almost all newer molecular phylogenies (see Chapter 40).

Interesting subjects for future research

1. Early development of species with various types of development
2. Differentiation of the coelomic sacs
3. Physiology of moulting
4. Segmentation genes

References

Anderson, D.T. 1966. The comparative early embryology of the Oligochaeta, Hirudinea and Onychophora. *Proc. Linn. Soc. N. S. W.* **91**: 11–43.

Anderson, D.T. 1973. *Embryology and Phylogeny of Annelids and Arthropods (International Series of Monographs in pure and applied Biology, Zoology 50)*. Pergamon Press, Oxford.

Anderson, D.T. and Manton, S.M. 1972. Studies on the Onychophora VIII. The relationship between the embryos and the oviduct in the viviparous placental onychophorans *Epiperipatus trinidadensis* Bouvier and *Macroperipatus torquatus* (Kennel) from Trinidad. *Phil. Trans. R. Soc. Lond. B* **264**: 161–189.

Bartolomaeus, T. and Ruhberg, H. 1999. Ultrastructure of the body cavity lining in embryos of *Epiperipatus biolleyi* (Onychophora, Peripatidae)—a comparison with annelid larvae. *Invert. Biol.* **118**: 165–174.

Budd, G.E. 2001. Why are arthropods segmented? *Evol. Dev.* **3**: 332–342.

Dray, N., Tessmar-Raible, K., Le Gouar, M., *et al.* 2010. Hedgehog signaling regulates segment formation in the annelid *Platynereis*. *Science* **329**: 339–342.

Eriksson, B.J. and Budd, G.E. 2001. Onychophoran cephalic nerves and their bearing on our understanding of head segmentation and stem-group evolution of Arthropoda. *Arthropod Struct. Dev.* **29**: 197–209.

Eriksson, B.J., Tait, N.N., Budd, G.E. and Akam, M. 2009. The involvement of engrailed and wingless during

segmentation in the onychophoran *Euperipatoides kanangrensis* (Peripatopsidae: Onychophora) (Reid 1996). *Dev. Genes Evol.* **219**: 249–264.

Evans, R. 1901. On the Malayan species of Onychophora. Part II.—The development of *Eoperipatus weldoni*. *Q. J. Microsc. Sci., N. S.* **45**: 41–88.

Grimaldi, D.A., Engel, M.S. and Nascibene, P.C. 2002. Fossiliferous Cretaceous amber from Myanmar (Burma): Its rediscovery, biotic diversity, and paleontological significance. *Am. Mus. Novit.* **3361**: 1–71.

Hoffmann, K. 1997. Ecdysteroids in adult females of a 'walking worm' *Euperipatoides leuckartii* (Onychophora, Peripatopsidae). *Invertebr. Reprod. Dev.* **32**: 27–30.

Homberg, U. 2008. Evolution of the central complex in the arthropod brain with respect to the visual system. *Arthropod Struct. Dev.* **37**: 347–362.

Janssen, R., Eriksson, B.J., Budd, G., Akam, M. and Prpic, N.M. 2010. Gene expression patterns in an onychophoran reveal that regionalization predates limb segmentation in pan-arthropods. *Evol. Dev.* **12**: 363–372.

Kennel, J. 1885. Entwicklungsgeschichte von *Peripatus edwardsii* Blanch. und *Peripatus torquatus* n.sp. *Arb. Zool.-Zootom. Inst. Würzburg* **7**: 95–299.

Maas, A., Mayer, G., Kristensen, R.M. and Waloszek, D. 2007. A Cambrian micro-lobopodian and the evolution of arthropod locomotion and reproduction. *Chin. Sci. Bull.* **52**: 3385–3392.

Manton, S.F. 1949. Studies on the Onychophora VII. The early embryonic stages of *Peripatopsis*, and some general considerations concerning the morphology and phylogeny of the Arthropoda. *Phil. Trans. R. Soc. Lond. B* **233**: 483–580.

Manton, S.F. 1977. *The Arthropoda. Habits, Functional Morphology, and Evolution*. Oxford University Press, Oxford.

Marotta, R. and Ruhberg, H. 2004. Sperm ultrastructure of an oviparous and an ovoviviparous onychophoran species (Peripatopsidae) with some phylogenetic considerations. *J. Zool. Syst. Evol. Res.* **42**: 313–322.

Mayer, G. 2006a. Origin and differentiation of nephridia in the Onychophora provide no support for the Articulata. *Zoomorphology* **125**: 1–12.

Mayer, G. 2006b. Structure and development of onychophoran eyes: what is the ancestral visual organ in arthropods? *Arthropod Struct. Dev.* **35**: 231–245.

Mayer, G. and Harzsch, S. 2007. Immunolocalization of serotonin in Onychophora argues against segmental ganglia being an ancestral feature of arthropods. *BMC Evol. Biol.* **7**: 118.

Mayer, G. and Whitington, P.M. 2009. Neural development in Onychophora (velvet worms) suggests a step-wise evolution of segmentation in the nervous system of Panarthropoda. *Dev. Biol.* **335**: 263–275.

Mayer, G., Ruhberg, H. and Bartolomaeus, T. 2004. When an epithelium ceases to exist - an ultrastructural study on the fate of the embryonic coelom in *Epiperipatus biolleyi*

(Onychophora, Peripatidae). *Acta Zool. (Stockh.)* **85**: 163–170.

Nylund, A., Ruhberg, H., Tönneland, A. and Meidell, B. 1988. Heart ultrastructure in four species of Onychophora (Peripatopsidae and Peripatidae) and phylogenetic implications. *Zool. Beitr., N. F.* **32**: 17–30.

Pearse, V., Pearse, J., Buchsbaum, M. and Buchsbaum, R. 1987. *Living Invertebrates*. Blackwell Scientific Publ., Palo Alto, CA.

Reid, A.L. 1996. Review of the Peripatopsidae (Onychophora) in Australia. *Invertebr. Taxon.* **10**: 663–936.

Robson, E.A. 1964. The cuticle of *Peripatopsis moseleyi*. *Q. J. Microsc. Sci., N. S.* **105**: 281–299.

Sclater, W.L. 1888. On the early stages of the development of a South American species of *Peripatus*. *Q. J. Microsc. Sci., N. S.* **28**: 343–363.

Sedgwick, A. 1885. The development of *Peripatus capensis*. Part I. *Q. J. Microsc. Sci., N. S.* **25**: 449–466.

Sedgwick, A. 1886. The development of the Cape species of *Peripatus*. Part II. *Q. J. Microsc. Sci., N. S.* **26**: 175–212.

Sedgwick, A. 1887. The development of the Cape species of *Peripatus*. Part III. On the changes from stage A to stage F. *Q. J. Microsc. Sci., N. S.* **27**: 467–550.

Sedgwick, A. 1888. The development of the Cape species of *Peripatus*. Part IV. The changes from stage G to birth. *Q. J. Microsc. Sci., N. S.* **28**: 373–396.

Sheldon, L. 1889. On the development of *Peripatus novæ-Zelandiæ*. *Q. J. Microsc. Sci., N. S.* **29**: 283–294.

Storch, V. and Ruhberg, H. 1993. Onychophora. In F.W. Harrison (ed.): *Microscopic Anatomy of Invertebrates*, vol. 12, pp. 11–56. Wiley-Liss, New York.

Strausfeld, N.J., Strausfeld, C.M., Loesel, R., Rowell, D. and Stowe, S. 2006. Arthropod phylogeny: onychophoran brain organization suggests an archaic relationship with a chelicerate stem lineage. *Proc. R. Soc. Lond. B* **273**: 1857–1866.

Wedeen, C.J., Kostriken, R.G., Leach, D. and Whitington, P. 1997. Segmentally iterated expression of an engrailed-class gene in the embryo of an Australian onychophoran. *Dev. Genes Evol.* **270**: 282–286.

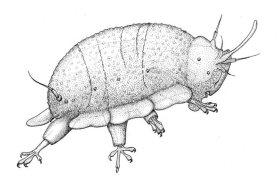

46

Phylum **Tardigrada**

Tardigrada, or water bears, is a small, easily recognized phylum of microscopic, often charmingly clumsy animals occurring in almost all types of habitats that are permanently or periodically moist—from the deep sea to soil, mosses, hot springs, and glaciers. Some species can go into anhydrobiosis, and seasonal changes between different morphologies, cyclomorphosis, are known in a number of species (Halberg *et al.* 2009b). So far about 1000 species have been described, but new species, both terrestrial and marine, are constantly being added. Morphological studies generally recognize two orders, Heterotardigrada and Eutardigrada, and this is supported by molecular analyses (Nelson 2002; Jørgensen *et al.* 2010). *Thermozodium* from a hot-sulphur spring in Japan seems intermediate between the two orders, but the type material and the type locality have now disappeared. The extraordinary abilities of the terrestrial species to withstand extreme conditions, cryptobiosis, must be seen as a specialization to special habitats and are probably not of importance for the understanding of the position of the phylum. Fossils from the Middle Cambrian strongly resemble living tardigrades, but have only three pairs of appendages (Müller *et al.* 1995). The microscopic anatomy was reviewed by Dewel *et al.* (1993).

The body always consists of a head, with or without various spines or other appendages, and four trunk segments, each with a pair of legs with a terminal group of sucking discs or claws (Fig. 46.1). The terminal or ventral mouth is a small circular opening, through which a peculiar telescoping mouth cone with stylets can be protruded. The complicated buccal tube, pharynx, and oesophagus are described in more detail below. The food seems to be bacteria, parts of larger plants, or animals, live or dead. Ingested food becomes surrounded by a peritrophic membrane and passes through the simple, endodermal midgut, which consists of cells with microvilli, to the short, ectodermal rectum. The legs have short, intrinsic muscles and consist of a proximal and a distal joint; the distal joint can be retracted into the proximal joint, and the legs cannot therefore be described as articulated. The ectoderm secretes a complex chitinous cuticle with lipid-containing layers; the type of chitin is unknown. The cuticle, which is heavily sclerotized in many terrestrial forms, is moulted periodically; the claws, the cuticle of the pharynx with stylets and stylet supports, and of the rectum are all shed at the moulting. The pharynx bulb is Y-shaped in cross section, with a constant pattern of radiating, myoepithelial cells, comprising a single sarcomere of a cross-striated muscle (Eibye-Jacobsen 1996, 1997). The more or less strongly calcified stylets (Bird and McClure 1997) are secreted by ectodermal glands, usually called salivary glands, but there is apparently no evidence for the secretion of digestive products.

Chapter vignette: *Wingstrandarctus corallinus*. (After Kristensen 1984.)

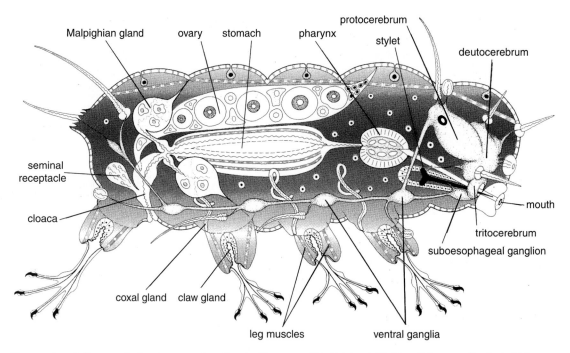

Figure 46.1. A diagram of a female ancestral tardigrade. (Courtesy of Dr Reinhardt Møbjerg Kristensen, Zoological Museum, University of Copenhagen.)

Stylets and stylet supports are formed as separate structures in separate lobes of the glands in *Batillipes* (Kristensen 1976), and this may apply to other forms too. The secretion of the stylets and their supports indicates that they are formed from modified legs.

Both the claws and the complicated toes, with cuticular 'tendons' connecting the suction pads, and terminal claws at the tips of the toes with the muscle attachment at the base, are cuticular structures secreted by the so-called foot glands (Kristensen 1976); these ectodermal structures have no glandular function and have nothing to do with the crural glands of the onychophorans.

The ectoderm is monolayered and is very thin on the body, where a fixed number of cells (eutely) are arranged in a symmetrical pattern that can be followed from species to species, even between heterotardigrades and eutardigrades (Marcus 1928).

The brain has traditionally been described as consisting of three parts, called protocerebrum, deutocerebrum, and tritocerebrum (Marcus 1929), but this view has been challenged by newer investigations

(Dewel and Dewel 1996; Wiederhöft and Greven 1996; Zantke *et al.* 2008). However, new studies using immunostaining of both a heterotardigrade and a eutardigrade lend full support to the description of Marcus (1929), although the homology of the various ganglia is uncertain (personal communication, Mr K.A. Halberg, Natural History Museum of Denmark, University Copenhagen). There is a pair of ventral nerves with a pair of ganglia with a commissure in each segment and with nerves to the legs.

Sense organs of the sensilla type are found on each segment in the arthrotardigrades. They consist of a few sensory cells with a modified cilium, surrounded by a trichogen cell and a tormogen cell.

There is a spacious body cavity that is mixocoel, although the embryological somites do not have a lumen (see below). The musculature consists of separate muscles often consisting of few, large cells arranged in a metameric pattern (Halberg *et al.* 2009a). All the muscles are cross-striated in the arthrotardigrades, whereas the eutardigrades have oblique cross-striated or smooth muscles, except the stylet muscles and the muscles of the pharyngeal bulb (see above). The muscles are attached to the cuticle through ectodermal cells with microfilaments and hemidesmosomes.

Excretory organs, called Malpighian tubules, are found in eutardigrades and *Thermozodium*, where they originate from the transition between midgut and rectum; their ultrastructure is strikingly similar to that of insects (Møbjerg and Dahl 1996).

An unpaired testis or ovarium (or an ovotestis) is situated dorsal to the gut with paired or unpaired gonoducts, ending either in a separate gonopore in front of the anus (in the heterotardigrades) or in the rectum (in the eutardigrades). Some species are parthenogenetic. The sperm is flagellate, but the head shows much variation, in some species with quite bizarre morphology (Rebecchi *et al.* 2000). It is transferred to the female during a copulation.

The large eggs are deposited in the exuvium during moulting. They are surrounded by a tough egg shell, which has made detailed embryological studies difficult. Earlier reports of enterocoely have not been substantiated, and the following description is based on the new studies using 4D microscopy of living embryos.

Hejnol and Schnabel (2005, 2006) studied the eutardigrade *Thulinia*, with some observations on the heterotardigrade *Echiniscoides*. The cleavage is total and equal, and the 64-cell stage is solid or with a narrow blastocoel. The cell lineages show much variation between the embryos studied. Two primary germ cells, each descending from one of the two first blastomeres or from one of the two sister blastomeres of the second cleavage, ingress first, followed by an invagination of four endodermal and eight mesodermal precursor cells. The endodermal cells divide and become organized in a median gut, and the mesodermal cells divide to form a pair of lateral bands. The blastopore closes, and a large anterior stomodaeal and a small proctodaeal invagination become established. The embryo now curves ventrally and becomes comma-shaped, with the large developing pharynx filling much of the head of the comma. The mesodermal bands divide into four somites on each side, with the nuclei at their periphery, but without coelomic cavities. The brain differentiates from the ectoderm dorsal to the pharynx, and the four ventral ganglia differentiate from the ventral ectoderm. The cells of the mesodermal somites disperse and become individual muscles. The differentiation of the limb buds was not followed.

Gabriel and Goldstein (2007) and Gabriel *et al.* (2007) studied the parthenogenetic eutardigrade *Hypsibius*. They observed a highly conserved cleavage pattern, and constructed a cell-lineage diagram, but the final fate of each blastomere was unfortunately not determined. They observed the ingression of two cells, which probably represent the primary germ cells, and a following epibolic gastrulation of endoderm and mesoderm. After gastrulation, the endomesoderm should form a tube, which should become divided into four pouches. Each pouch should give off a coelomic pouch on each side (and the four median parts should become connected again). This appears most unlikely, but the doubt can only be removed if the cell lineage of the crucial stages become established. *Engrailed* is expressed in dorsolateral ectodermal stripes along the posterior border of the four somites. Limb buds develop in line with the somites.

It appears that the marine arthrotardigrades is the group with most plesiomorphies, and that, for example, the malpighian tubules of the eutardigrades and the ability to go into cryptobiosis are adaptations to the terrestrial habitat (Renaud-Mornant 1982; Kristensen and Higgins 1984).

The monophyly of the Tardigrada seems unquestioned, but their relationships to other groups has been debated. Most authors regard them as panartropods, but a relationship with cycloneuralians has been proposed too, with the structure of the triradiate gut interpreted as a homologue of the pharynx of nematodes and loriciferans. However, the triradiate pharynx is simply the most efficient way of constructing an independent suction device, and triradiate pharynges, either myoepithelial (as in nematodes; Chapter 49) or with mesodermal musculature (as in kinorhynchs; Chapter 53), have obviously evolved in several independent lineages. The new information on the embryology with four somites that disperse into individual muscles in the body and limbs and the segmental gene expression (Gabriel and Goldstein 2007) are a clear indication about the arthropod relationship. Further similarities with the arthropods include the nervous system, which has paired ganglia in the head, and a ventral chain of ganglia with connectives. Many of the molecular phylogenies are in agreement with this (Dunn *et al.* 2008; Paps *et al.* 2009; Mallatt *et al.* 2010; Rota-Stabelli *et al.* 2010), although some are in favour of the cycloneuralian relationship (e.g. Meldal *et al.* 2007; Roeding *et al.* 2007; Lartillot and Philippe 2008; Hejnol *et al.* 2009).

Interesting subjects for future research

1. Embryology of all major groups
2. Type of the chitin in the cuticle

References

Bird, A.F. and McClure, S.G. 1997. Composition of the stylets of the tardigrade, *Macrobiotus* cf. *pseudohufelandi*. *Trans. R. Soc. S. Aust.* **121**: 43–50.

Dewel, R.A. and Dewel, W.C. 1996. The brain of *Echiniscus viridissimus* Peterfi, 1956 (Heterotardigrada): a key to understanding the phylogenetic position of tardigrades and the evolution of the arthropod head. *Zool. J. Linn. Soc.* **116**: 35–49.

Dewel, R.A., Nelson, D.R. and Dewel, W.C. 1993. Tardigrada. In F.W. Harrison (ed.): *Microscopic Anatomy of Invertebrates*, vol. 12, pp. 143–183. Wiley-Liss, New York.

Dunn, C.W., Hejnol, A., Matus, D.Q., *et al.* 2008. Broad phylogenomic sampling improves resolution of the animal tree of life. *Nature* **452**: 745–749.

Eibye-Jacobsen, J. 1996. On the nature of pharyngeal muscle cells in the Tardigrada. *Zool. J. Linn. Soc.* **116**: 123–138.

Eibye-Jacobsen, J. 1997. Development, ultrastructure and function of the pharynx of *Halobiotus crispae* Kristensen, 1982 (Eutardigrada). *Acta Zool. (Stockh.)* **78**: 329–347.

Gabriel, W.N. and Goldstein, B. 2007. Segmental expression of Pax3/7 and engrailed homologs in tardigrade development. *Dev. Genes Evol.* **217**: 421–433.

Gabriel, W.N., McNuff, R., Patel, S.K., *et al.* 2007. The tardigrade *Hypsibius dujardini*, a new model for studying the evolution of development. *Dev. Biol.* **312**: 545–559.

Halberg, K.A., Persson, D., Møbjerg, N., Wanninger, A. and Kristensen, R.M. 2009a. Myoanatomy of the marine tardigrade *Halobiotus crispae* (Eutardigrada: Hypsibiidae). *J. Morphol.* **270**: 996–1013.

Halberg, K.A., Persson, D., Ramløv, H., *et al.* 2009b. Cyclomorphosis in Tardigrada: adaptation to environmental constraints. *J. Exp. Biol.* **212**: 2803–2811.

Hejnol, A. and Schnabel, R. 2005. The eutardigrade *Thulinia stephaniae* has an indeterminate development and the potential to regulate early blastomere ablations. *Development* **132**: 1349–1361.

Hejnol, A. and Schnabel, R. 2006. What a couple of dimensions can do for you: Comparative developmental studies using 4D microscopy—examples from tardigrade development. *Integr. Comp. Biol.* **46**: 151–161.

Hejnol, A., Obst, M., Stamatakis, A., *et al.* 2009. Assessing the root of bilaterian animals with scalable phylogenomic methods. *Proc. R. Soc. Lond. B* **276**: 4261–4270.

Jørgensen, A., Faurby, S., Hansen, J.G., Møbjerg, N. and Kristensen, R.M. 2010. Molecular phylogeny of Arthrotardigrada (Tardigrada). *Mol. Phylogenet. Evol.* **54**: 1006–1015.

Kristensen, R.M. 1976. On the fine structure of *Batillipes noerrevangi* Kristensen 1976. 1. Tegument and moulting cycle. *Zool. Anz.* **197**: 129–150.

Kristensen, R.M. 1984. On the biology of *Wingstrandarctus corallinus* nov.gen. et sp., with notes on the symbiontic bacteria in the subfamily Florarctinae (Arthrotardigrada). *Vidensk. Medd. Dan. Naturhist. Foren.* **145**: 201–218.

Kristensen, R.M. and Higgins, R.P. 1984. A new family of Arthrotardigrada (Tardigrada: Heterotardigrada) from the Atlantic coast of Florida, U.S.A. *Trans. Am. Microsc. Soc.* **103**: 295–311.

Lartillot, N. and Philippe, H. 2008. Improvement of molecular phylogenetic inference and the phylogeny of the Bilateria. *Phil. Trans. R. Soc. Lond. B* **363**: 1463–1472.

Mallatt, J., Craig, C.W. and Yoder, M.J. 2010. Nearly complete rRNA genes assembled from across the metazoan animals: Effects of more taxa, a structure-based alignment, and paired-sites evolutionary models on phylogeny reconstruction. *Mol. Phylogenet. Evol.* **55**: 1–17.

Marcus, E. 1928. Zur vergleichenden Anatomie und Histologie der Tardigraden. *Zool. Jahrb., Allg. Zool.* **45**: 99–158.

Marcus, E. 1929. Tardigrada. *Bronn's Klassen und Ordnungen des Tierreichs*, 5. Band, 4. Abt., 3. Buch, pp. 1–608. Akademische Verlagsgesellschaft, Leipzig.

Meldal, B.H.M., Debenhamb, N.J., De Ley, P., *et al.* 2007. An improved molecular phylogeny of the Nematoda with special emphasis on marine taxa. *Mol. Phylogenet. Evol.* **42**: 622–636.

Møbjerg, N. and Dahl, C. 1996. Studies on the morphology and ultrastructure of the Malpighian tubules of *Halobiotus crispae* Kristensen, 1982 (Eutardigrada). *Zool. J. Linn. Soc.* **116**: 85–99.

Müller, K.J., Walossek, D. and Zakharov, A. 1995. 'Orsten' type phosphatized soft-integument preservation and a new record from the Middle Cambrian Kuonamka Formation in Siberia. *Neues Jahrb. Geol. Palaeontol. Abh.* **197**: 101–118.

Nelson, D.R. 2002. Current status of the Tardigrada: evolution and ecology. *Integr. Comp. Biol.* **42**: 652–659.

Paps, J., Baguñá, J. and Riutort, M. 2009. Lophotrochozoa internal phylogeny: new insights from an up-to-date analysis of nuclear ribosomal genes. *Proc. R. Soc. Lond. B* **276**: 1245–1254.

Rebecchi, L., Guidi, A. and Bertolani, R. 2000. Tardigrada. In K.G. Adiyodi and R.G. Adiyodi (eds): *Reproductive Biology of Invertebrates*, vol. 9B, pp. 267–291. Wiley, Chicester.

Renaud-Mornant, J. 1982. Species diversity in marine tardigrades. In D.R. Neslon (ed.): *Proceedings of the Third International Symposium on the Tardigrada*, pp. 149–177. East Tennessee State University Press, Johnson City, TN.

Roeding, F., Hagner-Holler, S., Ruhberg, H., *et al.* 2007. EST sequencing of Onychophora and phylogenomic analysis of Metazoa. *Mol. Phylogenet. Evol.* **45**: 942–951.

Rota-Stabelli, O., Campbell, L., Brinkmann, H., *et al.* 2010. A congruent solution to arthropod phylogeny: phylogenomics, microRNAs and morphology support monophyletic Mandibulata. *Proc. R. Soc. Lond. B* **278**: 298–306.

Wiederhöft, H. and Greven, H. 1996. The cerebral ganglia of *Milnesium tardigradum* Doyère (Apochela, Tardigrada): three dimensional reconstruction and notes on their ultrastructure. *Zool. J. Linn. Soc.* **116**: 71–84.

Zantke, J., Wolff, C. and Scholtz, G. 2008. Three-dimensional reconstruction of the central nervous system of *Macrobiotus hufelandi* (Eutradigrada, Parachela): implications for the phylogenetic position of the Tardigrada. *Zoomorphology* **127**: 21–36.

47

CYCLONEURALIA

Nematoda, Nematomorpha, Priapula, Kinorhyncha, and Loricifera have an anterior mouth, a cylindrical pharynx, and a collar-shaped, peripharyngeal brain (Fig. 47.1), and these characters are interpreted as apomorphies of a monophyletic group, which has been given the name Cycloneuralia, with reference to the shape of the brain (Ahlrichs 1995). In the previous edition of this book, I included the gastrotrichs in a group with the same name, but it now turns out that their brain does not have the characteristic cycloneuralian structure (Chapter 30), so they have been removed from the group. This makes the name synonymous with the concomitantly introduced name Introverta (Nielsen 1995), but Cycloneuralia has been used in most of the recent papers, so I have preferred to use it here. The names Nemathelminthes and Aschelminthes have been used for various selections of the phyla discussed here, but it seems best to abandon both of them.

This group seems morphologically well defined, but it is hardly ever shown as monophyletic in molecular phylogenetic analyses. The tardigrades are an ingroup of the ecdysozoans in some studies (Dunn *et al.* 2008). Most studies show scalidophorans and nematoids as separate clades in various positions relative to the arhropods, but the priapulans are basal in a number of trees (for example Hejnol *et al.* 2009). Paraphyly of the ecdysozoans, with the arthropods as an ingroup, has been suggested in a number of studies (for example Sørensen *et al.* 2008; Rota-Stabelli *et al.* 2010; Edgecombe *et al.* 2011), but many studies show poor taxon sampling, especially of the scalidophorans.

The anterior part of the body, which can be invaginated, is called the introvert (Fig. 47.2). This definition appears quite straightforward, but difficulties arise when the position of the mouth opening shall be defined to delimit the introvert. The buccal cavity and the pharynx with teeth can be everted, for example in priapulids, and it is difficult to find an anatomical criterion for the position of the mouth. I have chosen to emphasize the functional differences between the introvert and the buccal cavity plus pharynx: the introvert is used for penetration of a substratum, whereas the teeth in the buccal cavity and the pharynx are used for grasping and ingesting prey; this means that the spines/scalids of the introvert point away from the midgut when the introvert is invaginated, whereas the spines/scalids/jaws of the buccal cavity and pharynx point towards the midgut. This is of no help when the mouth cone of kinorhynchs and loriciferans is discussed, but can give a definition of the introvert in the nematodes.

If the above definition is followed, an introvert is well developed in priapulids, kinorhynchs, loriciferans, and larval nematomorphs; nematodes generally lack an introvert, but *Kinonchulus* (Riemann 1972) (Fig. 47.2) has an unmistakable introvert with six double rows of cuticular spines of varying length. The introvert is used for burrowing and grasping prey organisms, and most nematodes have apparently changed their feeding method to sucking with the myoepithelial pharynx.

All of the five phyla have compact cuticles that are moulted, but the chemical composition is not the same

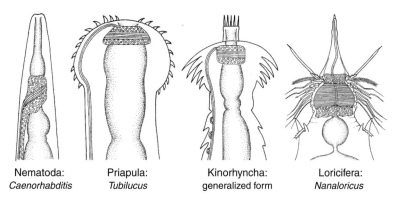

Nematoda:	Priapula:	Kinorhyncha:	Loricifera:
Caenorhabditis	*Tubilucus*	generalized form	*Nanaloricus*

Figure 47.1. Brain structure in the cycloneuralian phyla; left views except *Nanaloricus*, which is in dorsal view. Nematoda: *Caenorhabdites elegans* (based on White *et al.* 1986). Priapula: *Tubiluchus philippinensis* (based on Calloway 1975; Rehkämper *et al.* 1989). Kinorhyncha: generalized form (based on Hennig 1984; Kristensen and Higgins 1991). Loricifera: *Nanaloricus mysticus* (based on Kristensen 1991). The nematomorph brain is so modified that a meaningful comparison cannot be made (Chapter 54).

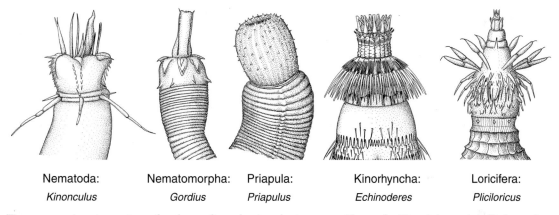

Nematoda:	Nematomorpha:	Priapula:	Kinorhyncha:	Loricifera:
Kinonculus	*Gordius*	*Priapulus*	*Echinoderes*	*Pliciloricus*

Figure 47.2. Anterior regions of cycloneuralians showing the introverts. Nematoda: *Kinonchulus sattleri.* (Redrawn from Riemann 1972.) Nematomorpha: larva of *Gordius aquaticus.* (Redrawn from Dorier 1930.) Priapula: *Priapulus horridus.* (Redrawn from Theel 1911.) Kinorhyncha: *Echinoderes aquilonius* (see the chapter vignette to Chapter 53). Loricifera: larva of *Pliciloricus gracilis.* (Redrawn from Higgins and Kristensen 1986.)

in all groups (Fig. 47.3). Nematoidea (Nematoda and Nematomorpha) have a thick layer of collagenous fibres in the inner layer of the cuticle, whereas the Scalidophora (Loricifera, Kinorhyncha, and Priapula) have cuticles with chitin. However, chitin is found in the pharyngeal cuticle of some nematodes (Neuhaus *et al.* 1997) and in the juvenile cuticle of some nematomorphs (Neuhaus *et al.* 1996). It is probable that the cuticle of the ancestor of the Cycloneuralia was chitinous, like that of their sister group, the Panarthropoda, and that the nematodes lost the cuticular chitin almost completely and reinforced their cuticle with thick layers of collagen. Carlisle (1959) could not detect chitin in the cuticle of newly moulted *Priapulus*, whereas a clear reaction was obtained in the exuviae, so the lack of chitin in the nematoid cuticle could be a paedomorphic trait. The moulting of the cuticle must be controlled by hormones, but the present knowledge of such compounds in cycloneuralians is limited to nematodes (Chapter 49), and the phylogenetic interpretation of the available information is most uncertain. The ectodermal cells appear to be com-

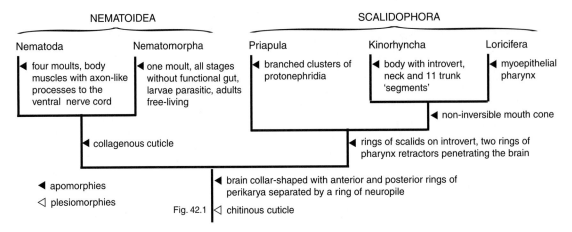

Figure 47.3. Phylogeny of the Cycloneuralia.

pletely without microvilli, and this is probably connected with the moulting.

The anterior mouth and the cylindrical pharynx surrounded by the collar-shaped brain are quite different from the ventral mouth and bilateral pharynx (often with paired jaws) and the bilateral brain of most of the other protostomes. Protostomian ontogeny indicates that the ventral mouth, bilateral pharynx, and bilateral brain with paired cephalic ganglia are ancestral to the group, and the morphology of the cycloneuralian mouth/pharynx/brain must therefore be regarded as apomorphic.

A triradiate sucking pharynx with radiating musculature is found in most of the cycloneuralian groups. It consists of myoepithelial cells in gastrotrichs, nematodes, and loriciferans, whereas the kinorhynchs have a thin epithelial covering of the mesodermal muscle cells; the priapulans have a more usual pharynx used in swallowing. A soft, triradiate pharynx functioning as a suction pump is also found in leeches (Harant and Grass 1959) and tardigrades (Chapter 46); also pycnogonids have a triradiate sucking pharynx (Dencker 1974), but radiating muscles attach to the outer wall of the proboscis. A few rotifers, such as *Asplanchna*, have a triradiate pharynx that is not a suction pump (Beauchamp 1965), and some onychophorans have a triradiate pharynx at early embryonic stages (Schmidt-Rhaesa *et al.* 1998). A soft, sucking pharynx may consist of a cylinder with radiating muscles that at contraction

expand the lumen; this structure is found, for example, in juvenile cyclorhagid kinorhynchs (Chapter 53), but it is probably a weak pump with an unstable shape. The triradiate structure is obviously a successful construction of a sucking pharynx. It appears to be the simplest shape of a subdivided cylinder that will function as a sucking organ; a pharynx with only two longitudinal bands of muscles (and no accessory structures) will at contraction become oval in cross-section without opening the lumen. Juveniles of the gastrotrich *Lepidodasys* have a circular pharynx, whereas the adults have the triradiate type (Ruppert 1982), which indicates that the radial type is the ancestral. A few nematodes have a round pharynx (Schmidt-Rhaesa *et al.* 1998) and this indicates that this type of pharynx may evolve from the triradiate type. Both Ruppert (1982) and Neuhaus (1994) discussed the pharynges of 'aschelminths' and concluded that the circular type must be ancestral in the gastrotrichs, so the triradiate pharynx cannot be ancestral in the cycloneuralians. This is supported by the different orientations of the pharynges with a dorsal median muscle in nematodes, loriciferans, and chaetonotoid gastrotrichs and a ventral median muscle in macrodasyoid gastrotrichs. Myoepithelial cells are undoubtedly an ancestral character within the eumetazoans (Chapter 10), and such cells occur here and there in the bilaterian phyla, so one cannot a priori say that a myoepithelium is plesiomorphic. It is possible that a myoepithelial pharynx is ancestral within the

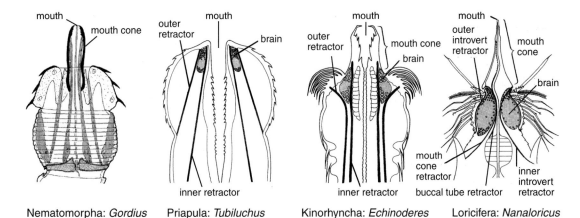

Figure 47.4. Longitudinal sections of introverts with brain and retractor muscles of a nematomorph larva and adults of the three cephalorhynch phyla. Nematomorpha: *Chordodes*. (Based on Müller *et al.* 2004.) Priapula: *Tubiluchus* sp. (Interpretation based on Calloway 1975 and Rehkämper *et al.* 1989.). Kinorhyncha: *Echinoderes capitata*. (Based on Nebelsick 1993.) Loricifera: *Nanaloricus mysticus*. (Based on Kristensen 1991 and personal communications from Dr R. M. Kristensen, University Copenhagen.)

Cycloneuralia, and secondarily lost in priapulans and kinorhynchs, but it may just as well have evolved independently in tardigrades, gastrotrichs, nematodes, and loriciferans.

The Lower Cambrian deposits have revealed very many fossils that must be classified as cycloneuralians, but which cannot with any certainty be interpreted as members of any of the living phyla (see for example Hou *et al.* 2004; Maas *et al.* 2007). The fossil record seems to indicate that only forms with a chitinous cuticle have become fossilized in the Cambrian, as no unquestionable nematodes and only larval nematomorphs have been reported.

The ancestor of the Cycloneuralia is difficult to visualize. It must have had the characteristic collar-shaped brain and the ventral nerve cord, and it probably had a chitinous cuticle that was moulted. The similarities between the introverts of the five groups, and especially between those of the nematomorph larvae and the scalidophorans (Fig. 47.1,2,4) indicate that it had an introvert with a mouth cone at some stage in the life cycle (Sørensen *et al.* 2008). It could have had a lorica, as had the centimetre-long *Siriloricus* from the Early Cambrian Sirius Passet (Peel 2010), and this was then lost in the lineage of the Nematoidea in connection with the loss of chitin. This would place the

Nematoidea as an ingroup of the Scalidophora, as suggested by Sørensen *et al.* (2008), but more molecular analyses of scalidophoran groups are needed before a more firm conclusion can be drawn.

References

Ahlrichs, W.H. 1995. *Ultrastruktur und Phylogenie von Seison nebaliae (Grube 1859) und Seison annulatus (Claus 1876).* Dissertation, Georg-August-University, Göttingen, Cuvillier Verlag, Göttingen.

Beauchamp, P. 1965. Classe des Rotifères. *Traité de Zoologie*, vol. 4(3), pp. 1225–1379. Masson, Paris.

Calloway, C.B. 1975. Morphology of the introvert and associated structures of the priapulid *Tubiluchus corallicola* from Bermuda. *Mar. Biol.* **31**: 161–174.

Carlisle, D.B. 1959. On the exuvia of *Priapulus caudatus* Lamarck. *Ark. Zool., 2. Ser.* **12**: 79–81.

Dencker, D. 1974. Das Skeletmuskulatur von *Nymphon rubrum* Hodge, 1862 (Pycnogonida: Nymphonidae). *Zool. Jahrb., Anat.* **93**: 272–287.

Dorier, A. 1930. Recherches biologiques et systématiques sur les Gordiacés. *Trav. Lab. Hydrobiol. Piscic. University Grenoble* **12**: 1–180.

Dunn, C.W., Hejnol, A., Matus, D.Q., *et al.* 2008. Broad phylogenomic sampling improves resolution of the animal tree of life. *Nature* **452**: 745–749.

Edgecombe, G.D., Giribet, G., Dunn, C.W., *et al.* 2011. Higher-level metazoan relationships: recent progress and remaining questions. *Org. Divers. Evol.* **11**: 151–172.

Harant, H. and Grassé, P.P. 1959. Classe des Annélides Achètes ou Hirudinées ou sangsues. *Traité de Zoologie*, vol. 5(1), pp. 272–287. Masson, Paris.

Hejnol, A., Obst, M., Stamatakis, A., *et al.* 2009. Assessing the root of bilaterian animals with scalable phylogenomic methods. *Proc. R. Soc. Lond. B* **276**: 4261–4270.

Hennig, W. 1984. *Taschenbuch der Zoologie, Band 2, Wirbellose I*. Gustav Fischer, Jena.

Higgins, R.P. and Kristensen, R.M. 1986. New Loricifera from southeastern United States coastal waters. *Smithson. Contr. Zool.* **438**: 1–70.

Hou, X.-G., Aldridge, R.J., Bergström, J., *et al.* 2004. *The Cambrian Fossils of Chengjiang, China*. Blackwell, Malden, MA.

Kristensen, R.M. 1991. Loricifera. In F.W. Harrison (ed.): *Microscopic Anatomy of Invertebrates*, vol. 4, pp. 351–375. Wiley-Liss, New York.

Kristensen, R.M. and Higgins, R.P. 1991. Kinorhyncha. In F.W. Harrison (ed.): *Microscopic Anatomy of Invertebrates*, vol. 4, pp. 377–404. Wiley-Liss, New York.

Maas, A., Huang, D., Chen, J., Waloszek, D. and Braun, A. 2007. Maotianshan-Shale nemathelminths—morphology, biology, and the phylogeny of Nemathelminthes. *Palaeogeogr. Palaeoclimatol. Palaeoecol.* **254**: 288–306.

Müller, M.C.M., Jochmann, R. and Schmidt-Rhaesa, A. 2004. The musculature of horsehair worm larvae (*Gordius aquaticus, Paragordius varius*, Nematomorpha): F-actin staining and reconstruction by cLSM and TEM. *Zoomorphology* **123**: 54–54.

Nebelsick, M. 1993. Nervous system, introvert, and mouth cone of *Echinoderes capitatus* and phylogenetic relationships of the Kinorhyncha. *Zoomorphology* **113**: 211–232.

Neuhaus, B. 1994. Ultrastructure of alimentary canal and body cavity, ground pattern, and phylogenetic relationships of the Kinorhyncha. *Microfauna Mar.* **9**: 61–156.

Neuhaus, B., Bresciani, J. and Peters, W. 1997. Ultrastructure of the pharyngeal cuticle and lectin labelling with wheat germ agglutinin-gold conjugate indicating chitin in the pharyngeal cuticle of *Oesophagostomum dentatum* (Strongylida, Nematoda). *Acta Zool. (Stockh.)* **78**: 205–213.

Neuhaus, B., Kristensen, R.M. and Lemburg, C. 1996. Ultrastructure of the cuticle of the Nemathelminthes and electron microscopical localization of chitin. *Verh. Dtsch. Zool. Ges.* **89(1)**: 221.

Nielsen, C. 1995. *Animal Evolution: Interrelationships of the Living Phyla*. Oxford University Press, Oxford.

Peel, J.S. 2010. A corset-like fossil from the Cambrian Sirius Passet Lagerstätte of North Greenland and its implications for cycloneuralian evolution. *J. Paleont.* **84**: 332–340.

Rehkämper, G., Storch, V., Alberti, G. and Welsch, U. 1989. On the fine structure of the nervous system of *Tubiluchus philippinensis* (Tubiluchidae, Priapulida). *Acta Zool. (Stockh.)* **70**: 111–120.

Riemann, F. 1972. *Kinonchulus sattleri* n.g. n.sp. (Enoplida, Tripyloidea), an aberrant freeliving nematode from the lower Amazoans. *Veroeff. Inst. Meeresforsch. Bremerhav.* **13**: 317–326.

Rota-Stabelli, O., Campbell, L., Brinkmann, H., *et al.* 2010. A congruent solution to arthropod phylogeny: phylogenomics, microRNAs and morphology support monophyletic Mandibulata. *Proc. R. Soc. Lond. B* **278**: 298–306.

Ruppert, E.E. 1982. Comparative ultrastructure of the gastrotrich pharynx and the evolution of myoepithelial foreguts in Aschelminthes. *Zoomorphology* **99**: 181–220.

Schmidt-Rhaesa, A., Bartolomaeus, T., Lemburg, C., Ehlers, U. and Garey, J.R. 1998. The position of the Arthropoda in the phylogenetic system. *J. Morphol.* **238**: 263–285.

Sørensen, M.V., Hebsgaard, M.B., Heiner, I., *et al.* 2008. New data from an enigmatic phylum: evidence from molecular data supports a sister-group relationship between Loricifera and Nematomorpha. *J. Zool. Syst. Evol. Res.* **46**: 231–239.

Theel, H. 1911. Priapulids and sipunculids dredged by the Swedish Antarctic expedition 1901–1903 and the phenomenon of bipolarity. *K. Sven. Naturvetenskapsakad. Handl.* **47(1)**: 1–36.

White, J.G., Southgate, E., Thomson, J.N. and Brenner, S. 1986. The structure of the nervous system of the nematode *Caenorhabditis elegans*. *Phil. Trans. R. Soc. Lond. B* **314**: 1–340.

48

NEMATOIDEA

The nematomorphs share several apomorphies with the nematodes, such as the cuticle with layers of crossing collagenous fibrils, the body wall with only longitudinal muscles, and the ectodermal longitudinal cords, and it appears unquestionable that the two groups together constitute a monophyletic group. The nematodes have highly characteristic muscles with an apical process to the nerve cords, and this is lacking in the nematomorphs. Further characters, such as the cuticle apparently without an epicuticle, the rod-shaped spermatozoa, and the rather different larval and adult stage, possibly with only one moult, clearly distinguish the nematomorphs from the nematodes. Nematodes have specialized their feeding mechanism by the evolution of a myoepithelial sucking pharynx, which has secondarily been lost in some of the endoparasites. The nematomorphs have retained the introvert in the larvae, and almost lost the gut in another line of specialization to parasitism. The retractile introvert, with spines and a non-inversible mouth cone of the larval nematomorphs, resembles the introverts of kinorhynchs and loriciferans (Fig. 47.1,2,4), and a molecular study has indicated a sister-group relationship (Sørensen *et al.* 2008) (see Chapter 47).

The molecular study of Bleidorn *et al.* (2002) of a number of nematomorphs and three nematodes (including *Mermis*) showed the two groups as sister groups, and similar results have been obtained by more recent studies (Mallatt and Giribet 2006; Meldal *et al.* 2007; Telford *et al.* 2008).

The monophyly of the Nematoidea seems unquestioned.

References:

Bleidorn, C., Schmidt-Rhaesa, A. and Garey, J.R. 2002. Systematic relationships of Nematomorpha based on molecular and morphological data. *Invert. Biol.* **121**: 357–364.

Mallatt, J. and Giribet, G. 2006. Further use of nearly complete 28S and 18S rRNA genes to classify Ecdysozoa: 37 more arthropods and a kinorhynch. *Mol. Phylogenet. Evol.* **40**: 772–794.

Meldal, B.H.M., Debenhamb, N.J., De Ley, P., *et al.* 2007. An improved molecular phylogeny of the Nematoda with special emphasis on marine taxa. *Mol. Phylogenet. Evol.* **42**: 622–636.

Sørensen, M.V., Hebsgaard, M.B., Heiner, I., *et al.* 2008. New data from an enigmatic phylum: evidence from molecular data supports a sister-group relationship between Loricifera and Nematomorpha. *J. Zool. Syst. Evol. Res.* **46**: 231–239.

Telford, M.J., Bourlat, S.J., Economou, A., Papillon, D. and Rota-Stabelli, O. 2008. The evolution of the Ecdysozoa. *Phil. Trans. R. Soc. Lond. B* **363**: 1529–1537.

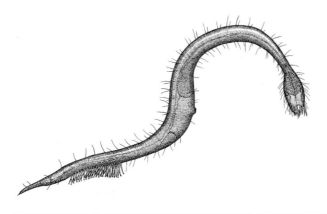

49

Phylum **Nematoda**

Nematoda is one of the most successful phyla today; about 30 000 species have been described, but it is believed that the number of living species should be counted in millions. Nematodes occur in almost all habitats, both aquatic and terrestrial, and many parasitize plants or animals and are of great economic importance. The tiny free-living, terrestrial nematode *Caenorhabditis elegans* is now one of the best studied animals, and a wealth of information about all aspects of the structure and biology of this animal can be found at the 'Wormatlas' (http://www.wormatlas.org) and 'Wormbook' (http://wormbook.org). The detail of anatomical, developmental, and genetic knowledge of this species perhaps biases perceptions of the phylum as a whole, as it is frequently implicitly assumed that what is true for *C. elegans* is true for all nematodes. The exceptions to the *C. elegans* 'rule' may actually be numerically dominant. Nematodes are difficult to recognize as fossils, and the earliest unquestionable fossil seems to be from the Devonian (Poinar *et al.* 2008).

Nematoda is a well-defined, monophyletic group, with Nematomorpha as the sister group (or a further outgroup) in most studies (Bleidorn *et al.* 2002; Brusca and Brusca 2003; Dunn *et al.* 2008; Edgecombe *et al.* 2011). Traditional, morphology-based nematode systematics has divided the phylum into Adenophorea, without caudal phasmids and with a simple, non-cuticularized excretory organ, and Secernentia, with

phasmids and a more complex excretory organ with a cuticularized duct (Brusca and Brusca 2003). However, studies of small-subunit rDNA sequences show a much more complicated picture, with three major clades, Enoplia, Dorylaimia, and Chromadoria (which includes the Secernentia), with Dorylaimia and Chromadorea likely to be more closely related (Blaxter *et al.* 1998; Holterman *et al.* 2006; Meldal *et al.* 2007).

In spite of the variations in habitats, all nematodes are of a remarkably similar body plan. Almost all species are cylindrical with tapering ends and with a thick elastic cuticle, which may be smooth, often with a microrelief; a few types have ring- or scale-shaped thickenings, and others have various types of spines or bristles. The cuticle maintains a turgor in the body cavity, which functions as a hydrostatic skeleton with bands of longitudinal body-wall muscles as antagonists; ring muscles are absent in the body wall and locomotory cilia are completely absent. This very specialized body plan has obviously made the invasion of very different niches possible without strong modifications.

A most characteristic feature of some nematodes is that the numbers of cells/nuclei in most organs appear to be constant within species (Hyman 1951). The small *Caenorhabditis elegans* (length about 1.3 mm, diameter about 80 μm) has a total number of 558 somatic nuclei in the newly hatched hermaphrodite and 560 in the male, and 959 and 1031, respectively, in the adult (not

---

Chapter vignette: *Draconema cephalatum*. (Redrawn from Steiner 1919.)

counting the germ cells) (Kenyon 1985); the intestine comprises 34 cells and the nervous system 302 neurons in the adult hermaphrodite (Sulston *et al.* 1983). Much larger species, for example *Ascaris lumbricoides* have about the same arrangement and number of nuclei in the nervous system (about 250; Stretton *et al.* 1978) and the numbers in most other organ systems are also constant; however, especially the syncytial ectoderm shows much higher numbers of nuclei in the larger species (Cunha *et al.* 1999). Some species have tissues, such as epidermis, endoderm, and some of the muscles consisting of large syncytia with a more indefinite number of nuclei, whereas others have a normal, cellular ectoderm. Regenerative powers are very small, as cell division is completed at hatching.

The microscopic anatomy was reviewed by Wright (1991). The body wall consists of a cuticle, a layer of ectodermal cells (often called hypodermis), a basal membrane, and a layer of longitudinal body-wall muscle cells (Moerman and Fire 1997). The cuticle also lines the buccal cavity, the pharynx, and the rectum. The outermost layer of the cuticle is thin epicuticle, which is a double membrane resembling the cell membrane, but both its thickness and its freeze fracture pattern differ from those of a normal cell membrane. It is the first cuticular layer to be secreted during ontogeny and also during synthesis of the new cuticle at moulting. The main layer of the body cuticle comprises several zones that consist mainly of collagen (Johnstone 1994); chitin is lacking in the body cuticle, but the pharyngeal cuticle of some genera contains chitin (Neuhaus *et al.* 1997). The median zone shows wide variation and contains highly organized structures in many species. The inner zone contains more or less complex layers of fibrils with different orientations. A few parasitic genera have only a reduced or partial cuticle. The cuticle is moulted four times (three times in *Pristionchus* (Félix *et al.* 1999)) and growth is related to the moults, although a considerable part of the growth is related to gradual stretching of wrinkles in the newly formed cuticle, resulting in a sigmoidal growth curve (Howells and Blainey 1983).

Moulting is usually said to be associated with changes in the levels of steroid hormones, especially ecdysteroids, but these hormones have not been found in any free-living nematode (Frand *et al.* 2005), although an ortholog of the ecdysone receptor protein has been found in some parasitic species (Ghedin *et al.* 2007; Graham *et al.* 2010). The 'Halloween genes', that mediate the biosynthesis of 20-hydroxyecdysone in arthropods, are absent in the genome of *Caenorhabditis* (Lafont and Mathieu 2007). The biological function of the ecdysteroids in nematodes is still obscure. They seem to be involved in growth regulation, embryogenesis, vitellogenesis, and moulting (Franke and Käuser 1989). They have been found in many animals and plants, but among metazoans only arthropods are known to be able to synthesize these molecules (Lafont 1997; Lafont and Koolman 2009). Some other cuticle biosynthetic genes show a pulse of expression just before each moult (Frand *et al.* 2005).

The ectodermal cells are arranged in longitudinal rows, with the nuclei concentrated in a pair of lateral thickenings or cords; an additional dorsal and ventral cord are found in the larger species, which may also have nuclei between the cords. The cords divide the somatic musculature into longitudinal bands. The muscle cells are connected to the cuticle via hemidesmosomes and tonofibrils in the ectodermal cells. Each muscle cell has an axon-like extension that reaches to the motor neurons in the median longitudinal nerves (see below). There are oblique muscles in the posterior region of the male, and a few muscular cells are associated with the intestine and the rectum.

The terminal mouth is surrounded by various labial organs and sensilla (see below), and opens into a buccal cavity often with cuticular structures such as thorns, hooks, jaws, or spines. Many genera of the order Dorylaimida have a grooved tooth that is used to penetrate food organisms; this tooth is secreted by one of the ventrolateral myoepithelial cells of the pharynx. The groove is almost closed in some genera, and the tooth is a long hypodermic needle with the base surrounding the opening to the pharynx completely in others. The remarkable genus *Kinonculus* has a conspicuous introvert with six longitudinal double rows of short and long spines in front of the normal rings of sensilla; there is a large grooved or hollow tooth in contact with the dorsal side of the pharynx (Fig. 47.2).

The long pharynx (sometimes called the oesophagus) is cylindrical, often with one or two swellings and has a lumen that is Y-shaped in cross section (Hoschitz *et al.* 2001). The one-layered epithelium consists of myoepithelial cells that form a dorsal and two lateral thickenings, while epithelial cells with conspicuous tonofilaments attaching to hemidesmosomes line the bottoms of the grooves. All the cells secrete a cuticle, which may form various masticatory structures. Kenyon (1985) proposed that the ancestral pharynx had only the lateral muscle bands, and that the dorsal band is a later specialization. The nematode pharynx is definitely a suction organ, and it appears that a pharynx with only a pair of lateral muscle bands will become oval in cross section at contraction without opening the lumen. A pharynx with only two bands of muscles is therefore not a likely ancestral character.

The intestine is a straight tube of usually unciliated cells with a microvillous border, but Zmoray and Guttekova (1969) found numerous cilia between long microvilli in the gut of the soil nematode *Eudorylaimus*, but this report is usually ignored. There are generally no muscles around the intestine, except the four posterior cells mentioned below, and the food particles are apparently pressed through the intestine by the movements of the pharynx.

The nervous system has sometimes been characterized as orthogonal (Beklemischew 1960; Reisinger 1972), but this is misleading. The nervous system of *Caenorhabditis* (Fig. 49.1) has been mapped in very fine detail (Albertson and Thomson 1976; White *et al.* 1986; Altun *et al.* 2009; Oren-Suissa *et al.* 2010), and

other well-studied species appear to have very similar systems, even with individual cells being recognizable between species. There is a collar-shaped brain around the pharynx, a ventral longitudinal nerve cord, and a concentration of cells surrounding the rectum; the few nerve cells located outside this typical protostomian (gastroneuralian) central nervous system are sensory cells. The brain consists of an anterior and a posterior ring with concentrations of perikarya, and a middle zone of neuropil. The nerves along the lateral and dorsal ectodermal thickenings, which have been described from all nematodes investigated for more than a century, consist of bundles of axons from the sensory cells, from motor cell bodies in the brain, or the ventral nervous system; none of the axons extend directly from the brain along these nerves, which are thus of a character quite distinct from that of the ventral cord. The whole nervous system (with the exception of six cell bodies) is situated between the epidermis and its basal membrane, and the neuromuscular junctions with the mesodermal muscles connect the two types of cells across the basal membrane. This is possible because the muscle cells have the axon-like protrusions mentioned above. The ventral cord contains a row of 'ganglia' containing the cell bodies of the motor neurons of the body-wall muscles; their connections to the lateral and dorsal longitudinal nerves are via lateral commissures, which form species-constant patterns. In *Ascaris* there are five units of neurons each with 11 cells; each unit has six right-side and one left-side commissure. Similar structures are known from other genera. The units have sometimes been described as segments, but it must be emphasized that the repeating units are not formed

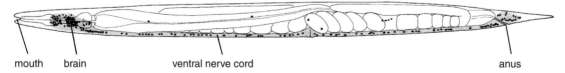

mouth   brain   ventral nerve cord   anus

Figure 49.1. Central nervous system (CNS) of *Caenorhabditis elegans*. The nuclei of the nervous cells in the left half of the animal are indicated as black dots. The CNS (anterior ganglion, dorsal ganglion, lateral ganglion, ventral ganglion, retrovesicular ganglion, ventral cord, pre-anal ganglion, dorso-rectal ganglion, and lumbar ganglion) is shaded. The few nuclei outside the CNS are exclusively of sensory cells. (Based on White *et al.* 1986.)

from teloblasts as are the segments of the annelids and arthropods (see below).

Various types of sensory organs have been described. Some sensory cells are internal, such as receptors just below the cuticle of the pharynx and the stretch receptors with a modified cilium in the lateral epithelial cords of enoplids. Other sense organs respond to exterior stimuli, such as chemoreceptors, mechanoreceptors, and eyespots. Sensilla are chemoreceptors (or perhaps sometimes mechanoreceptors), which typically consist of a nerve cell with one or a few short terminal, modified cilia (often called dendritic processes), a sheath cell, and a socket cell, but there may be more than one nerve cell in each sensillum. The distal extensions of the socket cell and the sheath cell surround a pit or another modified area of the cuticle, into which the short cilia protrude. Many of these sensory organs are situated on the flat surface, for example on the lips, while others may be associated with spicules, for example in the male copulatory apparatus. There are normally three rings of sensilla around the mouth, with four sensilla in the posterior ring and six in each of the anterior rings. A pair of large sensilla, the amphids, which are usually very conspicuous in the free-living species, is found laterally in the head or neck region. Each amphid contains 4–13 sensory cells, which in some species have many cilia. The amphidial pore or canal is large and varying in shape; funnel-shaped, circular, or spirally coiled. Most amphids have an additional unit lying deeply in the canal and consisting of an extended, distal part of the sheath cell, with strongly folded or microvillous cell membranes around the canal; some of the receptor cells may have a microvillous zone in the same area. Amphids are mechanosensory, chemosensory, and perhaps thermosensory in *Caenorhabditis*.

The body cavity is lined by ectodermal cells of the epithelial cords, mesodermal muscles, and endodermal intestine and is therefore a primary body cavity, which forms at a late stage of development (see below). It is spacious in the larger parasitic forms, but almost non-existent in small species. The excretory organ consists of 1–5 cells with considerable differences between the families. Many of the marine forms have one large glandular cell (called the ventral-gland or renette cell), but the system may be complicated by the presence of one or a pair of longitudinal tubular extensions. *Caenorhabditis* represents the most complicated type with four cells: a pore cell, a duct cell, a canal cell, and a fused pair of gland cells. The large glandular cell apparently secretes a fluid into its long, narrow lumen. Pulsation of the excretory system has been observed in several species, and the pulsations are correlated with the osmolarity of the surrounding fluid, so osmoregulation is one function of the system. It has been suggested that the excretory system secretes a 'moulting fluid' through the excretory pore into the space between the old and the new cuticle and that this fluid is necessary for the moulting. However, later observations indicate that the loosening of the old cuticle begins at the mouth rather than at the excretory pore, and laser ablation experiments with nuclei of the excretory cells have not prevented moulting (Singh and Sulston 1978). The role of the excretory system in moulting must therefore be characterized as uncertain.

The sac-shaped gonads consist of germinal cells surrounded by a basal membrane and open through a ventral pore. Some species are gonochoristic, but other species, such as *Caenorhabditis*, have protandric hermaphrodites and males. Fertilization is internal, and the males have a complicated copulatory apparatus. The sperm is aflagellate and amoeboid and its movement is based on a special protein, called major sperm protein, whereas actin seems to be unimportant (Wolgemuth *et al.* 2005); its final differentiation usually takes place within the female. The fertilized egg becomes surrounded by an egg shell, which contains chitin. The eggs are usually somewhat elongate, often with a slightly flattened or even concave side; in some species the polar bodies are formed at or near one end of the egg, but in other species at the flat/concave side (zur Strassen 1959). All species studied show a germ line (Schierenberg 1997) (as in other protostomes, see Chapter 22 and Fig. 21.1). The cells outside the germ line show chromosome diminution in a number of species, such as *Ascaris*, but not in others, such as *Caenorhabditis* (Tobler *et al.* 1992).

Table 49.1. Cell lineage of *Caenorhabditis elegans*. (Based on Sulston *et al.* 1983.) There is more variation in the cell lineage than this diagram shows. (See Schnabel *et al.* 1997.)

	S₁ (AB)	ectoderm:	body epithelium, epithelium of the mouth, epithelial cells in the pharynx (myoepithelial and marginal), epithelium of the rectum, nervous system, excretory system (4 cells)
		mesoderm:	one body-muscle cell, anal sphincter muscle, anal depressor muscle cell, left intestinal muscle cell, postembryonic mesoblast

- **S₁ (AB)**
  - ectoderm: body epithelium, epithelium of the mouth, epithelial cells in the pharynx (myoepithelial and marginal), epithelium of the rectum, nervous system, excretory system (4 cells)
  - mesoderm: one body-muscle cell, anal sphincter muscle, anal depressor muscle cell, left intestinal muscle cell, postembryonic mesoblast
- **S₂ (EMS)**
  - **MS**
    - ectoderm: epithelial cells in the pharynx (myoepithelial, gland), 6 nerve cells in the pharynx
    - mesoderm: body-wall muscle cells, right intestinal muscle cell, 'coelomocytes', 12 glia cells, somatic gonad cells
  - **E**
    - endoderm: intestine
- **P₂**
  - **S₃ (C)**
    - ectoderm: body-epithelium, 2 nerve cells near anus
    - mesoderm: body-wall muscle cells
  - **P₃**
    - **S₄ (D)** mesoderm: body-wall muscle cells
    - **P₄** germ cells

Grouping: Z → P₁ → {S₂ (EMS), P₂}; P₂ → {S₃ (C), P₃}; P₃ → {S₄ (D), P₄}. P₁ → {S₂ (EMS), P₂}. Z → {S₁ (AB), P₁}.

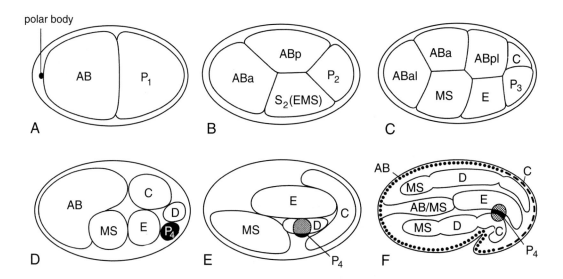

Figure 49.2. Embryology of *Caenorhabditis elegans*. Diagrams of embryos seen from the left. (A) 2-cell stage. (B) 4-cell stage after the shifting of the AB blastomeres. (C) 8-cell stage. (D) 26-cell stage; areas of blastomeres are indicated (there are 16 AB cells, 4 MS cells, 2 E cells, and 2 C cells). (E) 102-cell stage; the D, E and P₄ descendants have been invaginated. (F) More than 500 cells; AB cells now contribute to the anterior ectoderm and to the pharynx, E cells form the gut, some of the MS form some of the pharyngeal muscles, C cells contribute to the posterior ectoderm and MS, D and C cells form the longitudinal muscles. (A,B and D–F, redrawn from Schierenberg 1997; C, based on Hutter and Schnabel 1995.)

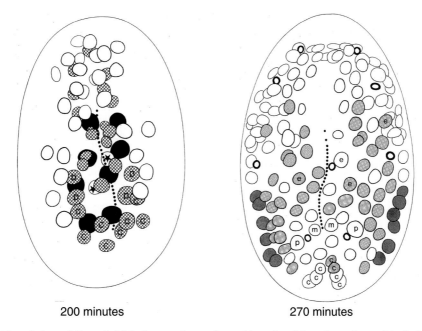

200 minutes   270 minutes

Figure 49.3. Ventral views of *Caenorhabditis elegans* embryos; the position of nuclei on the surface and in the blastoporal cleft is indicated. (Based on Sulston *et al.* 1983.)

Blastopore closure (200 min). The nuclei of the eight endodermal cells are black; the two germ line nuclei are marked with asterisks; the cross-hatched nuclei indicate the main body muscle precursors—the cells originating from S3 and S4 are marked with C and D respectively and the cells without letters are descendants of MS. The position of the blastoporal cleft is indicated by the dotted line.

Late gastrulation (270 min). The mesodermal cells (see above) are covered ventrally by cells that are the precursors of the ventral nervous system. All nuclei are of AB cells except the small posterior wedge of cells marked c. The nuclei of cells that will give rise to cells in the ventral nervous system (including ventral ganglion, retrovesicular ganglion, ventral cord, pre-anal ganglion, and lumbar ganglion) are shaded. The lightly shaded cells will invaginate as the last part of the gastrulation, bringing the darkly shaded lateral lines of cells in contact midventrally, where they become arranged in one line; these cells then divide into a nerve cell and an epithelial cell. The unshaded cell marked e becomes the H-shaped excretory cell; the uppermost shaded cell marked e divides to form the excretory duct cell and some ganglionic cells; the lowermost-shaded cells marked e divide to form the two excretory duct cells and some ganglionic cells. The two cells marked m are situated at the proctodaeum; they are precursors of the following mesodermal muscle cells: the AB body-wall muscle cell, the anal sphincter, the anal depressor, and the left intestinal muscle. The two cells marked p become the socket cells of the posterior sensilla called phasmids. The small thick rings indicate the results of programmed cell death.

Embryology of the large *Parascaris equorum* was studied by a number of authors more than a century ago (zur Strassen 1896, 1906; Boveri 1899; Müller 1903), but recent studies, especially of the small *Caenorhabditis elegans*, have set a new standard for the study of cell lineages (Sternberg and Horvitz 1982; Sulston *et al.* 1983; Schnabel *et al.* 1997) (Table 49.1). *Caenorhabites*, *Ascaris*, *Panagrellus*, and *Acrobeloides* represent three rhabditoid groups within the Chromadoria, and their cell lineages are remarkably similar (Sternberg and Horvitz 1982; Schulze and Schierenberg 2009).

The very well documented, normal embryology of *Caenorhabditis* will be described first (Figs. 49.2–3 and Table 49.1).

The anteroposterior axis of the embryo is determined by the entry point of the sperm (Goldstein and Hird 1996) and the zygote divides into an anterior and a posterior cell. The anterior cell S_I (AB, first somatic founder cell) divides longitudinally and, slightly later, the posterior cell (P$_1$, first germline cell) divides transversally into the anterior S$_2$(EMS) and the posterior P$_2$. Probably because of the narrowness of the egg shell, one of the emerging AB descendants occupies an anterior and the other a more posterior-dorsal position. The posterior cell gets contact with P$_2$, which in turn induces the cell to adopt the ABp fate. Another essential induction goes from P$_2$ to S$_2$(EMS). The ABa and ABp cells divide into right and left blastomeres, whereas the two other cells divide in anterior and posterior blastomeres: S$_2$(EMS) into the anterior MS and the posterior E, which becomes the endoderm; and P$_2$ into the ventral P$_3$ and the dorsal S$_3$(C). The MS cell divides obliquely into the anterior-left MSa, which gives rise to all left descendants, and the posterior-right MSp, which gives rise to right descendants. MSap and MSpp give rise to mainly mesodermal muscle cells, somatic gonad cells, and glia cells, whereas the MSaa and MSpa mainly form the ectodermal pharynx (with myoepithelial cells, gland cells, and motor neurons) and the midventral body muscles (Sulston *et al.* 1983; Kenyon 1985). A blastula is formed and gastrulation starts with movement of the two E cells to the interior of the embryo, eventually leading to an elongate blastopore surrounded by MS and D descendants; the P$_4$ cell (the primordial germ cell) lies at the posterior edge of the blastopore, but soon becomes internalized. The cells that will give rise to mesodermal muscles surround the blastopore (Fig. 49.3). The lateral blastopore lips become pressed together and fuse. After the fusion of the blastoporal lips, the surface of the embryo is covered by descendants of the S$_1$(AB) and S$_3$(C) cells, with the last-mentioned cells forming a band along the midline in the posterior part of the embryo; the stomodeal invagination, which develops into the pharynx, is composed of a wider variety of cells. In the second part of the embryogenesis (the morphogenetic phase), the embryo forms a ventral indentation so that parts of the dorsal side of the head and tail are seen in a ventral view. On both sides of the blastopore lie extensive areas, with cells that will give rise to neurons in the ventral nerve cord; when these cells migrate into the embryo the most lateral cells, which form a row of six on each side (Fig. 49.3), finally meet and become arranged in a single, midventral line. Each of these cells

divides postembryonically into an epidermal cell and an inner nerve cell, and this internalizes the last cells of the nerve cord. The cells of the two rows are descendants of the cells ABp(r/l)ap, and their cell lineages (see Fig. 49.4) show that these cells are not stem cells like the neuroblasts of, for example, the leeches (Chapter 25), and the two cells give rise to several other types of cells. The ring-shaped brain originates exclusively from S$_1$(AB) cells. The precursors of the four cells of the excretory organ are located in the field of nervous precursor cells (Fig. 49.3). Mixed embryological origin is shown, for example, by the longitudinal body-wall muscles that are composed of descendants from the cells MS, S$_4$(D), S$_3$(C), and S$_1$(AB); the ectodermal epithelium comprises large syncytial areas, and the dorsal syncytium is formed by the fusion of cells from S$_2$(AB) and S$_3$(C).

The embryology of *Romanomermis* (Schierenberg and Schulze 2008; Schulze and Schierenberg 2009) (Table 49.2), representing the Dorylaimia, resembles that of *Caenorhabditis* in some details, but important differences are found in the cleavage pattern. A speciality of *Romanomermis* is that coloured cytoplasm is segregated to the S$_2$ cell, which appears to give rise to the complete ectoderm. This cell divides to form a transverse ring of eight cells. This ring of cells then dupli-

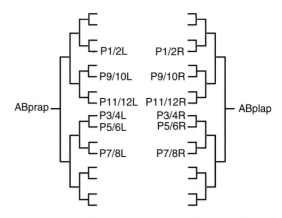

Figure 49.4. Cell lineage of the right and left row of lateral cells that meet in the ventral midline and contribute to the ventral cord in *Caenorhabditis elegans* (see Fig. 49.3). '1/2' indicates that the position in the midventral row of cells after complete blastopore closure can be either number 1 or 2. (Based on Sulston *et al.* 1983.)

Table 49.2. Cell lineage of *Romanomermis culicivorax*. (Based on Schulze and Schierenberg 2009.)

		$S_1 a$		neurons
	S_1	$S_1 pa$		mouth region, pharynx, ?neurons
	$S_1 p$	$S_1 pp$	$S_1 ppa$	pharynx
Z			$S_1 ppp$	gut
	S_2			epidermis, ?neurons
P_1	S_3			body muscles
	P_2	S_4		?somatic gonad
	P_3	P_4		germ cells

cates at least three times to form eight rings of cells, and their descendants finally cover the whole embryo.

Very different types of early embryology are found in species of the aquatic order Enoplia. The egg of *Tobrilus* divides equally a number of times and a coeloblastula is formed at the 32-cell stage. At the 64-cell stage, a few cells immigrate into the blastocoel forming a small archenteron, with cells adjacent to the 'blastopore' region subsequently developing into the pharynx (Schierenberg 2005). A variable pattern is seen in *Enoplus*, which does not show any distinct early cleavage pattern. However, a large endoderm precursor cell can be recognized at the 8-cell stage (Malakhov 1998; Voronov and Panchin 1998). The descendents of the other blastomeres appear to adopt variable fates and the origin of the germ cells could not be determined. Gastrulation and morphogenesis resemble those of *Caenorhabditis* and *Romanomermis*.

It is believed that the embryology of *Tobrilus*, with a coeloblastula and an embolic gastrulation, represents the ancestral type of within the nematodes, and this could indicate that *Tobrilus* represents a clade 'basal' to the remaining groups (Schierenberg and Schulze 2008).

Further development is through four (or sometimes three) moults, some of which take place inside the egg in some species.

In *Acroleboides*, early blastomeres can regulate, in that lost cells are replaced in a hierarchical manner, e.g. MS can take over the AB programme but not vice versa (Wiegner and Schierenberg 1999). Experimental and comparative analyses of nematode embryogenesis have revealed an unexpected degree of variability and the presence of an intricate network of interactions between the blastomeres (Hutter and Schnabel 1995; Schierenberg and Schulze 2008).

The nematodes are clearly a monophyletic group. The ground plan with a thick collagenous cuticle that is moulted and longitudinal muscle bands, together functioning in a hydrostatic skeleton, is not seen in any other phylum, except in the nematomorphs (Chapter 50), and the three separate rings of sensilla, with six sensilla on the lips, and 6+4 sensilla on the body, the highly specialized sperm, and the four moults seem to be very reliable apomorphies (Lorenzen 2007). Also the pair of large chemosensilla, the amphids, that are apparently always present in the cephalic region, may be a nematode apomorphy. A sister-group relationship with the nematomorphs is strongly indicated; this is further discussed in Chapter 50.

The highly determined cleavage type bears no resemblance to the spiral cleavage (see Fig. 23.2 and compare Table 49.1 with, for example, Fig. 22.9 and Table 23.1), but the more 'simple' cleavage pattern of *Tobrilus* is reminiscent of the embryology of equal-cleaving gastropods (Chapter 27). The mesoderm originates from a ring of cells around the blastopore in *Caenorhabditis* and this is quite different from mesoderm formation in the spiralians (Chapter 23). The ultrastructure of the sensilla resembles that of some arthropod sense organs, but this 'probably reflects evolution of convergent solutions to the problem of how to get a sensory neuron into or through an exoskeleton' (Ward *et al.* 1975), and it should be noted that the three cells of the nematode sensillum originate from different blastomeres, as opposed to the arthropod sensilla that differentiate from one cell (Chapter 44).

The origin of the tube-shaped gut, through fusion of the lateral blastopore lips, and the development of the ventral nerve cord along the fusion line, unequivocally define the nematodes as protostomians (gastroneuralians). Development is direct without any trace of a ciliated trochophore stage or of the apical, larval nervous system characteristic of the Spiralia, and the terminal mouth and cylindrical pharynx surrounded by the collar-shaped ganglion are cycloneuralian characters. The moulted cuticle consisting of collagen clearly determines their position on the phylogenetic tree (Fig. 47.3).

Studies on molecular phylogeny invariably place the nematodes in the Ecdysozoa (Hejnol *et al.* 2009; Pick *et al.* 2010), in some cases more specifically within the Cycloneuralia (Dunn *et al.* 2008), although there is not much consensus about the topology at the base of the Ecdysozoa. The tardigrades are sometimes included (Hejnol *et al.* 2009), but this is probably the result of long branch attraction.

Interesting subjects for future research

1. Ultrastructure of *Kinonchulus*

2. Ultrastructure and hormonal regulation of moulting and its evolution in the phylum

3. Reductive evolution of a regulative developmental system to the cell-lineage determined, mosaic system seen in Caenorhabditis

References

Albertson, D.G. and Thomson, J.N. 1976. The pharynx of *Caenorhabditis elegans*. *Phil. Trans. R. Soc. Lond. B* **275**: 299–325.

Altun, Z.F., Chen, B., Wang, Z.-W. and Hall, D.H. 2009. High resolution map of *Caenorhabditis elegans* gap junction proteins. *Dev. Dyn.* **238**: 1936–1950.

Beklemischew, W.N. 1960. *Die Grundlagen der vergleichenden Anatomie der Wirbellosen, 2 vols*. VEB Deutscher Verlag der Wissenschaften, Berlin.

Blaxter, M.L., De Ley, P., Garey, J.R., *et al.* 1998. A molecular evolutionary framework for the phylum Nematoda. *Nature* **392**: 71–75.

Bleidorn, C., Schmidt-Rhaesa, A. and Garey, J.R. 2002. Systematic relationships of Nematomorpha based on molecular and morphological data. *Invert. Biol.* **121**: 357–364.

Boveri, T. 1899. Die Entwickelung von *Ascaris megalocephala* mit besonderer Rücksicht auf die Kernverhältnisse. *Festschrift zum siebenzigsten Geburtstag von Carl von Kupffer*, pp. 383–430. Gustav Fischer, Jena.

Brusca, R.C. and Brusca, G.J. 2003. *Invertebrates, 2nd ed.* Sinauer, Sunderland, MA.

Cunha, A., Azevedo, R.B.R., Emmons, S.W. and Leroi, A.M. 1999. Variable cell number in nematodes. *Nature* **402**: 253.

Dunn, C.W., Hejnol, A., Matus, D.Q., *et al.* 2008. Broad phylogenomic sampling improves resolution of the animal tree of life. *Nature* **452**: 745–749.

Edgecombe, G.D., Giribet, G., Dunn, C.W., *et al.* 2011. Higher-level metazoan relationships: recent progress and remaining questions. *Org. Divers. Evol.* **11**: 151–172.

Félix, M.A., Hill, R.J., Schwarz, H., *et al.* 1999. *Pristionchus pacificus*, a nematode with only three juvenile stages, displays major heterochronic changes relative to *Caenorhabditis elegans*. *Proc. R. Soc. Lond. B* **266**: 1617–1621.

Frand, A.R., Russel, S. and Ruvkun, G. 2005. Functional genomic analysis of *C. elegans* molting. *PLoS Biol.* **3(10)**: e312.

Franke, S. and Käuser, G. 1989. Occurrence and hormonal role of ecdysteroids in non-arthropods. In J. Koolman (ed.): *Ecdysone*, pp. 296–307. Georg Thieme, Stuttgart.

Ghedin, E., Wang, S., Spiro, D., *et al.* 2007. Draft genome of the filarial nematode parasite *Brugia malayi*. *Science* **317**: 1756–1760.

Goldstein, B. and Hird, S.N. 1996. Specification of the anteroposterior axis in *Caenorhabditis elegans*. *Development* **122**: 1467–1474.

Graham, L.D., Kotze, A.C., Fernley, R.T. and Hill, R.J. 2010. An ortholog of the ecdysone receptor protein (EcR) from the parasitic nematode *Haemonchus contortus*. *Mol. Biochem. Parasitol.* **171**: 104–107.

Hejnol, A., Obst, M., Stamatakis, A., *et al.* 2009. Assessing the root of bilaterian animals with scalable phylogenomic methods. *Proc. R. Soc. Lond. B* **276**: 4261–4270.

Holterman, M., van der Wurff, A., van den Elsen, S., *et al.* 2006. Phylum-wide analysis of SSU rDNA reveals deep phylogenetic relationships among nematodes and accelerated evolution toward crown clades. *Mol. Biol. Evol.* **23**: 1792–1800.

Hoschitz, M., Bright, M. and Ott, J.A. 2001. Ultrastructure and reconstruction of the pharynx of *Leptonemella juliae* (Nematoda, Adenophorea). *Zoomorphology* **121**: 95–107.

Howells, R.E. and Blainey, l.J. 1983. The moulting process and the phenomenon of intermoult growth in the filarial nematode Brugia pahangi. *Parasitology* **87**: 493–505.

Hutter, H. and Schnabel, R. 1995. Specification of anterior-posterior differences within the AB lineage in the *C. elegans* embryo: a polarising induction. *Development* **121**: 1559–1568.

Hyman, L.H. 1951. *The Invertebrates, vol. 3. Acanthocephala, Aschelminthes, and Entoprocta. The Pseudocoelomate Bilateria.* McGraw-Hill, New York.

Johnstone, I.L. 1994. The cuticle of the nematode *Caenorhabditis elegans*: a complex collagen structure. *BioEssays* **16**: 171–178.

Kenyon, C. 1985. Cell lineage and the control of *Caenorhabditis elegans* development. *Phil. Trans. R. Soc. Lond. B* **312**: 21–38.

Lafont, R. 1997. Ecdysteroids and related molecules in animals and plants. *Arch. Insect Biochem. Physiol.* **35**: 3–20.

Lafont, R. and Koolman, J. 2009. Diversity of ecdysteroids in animal species. In G. Smagghe (ed.): *Ecdysone: Structures and Functions*, pp. 47–71. Springer, Dordrecht.

Lafont, R. and Mathieu, M. 2007. Steroids in aquatic invertebrates. *Ecotoxicology* **16**: 109–130.

Lorenzen, S. 2007. Nematoda. In W. Westheide and R. Rieger (eds): *Spezielle Zoologie. Teil 1: Einzeller und Wirbellose Tiere*, 2nd ed., pp. 732–751. Elsevier, München.

Malakhov, V.V. 1998. Embryological and histological peculiarities of the order Enoplida, a primitive group of nematodes. *Russ. J. Nematol.* **6**: 41–46.

Meldal, B.H.M., Debenhamb, N.J., De Ley, P., *et al.* 2007. An improved molecular phylogeny of the Nematoda with special emphasis on marine taxa. *Mol. Phylogenet. Evol.* **42**: 622–636.

Moerman, D.G. and Fire, A. 1997. Muscle: structure, function, and development. In D.L. Riddle (ed.): *C. elegans II*, pp. 417–470. Cold Spring Harbor Laboratory Press, Plainview, NY.

Müller, H. 1903. Beitrag zur Embryonalentwicklung der *Ascaris megalocephala. Zoologica (Stuttg.)* **17**: 1–30.

Neuhaus, B., Bresciani, J. and Peters, W. 1997. Ultrastructure of the pharyngeal cuticle and lectin labelling with wheat germ agglutinin-gold conjugate indicatimng chitin in the pharyngeal cuticle of *Oesophagostomum dentatum* (Strongylida, Nematoda). *Acta Zool. (Stockh.)* **78**: 205–213.

Oren-Suissa, M., Hall, D.H., Treinin, M., Shemer, G. and Podbilewicz, B. 2010. The fusogen EFF-1 controls sculpting of mechanosensory dendrites. *Science* **328**: 1285–1288.

Pick, K.S., Philippe, H., Schreiber, F., *et al.* 2010. Improved phylogenomic taxon sampling noticeably affects nonbilaterian relationships. *Mol. Biol. Evol.* **27**: 1983–1987.

Poinar, G., Kerp, H. and Hass, H. 2008. *Palaeonema phyticum* gen. n., sp. n. (Nematoda: Palaeonematidae fam. n.), a Devonian nematode associated with early land plants. *Nematology* **10**: 9–14.

Reisinger, E. 1972. Die Evolution des Orthogons der Spiralier und das Archicoelomatenproblem. *Z. Zool. Syst. Evolutionsforsch.* **10**: 1–43.

Schierenberg, E. 1997. Nematodes, the roundworms. In S.E. Gilbert and A.M. Raunio (eds): *Embryology. Constructing the Organism*, pp. 131–148. Sinauer Associates, Sunderland, MA.

Schierenberg, E. 2005. Unusual cleavage and gastrulation in a freshwater nematode: developmental and phylogenetic implications. *Dev. Genes Evol.* **215**: 103–108.

Schierenberg, E. and Schulze, J. 2008. Many roads lead to Rome: different ways to construct a nematode. In A.

Minelli and G. Fusco (eds): *Evolving Pathways. Key Themes in Evolutionary Developmental Biology*, pp. 261–280. Cambridge University Press, Cambridge.

Schnabel, R., Hutter, H., Moerman, D. and Schnabel, H. 1997. Assessing normal embryogenesis in *Caenorhabditis elegans* using a 4D microscope: variability of development and regional specification. *Dev. Biol.* **184**: 234–265.

Schulze, J. and Schierenberg, E. 2009. Embryogenesis of *Romanomermis culicivorax*: An alternative way to construct a nematode. *Dev. Biol.* **334**: 10–21.

Singh, R.N. and Sulston, J.E. 1978. Some observations on moulting in *Caenorhabditis elegans*. *Nematologica* **24**: 63–71.

Steiner, G. 1919. Untersuchungen über den allgemeinen Bauplan des Nematodenkörpers. *Zool. Jahrb., Anat.* **43**: 1–96.

Sternberg, P.W. and Horvitz, H.R. 1982. Postembryonal nongonadal cell lineages of the nematode *Panagrellus redivivus*: Description and comparison with those of *Caenorhabditis elegans. Dev. Biol.* **93**: 181–205.

Stretton, A.O.V., Fishpool, R.M., Southgate, E., *et al.* 1978. Structure and physiological activity of the motoneurons of the nematode Ascaris *Proc. Natl. Acad. Sci. USA* **75**: 3493–3497.

Sulston, J.E., Schierenberg, E., White, J.G. and Thomson, J.N. 1983. The embryonic cell lineage of the nematode *Caenorhabditis elegans. Dev. Biol.* **100**: 64–119.

Tobler, H., Etter, A. and Müller, F. 1992. Chromatin diminution in nematode development. *Trends Genet.* **8**: 427–432.

Voronov, D.A. and Panchin, Y.V. 1998. Cell lineage in marine nematode *Enoplus brevis. Development* **125**: 143–150.

Ward, S., Thompson, N., White, J.G. and Brenner, S. 1975. Electron microscopical reconstruction of the anterior sensory anatomy of the nematode *Caenorhabditis elegans. J. Comp. Neurol.* **160**: 313–338.

White, J.G., Southgate, E., Thomson, J.N. and Brenner, S. 1986. The structure of the nervous system of the nematode *Caenorhabditis elegans. Phil. Trans. R. Soc. Lond. B* **314**: 1–340.

Wiegner, O. and Schierenberg, E. 1999. Regulative development in a nematode embryo: a hierarchy of cell fate transformations. *Dev. Biol.* **215**: 1–12.

Wolgemuth, C.W., Miao, L., Vanderlinde, O., Roberts, T. and Oster, G. 2005. MSP dynamics drives nematode sperm locomotion. *Biophys. J.* **88**: 2462–2471.

Wright, K.A. 1991. Nematoda. In F.W. Harrison (ed.): *Microscopic Anatomy of Invertebrates*, vol. 4, pp. 111–195. Wiley-Liss, New York.

Zmoray, I. and Guttekova, A. 1969. Ecological conditions for occurrence of cilia in intestines of nematodes. *Biologia (Bratisl.)* **24**: 97–112.

zur Strassen, O. 1896. Embryonalentwicklung der *Ascaris megalocephala. Arch Entwicklungsmech. Org.* **3**: 27–105.

zur Strassen, O. 1906. Die Geschichte der T-Riesen von *Ascaris megalocephala* als Grundlage zu einer Entwicklungsmechanik dieser Species. *Zoologica (Stuttg.)* **17(40)**: 1–342.

zur Strassen, O. 1959. Neue Beiträge zur Entwicklungsgeschichte der Nematoden. *Zoologica (Stuttg.)* **38(107)**: 1–142.

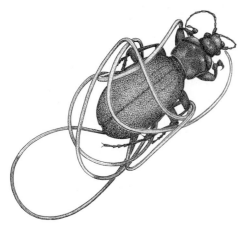

Phylum **Nematomorpha**

Nematomorpha, or hair worms, is a small phylum of nematode-like parasites that spend the larval stage in the body cavities of arthropods, while the adult sexual stage is free and non-feeding, but can nevertheless live for several months. About 325 species have been described, representing two orders: the marine Nectonematoidea, with the only genus *Nectonema* with five species (Poinar and Brockerhoff 2001), and the limnic-terrestrial Gordioidea. *Nectonema* larvae parasitize decapod crustaceans, and the adult stage is pelagic with dorsal and ventral double rows of swimming bristles. The gordioid larvae are found in insects, and the adults crawl in or cling to the vegetation. Some Cambrian fossils resemble nematomorphs, but most of these forms are probably cycloneuralian stem groups (see Chapter 47). The oldest unequivocal record is of a gordioid from the Early Cretaceous amber (Poinar and Buckley 2006).

The adult body is an extremely slender cylinder, in some species more than a metre long. The anterior end is rounded without any appendages in gordioids, with a small mouth opening at the ventral side in some species, whereas other species lack a mouth; the pharynx or oesophagus is a solid strand of cuticle in most species. *Nectonema* has a minute buccal cavity with a pair of teeth leading to a narrow, cuticularized pharynx/oesophagus, which shows only one cell in cross section. The intestine is a narrow tube of monolayered

epithelial cells with microvilli; it opens into a cuticle-lined cloaca. It is clear that nutrient uptake does not take place through the digestive system, but the intestine is involved in storage of substances taken up through the cuticle.

The microscopic anatomy was reviewed by Bresciani (1991) and Schmidt-Rhaesa (2005). The body wall consists of a monolayered ectoderm, covered by a cuticle with several layers, and a layer of longitudinal muscle cells. The larval cuticle consists of a thin outer glycocalyx, a thin osmiophilic layer, a homogeneous layer, and a fibrillar layer (Schmidt-Rhaesa 1996b, 2005). The adult cuticle comprises a thin outer osmiophilic zone and a thick inner zone of thick collagenous fibres in layers of alternating orientation parallel to the surface (Schmidt-Rhaesa 2005; Schmidt-Rhaesa and Gerke 2006). The outer layer is often called epicuticle, but it does not have the trilaminar structure found, for example, in nematodes. Chitin has been detected in the cuticle of juveniles, but not in adults of *Nectonema* (Neuhaus *et al.* 1996). The natatory bristles of *Nectonema* consist of parallel fibrils and are loosely attached to the outer layer of the body cuticle (Schmidt-Rhaesa 1996b). The ectoderm is thin except for a dorsal longitudinal thickening or cord; *Nectonema* has both a dorsal and a ventral cord. The layer of longitudinal muscles is interrupted at the cords.

Chapter vignette: *Polygordius aquaticus* just emerged from its host. (Redrawn from Bresciani 1991.)

The gordioid nervous system (Schmidt-Rhaesa 1996a) comprises a brain surrounding the pharynx and having a large, dorsal organ that has been interpreted as an eye, a ventral nerve cord, and an anal ganglion. The whole nervous system is basiepithelial, but the ventral nerve cord moves into the mesenchyme surrounded by a mid-ventral extension of the ectodermal basal membrane (Schmidt-Rhaesa 2005). The larval brain of *Nectonema* surrounds the pharynx, but the dorsal part is reduced in the adult, which has four giant nerve cells with axons to the ventral cord (Schmidt-Rhaesa 1996a,b). Axons with neurosecretory vesicles in contact with the basal side of the muscles have been observed in *Nectonema* (Schmidt-Rhaesa 1996a); in *Gordius*, thin extensions from the basal side of the muscle cells along the basement membrane could form the connection to the nerve cord (Schmidt-Rhaesa 1998).

The muscles are very unusual, with a peripheral layer of very thick paramyosin filaments with a unique organization and thin (?actin) filaments (Schmidt-Rhaesa 1998). The muscle cells are completely surrounded by extracellular matrix and are anchored to the cuticle via hemidesmosomes and tonofibrils in the ectoderm. The thick, fibrous cuticle and the longitudinal muscles work together as a hydrostatic skeleton around the pseudocoel.

The gordioids have the primary body cavity almost completely filled with a mesenchymatous tissue; narrow pseudocoelomic canals surround the intestine and the gonads. *Nectonema* has a more spacious pseudocoel that is divided into a small cephalic chamber and a long body cavity by a septum (Schmidt-Rhaesa 1996b).

Haemal systems and excretory organs are absent.

Paired, sac-shaped gonads are situated in the pseudocoel, surrounded by a continuous epithelium, and open into the cloaca. Ripe sperm in the testes of *Gordius* are rod-shaped, but the shape changes when the sperm is deposited on the female and enters the receptacle, where it finally becomes filiform with an acrosome at the tip of a flexible anterior part and a cylindrical nucleus in the rod-shaped posterior end (Schmidt-Rhaesa 1997; De Villalobos *et al.* 2005).

The gordioid eggs are fertilized in the uterus and deposited in strings. Cleavage is total and usually equal, and there is much variation in the arrangement of the blastomeres (Inoue 1958). A coeloblastula is formed, and in *Gordius* it soon becomes filled with a compact mass of cells (Mühldorf 1914; Malakhov and Spiridonov 1984). A thick-walled, shallow, anterior invagination becomes the introvert, and a more thin, posterior invagination becomes the gland (see below) and the intestine (Montgomery 1904; Inoue 1958). The cuticular lining of the anterior invagination indicates that it represents the stomodaeum. The full-grown larva (Zapotosky 1974, 1975; Müller *et al.* 2004; Jochmann and Schmidt-Rhaesa 2007) has an anterior, preseptal region with the introvert, and a posterior, postseptal region; both regions have an annulated epidermis with a thick cuticle. The two body regions are separated by the septum, which is a diaphragm of six mesodermal cells surrounded by a thickened basal membrane only pierced by the pharynx (oesophagus) cell. The introvert carries three rings each with six cuticular spines and a cylindrical mouth cone (proboscis), which can be withdrawn but not inverted (Fig. 47.4). The mouth cone has a smooth cuticle with three longitudinal cuticular rods, which fuse anteriorly around the narrow mouth opening. Protrusion of the introvert and mouth cone appears to be caused by contraction of parietal muscles from the septum to the body wall just behind the introvert; retractor muscles from the septum insert on the introvert in front of the anterior ring of spines and inside the mouth cone. One epithelial cell, containing a contorted cuticular canal, extends from the mouth opening to the septum, where the canal continues through two postseptal cells to a gland consisting of eight large cells; the function of gland + duct has not been demonstrated. The postseptal region contains the gland, an 'intestine' with a short rectum, six longitudinal muscles, and a number of undifferentiated cells. The intestine contains a number of granules of unknown function, but Dorier (1930) observed that the content is given off and suggested that it is used in cyst formation. Mühldorf (1914) and Malakhov and Spiridonov (1984) observed a ventral double row of ectodermal nuclei that were supposed to represent the ventral nervous cord. The larva hatches and must penetrate the body wall or gut of a host by means of the introvert and mouth cone in order to develop further; this may happen directly or, perhaps more normally, after a period of encystment, where the cysts become ingested by the host (Dorier 1930) or by an intermediate host that is in turn swallowed by the final host (Hanelt *et al.* 2005). The following development takes place in the body cavity of the arthropod. The introvert disappears, and the preseptal region diminishes while the postseptal region grows enormously, so that the preseptal region of the adult is represented only by the anterior, hemispherical calotte. The sexually mature specimens break out of the host, usually through the anal region (Dorier 1930).

Huus (1931) observed copulation and subsequent spawning of eggs of *Nectonema*, but the eggs did not develop. The youngest larva observed was already inside the crustacean host; it was cylindrical with a short introvert with two rings of six hooks each and a pair of anterior spines. The drawings show a gut without a lumen in the anterior end, but a somewhat older stage, still with hooks on the introvert, showed the thin, curved cuticle-lined oesophagus observed in the adults. Schmidt-Rhaesa (1996b) observed stages of different age inside the host. The youngest stage had the septum in contact with the anterior body wall, so the cerebral cavity develops later. The adult cuticle including the natatory bristles underneath the larval cuticle was described; only one moult was observed.

The nematomorphs are most probably a monophyletic group. They have been interpreted as specialized nematodes, perhaps derived from the parasitic mermithoids, which have a similar life cycle with juveniles parasitizing arthropods and free-living adults that do not feed, and with a thin pharynx without connection to the intestine. However, the mermithoids have the usual nematode sensillae and amphid, ovaria with an anterior and a posterior branch extending from the median genital opening, the four moults, and no trace of an introvert, so this possibility does not seem convincing (Lorenzen 1985; Schmidt-Rhaesa 1996b).

Molecular phylogenetic studies support the monophyly of the phylum and its sister-group relationship with the nematodes (Chapter 48).

Interesting subjects for future research

1. Embryology
2. Number of moults in *Nectonema*
3. Structure of the brain and the putative eye

References

Bresciani, J. 1991. Nematomorpha. In F.W. Harrison (ed.): *Microscopic Anatomy of Invertebrates*, vol. 4, pp. 197–218. Wiley-Liss, New York.

De Villalobos, C., Restelli, M., Schmidt-Rhaesa, A. and Zanca, F. 2005. Ultrastructural observations of the testicular epithelium and spermatozoa of *Pseudochordodes bedriagae* (Gordiida, Nematomorpha). *Cell Tissue Res.* **321**: 251–255.

Dorier, A. 1930. Recherches biologiques et systématiques sur les Gordiacés. *Trav. Lab. Hydrobiol. Piscic. University Grenoble* **12**: 1–180.

Hanelt, B., Thomas, F. and Schmidt-Rhaesa, A. 2005. Biology of the phylum Nematomorpha. *Adv. Parasitol.* **59**: 243–305.

Huus, J. 1931. Über die Begattung bei *Nectonema munidae* Br. und über den Fund der Larve von dieser Art. *Zool. Anz.* **97**: 33–37.

Inoue, I. 1958. Studies on the life history of *Chordodes japonensis*, a species of Gordiacea. I. Development and structure of the larva. *Jap. J. Zool.* **12**: 203–218.

Jochmann, R. and Schmidt-Rhaesa, A. 2007. New ultrastructural data from the larva of *Paragordius varius* (Nematomorpha). *Acta Zool. (Stockh.)* **88**: 137–144.

Lorenzen, S. 1985. Phylogenetic aspects of pseudocoelomate evolution. In S. Conway Morris (ed.): *The Origins and Relationships of Lower Invertebrates*, pp. 210–223. Oxford University Press, Oxford.

Malakhov, V.V. and Spiridonov, S.E. 1984. The embryogenesis of *Gordius* sp. from Turkmenia, with special reference to the position of the Nematomorpha in the animal kingdom. *Zool. Zh.* **63**: 1285–1296.

Montgomery, T.H. 1904. The development and structure of the larva of *Paragordius*. *Proc. Acad. Sci. Nat. Phila.* **56**: 738–755.

Mühldorf, A. 1914. Beiträge zur Entwicklungsgeschichte und zu den phylogenetischen Beziehungen der Gordiuslarve. *Z. Wiss. Zool.* **111**: 1–75.

Müller, M.C.M., Jochmann, R. and Schmidt-Rhaesa, A. 2004. The musculature of horsehair worm larvae (*Gordius aquaticus*, *Paragordius varius*, Nematomorpha): F-actin staining and reconstruction by cLSM and TEM. *Zoomorphology* **123**: 54–54.

Neuhaus, B., Kristensen, R.M. and Lemburg, C. 1996. Ultrastructure of the cuticle of the Nemathelminthes and electron microscopical localization of chitin. *Verh. Dtsch. Zool. Ges.* **89(1)**: 221.

Poinar, G.J. and Brockerhoff, A.M. 2001. *Nectonema zealandica* n. sp. (Nematomorpha: Nectonematoidea) parasitising the purple rock crab *Hemigrapsus edwardsi* (Brachyura: Decapoda) in New Zealand, with notes on the prevalence of infection and host defence reactions. *Syst. Parasitol.* **50**: 149–157.

Poinar, G.J. and Buckley, R. 2006. Nematode (Nematoda: Mermithidae) and hairworm (Nematomorpha: Chordodidae) parasites in Early Cretaceous amber. *J. Invertebr. Pathol.* **93**: 36–41.

Schmidt-Rhaesa, A. 1996a. The nervous system of *Nectonema munidae* and *Gordius aquaticus*, with implications for the ground pattern of the Nematomorpha. *Zoomorphology* **116**: 133–142.

Schmidt-Rhaesa, A. 1996b. Ultrastructure of the anterior end in three ontogenetic stages of *Nectonema munidae* (Nematomorpha). *Acta Zool. (Stockh.)* **77**: 267–278.

Schmidt-Rhaesa, A. 1997. Ultrastructural observations of the male reproductive system and spermatozoa of *Gordius aquaticus* L., 1758. *Invertebr. Reprod. Dev.* **32**: 31–40.

Schmidt-Rhaesa, A. 1998. Muscular ultrastructure in *Nectonema munidae* and *Gordius aquaticus* (Nematomorpha). *Invertebr. Biol.* **117**: 37–44.

Schmidt-Rhaesa, A. 2005. Morphogenesis of *Paragordius varius* (Nematomorpha) during the parasitic phase. *Zoomorphology* **124**: 33–46.

Schmidt-Rhaesa, A. and Gerke, S. 2006. Cuticular ultrastructure of *Chordodes nobilii* Camerano, 1901, with a comparison of cuticular ultrastructure in horsehair worms (Nematomorpha). *Zool. Anz.* **245**: 269–276.

Zapotosky, J.E. 1974. Fine structure of the larval stage of *Paragordius varius* (Leidy, 1851) (Gordioidea: Paragordiidae). I. The preseptum. *Proc. Helminthol. Soc. Wash.* **41**: 209–221.

Zapotosky, J.E. 1975. Fine structure of the larval stage of *Paragordius varius* (Leidy, 1851) (Gordioidea: Paragordiidae). II. The postseptum. *Proc. Helminthol. Soc. Wash.* **42**: 103–111.

51

SCALIDOPHORA

Priapula, Kinorhyncha, and Loricifera are now usually treated together as a monophyletic group called Scalidophora (Lemburg 1995; Ehlers *et al.* 1996; Schmidt-Rhaesa 1996; Neuhaus and Higgins 2002). The three phyla share a number of characters that can best be interpreted as apomorphies: (1) rings of scalids on introvert; (2) flosculi; and (3) two rings of introvert retractors attached through the collar-shaped brain (Fig. 47.3). The variation between the three groups is so small compared with the enormous variation in phyla such as molluscs and urochordates, that they might well be treated as three classes of a phylum Scalidophora.

Malakhov (1980) introduced the name Cephalorhyncha for a group consisting of priapulans, kinorhynchs, and nematomorphs; the loriciferans were added later (Adrianov *et al.* 1990; Malakhov and Adrianov 1995). However, the similarity between, for example, the proboscis with armature in the embryological stages of kinorhynchs and nematomorphs (Adrianov and Malakhov 1994) can be interpreted as a shared character of all the cycloneuralians, and the collagenous nature of the nematomorph cuticle (Chapter 50) indicates a closer relationship with the Nematoda.

The cuticle consists of three layers, a trilaminar epicuticle without chitin, an exocuticle consisting mainly of proteins, but containing some chitin in kinorhynchs, and a chitin-containing endocuticle (Schmidt-Rhaesa 1998). The epicuticle resembles that of nematodes and nematomorphs, which together are regarded as the sister group of the Scalidophora; the body cuticle of these two groups is generally collagenous without chitin, but chitin is found in the pharynx of some nematodes and also in their egg shells (Chapter 49) and in nematomorph larvae (Chapter 50).

Characteristic chitinous spines or scales around the introvert, called scalids, form variations over a pentagonal pattern in kinorhynchs, a hexagonal pattern in loriciferans, and is more variable in priapulans; they contain cilia from one or more monociliate sensory cells and are not observed in any other cycloneuralian group. Their shape is highly variable, as indicated by the names of the different types, spinoscalids, trichoscalids, clavoscalids etc., but their essential structure is identical (Chapters 52–54). Nematodes and nematomorphs have cuticular sense organs called sensilla that consist of three cell types (Chapter 49); they may protrude as papillae or spines, but their special structure with the three cell types does not resemble that of the scalids, and their cuticular parts are collagenous. Sensory pits or flosculi, with a sheath cell surrounding one or more sensory cells with a cilium surrounded by microvilli, are also an autapomorphy of the Scalidophora (Lemburg 1995).

The retractor muscles of the introvert show a highly characteristic pattern. There are two rings of muscles that both attach to the ectoderm around the mouth or at the base of the mouth cone; the inner ring of muscles runs along the inner side of the brain ring, in some cases penetrating its anterior part, and the outer ring runs along the outside of the brain (Fig. 47.4). Anterior-muscle attachments penetrating the brain with elongate, tonofibril-containing ectodermal cells (sometimes called tanycytes) have been observed in all three phyla (Kristensen and Higgins 1991), but their position has only been documented in a few species.

Several priapulan-like fossils are known from Chienjiang (Maas *et al.* 2007) and Burgess Shale (Briggs *et al.* 1994). The Middle Cambrian *Orstenoloricus* is interpreted as a stem-group loriciferan (Maas *et al.* 2009), and loricate larvae of priapulan-loriciferan stem group(s) have been reported from Sirius Passet (Peel 2010) and Australian Middle Cambrian (Maas *et al.* 2009). No kinorhynch fossils have been reported. A number of Cambrian fossils earlier classified as nematomorphs, based on similarities with larvae of living nematomorphs, are now interpreted as priapulans (Maas *et al.* 2007).

The interrelationships of the three phyla have been discussed by several recent papers based on morphology (for example Neuhaus and Higgins 2002; Lorenzen 2007). A sister-group relationship between priapulans and loriciferans is supported by the presence of a lorica in several larval stages (Maas *et al.* 2007; Maas *et al.* 2009), and the Cambrian *Sirilorica* was interpreted as a stem group of these two phyla (Peel 2010). However, Kristensen (1983) and Higgins and Kristensen (1986) emphasize the presence of a non-inversible mouth cone and detailed similarities between several types of scalids in loriciferans and kinorhynchs and regard these two phyla as sister groups. The loriciferan pharynx has a myoepithelium, whereas kinorhynchs and priapulans have exclusively mesodermal musculature. The myoepithelial pharynx has often been interpreted as a symplesiomorphy of the aschelminths, but it appears more probable that this type of suction pump has evolved independently a number of times.

Kinorhynchs and priapulans show up as sister groups in several molecular analyses (Todaro *et al.* 2006; Gerlach *et al.* 2007; Dunn *et al.* 2008; Sørensen *et al.* 2008; Paps *et al.* 2009a, 2009b), but loriciferans are only very rarely included. The 18S rRNA studies of Sørensen *et al.* (2008) showed the Loricifera as sister group of the Nematomorpha and quite distant from the Kinorhyncha, but this finds no support from morphology. A study of mitochondrial and nuclear genes of *Priapulus* (Webster *et al.* 2006) indicated that the priapulan genome is highly conserved within the ecdysozoans. A study of mitochondrial gene order of *Priapulus* (Webster *et al.* 2007) showed highest similarity with chelicerates (Xiphosura and Acari), but the mitochondrial analyses often give unexpected results.

The non-inversible mouth cone of kinorhynchs and loriciferans indicates a sister-group relationship (Fig. 47.3), but the three groups may also be seen as an unresolved trichotomy.

References

Adrianov, A.V. and Malakhov, V.V. 1994. *Kinorhyncha: Structure, Development, Phylogeny and Taxonomy.* Nauka, Moscow.

Adrianov, A.V., Malakhov, V.V. and Yushin, V.V. 1990. Loricifera—a new taxon of marine invertebrates. *Sov. J. Mar. Biol.* **15**: 136–138.

Briggs, D.E.G., Erwin, D.H. and Collier, F.J. 1994. *The Fossils of the Burgess Shale.* Smithsonian Institution Press, Washington.

Dunn, C.W., Hejnol, A., Matus, D.Q., *et al.* 2008. Broad phylogenomic sampling improves resolution of the animal tree of life. *Nature* **452**: 745–749.

Ehlers, U., Ahlrichs, W., Lemburg, C. and Schmidt-Rhaesa, A. 1996. Phylogenetic systematization of the Nemathelminthes (Aschelminthes). *Verh. Dtsch. Zool. Ges.* **89**: 8.

Gerlach, D., Wolf, M., Dandekar, T., *et al.* 2007. Deep metazoan phylogeny. *In Silico Biol.* **7**: 0015.

Higgins, R.P. and Kristensen, R.M. 1986. New Loricifera from southeastern United States coastal waters. *Smithson. Contr. Zool.* **438**: 1–70.

Kristensen, R.M. 1983. Loricifera, a new phylum with Aschelminthes characters from the meiobenthos. *Z. Zool. Syst. Evolutionsforsch.* **21**: 163–180.

Kristensen, R.M. and Higgins, R.P. 1991. Kinorhyncha. In F.W. Harrison (ed.): *Microscopic Anatomy of Invertebrates,* vol. 4, pp. 377–404. Wiley-Liss, New York.

Lemburg, C. 1995. Ultrastructure of sense organs and receptor cells of the neck and lorica of the *Halicryptus spinulosus* larva (Priapulida). *Microfauna Mar.* **10**: 7–30.

Lorenzen, S. 2007. Nemathelminthes i.e.s. In W. Westheide and R. Rieger (eds): *Spezielle Zoologie. Teil 1: Einzeller und Wirbellose Tiere,* 2nd ed., pp. 725–763. Elsevier/Spektrum, München.

Maas, A., Huang, D., Chen, J., Waloszek, D. and Braun, A. 2007. Maotianshan-Shale nemathelminths—morphology, biology, and the phylogeny of Nemathelminthes. *Palaeogeogr. Palaeoclimatol. Palaeoecol.* **254**: 288–306.

Maas, A., Waloszek, D., Haug, J.T. and Müller, K.J. 2009. Loricate larvae (Scalidophora) from the Middle Cambrian of Australia. *Memoirs of the Association of Australasian Palaeontologists* **37**: 281–302.

Malakhov, V.V. 1980. Cephalorhyncha, a new type of animal kingdom uniting Priapulida, Kinorhyncha, Gordiacea, and a system of Aschelminthes worms. *Zool. Zh.* **59**: 485–499.

Malakhov, V.V. and Adrianov, A.V. 1995. *Cephalorhyncha—a new phylum of the Animal Kingdom (In Russian, English summary)*. KMK Scientific Press Ltd, Moscow.

Neuhaus, B. and Higgins, R.P. 2002. Ultrastructure, biology, and phylogenetic relationships of Kinorhyncha. *Integr. Comp. Biol.* **42**: 619–632.

Paps, J., Baguñá, J. and Riutort, M. 2009a. Lophotrochozoa internal phylogeny: new insights from an up-to-date analysis of nuclear ribosomal genes. *Proc. R. Soc. Lond. B* **276**: 1245–1254.

Paps, J., Baguñà, J. and Riutort, M. 2009b. Bilaterian phylogeny: A broad sampling of 13 nuclear genes provides a new Lophotrochozoa phylogeny and supports a paraphyletic basal Acoelomorpha. *Mol. Biol. Evol.* **26**: 2397–2406.

Peel, J.S. 2010. A corset-like fossil from the Cambrian Sirius Passet Lagerstätte of North Greenland and its implications for cycloneuralian evolution. *J. Paleont.* **84**: 332–340.

Schmidt-Rhaesa, A. 1996. *Zur Morphologie, Biologie und Phylogenie der Nematomorpha.* Cuvillier, Göttingen.

Schmidt-Rhaesa, A. 1998. Phylogenetic relationships of the Nematomorpha—a discussion of current hypotheses. *Zool. Anz.* **236**: 203–216.

Sørensen, M.V., Hebsgaard, M.B., Heiner, I., *et al.* 2008. New data from an enigmatic phylum: evidence from molecular data supports a sister-group relationship between Loricifera and Nematomorpha. *J. Zool. Syst. Evol. Res.* **46**: 231–239.

Todaro, M.A., Telford, M.J., Lockyer, A.E. and Littlewood, D.T.J. 2006. Interrelationships of the Gastrotricha and their place among the Metazoa inferred from 18SrRNA genes. *Zool. Scr.* **35**: 251–259.

Webster, B.L., Copley, R.R., Jenner, R.A., *et al.* 2006. Mitogenomics and phylogenomics reveal priapulid worms as extant models of the ancestral ecdysozoan. *Evol. Dev.* **8**: 502–510.

Webster, B.L., Mackenzie-Dodds, J.A., Telford, M.J. and Littlewood, D.T.J. 2007. The mitochondrial genome of *Priapulus caudatus* Lamarck (Priapulida: Priapulidae). *Gene* **389**: 96–105.

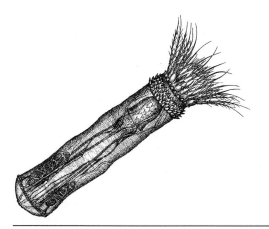

52

Phylum **Priapula**

Priapula is a small phylum comprising a few well-known, marine, macrobenthic genera, such as *Priapulus* and *Halicryptus*, and three meiobenthic genera, *Tubiluchus*, *Meiopriapulus*, and *Maccabeus*, which have been described within the last decades. A total of 20 living species are now recognized; they are placed in three families, sometimes with *Maccabeus* placed in a separate order (Adrianov and Malakhov 1996). A number of genera from the Lower Cambrian bear striking resemblance to some of the living genera with a morphological variation of about the same magnitude as that of the living species (Hou *et al.* 2004). The Cambrian-Silurian palaeoscolecids are now regarded as stem-group priapulans (Harvey *et al.* 2010).

The body consists of a cylindrical trunk and a large introvert; some genera have one or two caudal appendages. The microscopical anatomy was reviewed by Storch (1991). *Maccabeus* (see the chapter vignette) has a double circle of spiny tentacles and a ring of setose spines around the mouth. The introvert carries many rings of different types of scalids mostly with a monociliate sensory cell (Storch *et al.* 1995); the scalids point posteriorly when the introvert is everted. Sensory organs, called flosculi and tubuli, consisting of one or a few monociliate sensory cells surrounded by a sheath cell with cuticular papillae, are found in various parts of the body (Lemburg 1995a,b).

The introvert can be everted by the contraction of the trunk muscles, with the body cavity functioning as a hydrostatic skeleton. Retraction of the introvert is by contraction of two sets of muscles, an outer ring extending from the mouth region to the posterior limit of the introvert, and an inner ring from the mouth region to the mid-region of the trunk (see below and Fig. 47.4). Burrowing through the substratum is by eversion and contraction of the introvert combined with peristaltic movements of the whole body (Hammond 1970). The caudal appendages are extensions of the body wall with musculature and body cavity; they are believed to have respiratory functions.

The ectoderm is monolayered and covered by a chitinous cuticle, consisting of an electron-dense epicuticle, an electron-dense exocuticle, and a fibrillar, electron-lucent endocuticle in *Halicryptus* and *Priapulus* (Saldarriaga *et al.* 1995; Lemburg 1998); the inner zone has a layer of crossing fibres of unknown chemical composition in *Meiopriapulus*. Chitin is found in the endocuticle and in small amounts in the exocuticle of the larval lorica of *Priapulus* (Lemburg 1998). The chitin is of the β-type (Shapeero 1962). The mouth opening is usually surrounded by a narrow field of buccal papillae with many sensory cells and opens into a pharynx with circles of cuticular teeth that lack the cilia found in the scalids; the teeth point towards the midgut when the pharynx is not everted. The macrobenthic species have a large muscular pharynx and a very short oesophagus with a sphincter muscle just in front of the midgut. *Meiopriapulus* and *Tubiluchus* have been described as having a mouth cone and lacking

Chapter vignette: *Maccabeus tentaculatus*. (Redrawn from Por 1972.)

pharyngeal teeth, but as the epithelium with the teeth can be inverted, it is rather the pharynx that can be everted; a non-inversible mouth cone like that of kinorhynchs and loriciferans is not present. *Meiopriapulus* and *Tubiluchus* have the pharynx, followed by a long oesophagus with longitudinal folds and a muscular, gizzard-like swelling, with an anterior ring of small cuticular projections and a posterior ring of long, comb-like plates. The pharynx has longitudinal, circular, and oblique muscles. The larger species are carnivores, which grasp annelids and other prey organisms and ingest them through the action of the pharyngeal teeth (van der Land 1970). *Tubiluchus* and *Meiopriapulus* scrape bacteria and other small organisms from sediment particles (Storch *et al.* 1989; Higgins and Storch 1991), whereas *Maccabeus* apparently swallows larger prey caught by the pharynx, which is then retracted by the unique pharynx retractors to the posterior end of the body (Por and Bromley 1974). It appears that none of the species use the pharynx as a suction pump.

The midgut is a long tube of cells with microvilli; it is surrounded by longitudinal and circular muscles. The rectum has a folded cuticle.

The central nervous system consists of a collar-shaped brain, with an anterior and a posterior ring of perikarya separated by a zone of neuropil, an unpaired ventral cord, and a caudal ganglion ventral to the anus; all these structures are intraepithelial. A number of longitudinal nerves and a series of annular nerves extend along the body. The sense organs of the intro-

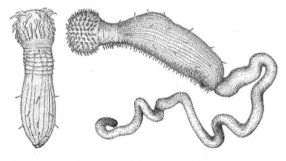

Figure 52.1. Larva and adult male of *Tubiluchus corallicola*; only the male has the cuticular spines on the ventral side of the abdomen. (Redrawn from van der Land 1970 and Calloway 1975.)

vert are innervated from the brain (Rothe and Schmidt-Rhaesa 2010).

There is a rather spacious body cavity filled with a fluid containing erythrocytes and amoebocytes (Schreiber and Storch 1992). There is no peritoneal cell layer in the larger species, and the isolated cells observed at the inner side of the body wall are amoebocytes. *Meiopriapulus* has a coelomic muscular collar around the gut in the introvert.

The body wall of the trunk and introvert has an outer layer of ring muscles and an inner layer of longitudinal muscles. There is a ring of short, outer introvert-retractor muscles extending from the anterior part of the brain to the base of the introvert, and eight long, inner retractor muscles extending from the posterior part of the brain to the middle region of the body wall (Fig. 47.4). The retractor muscles are apposed to the collar-shaped brain and attached to elongate epidermal cells (tanycytes). These cells penetrate the brain and attach to the striated retractor muscles at the Z-bands. The pharynx is surrounded by longitudinal, oblique, and circular muscles. *Maccabeus* has two rings of muscles between the oesophagus and the body wall.

Priapulus has large, branched protonephridia with monociliate terminal cells, which make sinuous weirs between each other. *Tubiluchus* has biciliate terminal cells. The protonephridia and the sac-shaped gonads have common ducts, with a pair of openings lateral to the anus in *Priapulus* and *Tubiluchus*, but with openings in the rectum in *Meiopriapulus*. There are myoepithelial cells in the urogenital organs of *Halicryptus*, particularly at the mesenteria.

Priapulus is a free spawner (Lang 1953), whereas *Tubiluchus* has internal fertilization (Alberti and Storch 1988). The sperm of most genera is of the primitive type with a rounded head, whereas *Tubiluchus* has sperm with an elongate head with a peculiar, spirally coiled acrosome (Storch *et al.* 2000). The first cleavages of *Priapulus* are radial and highly stereotypical. A coeloblastula is formed, and gastrulation is through ingression of the four cells at the blastoporal pole, first two larger cells and then two smaller cells (Wennberg *et al.* 2008). The origin of the mesoderm is unknown. All species except *Meiopriapulus* develop through a characteristic loricate larval stage, with a ring of longitudinal chitinous shields on the trunk (Higgins *et al.* 1993; see Fig. 52.1). The newly hatched larvae of *Halicryptus* and *Priapulus* lack a lorica, but it develops in the later larval stages (Storch and Higgins 1991; Janssen *et al.* 2009; Wennberg *et al.* 2009). All species go through a series of moults, both in the larval stage with the lorica and in the adult stage. The formation of the various cuticular layers during moulting closely resembles that observed in arthropods, tardigrades, and nematodes (Lemburg 1998). The rich Burgess Shale fauna (Hou *et al.* 2004) shows that the living priapulans are the few surviving representatives of a much larger group. Their close relationships with kinorhynchs and loriciferans are indicated by the similarities in the cuticle with scalids, the structure of the introvert with the two rings of retractor muscles (Fig. 47.1), and the structure of the brain with tanycytes.

It is difficult to point to strong apomorphies of the Priapula, but the complicated protonephridia with clusters of terminals is probably one (Neuhaus and Kristensen 2007). Lemburg (1999) adds the presence of a special cuticular layer (exocuticula III) in the larvae, but this needs further study.

The interrelationships of priapulans, kinorhynchs, and loriciferans are discussed in Chapter 51.

Interesting subjects for future research

1. Embryology and postembryonal development of several species
2. Moulting

References

Adrianov, A.V. and Malakhov, V.V. 1996. *Priapulida (Priapulida): structure, development, phylogeny, and classification (In Russian, English summary).* KMK Science Press, Moscow.

Alberti, G. and Storch, V. 1988. Internal fertilization in a meiobenthic priapulid worm: *Tubiluchus philippinensis* (Tubiluchidae, Priapulida). *Protoplasma* **143**: 193–196.

Calloway, C.B. 1975. Morphology of the introvert and associated structures of the priapulid *Tubiluchus corallicola* from Bermuda. *Mar. Biol.* **31**: 161–174.

Hammond, R.A. 1970. The burrowing of *Priapulus caudatus*. *J. Zool.* **162**: 469–480.

Harvey, T.H.P., Dong, X. and Donoghue, P.C.J. 2010. Are palaeoscolecids ancestral ecdysozoans? *Evol. Dev.* **12**: 177–200.

Higgins, R.P. and Storch, V. 1991. Evidence for direct development in *Meiopriapulus fijiensis* (Priapulida). *Trans. Am. Microsc. Soc.* **110**: 37–46.

Higgins, R.P., Storch, V. and Shirley, T.C. 1993. Scanning and transmission electron microscopical observations on the larvae of *Priapulus caudatus* (Priapulida). *Acta Zool. (Stockh.)* **74**: 301–319.

Hou, X.-G., Aldridge, R.J., Bergström, J., *et al.* 2004. *The Cambrian Fossils of Chengjiang, China.* Blackwell, Malden, MA.

Janssen, R., Wennberg, S.A. and Budd, G.E. 2009. The hatching larva of the priapulid worm *Halicryptus spinulosus*. *Front. Zool.* **6**: 8.

Lang, K. 1953. Die Entwicklung des Eies von *Priapulus caudatu* Lam. und die systematische Stellung der Priapuliden. *Ark. Zool.*, 2. Ser. **5**: 321–348.

Lemburg, C. 1995a. Ultrastructure of sense organs and receptor cells of the neck and lorica of the *Halicryptus spinulosus* larva (Priapulida). *Microfauna Mar.* **10**: 7–30.

Lemburg, C. 1995b. Ultrastructure of the introvert and associated structures of the larvae of *Halicryptus spinulosus* (Priapulida). *Zoomorphology* **115**: 11–29.

Lemburg, C. 1998. Electron microscopical localization of chitin in the cuticle of *Halicryptus spinulosus* and *Priapulus caudatus* (Priapulida) using gold-labelled wheat germ agglutinin: phylogenetic implications for the evolution of the cuticle within the Nemathelminthes. *Zoomorphology* **118**: 137–158.

Lemburg, C. 1999. *Ultrastrukturelle Untersuchungen an den Larven von Halicryptus spinulosus und Priapulus caudatus.* Cuvillier, Göttingen.

Neuhaus, B. and Kristensen, R.M. 2007. Ultrastructure of the protonephridia of larval *Rugiloricus* cf. *cauliculus*, male *Armorloricus elegans*, and female *Nanaloricus mysticus* (Loricifera). *J. Morphol.* **268**: 357–370.

Por, F.D. 1972. Priapulida from deep bottoms near Cyprus. *Isr. J. Zool.* **21**: 525–528.

Por, F.D. and Bromley, H.J. 1974. Morphology and anatomy of *Maccabeus tentaculatus* (Priapulida: Seticoronaria). *J. Zool.* **173**: 173–197.

Rothe, B.H. and Schmidt-Rhaesa, A. 2010. Structure of the nervous system in *Tubiluchus troglodytes* (Priapulida). *Invertebr. Biol.* **129**: 39–58.

Saldarriaga, J.F., Voss-Foucart, M.F., Compère, P., *et al.* 1995. Quantitative estimation of chitin and proteins in the cuticle of five species of Priapulida. *Sarsia* **80**: 67–71.

Schreiber, A. and Storch, V. 1992. Free cells and blood proteins of *Priapulus caudatus* Lamarck (Priapulida). *Sarsia* **76**: 261–266.

Shapeero, W.L. 1962. The epidermis and cuticle of *Priapulus caudatus*. *Trans. Am. Microsc. Soc.* **81**: 352–355.

Storch, V. 1991. Priapulida. In F.W. Harrison (ed.): *Microscopic Anatomy of Invertebrates*, vol. 4, pp. 333–350. Wiley-Liss, New York.

Storch, V. and Higgins, R.P. 1991. Scanning and transmission electron microscopic observations on the larva of *Halicryptus spinulosus* (Priapulida). *J. Morphol.* **210**: 175–194.

Storch, V., Higgins, R.P. and Morse, M.P. 1989. Internal anatomy of *Meiopriapulus fijiensis* (Priapulida). *Trans. Am. Microsc. Soc.* **108**: 245–261.

Storch, V., Higgins, R.P., Anderson, P. and Svavarsson, J. 1995. Scanning and transmission electron microscopic analysis of the introvert of *Priapulopsis australis* and *Priapulopsis bicaudatus* (Priapulida). *Invertebr. Biol.* **114**: 64–72.

Storch, V., Kempendorf, C., Higgins, R.P., Shirley, T.C. and Jamieson, B.G.M. 2000. Priapulida. In K.G. Adiyodi, R.G. Adiyodi and B.G.M. Jamieson (eds): *Reproductive Biology of Invertebrates*, vol. 9B, pp. 1–19. John Wiley, Chichester.

van der Land, J. 1970. Systematics, zoogeography, and ecology of the Priapulida. *Zool. Verh. (Leiden)* **112**: 1–118.

Wennberg, S.A., Janssen, R. and Budd, G.E. 2008. Early embryonic development of the priapulid worm *Priapulus caudatus*. *Evol. Dev.* **10**: 326–338.

Wennberg, S.A., Janssen, R. and Budd, G.E. 2009. Hatching and earliest larval stages of the priapulid worm *Priapulus caudatus*. *Invertebr. Biol.* **128**: 157–171.

Phylum **Kinorhyncha**

Kinorhyncha, or mud dragons, is a small phylum comprising about 200 benthic, marine species, which are almost all less than one millimetre long. Two orders are generally recognized, but the group is actually very homogeneous (Adrianov and Malakhov 1999b; Neuhaus and Higgins 2002)

The body consists of an introvert, a neck, and 11 trunk segments, each with a dorsal and one or a pair of ventral cuticular plates. The introvert can be retracted into the anterior part of the trunk; there is a closing structure consisting of a ring of small plates at the neck in the Cyclorhagida, and of two to four dorsal and two to four ventral platelets in the Homalorhagida.

The microscopic anatomy has been reviewed by Kristensen and Higgins (1991), Neuhaus (1994), and Neuhaus and Higgins (2002). The anterior end of the fully extended introvert carries a mouth cone with a ring of nine oral styles surrounding the mouth (Fig. 47.1); the mouth cone is retracted, but not inverted when the introvert retracts. The median part of the introvert carries several rings of scalids with locomotory and sensory functions. The sockets of the scalids contain a number of monociliate sensory cells, with the cilia extending to the tip of the scalids. The introvert is everted by contraction of the body musculature, with the narrow body cavity functioning as a hydrostatic skeleton. Retraction is by contraction of two sets of muscles between the introvert and a number of trunk segments (see below).

The mouth leads to a short buccal cavity (inside the mouth cone) with three or four pentamerous rings of cuticular pharyngeal styles, each with a ciliated receptor cell and a terminal pore, followed by a zone with numerous cuticular fibres. The muscular pharynx is circular or slightly nine lobed in cross-section in the Cyclorhagida, and rounded with a tri-radiate lumen and a ventral and a pair of lateral muscular swellings in the Homalorhagida. The ectoderm of the oral cavity bears several rings of scalids of different types. Strong squeezing may rupture some of the pharyngeal muscles and force an eversion of the pharyngeal wall with the scalids. The muscular pharynx bulb has a layer of ectoderm with a thin cuticle, and the mesodermal musculature comprises both radial and circular muscle cells. The feeding biology is poorly studied, but many of the cyclorhagids swallow diatoms, whereas most of the homalorhagids ingest detritus by opening the mouth and sucking the material with the pharyngeal bulb.

The midgut is a tube of large endodermal cells with microvilli surrounded by a thin layer of longitudinal and circular, mesodermal muscle cells. There is a cuticle-lined rectum.

The monolayered ectoderm is generally covered by a compact, chitinous cuticle without microvilli, but

---

Chapter vignette: *Echinoderes aquilonius*. Disko, Greenland, July 1988. (Drawn from a SEM preparation courtesy of Dr R. M. Kristensen, University Copenhagen, Denmark.)

microvilli have been observed in the cuticle of the pharyngeal crown surrounding the anterior rim of the pharynx bulb and in the cuticular plates of the body. The plates of the trunk have large, ventral apodemes for the attachment of the longitudinal muscles between the segments. Various spines with associated glands occur on the cuticular plates, especially in the posterior end. The central nervous system comprises a collar-shaped brain surrounding the anterior part of the pharynx at the base of the mouth cone. Three regions can be recognized; an anterior region with ten clusters of perikarya, a middle region consisting mainly of neuropil, and a posterior ring with eight or ten clusters of perikarya. The anterior ganglia innervate the mouth cone, which has a small nerve ring with ten small ganglia and the scalids on the introvert. There is a paired, midventral, intraepithelial cord with ganglionic swellings and commissures in each segment, and ending in a ventral anal ganglion; the ventral nerve cord appears to be exclusively motor. Three pairs of longitudinal nerves with segmental ganglia are situated laterally and dorsally. The different types of scalids all have cilia extending to the tip and are believed to be either chemoreceptors or mechanoreceptors, while at the same time functioning in locomotion and perhaps food manipulation. The sensory organs called flosculi, which consist of a monociliate sensory cell with a circle of microvilli surrounded by a cell with a ring of cuticular papillae, and sensory spots, which consist of two different types of monociliate sensory cells surrounded by a sheath cell with an oval field of cuticular papillae, occur in species-specific positions. The mid-dorsal spine on the fourth segment of *Echinoderes capitatus* contains a multiciliary receptor cell. A very special cephalic sensory organ, consisting of two cells surrounding a cavity with a highly modified cilium, has been found in *Pycnophyes*.

The muscles of the body wall are cross striated and attach to the cuticle by way of ectodermal cells with tonofilaments and hemidesmosomes. The longitudinal muscles between the segmental cuticular plates and the dorsoventral muscles are segmentally arranged (Müller and Schmidt-Rhaesa 2003). The narrow primary body cavity contains amoebocytes. A ring of 16 outer retractor muscles extend from the epithelium of the introvert

and along the outer side of the brain, whereas a ring of 12 inner retractor muscles extend from the base of the mouth cone and along the inner side of the brain; both sets of muscles attach to the epithelium of a number of the anterior segments (see also Fig. 47.4). The presence of tanycytes in the brain has been reported, but without indication of their precise position.

A pair of protonephridia opens at the ninth segment. Each consists of a tube with a number of biciliate cells and a microvillous weir.

A pair of sac-shaped gonads opens in a posteroventral pore. Spermatophores have been observed in some genera (Brown 1983), and the presence of seminal receptacles in all species investigated indicates internal fertilization. This fits well with the aberrant type of sperm with a large, sausage-shaped head and a very short cilium (Adrianov and Malakhov 1999a,b). The fertilized eggs of *Echinoderes* are deposited singly and the female actively covers the egg with detritus (Kozloff 2007). The 18-cell stage shows four internal, probably endodermal cells; the orientation could not be recognized. The later embryos curve ventrally and the pharyngeal structures develop. The embryos hatch as small, nine-segmented juveniles (Kozloff 1972). The post-embryonic development of representatives of both orders shows a series of five to six moults, which gradually transform the juvenile into an adult (Higgins 1974; Neuhaus 1995; Sørensen *et al.* 2010).

The 'segmented' musculature and nervous system corresponding to the nine rings of cuticular plates of the trunk clearly demonstrate the monophyly of the group. The arrangement superficially resembles that found in the arthropods (Chapter 41) and some nematodes (Chapter 46), but there is nothing to indicate that the segments arise from a posterior teloblastic zone; a similar 'pseudosegmentation' is seen in the muscles with associated nerve cells in bdelloid rotifers (Chapter 34). The whole structure of the radially symmetrical introvert, of the pharynx with teeth or scalids, and of the collar-shaped brain surrounding the pharynx make thoughts about a closer relationship between the kinorhynchs and arthropods highly improbable. The large introvert with different types of scalids, the chitinous cuticle, the two rings of introvert retractor muscles, and the presence of tanycytes are characters shared with priapulans and loriciferans demonstrating the monophyly of the Scalidophora. The presence of a non-inversible mouth cone with cuticular ridges and spines indicates a sister-group relationship with the loriciferans (Chapter 51).

Interesting subjects for future research

1. Embryology
2. Moulting hormones and secretion of the new cuticle at moulting

References

Adrianov, A.V. and Malakhov, V. 1999a. Kinorhyncha. In K.G. Adiyodi and R.G. Adiyodi (eds): *Reproductive Biology of Invertebrates*, vol. 9A, pp. 193–211. Wiley, Chichester.

Adrianov, A.V. and Malakhov, V.V. 1999b. *Cephalorhyncha of the World Ocean (In Russian, English summary)*. KMK Scientific Press Ltd, Moscow.

Brown, R. 1983. Spermatophore transfer and subsequent sperm development in a homalorhagid kinorhynch. *Zool. Scr.* **12**: 257–266.

Higgins, R.P. 1974. Kinorhyncha. In A.C. Giese (ed.): *Reproduction of Marine Invertebrates*, vol. 1, pp. 507–518. Academic Press, New York.

Kozloff, E.N. 1972. Some aspects of development in *Echinoderes* (Kinorhyncha). *Trans. Am. Microsc. Soc.* **91**: 119–130.

Kozloff, E.N. 2007. Stages of development, from first cleavage to hatching, of an *Echinoderes* (Phylum Kinorhyncha: Class Cyclorhagida). *Cah. Biol. Mar.* **48**: 199–206.

Kristensen, R.M. and Higgins, R.P. 1991. Kinorhyncha. In F.W. Harrison (ed.): *Microscopic Anatomy of Invertebrates*, vol. 4, pp. 377–404. Wiley-Liss, New York.

Müller, M.C.M. and Schmidt-Rhaesa, A. 2003. Reconstruction of the muscle system in *Antygomonas* sp. (Kinorhyncha, Cyclorhagida) by means of phalloidin labelling and cLSM. *J. Morphol.* **256**: 103–110.

Neuhaus, B. 1994. Ultrastructure of alimentary canal and body cavity, ground pattern, and phylogenetic relationships of the Kinorhyncha. *Microfauna Mar.* **9**: 61–156.

Neuhaus, B. 1995. Postembryonic development of *Paracentrophyes praedictus* (Homalorhagida): neoteny questionable among the Kinorhyncha. *Zool. Scr.* **24**: 179–192.

Neuhaus, B. and Higgins, R.P. 2002. Ultrastructure, biology, and phylogenetic relationships of Kinorhyncha. *Integr. Comp. Biol.* **42**: 619–632.

Sørensen, M.V., Accogli, G. and Hansen, J.G. 2010. Postembryonic development of *Antygomonas incomitata* (Kinorhyncha: Cyclorhagida). *J. Morphol.* **271**: 863–882.

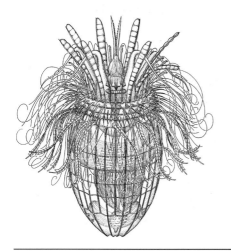

54

Phylum **Loricifera**

The newly discovered phylum Loricifera now comprises nine described genera with a total of about 30 species, but new types continue to be discovered. The microscopic animals, body length only 100–500 μm for most species with one species reaching about 800 μm, occur in marine sediments of all types, from coarse, shallow-water sediments to fine, deep-sea mud. Even completely anoxic sediments are inhabited by a number of species (Danovaro *et al.* 2010).

The body consists of a trunk with an exoskeleton called the lorica and an introvert that can be invaginated into the anterior part of the trunk; the mouth is situated at the tip of a non-inversible mouth cone (Figs. 47.1,2,4 and 54.1). The anatomy was reviewed by Kristensen (1991a, 2002).

The complicated mouth cone has stiffening stylets or ridges. The introvert proper carries nine circles of scalids of different types, totalling 285 in males of *Nanaloricus*. The scalids contain several monociliate sensory cells; the spinoscalids have intrinsic muscles. The trichoscalids on the neck contain one cilium, arise from a socket in one or two small cuticular plates, and can be moved by a pair of small muscles at the socket. The trichoscalids of *Pliciloricus* are so thin that they may act as locomotory cilia (see the chapter vignette). Other types are flattened, clavate, shaped like a pea-pod, spiny or hairy, branched, or articulated. The neck may have rows of cuticular plates with additional scalids.

The trunk is covered by the lorica that consists of 6–30 longitudinal cuticular plates. The plates are rather stiff in *Nanaloricus*, but those of *Pliciloricus* and *Rugiloricus* are very thin and may form longitudinal folds. The posterior end of many species shows a number of smaller plates surrounding the anus.

In *Nanaloricus* and *Armoloricus* (Kristensen and Gad 2004), the tip of the mouth cone continues in a hexagonal, telescoping mouth tube with six rows of small cuticular teeth, which in turn continues in a long, flexible, heavily cuticularized, and therefore non eversible, buccal canal. The buccal canal passes through the collar-shaped brain and ends in a triradiate, myoepithelial pharynx bulb, followed by a short oesophagus. The pharynx lumen is hexaradiate with weak radiating muscles and large gland cells. The midgut consists of large cells with microvilli, and the rectum has a thin cuticle. *Rugiloricus* (Heiner 2008) has a less complicated mouth cone without an eversible mouth tube; the mid-ventral pair of anterior scalids are more or less fused and have the shape of serrated stylets. The ectoderm is monolayered and covered by a compact cuticle, which consists of a trilaminate epicuticle, one to three amorphous layer(s), and a basal fibrillar layer that contains chitin (Neuhaus *et al.* 1997).

The central nervous system is intraepithelial and comprises a large brain surrounding the gut in the anterior part of the introvert, and a number of longitudinal

Chapter vignette: *Pliciloricus enigmaticus*. (Modified from Higgins and Kristensen 1986; courtesy of Dr R.M. Kristensen.)

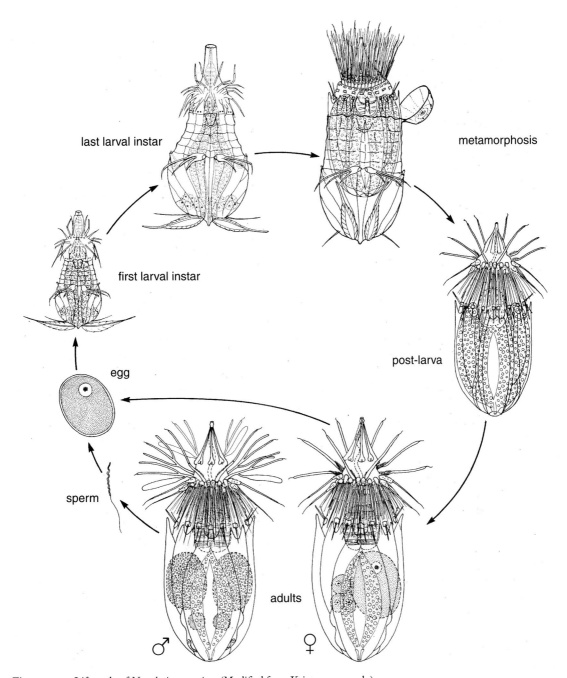

last larval instar

metamorphosis

first larval instar

post-larva

egg

sperm

adults

♂   ♀

Figure 54.1. Life cycle of *Nanaloricus mysticus*. (Modified from Kristensen 1991b.)

nerves. The brain is collar-shaped with anterior and posterior concentrations of ganglionic cell bodies and a median zone of neuropil; it is insunk from the epithelium but surrounded by an extension of the basal membrane, as shown by the ring-shaped connection between the two structures. The cells in the forebrain are concentrated in a ring of eight ganglia that innervate the scalids and the mouth cone; the cells in the hindbrain are arranged in a ring of ten ganglia connected with ten longitudinal nerve cords. The ventral pair of nerve cords is larger than the others and pass through a subpharyngeal ganglion and a number of double ganglia to a paired caudal ganglion. The brain is surrounded by circular muscles, and special epidermal cells with long tonofilament bundles attach to the circular muscles and to the head retractors.

The muscular system consists of a high number of small muscles, many comprising only one or two myomeres, but not all of the muscles have been described. A varying number (15–24) of outer introvert retractors extend from the base of the mouth cone through the brain to the body wall at the transition between the introvert and neck. Five inner introvert retractors extend from the base of the mouth cone to the posterior part of the neck. The adult *Nanaloricus* has a system of six muscles situated in and behind the brain, with long tendons to the mouth region; these muscles can retract the mouth tube (Fig. 47.4). Small ring muscles are situated along some of the rings of scalids and between the plates of the lorica. *Nanaloricus*, with only six plates in the lorica, has more individualized muscles between the plates.

The sexes are separate, and there are obvious differences between the numbers and types of scalids between the sexes. There is a pair of gonads with protonephridia opening into the gonadal lumen. The terminal cells are monociliate (Neuhaus and Kristensen 2007). The sperm has a rod-shaped head and a long cilium (Heiner *et al.* 2009).

The embryology has not been described. *Nanaloricus*, which is apparently one of the most specialized genera, has a rather simple life cycle (Fig. 54.1). Its first postembryonic stage is the so-called Higgins larva, which has the general anatomy of the adult, but with fewer scalids. Its most characteristic structure is a pair of large, pointed moveable spines, situated in the posterior end; these appendages are wide and flipper-like and can be used in swimming. Other genera, such as *Armorloricus* (Heiner and Kristensen 2009) and especially *Pliciloricus* (Heiner 2008), have more complicated life cycles with various types of paedogenetic, viviparous development often inside an exuvium. There are varying numbers of moults, and the whole life cycle is apparently still incompletely known in several species. The exuvia comprise all the cuticularized regions and show the armature of the pharynx/oesophagus especially clearly.

The feeding biology is unknown, but either a very narrow mouth, surrounded by spines is present, or there is an eversible mouth tube with small teeth, and the pharyngeal bulb, which seems ideal for sucking, suggesting that at least the adults pierce other animals and suck their body fluids. The larvae have larger mouth zones, but probably also feed by sucking. Three species have been found in a permanently anoxic environment (Danovaro *et al.* 2010); they have mitochondria resembling hydrogenosomes associated with putative methanogenic archaeans (Mentel and Martin 2010).

The Loricifera are definitely a monophyletic group. The life cycle with the Higgins larva, the mouth cone with the very narrow mouth, the myoepithelial pharynx, and the scalids with intrinsic muscles separate the group from priapulans and kinorhynchs, which must be regarded as the sister groups. The non-inversible mouth cone and the structure of the brain indicate a sister-group relationship with the kinorhynchs. The sister-group relationship with nematomorphs indicated in a moleular analysis (Sørensen *et al.* 2008) is discussed in Chapter 47. Warwick (2000) interpreted the loriciferans as progenetic priapulans, but this idea has not found any support.

Interesting subjects for future research

1. Embryology
2. Feeding biology of larvae and adults

References

Danovaro, R., Dell'Anno, A., Pusceddu, A., *et al.* 2010. The first metazoa living in permanently anoxic conditions. *BMC Biology* **8**: 30.

Heiner, I. 2008. *Rugiloricus bacatus* sp. nov. (Loricifera—Pliciloricidae) and a ghost-larva with paedogenetic reproduction. *Syst. Biodivers.* **6**: 225–247.

Heiner, I. and Kristensen, R.M. 2009. *Urnaloricus gadi* nov. gen. et nov. sp. (Loricifera, Urnaloricidae nov. fam.), an aberrant Loricifera with a viviparous pedogenetic life cycle. *J. Morphol.* **270**: 129–153.

Heiner, I., Sørensen, K.J.K. and Kristensen, R.M. 2009. The spermiogenesis and the early spermatozoa of *Armorloricus elegans* (Loricifera, Nanaloricidae). *Zoomorphology* **128**: 285–304.

Higgins, R.P. and Kristensen, R.M. 1986. New Loricifera from southeastern United States coastal waters. *Smithson. Contr. Zool.* **438**: 1–70.

Kristensen, R.M. 1991a. Loricifera. In F.W. Harrison (ed.): *Microscopic Anatomy of Invertebrates*, Vol. 4, pp. 351–375. Wiley-Liss, New York.

Kristensen, R.M. 1991b. Loricifera: a general biological and phylogenetic overview. *Verh. Dtsch. Zool. Ges.* **84**: 231–246.

Kristensen, R.M. 2002. An introduction to Loricifera, Cycliophora, and Micrognathozoa. *Integr. Comp. Biol.* **42**: 641–651.

Kristensen, R.M. and Gad, G. 2004. *Armorloricus*, a new genus of Loricifera (Nanaloricidae) from Trezen ar Skoden (Roscoff, France). *Cah. Biol. Mar.* **45**: 121–156.

Mentel, M. and Martin, W. 2010. Anaerobic animals from an ancient, anoxic ecological niche. *BMC Biology* **8**: 32.

Neuhaus, B. and Kristensen, R.M. 2007. Ultrastructure of the protonephridia of larval *Rugiloricus* cf. *cauliculus*, male *Armorloricus elegans*, and female *Nanaloricus mysticus* (Loricifera). *J. Morphol.* **268**: 357–370.

Neuhaus, B., Kristensen, R.M. and Peters, W. 1997. Ultrastructure of the cuticle of Loricifera and demonstration of chitin using gold-labelled wheat germ agglutinin. *Acta Zool. (Stockh.)* **78**: 215–225.

Sørensen, M.V., Hebsgaard, M.B., Heiner, I., *et al.* 2008. New data from an enigmatic phylum: evidence from molecular data supports a sister-group relationship between Loricifera and Nematomorpha. *J. Zool. Syst. Evol. Res.* **46**: 231–239.

Warwick, R.M. 2000. Are loriciferans paedomorphic (progenetic) priapulids? *Vie Milieu* **50**: 191–193.

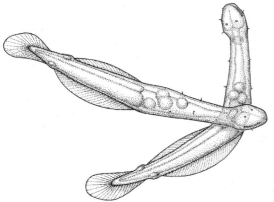

55

Phylum **Chaetognatha**

Chaetognatha, or arrowworms, is a small phylum of about 120 marine, mainly holopelagic 'worms'; some species, mainly of *Spadella* and *Paraspadella*, are benthic, and it appears that many deep-water forms remain to be described. The Lower Cambrian (Chengjiang) *Protosagitta* (Hu *et al.* 2007) is unquestionably a chaetognath. The monophyly of the group has apparently not been questioned, but the position of the group has been much debated. Traditional morphology based systematics has usually placed the group in the Deuterostomia because of the fate of the blastopore, but the nervous system and its ontogeny is definitely protostomian, and this is also indicated in most of the molecular phylogenies (see below).

The usually quite transparent body is cylindrical, with a rounded head and a tapering tail; there are one or two pairs of completely hyaline lateral fins and a large, horizontal tail fin. The microscopic anatomy was reviewed by Shinn (1997). The head has a ventral mouth surrounded by various teeth and paired lateral groups of large spines (hooks) of α-chitin used in grasping prey organisms, mainly copepods. The prey is apparently poisoned by a tetrodotoxin produced by bacteria somewhere in the head, but their exact location is unknown. An ectodermal fold with mesodermal muscles, the hood, can enwrap the head almost completely giving it a streamlined shape, and it can be retracted rapidly to expose the teeth and spines; ven-

trally it originates at the posterior border of the head, laterally the attachment curves forward, and the dorsal attachment forms an inverted V that reaches almost to the anterior tip of the head.

The ectoderm is monolayered with a cuticle at the inner side of the hood and on the anterior and ventral sides of the head, whereas the remaining parts of the body have a multilayered ectoderm. The cuticle with teeth and spines form a continuous structure. The teeth and spines are complicated structures formed by several cells; they have a high content of crystalline α-chitin with very little protein and are impregnated with zinc and silicon. The multilayered epithelium consists of an outer layer of polygonal cells covering two or more layers of interdigitating cells, with abundant bundles of tonofilaments. The fins are extensions of the basal membrane covered by the multilayered epithelium and stiffened by elongate fin-ray cells with a paracrystalline body of filaments with aligned substructures.

Two ectodermal organs of unknown function are located anteriorly on the dorsal side: the corona and the retrocerebral organ. The corona is an oval band of monociliate cells with a peculiar subsurface canal surrounding a glandular epithelium; the ciliary cells appear to be sensory and innervated by a pair of coronal nerves, but the function of the organ is totally unknown. The retrocerebral organ consists of a pair of sacs located in the posterior side of the cerebral

Chapter vignette: Two courting *Spadella cephaloptera*. (Modified from G. Thorson's Christmas card 1966.)

ganglion, with fine ducts opening in a common anterior pore. The sacs consist of large cells with numerous intertwined microvilli with ciliary basal bodies and a core filament.

The mouth opens into a pharynx, or oesophagus, that leads to a tubular intestine, which has a pair of anterior diverticula in some species, and leads further to a short rectum; absorptive cells of the intestine and all rectal cells are multiciliate.

The nervous system (Harzsch and Müller 2007; Harzsch and Wanninger 2010; Rieger *et al.* 2010) comprises a number of ganglia, notably the dorsal cerebral ganglion, a pair of lateral vestibular ganglia sending nerves to the muscles of the spines, and a large ventral ganglion; paired nerves connect the cerebral ganglion with the vestibular ganglia and the ventral ganglion. The cerebral ganglion comprises a central area of neuropil with lateral and dorsal areas of nuclei. It sends paired nerves to the esophageal ganglia, various sense organs, and the ventral ganglion. The ventral ganglion is a longitudinally elongate, almost rectangular structure situated at the median part of the ventral side of the body. The reported absence of the ventral ganglion in *Bathybelos* (Bieri and Thuesen 1990) requires confirmation. The ganglion has a median zone of neuropil and lateral zones of nuclei, about 12 pairs of lateral nerves, and a pair of posterior nerves, which pass beyond the anus. Identifiable RFamide-positive neurons are found in a repeated pattern along the ganglion. The nervous structures are all situated outside the basal membrane or internalized, but are still surrounded by the basal membrane, and the innervation of muscles is through the basal membrane, in some cases via thin-walled pits in the membrane. Sensory structures comprise paired eyes, each with numerous ciliary sensory cells and one pigment cell, and vibration sensitive ciliary fence organs with monociliate sensory cells.

The main body cavities are surrounded by a mesodermal epithelium and thus fall within the usual definition of coeloms. The head contains one narrow cavity and there is a pair of lateral cavities in the body; the lateral cavities are divided by a transverse septum at the level of the anus into an anterior trunk coelom containing the ovaries and a posterior tail coelom containing the testes. The body musculature consists of four areas of longitudinal 'primary' muscles, dorsally and ventrally separated by the basal membrane of the mesenteries and narrow bands of 'secondary' muscles, and laterally by narrow bands of 'secondary' muscles. The primary muscles are cross-striated and consist of two types of muscle fibres. The secondary muscles likewise consist of cross-striated fibres, but the ultrastructure is unique with two alternating types of sarcomeres. The body musculature (in the adults) and the gut is covered by a peritoneum, which forms a dorsal and a ventral mesentery. Some of the epithelial cells associated with the secondary muscles contain myofilaments and carry single cilia, and also the peritoneal cells of the gut are myoepithelial. The tail coeloms are subdivided by incomplete longitudinal mesenteries, which may be partially ciliated.

There is a narrow haemal cavity around the gut with a larger longitudinal canal on the dorsal side and lateral sinuses at the level of the anus. The haemal fluid does not contain pigments. Nephridia have not been observed.

Both testes and ovaries are elongate bodies covered by the peritoneum and attached to the lateral body wall. Clusters of spermatogonia break off from the testes and circulate in the tail coeloms during differentiation. Mature spermatozoa move into a pair of seminal vesicles and are passed in a loosely organized spermatophore to the body surface of the partner during a pseudocopulation; the spermatozoa then migrate to the female gonopore (Pearre *et al.* 1991). An oviduct, consisting of an inner, syncytial layer and an outer, cellular layer, lies laterally in the ovary and opens on a small dorsolateral papilla at the level of the anus; the inner part of the oviduct functions as a seminal receptacle. The ripe eggs are surrounded by a thick membrane with a small micropyle occupied by a number of accessory cells, one of which appears to form a canal for the penetration of the sperm from the oviduct (Goto 1999). The fertilized eggs enter the oviduct and are shed free in the water, or attached in small packets near the gonopore or to objects on the bottom (Kapp 1991). The sperm is very unusual, with the centrioles at the anterior end, and the nucleus and one mitochondrion forming a slim structure along most of the axonene.

The development of *Sagitta* and *Spadella* has been studied with classical methods by a number of authors (Hertwig 1880; Doncaster 1902; Burfield 1927; John 1933) and with blastomere marking by Shimotori and Goto (1999, 2001). No important difference has been reported between the genera and species, but the reports from the two types of study on the origin of the ventral ganglion are difficult to reconcile.

After fertilization and formation of the polar bodies, a small germ granule (Carré *et al.* 2002) develops in the cytoplasm near the blastoporal pole, and this body remains undivided during the first five cleavages, situated in one of the cells at the blastoporal pole (Fig. 55.1). Cleavage is total and equal, and early development is rather easy to follow because of the transparency of the embryos. Shimotori and Goto (1999, 2001) marked cells of 2- and 4-cell stages, and showed that the two first cleavages contain the primary, apical-blastoporal axis. The 4-cell stage shows a spiral-cleavage-like configuration, with two opposite blastomeres having a contact zone at the apical pole, while the other pair has a contact zone at the blastoporal pole. If the blastomeres are named in accordance with the spiral terminology, with the descendents of the B cell covering the dorsal side of the juvenile, the germ granule is situated in the D cell. A blastula with a narrow blastocoel develops; one side flattens at the stage of about 64 cells, and a typical invagination gastrula is formed. The germ granule divides and the material becomes distributed to two daughter blastomeres, which can also be recognized by their large nuclei. These cells are the primordial germ cells; they are situated at the bottom of the archenteron, but they soon detach from its wall, move into the archenteron, and divide once. The anterolateral parts of the archenteron wall form a pair of folds that grow towards the blastopore carrying the germ cells at the tips. The anterior part of the archenteron thus becomes divided into a median gut compartment and a pair of lateral mesodermal sacs. The blastopore closes and a stomodaeal invagination develops from the opposite pole; its position relative to the apical pole has not been

ascertained. The anterior parts of the mesodermal sacs become pinched off and fuse to form the head coelom. The embryo now elongates and curves inside the egg membrane, and all cavities inside the embryo become very narrow. The gut becomes a compact, flat, median column bordered by the lateral mesodermal masses, which meet along the midline at the posterior end of the embryo. The brain ganglion develops from a thickening of the dorsal ectoderm of the head, and the ventral ganglion develops as a pair of lateral ectodermal thickenings, which fuse ventrally some time after the hatching. The cell-lineage studies show that the ectoderm develops from the A, B and C cells, with the A- and C-cell descendents covering the left and right side of the ventral and dorsal sides of the head region. The ectoderm of the trunk develops from the B cell. The ventral part of the ventral ganglion should develop from the A and C cells and the laterodorsal parts from the D cell. The mesoderm develops from all four quadrants and the gut from the B and D cells. The longitudinal muscles originate from all four quadrants and the germinal cells from the D cell. Unexpectedly, the inner epidermis layer should originate from the D cell.

The newly hatched juvenile (Shinn and Roberts 1994) looks completely compact, but the coelomic cavities expand and the lumen of the gut develops as a slit; the lateral mesodermal sacs are quite thin-walled except at the two paired, longitudinal lateral muscles. The mesoderm with the germ cells develops a fold that finally divides each lateral cavity into an anterior and a posterior compartment, each with one of the germ cells at the lateral wall; the anterior germ cells become incorporated into the ovaries and the posterior ones into the testes. The anus breaks through at the level of the septa between the lateral cavities; its position relative to that of the closed blastopore is unknown.

This phylum is regarded as one of the most isolated groups among the metazoans. Both the multilayered epithelium, the structure of the fins, and the corona can be mentioned as autapomorphies of the phylum. Its phylogenetic position has been a matter of much discussion, and attempts have been made to relate them to almost any other metazoan phylum,

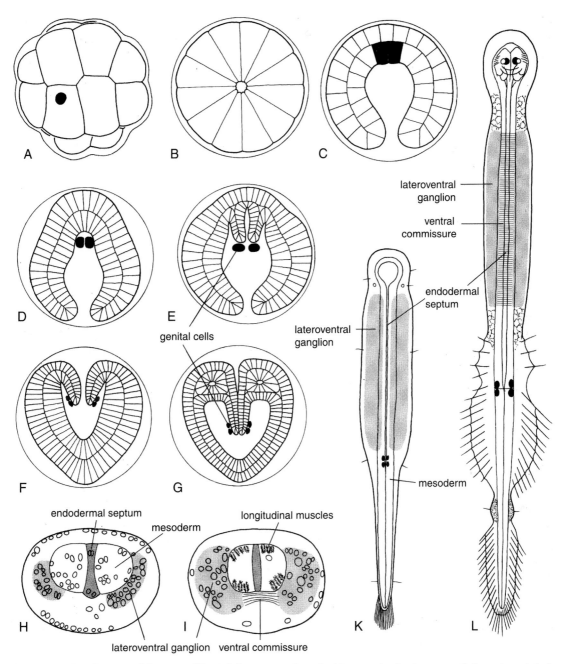

Figure 55.1. Development of *Sagitta* sp. (A) 16-Cell stage seen from the blastoporal pole; the germ cell determinant is indicated in black. (B) Blastula. (C) Gastrula; the germ cells (black) are still situated in the endoderm. (D) Late gastrula; the germ cells are situated in the archenteron. (E) Early stage of mesoderm formation. (F) Development of the mouth. (G) Development of the anterior pair of coelomic pouches. (H) Transverse section of a stage like that in (G), showing the lateroventral ganglionic cells (grey). (I) Transverse section of a stage like that in (L), showing the ventral ganglion. (K) Newly hatched juvenile. (L) Juvenile four days after hatching, with brain and ventral ganglion. (A and C redrawn from Elpatiewsky 1910; B and D–G redrawn from Burfield 1927; H–L redrawn from Doncaster 1902.)

both protostomian and deuterostomian (Ghirardelli 1968; Bone *et al.* 1991).

The differentiation of the archenteron resembles the enterocoely of deuterostomes (Chapter 56), but also that of brachiopods (Chapter 41), and this is the reason why so many earlier authors placed the Chaetognatha in the Deuterostomia. However, a more specific position or sister group has not been pointed out, and most other characters point to protostomian relationships. The lateral body cavities of the chaetognaths are definitely coeloms, but their homology with coeloms of other groups is unproven. It is often stated that chaetognaths have three pairs of coeloms like the 'archicoelomates' (with protocoel, mesocoel, and metacoel), but there are no specific similarities between the coeloms of the two groups. I regard the coelomic cavities of the chaetognaths an autapomorphy of the phylum.

The multilayered epithelium of the body has been interpreted as a deuterostome character, because similar epithelia are well known from the vertebrates, but to my knowledge nobody has proposed that chaetognaths and vertebrates should be sister groups. A multilayered epithelium of the body has also been described in the polychaete *Travisia* (Chapter 25), so this character seems unreliable.

Strong arguments for a protostomian relationship are found in the nervous system that is decidedly protostomian (gastroneuralian) both in embryos and adults. The development of a longitudinal, midventral ganglion by the fusion of a pair of longitudinal neural folds is one of the key characters of the Protostomia (Chapter 22). The cerebral ganglion has not been related to the apical pole, but it develops from the epithelium of the dorsal side of the head, and becomes connected to the ventral nerve cord through a pair of circumoesophageal commissures. The studies of marked blastomeres have demonstrated the existence of four quadrants, with one containing the primordial germ cell, resembling that of both spiralians and ecdysozoans (Fig. 21.1), but there is no sign of a spiral pattern.

Molecular phylogeny has placed Chaetognatha in many different positions: 'between' Protostomia and Deuterostomia (for example Giribet *et al.* 2009); as sister group of Lophotrochozoa + Ecdysozoa (for example Lartillot and Philippe 2008; Witek *et al.* 2009); as sister group of Protostomia (for example Helfenbein *et al.* 2004; Marlétaz *et al.* 2006; Marlétaz and Le Parco 2008); as sister group of Lophotrochozoa (for example Matus *et al.* 2006; Dunn *et al.* 2008; Hejnol *et al.* 2009); as an ingroup of the Lophotrochozoa (for example Papillon *et al.* 2004; Helmkampf *et al.* 2008b); as sister group of the Ecdysozoa (for example Giribet *et al.* 2000); and as an ingroup of Ecdysozoa (for example Helmkampf *et al.* 2008a; Paps *et al.* 2009); to my knowledge they have never been placed with the deuterostomes.

The Hox genes show a peculiar pattern with posterior genes 'intermediate' between protostome and deuterostome genes (Papillon *et al.* 2003), indicating a basal bilaterian position of the chaetognaths, and (Matus *et al.* 2007) found both spiralian and ecdysozoan 'motifs' in some of the genes. The position in the protostomes is also supported by the protostomian quadrant-type cleavage pattern, where the cell that will give rise to the germ cells can be identified from the early cleavage stages.

The combination of information from structure, embryology, and the Hox genes indicates that the Chaetognatha is a basal protostomian group. But whether it is the sister group of all the protostomes, of the spiralians, or of the ecdysozoans remains to be seen. I have decided to place them in a trichotomy at the base of the Protostomia (Fig. 22.1), in agreement with the phylogenomic study of Edgecombe *et al.* (2011).

Interesting subjects for future research

1. Embryology and cell lineage; origin of mouth and anus, differentiation of the head coelom
2. Ultrastructure of the spines

References

Bieri, R. and Thuesen, E.V. 1990. The strange worm *Bathybelos*. *Am. Sci.* **78**: 542–549.

Bone, Q., Kapp, H. and Pierrot-Bults, A.C. 1991. Introduction and relationships of the group. In Q. Bone, H. Kapp and A.C. Pierrot-Bults (eds): *The Biology of Chaetognaths*, pp. 1–4. Oxford University Press, Oxford.

Burfield, S.T. 1927. L.M.B.C. Memoir 28: *Sagitta*. *Proc. Trans. Liverpool Biol. Soc.* **41**(**Appendix 2**): 1–101.

Carré, D., Djediat, C. and Sardet, C. 2002. Formation of a large Vasa-positive germ granule and its inheritance by germ cells in the enigmatic chaetognaths. *Development* **129**: 661–670.

Doncaster, L. 1902. On the development of *Sagitta*; with notes on the anatomy of the adult. *Q. J. Microsc. Sci., N. S.* **46**: 351–395.

Dunn, C.W., Hejnol, A., Matus, D.Q., *et al.* 2008. Broad phylogenomic sampling improves resolution of the animal tree of life. *Nature* **452**: 745–749.

Edgecombe, G.D., Giribet, G., Dunn, C.W., *et al.* 2011. Higher-level metazoan relationships: recent progress and remaining questions. *Org. Divers. Evol.* **11**: 151–172

Elpatiewsky, W. 1910. Die Urgeschlechtszellenbildung bei *Sagitta. Anat. Anz.* **35**: 226–239.

Ghirardelli, E. 1968. Some aspects of the biology of the chaetognaths. *Adv. Mar. Biol* **6**: 271–375.

Giribet, G., Distel, D.L., Polz, M., Sterrer, W. and Wheeler, W.C. 2000. Triploblastic relationships with emphasis on the acoelomates and the position of Gnathostomulida, Cycliophora, Plathelminthes, and Chaetognatha: a combined approach of 18S rDNA sequences and morphology. *Syst. Biol.* **49**: 539–562.

Giribet, G., Dunn, C.W., Edgecombe, G.D., *et al.* 2009. Assembling the spiralian tree of life. In M.J. Telford and D.T.J. Littlewood (eds): *Animal Evolution. Genomes, Fossils, and Trees*, pp. 52–64. Oxford University Press, Oxford.

Goto, T. 1999. Fertilization process in the arrow worm *Spadella cephaloptera* (Chaetognatha). *Zool. Sci. (Tokyo)* **16**: 109–114.

Harzsch, S. and Müller, C.H.G. 2007. A new look at the ventral nerve center of *Sagitta*: implications for the phylogenetic position of Chaetognatha (arrow worms) and the evolution of the bilaterian nervous system. *Front. Zool.* **4**: 14.

Harzsch, S. and Wanninger, A. 2010. Evolution of invertebrate nervous systems: the Chaetognatha as a case study. *Acta Zool. (Stockh.)* **91**: 35–43.

Hejnol, A., Obst, M., Stamatakis, A., *et al.* 2009. Assessing the root of bilaterian animals with scalable phylogenomic methods. *Proc. R. Soc. Lond. B* **276**: 4261–4270.

Helfenbein, K.G., Fourcade, H.M., Vanjani, R.G. and Boore, J.L. 2004. The mitochondrial genome of *Paraspadella gotoi* is highly reduced and reveals that chaetognaths are sister group to protostomes. *Proc. Natl. Acad. Sci. USA* **191**: 10639–10643.

Helmkampf, M., Bruchhaus, I. and Hausdorf, B. 2008a. Multigene analysis of lophophorate and chaetognath phylogenetic relationships. *Mol. Phylogenet. Evol.* **46**: 206–214.

Helmkampf, M., Bruchhaus, I. and Hausdorf, B. 2008b. Phylogenomic analyses of lophophorates (brachiopods, phoronids and bryozoans) confirm the Lophotrochozoa concept. *Proc. R. Soc. Lond. B* **275**: 1927–1933.

Hertwig, O. 1880. Die Chaetognathen. Eine Monographie. *Jena. Z. Naturw.* **14**: 196–311.

Hu, S., Steiner, M., Zhu, M., *et al.* 2007. Diverse pelagic predators from the Chengjiang Lagerstätte and the establishment of modern-style ecosystems in the Early Cambrian. *Palaeogeogr. Palaeoclimatol. Palaeoecol.* **254**: 307–316.

John, C.C. 1933. Habits, structure, and development of *Spadella cephaloptera. Q. J. Microsc. Sci., N. S.* **75**: 625–696.

Kapp, H. 1991. Morphology and anatomy. In Q. Bone, H. Kapp and A.C. Pierrot-Bults (eds): *The Biology of Chaetognaths*, pp. 5–17. Oxford University Press, Oxford.

Lartillot, N. and Philippe, H. 2008. Improvement of molecular phylogenetic inference and the phylogeny of the Bilateria. *Phil. Trans. R. Soc. Lond. B* **363**: 1463–1472.

Marlétaz, F. and Le Parco, Y. 2008. Careful with understudied phyla: the case of the chaetognath. *BMC Evol. Biol.* **8**: 251.

Marlétaz, F., Martin, E., Perez, Y., *et al.* 2006. Chaetognath phylogenomics: a protostome with deuterostome-like development. *Curr. Biol.* **16**: R577–R578.

Matus, D.Q., Copley, R.R., Dunn, C.W., *et al.* 2006. Broad taxon and gene sampling indicate that chaetognaths are protostomes. *Curr. Biol.* **16**: R575–R576.

Matus, D.Q., Halanych, K.M. and Martindale, M.Q. 2007. The *Hox* gene complement of a pelagic chaetognath. *Integr. Comp. Biol.* **47**: 854–864.

Papillon, D., Perez, Y., Caubit, X. and Le Parco, Y. 2004. Identification of chaetognaths as protostomes is supported by the analysis of their mitochondrial genome. *Mol. Biol. Evol.* **21**: 2122–2129.

Papillon, D., Perez, Y., Fasano, L., Le Parco, Y. and Caubit, X. 2003. *Hox* gene survey in the chaetognath *Spadella cephaloptera*: evolutionary implications. *Dev. Genes Evol.* **213**: 142–148.

Paps, J., Baguñà, J. and Riutort, M. 2009. Bilaterian phylogeny: A broad sampling of 13 nuclear genes provides a new Lophotrochozoa phylogeny and supports a paraphyletic basal Acoelomorpha. *Mol. Biol. Evol.* **26**: 2397–2406.

Pearre, S., Jr, Kapp, H. and Pierrot-Bults, A.C. 1991. Growth and reproduction. In Q. Bone, H. Kapp and A.C. Pierrot-Bults (eds): *The Biology of Chaetognaths*, pp. 61–75. Oxford University Press, Oxford.

Rieger, V., Perez, Y., Müller, C.H.G., *et al.* 2010. Immunohistochemical analysis and 3D reconstruction of the cephalic nervous system in Chaetognatha: insights into the evolution of an early bilaterian brain? *Invertebr. Biol.* **129**: 77–104.

Shimotori, T. and Goto, T. 1999. Establishment of axial properties in the arrow worm embryo, *Paraspadella gotoi* (Chaetognatha): developmental fate of the first two blastomeres. *Zool. Sci. (Tokyo)* **16**: 459–469.

Shimotori, T. and Goto, T. 2001. Developmental fates of the first four blastomeres of the chaetognath *Paraspadella gotoi*: relationship to protostomes. *Dev. Growth Differ.* **43**: 371–382.

Shinn, G.L. 1997. Chaetognatha. In F.W. Harrison (ed.): *Microscopic Anatomy of Invertebrates*, vol. 15, pp. 103–220. Wiley-Liss, New York.

Shinn, G.L. and Roberts, M.E. 1994. Ultrastructure of hatchling chaetognaths (*Ferosagitta hispida*): epithelial arrangement of the mesoderm and its phylogenetic implications. *J. Morphol.* **219**: 143–163.

Witek, A., Herlyn, H., Ebersberger, I., Welch, D.B.M. and Hankeln, T. 2009. Support for the monophyletic origin of Gnathifera from phylogenomics. *Mol. Phylogenet. Evol.* **53**: 1037–1041.

56

DEUTEROSTOMIA

Deuterostomia is a well-delimited group of bilateral animals. It is accepted in all major morphological studies and textbooks and is also supported in almost all molecular studies. As indicated by the name, the deuterostomes typically have a blastopore that becomes the anus in the adult, while the mouth is a new opening from the bottom of the archenteron. This type of development can actually be followed in a number of species in echinoderms and enteropneusts, whereas the chordates show a highly specialized blastopore fate.

The Deuterostomia falls into two well-defined groups, Ambulacraria (Chapter 57), comprising enteropneusts, pterobranchs, and echinoderms, and Chordata (Chapter 62), comprising cephalochordates, urochordates, and vertebrates (Fig. 56.1). The Hox genes show a characteristic deuterostome signature (Fig. 21.3); the chordates have a long chordate signature with *Hox9–13(15)*, whereas the ambulacrarians have a shorter ambulacrarian signature with *Hox9/10* and three *Hox11/13* genes.

The ambulacrarians appear to have retained a long series of ancestral characters that have been lost in the chordates, which could indicate that the chordates are an ingroup of the ambulacrarians, but both the Hox genes and a number of other ambulacrarian apomorphies (see Chapter 57) exclude this possibility.

Echinoderms with planktotrophic larvae have a blastopore that directly becomes the anus, and the larval mouth is formed as a new opening from the bottom of the archenteron, which bends ventrally and

becomes connected to a shallow stomodaeum. The anterior coelomic sac (protocoel) usually pinches off from the bottom of the archenteron before the hydropore is formed (see below), but in the holothurians *Labidoplax* and *Stichopus*, the archenteron first bends dorsally and forms the hydropore, with the protocoel subsequently becoming pinched off, and then the archenteron bends ventrally to form the new mouth (Chapter 57). The significance of this is not known. Species with large yolk-rich eggs have modified development, but the two types of development can be found even within the same genus (Chapter 58), and corresponding areas of the eggs give rise to identical adult structures in the two types. Later stages are complicated in several of the major groups; a new adult mouth is formed at metamorphosis in asteroids and echinoids, and the adult ophiuroids lack the anus. There are no reports of a development where the blastopore becomes the mouth and no signs of amphistomy, i.e. the division of the blastopore into mouth and anus.

The enteropneusts show limited variation, and the blastopore becomes the anus in all species studied (Chapter 60).

The chordates do not have primary larvae, and their development is complicated, owing to early development of the neural folds, which enclose the blastopore at the posterior end forming the neurenteric canal. This usually closes the blastopore and a new anus is formed, but the blastopore actually appears to become further invaginated and transformed into the adult anus in

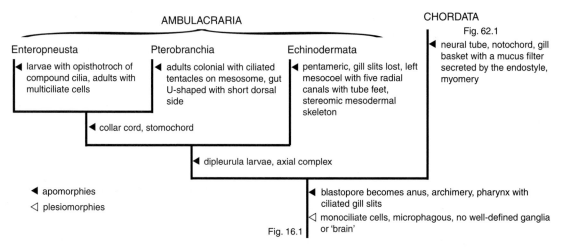

Figure 56.1. Phylogeny of the Deuterostomia.

anurans (Chapter 65). These modifications of the developmental stages must be related to the lecithotrophic development or viviparity in urochordates and vertebrates, and to a very early differentiation of the adult feeding structures in amphioxus (Chapter 63).

The first cleavage is median in most groups, i.e. the first cleavage separates the right and left halves of the embryo. The echinoderms show some variation, with a 'normal' right-left first division in several species, and oblique cleavage planes dividing the zygote into oral, right, aboral, and left parts in some echinoids (Chapter 58). Studies of early development of enteropneusts, urochordates, and cephalochordates (Chapters 60, 63, 64) have demonstrated that the first cleavage is median. The vertebrates show much variation in connection with yolk-rich eggs or placentally nourished embryos, but a first median cleavage is well-documented, for example in anurans (Chapter 65). This pattern of the two first cleavages being median and transversal is different from the predominant pattern in the protostomes, which have the first two cleavages dividing the embryo roughly into four quadrants: anterior, right, posterior, and left cells, with the D-cell having a special fate (Fig. 21.1). There are many species with aberrant cleavage patterns related to a large amount of yolk or placental nourishment, just as in the protostomes, and the urochordates have a highly specialized bilateral cleavage (Chapter 64).

The degree of determination varies somewhat between the deuterostome phyla, but each cell of the 2-cell stage is usually competent of developing into a complete embryo, and this is also the case for cells of the 4-cell stage in several species; half-gastrulae of echinoderms are able to regenerate completely (Chapter 58). Only the urochordate embryos are highly determined (Chapter 64).

Cleavage leads to the formation of blastula and gastrula in all the forms with small eggs developing into free larvae, and also in many species with yolk-rich eggs.

An apical organ, with sensory cells with long cilia at the position of the polar bodies, is found in most of the ambulacrarian larvae. There is no trace of cerebral ganglia like those of the spiralians (Chapter 23; Fig. 22.10). The apical organs are slightly thickened epithelial areas, with basiepithelial neuropil and single neuronal connections to other areas. They degenerate at metamorphosis. Chordate larvae/embryos have no apical organ.

The dipleurula larva (Fig. 21.2) is probably ancestral in the Ambulacraria, whereas the chordates do not have primary larvae (Chapters 57 and 62).

The origin of the mesoderm and the type of coelom formation has been given high rank in earlier phylogenetic studies. A hypothetic ancestor, in the shape of a creeping bilateral organism with three pairs of coelomic sacs, of which the two first pairs are connected to each other on each side through a common coelomoduct, has had a central position in many discussions of the phylogeny of the deuterostomes (see below).

Mesoderm and coelom formation shows an enormous intraphyletic variation both in enteropneusts and echinoderms (Fig. 57.2), but the mesoderm is in all cases formed from the endoderm (archenteron), never from the blastopore lips as in the protostomes (Chapter 22). The mesoderm is given off as coelomic pouches, enterocoely, in most ambulacrarians and chordates, but in the shape of solid outgrowths or diffuse delamination or ingression in a few species where the cavities accordingly arise through schizocoely (Fig. 57.2). The mesoderm formation observed in many echinoids and asteroids, viz. a pocket pinched off from the apical end of the archenteron that then gives off mesocoel + metacoel on each side, is often regarded as typical of the deuterostomes, but it is difficult to infer an ancestral pattern.

Coelomoduct development is poorly studied. It appears that a coelomopore is formed when an extension from a coelomic sac contacts the ectoderm, which may sometimes form a small invagination, resembling a stomodaeum or proctodaeum (for example in larvae of enteropneusts and echinoderms, see Ruppert and Balser 1986).

The circulatory system (Malakhov 1977; Nübler-Jung and Arendt 1996) of enteropneusts comprises a dorsal longitudinal vessel, in which the blood flows anteriorly to the heart/glomerulus complex and posteriorly from this through vessels around the oesophagus to a ventral longitudinal vessel. Amphioxus and the chordates have one or a pair of ventral longitudinal vessels with an anteriorly directed flow propelled by a ventral heart, and dorsal vessels with a posterior flow of the blood. If the chordate groups are flipped over, the circulation pattern becomes the same as in the enteropneusts.

Comparisons between chordates and their sister group, the ambulacrarians, and especially with enteropneusts, reveal both a number of similarities and a number of differences.

The ambulacrarians lack a notochord. The notochord develops from the roof of the archenteron, whereas the 'stomochord' of the hemichordates develops from the pharynx. The two organs have both different structure and origin, so they cannot be homologous (see Chapter 59).

The hemichordates have a collar chord, which has been homologized with the neural tube of the chordates, but this is not supported by studies of the nerve cells (Chapter 59).

The enteropneusts have a pharynx with gill slits of a structure similar to that of the cephalochordates and tunicates, but without a mucus filter. Another difference is that the tongue bars develop from the dorsal side in the enteropneust, but from the 'dorsal' (neural) side in the chordates. The homology of these structures is partially based on the assumption that the ciliary filtering system of the enteropneusts can be modified to a mucociliary filter system. A parallel example of such a transformation is seen in the prosobranch *Crepidula*, which has modified the respiratory gill to a filter-feeding structure with long, parallel gill bars, resembling that of the bivalves, but filtering with a mucus net secreted by an endostyle-like gland at the base of the gill (Werner 1953). Studies of both morphology and gene expressions (Cannon *et al.* 2009) support the generally accepted homology of the ciliated gill slits of the enteropneusts, the mucociliary filter-feeding gill slits of cephalochordates and tunicates, and the respiratory gill slits of the vertebrates. The homology of the gill pores of *Cephalodiscus* has not been studied. Ruppert *et al.* (1999) searched the branchial gut of the enteropneust *Schizocardium* for structures resembling the chordate endostyle and found that the epibranchial gland (which is situated along the dorsal side of the branchial gut) consists of longitudinal rows of different cell types highly reminiscent of the cell types found in the endostyle of amphioxus (Fig. 56.2).

The complex of gill slits and endostyle forms a beautiful evolutionary transformation series. The first step is the ciliary filter-feeding slits of the enteropneusts, which lack the mucus net secreted by an endostyle, but that have iodine-binding cells in several epithelia (Chapter 60). The second step is the mucociliary filter-feeding systems of cephalochordates and

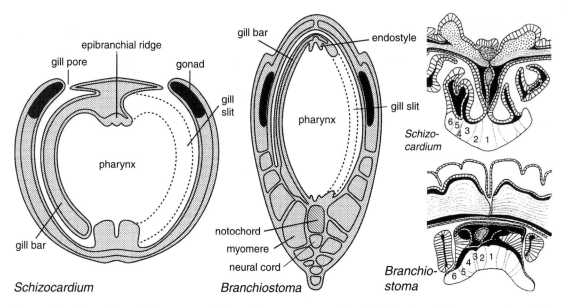

Figure 56.2. Comparisons of pharynx morphology of enteropneusts and cephalochordates. Left: cross-sections of the branchial zone of the enteropneust *Schizocardium* and the cephalochordate *Brachiostoma*. The sections are based on van der Horst (1927–1939) and Ruppert (1997), respectively, but modified to show sections that pass through a gill bar on one side and a gill slit on the other. Right: details of the hypobranchial ridge of *Schizocardium* and the endostyle of *Branchiostoma*; the longitudinal rows of epithelial cells with similar morphology are numbered. The sections of *Branchiostoma* are in both cases dorsoventrally inverted.

urochordates (Chapters 63, 64), which have an endostyle with bands of cells secreting the mucus filter, and of cells that bind iodine and secrete iodinated proteins into the net and into the blood stream. The third step is the mucus filter-feeding system, with the water pumped through the filter by contractions of the pharynx as seen in the ammocoetes larva (Chapter 65), which has an endostyle with only little importance for the secretion of the filter, but with iodine-binding and secreting cells. The fourth step is the exclusively respiratory gill system of adult lampreys and many other aquatic vertebrates, and the transformation of the endostyle into the endocrine thyroid, which accumulates iodine and secretes thyroxine (Chapter 65).

Adults of the ambulacrarians have a weak central nervous system without a brain-like centre. The development of the central nervous system of the chordates through the fusion of the neural folds and the complete internalization of the neural plate is described in all textbooks (Chapter 62).

The ancestral deuterostome was in all probability pelago-benthic with a dipleurula larva, and a benthic, vermiform adult with an archimeric body plan with a foregut with gill slits and a nervous system without a centralized brain (see for example Cameron 2005; Swalla and Smith 2008), but other details of the adult morphology are more difficult to infer.

References

Cameron, C.B. 2005. A phylogeny of the hemichordates based on morphological characters. *Can. J. Zool.* **83**: 196–215.

Cannon, J.T., Rychel, A.L., Eccleston, H., Halanych, K.M. and Swalla, B.J. 2009. Molecular phylogeny of Hemichordata, with updated status of deep-sea enteropneusts. *Mol. Phylogenet. Evol.* **52**: 17–24.

Malakhov, V.V. 1977. The problem of the basic structural plan in various groups of Deuterostomia. *Zh. Obshch. Biol.* **38**: 485–499.

Nübler-Jung, K. and Arendt, D. 1996. Enteropneusts and chordate evolution. *Curr. Biol.* **6**: 352–353.

Ruppert, E.E. 1997. Cephalochordata (Acrania). In F.W. Harrison (ed.): *Microscopic Anatomy of Invertebrates*, vol. 15, pp. 349–504. Wiley-Liss, New York.

Ruppert, E.E. and Balser, E.J. 1986. Nephridia in the larvae of hemichordates and echinoderms. *Biol. Bull.* **171**: 188–196.

Ruppert, E.E., Cameron, C.B. and Frick, J.E. 1999. Endostyle-like features of the dorsal epibranchial ridge of an enteropneust and the hypothesis of dorsal-ventral axis inversion in chordates. *Invertebr. Biol.* **118**: 202–212.

Swalla, B.J. and Smith, A.B. 2008. Deciphering deuterostome phylogeny: molecular, morphological and palaeontological perspectives. *Phil. Trans. R. Soc. Lond. B* **363**: 1557–1568.

van der Horst, C.J. 1927–1939. Hemichordata. *Bronn's Klassen und Ordnunges des Tierreichs, 4. Band, 4. Abt., 2. Buch, 2. Teil*, pp. 1–737. Akademische Verlagsgesellschaft, Leipzig.

Werner, B. 1953. Über den Nahrungserwerb der Calyptraeidae (Gastropoda Prosobranchia). *Helgol. Wiss. Meeresunters.* **4**: 260–315.

57

AMBULACRARIA

Echinoderms and enteropneusts (and later on ptero-branchs) have almost unanimously been placed together in the Ambulacraria, ever since Metschnikoff (1881) emphasized the similarities between their coelomic systems and their larvae. Also the molecular studies strongly support the group (for example Cannon *et al.* 2009; Hejnol *et al.* 2009; Edgecombe *et al.* 2011). The Hox genes show an 'ambulacrarian signature' (Fig. 21.3). Enteropneusts and pterobranchs are either sister groups, or the pterobranchs are an ingroup of the enteropneusts (see Chapter 59).

The ambulacrarians show archimery, i.e. a division of the body into three regions, prosome, meso-some, and metasome, which cannot always be recognized externally, but that have well-defined coelomic compartments, protocoel, mesocoel, and metacoel, respectively.

The prosome is an externally well-defined region in pterobranchs and enteropneusts, but it is completely integrated in the body in the echinoderms. The proto-coel is well-developed in the embryos/larvae and becomes integrated into the axial complex (see below).

The mesosome of the pterobranchs carries tentacles with mesocoelomic canals. In the enteropneusts, the mesosome has the shape of a collar with a pair of simple coelomic sacs. The echinoderm mesosome is completely integrated in the body, but its left sac is highly modified, forming the hydrocoel or water vascular system.

The metasome is the main part of the body in all the phyla, but it cannot be distinguished externally in the echinoderms. The metacoel is paired in all the three phyla. The gonads are situated in the peritoneum, and the gametes are shed through separate ducts.

The most conspicuous ambulacrarian apomorphies are the axial organ (Fig. 57.1), which has no counterpart in the chordates, and the dipleurula larva, which, however, could be ancestral in the deuterostomes (see Chapters 56 and 62).

The ontogenetic origin and the structure/function of the axial organ in echinoderms and enteropneusts are so similar that their homology can hardly be questioned. The adult structure is similar in the

Figure 57.1. Morphology of the anterior coelomic cavities in larvae and adults of ambulacrarians with emphasis on the transformations of parts of the larval coelomic sacs into the adult's axial complex: The larvae are drawn on the basis of photographs in Ruppert and Balser (1986), while the adults are diagrammatic median sections with some of the structures shifted slightly to get the necessary details into the plane of the drawing. Pterobranchs: adult *Cephalodiscus gracilis* (based on Lester 1985 and Dilly *et al.* 1986); enteropneusts: larva of *Schizocardium brasiliense*, adult of *Saccoglossus kowalevskii* (based on Balser and Ruppert 1990); echinoderms: larva of *Asterias forbesi*, adult of *Asterias* (based on Nichols 1962 and Ruppert and Balser 1986). Mesocoel, blue; renal chamber (protocoel), green; pericardium, black; blood, red. The lines with dots indicate a coelomic layer consisting of podocytes. The small arrows indicate the direction of the presumed or proven ultrafiltration of primary urine. See Colour Plate 8.

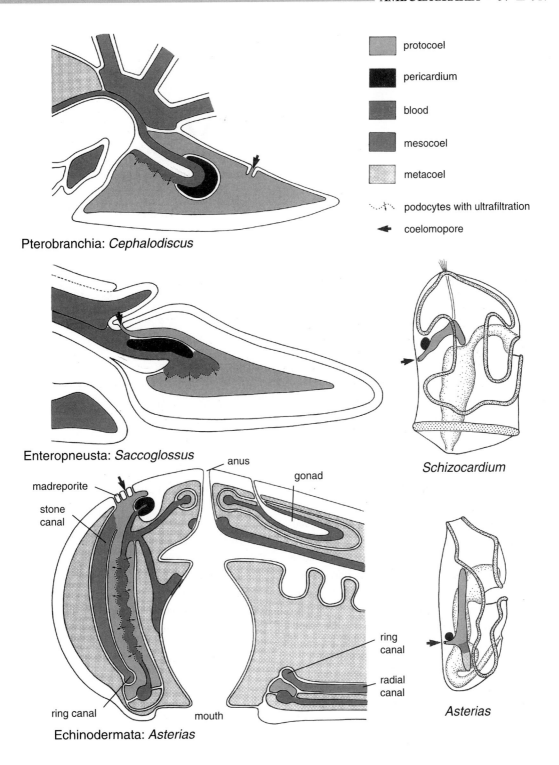

protocoel

pericardium

blood

mesocoel

metacoel

⋅⋅⋅⋅ podocytes with ultrafiltration

◂ coelomopore

Pterobranchia: *Cephalodiscus*

Enteropneusta: *Saccoglossus*

Schizocardium

anus

gonad

madreporite

stone canal

ring canal

radial canal

Asterias

ring canal

mouth

Echinodermata: *Asterias*

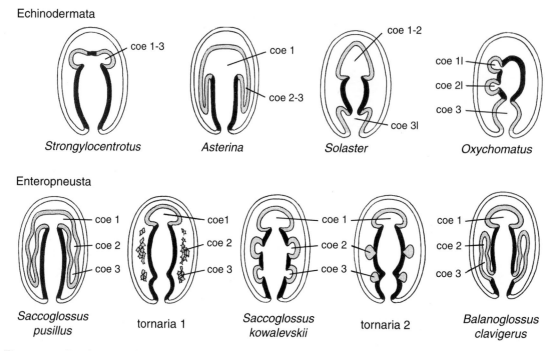

Figure 57.2. Development of the coelomic cavities in echinoderms and enteropneusts; the diagrams show some features that do not occur simultaneously in the embryos: the protocoel is usually pinched off before the meso- and metacoels develop, and the blastopore is usually closed at an earlier stage; the blastopore is, in most cases, not in the plane of the diagram. Echinodermata: *Strongylocentrotus lividus* (based on Ubisch 1913); *Solaster endeca* (based on Gemmill 1912); *Asterina gibbosa* (based on Ludwig 1882); and *Oxychomatus japonicus* (based on Holland 1991).—Enteropneusta: *Saccoglossus pusillus* (based on Davis 1908); tornaria larva 1 (based on Dawydoff 1944, 1948); *Saccoglossus kowalevskii* (based on Bateson 1884); tornaria larva 2 (based on Dawydoff 1944, 1948); and *Balanoglossus clavigerus* (based on Stiasny 1914). Ectoderm, white; mesoderm, grey; endoderm, black; coe 1, protocoel; coe 2, mesocoel; coe 3, metacoel; l, left.

pterobranchs, but the ontogeny has not been studied. The complex is situated in the prosome and comprises a heart consisting of a blood vessel (without an endothelium), partially surrounded by a pericardium (the right protocoel), and a specialized area with podocytes along the efferent branch. Primary urine is believed to be formed through ultrafiltration from the blood vessel through the basal membrane covered with podocytes to the left protocoel, which functions as a nephridial capsule. The urine flows from the protocoel through the short coelomoduct to a median coelomopore/nephridiopore. The prosome cannot be recognized in the adult echinoderms, but the ontogenetic origin of the respective components of their axial organ clearly demonstrates its homology with those of the other two phyla. The pericardium can be recognized as a pulsatile vesicle already in the larvae of echinoderms and enteropneusts (Ruppert and Balser 1986). The origin of the vesicle is not clear in all cases, but development from the protocoel has been indicated in some cases.

The 'text-book truth' about the development of the deuterostome mesoderm is that it is through enterocoely, i.e. coelomic pockets formed from the archenteron, typically three pairs. However, this needs a good deal of modification. It appears that the mesoderm always derives from the wall of the archenteron, but the coelomic sacs develop from compact extensions from the archenteron, or from scattered mesodermal cells in some tornariae (Fig. 57.2). There is considerable variation in the origin of the three pairs of coeloms.

The ambulacrarian nervous systems are of an unusual type, without a brain or well-defined ganglia. This is very obvious in the echinoderms, but the collar cord of the hemichordates has sometimes been compared with the neural tube of the chordates. It contains concentrations of neurons, but so do the longitudinal nerve cords (Chapter 60).

The ancestral ambulacrarian larva is usually called 'dipleurula' (a name introduced by Semon 1888), but several authors around the turn of the last century (review in Holland 1988), most prominently Bather (1900), unfortunately used the name for a hypothetical benthic ancestor of the echinoderms, an organism characterized by the presence of three pairs of coelomic sacs, and this usage of the term has continued too (see for example Hyman 1955; Gruner 1980). Ubaghs (1967) used the term for both the larva and the adult. Here I use the term 'dipleurula' for the ancestral larval type with a peri-oral, upstream-collecting ciliary ring, the neotroch, which is formed on monociliate cells (Fig. 21.2).

The planktotrophic larvae of echinoderms and enteropneusts are variations over the dipleurula theme, with the neotroch pulled out into sometimes quite bizarre shapes, in the echinoderms on arms with or without calcareous skeletal rods, and in the enteropneusts in meandering shapes on the blown-up body. The enteropneust larva, tornaria, has a large ring of compound cilia on multiciliate cells, the opisthotroch, at the edge of the expanded anal region (Fig. 60.1). The neotroch is an upstream-collecting ciliary band, which separates particles from the water current set up by the cilia through localized ciliary reversals (Strathmann 2007). The ciliary bands on the tentacles of the adult pterobranchs have a similar structure and the function is probably similar too. It is clearly different from the ciliary-sieving mechanism of the brachiozoans, which likewise collect particles on the upstream-side of the band, but through the mechanical sieve formed by the rather stiff laterofrontal cilia (Chapter 39). The neotroch is used both in particle collection and locomotion in echinoderm and the youngest enteropneust larvae, but the locomotory function is taken over by special regions of the neotroch of certain echinoid larvae and by the opisthotroch in the enteropneust tornaria larvae.

References

Balser, E.J. and Ruppert, E.E. 1990. Structure, ultrastructure, and function of the preoral heart-kidney in *Saccoglossus kowalevskii* (Hemichordata, Enteropneusta) including new data on the stomochord. *Acta Zool. (Stockh.)* **71**: 235–249.

Bateson, W. 1884. The early stages of the development of *Balanoglossus* (sp. incert.). *Q. J. Microsc. Sci., N. S.* **24**: 208–236.

Bather, F.A. 1900. The Echinoderma. In A.R. Lankester (ed.): *A Treatise on Zoology*, vol. 3, Adam & Charles Black, London.

Cannon, J.T., Rychel, A.L., Eccleston, H., Halanych, K.M. and Swalla, B.J. 2009. Molecular phylogeny of Hemichordata, with updated status of deep-sea enteropneusts. *Mol. Phylogenet. Evol.* **52**: 17–24.

Davis, B.M. 1908. The early life-history of *Dolichoglossus pusillus* Ritter. *University Calif. Publ. Zool.* **4**: 187–226.

Dawydoff, C. 1944. Formation des cavités coelomiques chez les tornaria du plancton indochinois. *C. R. Hebd. Seanc. Acad. Sci. Paris* **218**: 427–429.

Dawydoff, C. 1948. Classe des Entéropneustes + Classe des Ptérobranches. *Traité de Zoologie*, vol. 11, pp. 369–489. Masson, Paris.

Dilly, P.N., Welsch, U. and Rehkämper, G. 1986. Fine structure of heart, pericardium and glomerular vessel in *Cephalodiscus gracilis* M'Intosh, 1882 (Pterobranchia, Hemichordata). *Acta Zool. (Stockh.)* **67**: 173–179.

Edgecombe, G.D., Giribet, G., Dunn, C.W., *et al.* 2011. Higher-level metazoan relationships: recent progress and remaining questions. *Org. Divers. Evol.* **11**: 151–172

Gemmill, J.F. 1912. The development of the starfish *Solaster endeca* Forbes. *Trans. Zool. Soc. Lond.* **20**: 1–71.

Gruner, H.-E. 1980. Einführung. In H.-E. Gruner (ed.): *Lehrbuch der Speziellen Zoologie (4th ed.) (begründet von A. Kaestner)*, 1. Band, 1. Teil, pp. 15–156. Gustav Fischer, Stuttgart.

Hejnol, A., Obst, M., Stamatakis, A., *et al.* 2009. Assessing the root of bilaterian animals with scalable phylogenomic methods. *Proc. R. Soc. Lond. B* **276**: 4261–4270.

Holland, N.D. 1988. The meaning of developmental asymmetry for echinoderm evolution: a new interpretation. In C.R.C. Paul (ed.): *Echinoderm Phylogeny and Evolutionary Biology*, pp. 13–25. Oxford University Press, Oxford.

Holland, N.D. 1991. Echinodermata: Crinoidea. In A.C. Giese, J.S. Pearse and V.B. Pearse (eds): *Reproduction of Marine Invertebrates*, vol. 6, pp. 247–299. Boxwood Press, Pacific Grove, CA.

Hyman, L.H. 1955. *The Invertebrates, vol. 4. Echinodermata.* McGraw-Hill, New York.

Lester, S.M. 1985. *Cephalodiscus* sp. (Hemichordata: Pterobranchia): observations of functional morphology, behavior and occurrence in shallow water around Bermuda. *Mar. Biol.* **85**: 263–268.

Ludwig, H. 1882. Entwicklungsgeschichte der *Asterina gibbosa* Forbes. *Z. Wiss. Zool.* **37**: 1–98.

Metschnikoff, E. 1881. Über die systematische Stellung von *Balanoglossus* [2]. *Zool. Anz.* **4**: 154–157.

Nichols, D. 1962. *Echinoderms*. Hutchinson University Library, London.

Ruppert, E.E. and Balser, E.J. 1986. Nephridia in the larvae of hemichordates and echinoderms. *Biol. Bull.* **171**: 188–196.

Semon, R. 1888. Die Entwicklung der *Synapta digitata* und die Stammesgeschichte der Echinodermen. *Jena. Z. Naturw.* **22**: 1–135.

Stiasny, G. 1914. Studien über die Entwicklung des *Balanoglossus clavigerus* Delle Chiaje. II. Darstellung der weiteren Entwicklung bis zur Metamorphose. *Mitt. Zool. Stn. Neapel* **22**: 255–290.

Strathmann, R.R. 2007. Time and extent of ciliary response to particles in a non-filtering feeding mechanism. *Biol. Bull.* **212**: 93–103.

Ubaghs, G. 1967. General characters of echinoderms. In R.C. Moore (ed.): *Treatise of Invertebrate Paleontology, Part S*, vol. 1, pp. 3–60. Lawrence, KS.

Ubisch, L. 1913. Die Entwicklung von *Strongylocentrotus lividus* (*Echinus microtuberculatus, Arbacia pustulosa*). *Z. Wiss. Zool.* **106**: 409–448.

58

Phylum **Echinodermata**

The living Echinodermata is one of the most well-defined animal phyla, characterized by the unique pentameric body plan, with the left mesocoel (hydrocoel) specialized as the water-vascular system with a perioral ring and (usually) five radial canals with tube feet (podia). The pentameric symmetry can be recognized in all adult, living echinoderms, even though some holothurians and echinoids externally appear bilateral and some asteroids show a higher number of arms. The pentamery is secondary, as indicated by the bilaterality of early larval stages, where the pentameric symmetry develops through the very different growth of the mesocoel of the two sides (Smith 2008). Abnormal embryos with equal growth of the two sides develop into 'Janus-juveniles' with two oral sides (Herrmann 1981); there is no indication of a fusion of the mesocoels of the two sides as proposed by Morris (1999). The stereomic, calcareous, mesodermal skeleton is another characteristic of all echinoderms (Smith 2005), although it is inconspicuous, for example in some holothurians; similar mesodermal skeletal elements are found in some articulate brachiopods (Chapter 41). Also the mutable connective tissue is an echinoderm apomorphy. A peculiarity is that the echinoderms lack a brain and a well-defined anterior end; there are nerve cords around the mouth and along the ambulacra, but there is no central, coordinating nervous centre. This highly derived body plan is associated with co-options of Hox genes into new functions (see below).

All echinoderms are marine, and almost all are benthic in the adult stage, but planktotrophic larvae are known in almost all groups, and the pelago-benthic life cycle appears to be ancestral (Nielsen 1998; McEdward and Miner 2001). About 6600 living species are recognized, and there is a very extensive fossil record comprising about 13 000 species.

There was an early radiation in the Cambrian, and a number of extinct classes are known from the Cambrian-Ordovician; some of Cambrian forms had ambulacral plates indicating the presence of a water-vascular system with tube feet (Smith 2005; Sumrall and Wray 2007). The living classes (except the holothuroids) are known from the early Ordovician; all the other classes of the Cambrian radiation died out during the Late Palaeozoic, and only a few genera of the living classes survived the Permian-Triassic extinction (Smith 2004). There is no consensus about the phylogeny of the extinct groups (Mooi 2001), but studies on morphology, fossils, and molecules agree that the living forms represent two clades, Pelmatozoa with the crinoids, and Eleuterozoa with the remaining four classes. The interrelationships between the eleuterozoan classes is more uncertain, but the asteroids appear to be the sister group of the remaining three classes (Smith 2005; Swalla and Smith 2008; Janies

Chapter vignette: The stalked crinoid *Cenocrinus asterias*. (Based on Rasmussen 1977.)

et al. 2011). The small deep-sea echinoderm *Xyloplax* (Rowe *et al.*, 1988, 1994) was placed in a class of its own, the Concentricycloidea; however, the authors regarded the class as a sister group of certain valvatid asteroids, which is incompatible with the cladistic method adopted here. Subsequent authors have placed it within the asteroids (Janies and Mooi 1998; Janies *et al.* 2011).

The body is of a quite different shape in the five living classes, flower-shaped, star-shaped, globular, or worm-like, and there is considerable variation in the shape of the gut, which may be a straight tube between mouth and anus, coiled with several loops, a large sack without an anus, or a small sack with radial extensions. The microscopic anatomy was reviewed by Harrison (1994). All epithelia are monolayered, and one cilium is found on each cell both in ectoderm, endoderm, peritoneum, and in some myocytes.

The calcareous endoskeleton is formed in the mesoderm and has a lattice-like structure, called stereom. Each ossicle consists of numerous microcrystals with parallel orientation; they are secreted by primary mesenchyme cells in many of the larvae (see below) and some of the adult skeletal plates. The plates make firm contact, for example, in the test of the sea-urchins, but elsewhere they mostly form more loose connections or joints held together by collagenous material or muscles. The secretion of the plates is organized through a set of genes that are quite different from those that are involved in skeleton formation in vertebrates (Bottjer *et al.* 2006). There is an unusual type of connective tissue, catch or mutable connective tissue, that can change its stiffness drastically by nervous control through neuropeptides (Birenheide *et al.* 1998; Santos *et al.* 2005).

The presence of prosome, mesosome, and metasome cannot be recognized externally, but the development of the coelomic compartments from coelomic pouches of the bilateral larvae clearly reveals the archimery (Fig. 57.2). The most instructive illustrations of the structure and origin of the coelomic cavities are still to be found in Delage and Hérouard (1903).

The left protocoel (axocoel) is connected to the exterior through the madreporite in most groups. The coelomic cavity just below the madreporite, the madreporic chamber, is in open connection with the stone canal that is a part of the left mesocoel (see below). A canal from the madreporic chamber to a perioral ring is specialized as the axial gland or axial complex, which is a nephridium in most groups (Fig. 57.1). It consists of a haemal space lined by a basement membrane covered by peritoneal podocytes of the axocoel. Several elements in the axial complex contain muscle cells, and various parts of the organ have been observed to pulsate. It is believed that primary urine is formed through pressure ultrafiltration from the blood through the basement membrane and between the podocytes to the axocoel; the urine may become modified in the axial complex on the way to the madreporite that functions as a nephridiopore (Ruppert and Balser 1986). The axial complex is clearly recognized in asteroids, ophiuroids, and echinoids. In crinoids and many holothurians, the axial canal has lost contact with the madreporic chamber and opens into the metacoel; these groups use other organs/tissues for excretion, and excretion for example from gills, papulae, respiratory trees, and especially from the gut are known from many species of all classes.

The right protocoel and mesocoel degenerates early in development except for a small contractile sac, the dorsal sac, that surrounds a blind haemal space (MacBride 1896).

The left mesocoel (hydrocoel) forms the water-vascular system, with the stone canal along the axial complex from the madreporic chamber to a circumoral ring canal; radial canals extend from the ring canal along the body wall, giving rise to double rows of tube feet. The zones with radial canals are called radial areas or ambulacra, and the areas between them are called interradial or interambulacral. Mooi and David (1997, 1998, 2008) have pointed out that the surface can be divided into axial areas, associated with the growth of the ambulacra, and interaxial areas with no special growth pattern.

The metacoel consists of the larger, perivisceral coeloms, which surround gut, gonads, most of the other coelomic canals, and various narrow coelomic canals along the ambulacra. The asteroids have the most complicated system of metacoelomic canals comprising both an oral ring with radial canals and an aboral ring with extensions surrounding the gonads (Fig. 57.1). The left somatocoel is very spacious in echinoids, but both somatocoels are very narrow in the ophiuroids.

The above-mentioned haemal system is partially a spongiose mass, and it is clear that other organs are involved in transport of nutrients and oxygen. Peritoneal cilia in several of the coelomic compartments create circulation, for example around the gonads, in the coelomic fluid of the tube feet, and in gill-like structures with extensions of various coelomic compartments.

The nervous system is complicated and quite unusual in that a coordinating centre that could be called a brain is lacking (Cobb 1988, 1995; Heinzeller and Welsch 2001). There are ring nerves around the oesophagus and radial nerves along the ambulacra. The ring and radial nerves are internalized by a neurulation-like process in ophiuroids, echinoids, and holothuroids; the infolding, which forms the so-called epineural canals, can be directly observed during development (Ubisch 1913). These nerves consist of an ectodermal, intraepithelial part, called the ectoneural nerve, and a mesodermal part, called the hyponeural nerve, separated by a basement membrane. The hyponeural nerve is differentiated from the peritoneum of the narrow radial oral canal of the metacoel. In types with epineural canals, the ectoneural nerve is situated in the aboral ectodermal lining of the canal. The ectoneural part of the nervous system is apparently mainly sensory, while the hyponeural is motory; the communication between the two systems is via synapses across the basement membrane (Cobb 1995) or through short nerves between the two (Mashanov et al. 2006). Some muscle cells of podia and pedicellariae have a long thin extension reaching to the basement membrane just opposite to ectoneural nerve endings. Chemical synapses are rare and apparently difficult to identify, but have been described from asteroids, ophiuroids, and holothurians (Mashanov et al. 2006). Gap junctions or the pannexin/innexin and connexin genes have not been observed in any

echinoderm (Burke et al. 2006), so the electrical conduction between the cells must be through other structures (Garré and Bennett 2009).

The gonads are formed from mesodermal elements of the metacoel (MacBride 1896). The primordial germ cells in holothurians are epithelial with an apical cilium; the apical pole becomes the apical pole of the egg, and ultimately the apical pole of the larva (Frick and Ruppert 1996). The primary axis of the egg is thus maternally determined, as also shown by the classical experiments with transverse bisections of eggs and embryos of various stages (Hörstadius 1939a). Separate gonoducts open through pores in the genital plates (including the madreporite). Most species spawn the gametes freely in the water, where fertilization takes place, but a few have brood protection. The unfertilized egg has a jelly coat with a canal at the apical pole where the polar bodies will be given off (Boveri 1901). The sperm is of the primitive type with globular to conical heads (Jamieson 2000).

Reviews of the development of all five classes can be found in Chapters 4–8 in Giese et al. (1991). The oral-aboral axis becomes specified at the first cleavage in Strongylocentrotus (Cameron et al. 1989), but the axis is labile during the early stages as shown by the considerable powers of regeneration, as shown both by classical experiments with blastomere manipulation (Hörstadius 1939a) and modern experiments using marker gene expression (Davidson et al. 1998). Isolated blastomeres of the 2-cell stage develop into sexually mature adults in Strongylocentrotus (Cameron et al. 1996), and later embryological stages also regenerate when cut along the primary axis. Embryos and larvae of the starfish Asterina and Pisaster show similar patterns of regeneration (Maruyama and Shinoda 1990; Vickery and McClintock 1998). The oral-aboral axis becomes specified at a later stage in other sea urchins (Kominami 1983), and Henry et al. (1992) concluded that the cleavage pattern has become dissociated from axis specification in several cases. Fate mapping of blastomeres of the indirect-developing Ophiopholis aculeata shows an oblique pattern (Primus 2005). The first cleavage separates a dorsal and ventral blastomere, with endoderm developing exclusively

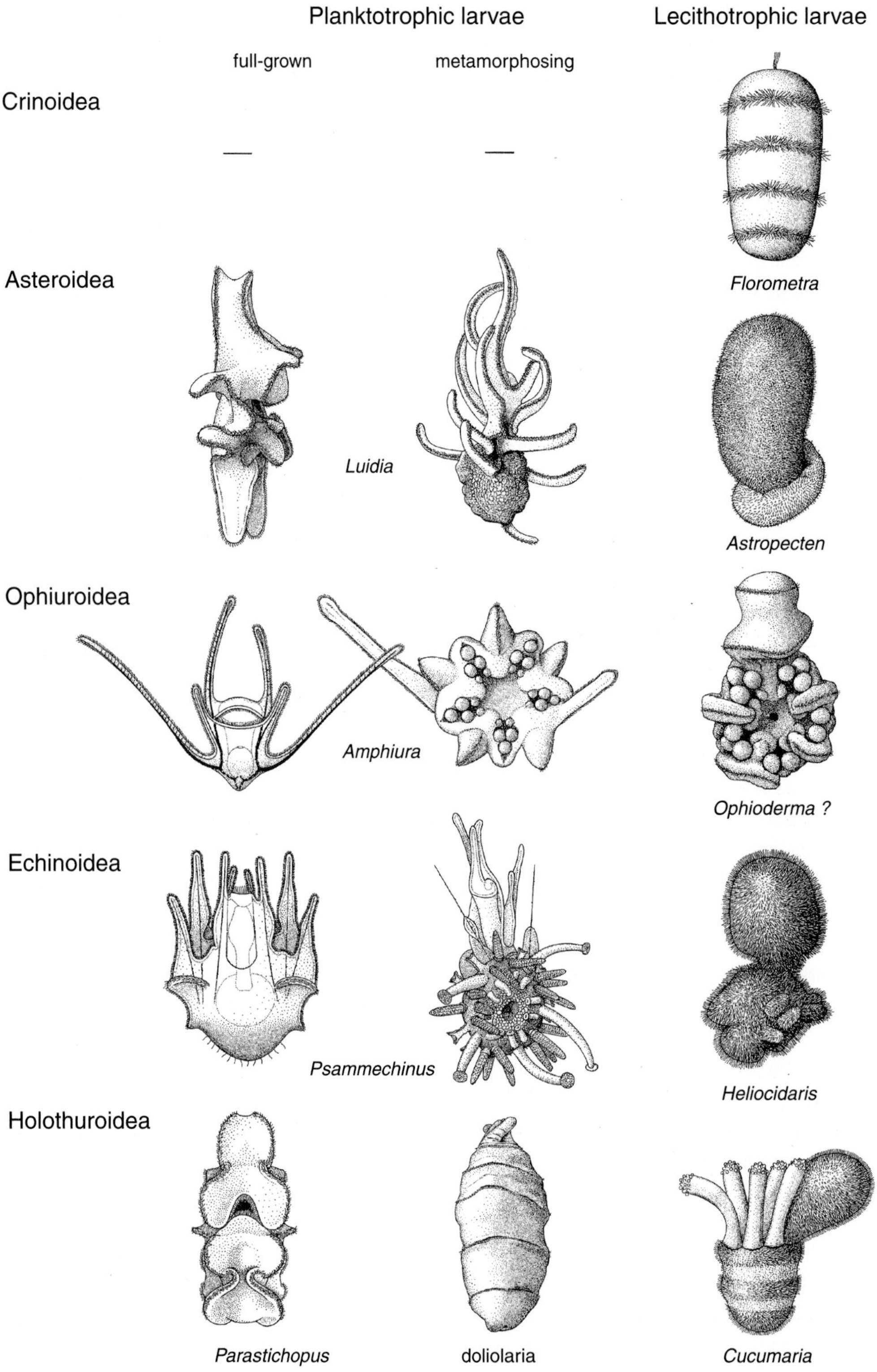

Planktotrophic larvae

full-grown   metamorphosing

Lecithotrophic larvae

Crinoidea

Florometra

Asteroidea

Luidia

Astropecten

Ophiuroidea

Amphiura

Ophioderma ?

Echinoidea

Psammechinus

Heliocidaris

Holothuroidea

Parastichopus   doliolaria   *Cucumaria*

and the mesoderm almost exclusively from the ventral blastomere. Experiments with isolated halves of early cleavage stages showed rather limited powers of regulation.

The first two cleavages follow the main axis of the egg so that four cells of equal size are formed, but what determines the planes of these cleavages is unknown. The oral-aboral axis is specified before the first cleavage in the direct developing sea urchin *Heliocidaris erythrogramma* but with some variations. The plane of the first cleavage is median in *Heliocidaris*, transverse in *Lytechinus*, oblique in *Strongylocentrotus*, and in any of these positions in *Hemicentrotus* (Wray 1997). The body axes of echinoid and asteroid larvae are problematic because the adult 'dorsal-ventral' axis is almost perpendicular to the larval oral-aboral axis, with a new mouth developing from an adult rudiment (imaginal disc) on the left side of the larva; ophiuroid development and metamorphosis seems easier to understand because its larval mouth becomes the adult mouth.

The embryology of echinoderms has been studied in detail by many authors over more than a century (see for example Delage and Hérouard 1903), and the classical studies of echinoids, such as *Echinus*, *Strongylocentrotus*, and *Paracentrotus* (MacBride 1903; Ubisch 1913), and asteroids, such as *Asterina* and *Asterias* (MacBride 1896; Gemmill 1914), form a good background for the modern studies especially of *Strongylocentrotus* and *Heliocidaris* (reviews in Wray 1997; Davidson *et al.* 1998; Arenas-Mena *et al.* 2000). Larval development and metamorphosis of *Psammechinus* are beautifully illustrated by Czihak (1960). The development of echinoids and asteroids

with planktotrophic larvae is especially well studied, and the description of the development will first deal with the echinoids, exemplified by *Strongylocentrotus purpuratus*, then with the other classes, and finally with types with lecithotrophic larvae and direct development (Fig. 58.1).

In *S. purpuratus*, the first two cleavages divide the embryo into lateral, anterior, and posterior quadrants, and the third cleavage is equatorial, dividing the embryo into an apical (animal) and a blastoporal (vegetal) half. The fourth cleavage divides the apical tier into a ring of eight cells, the blastoporal tier into an upper tier of four large macromeres, and a lower tier of four small micromeres; the micromeres divide again to form a central area of small micromeres surrounded by a ring of larger micromeres. A ciliated coeloblastula is formed after several cleavages, with a tuft of longer cilia at the apical pole indicating the first nervous centre, the apical organ (Byrne *et al.* 2007). The fates of the blastomeres have been followed in great detail and a fate map has been constructed (Davidson *et al.* 1998; Fig. 58.2). A complicated system of signalling, between the blastomeres of the late blastula to early gastrula stages, and its genetic background have been described by Davidson (1999). The main part of the blastoporal pole invaginates and gives rise to endoderm and mesoderm, with the most 'vegetal' part of the macromeres and the small micromeres giving rise to coelomic pouches and mesenchyme, whereas the larger micromeres become the skeletogenic cells. A band of narrow cells with long cilia, the neotroch, develops around the prospective oral field, which occupies the anterior and parts of the

Figure 58.1. Planktotrophic and lecithotrophic larval types of the five echinoderm classes. Crinoidea: lecithotrophic larva of *Florometra serratissima* (based on Lacalli and West 1986; planktotrophic larvae have not been reported). Asteroidea: full-grown bipinnaria larva and brachiolaria larva with developing sea star of *Luidia* sp. (plankton off Nassau, Bahamas, October 1990); lecithotrophic larva of *Astropecten latespinosus* (redrawn from Komatsu *et al.* 1988). Ophiuroidea: ophiopluteus of *Amphiura filiformis* (redrawn from Mortensen 1931), metamorphosing larva of *Amphiura* sp. (plankton off Kristineberg Marine Biological Station, Sweden, October 1984); lecithotrophic ophiuroid larva (possibly *Ophioderma squamulosa*, see Mortensen 1921) (plankton off San Salvador Island, Bahamas, October 1990). Echinoidea: full-grown echinopluteus larva and larva with developing sea urchin of *Psammechinus miliaris* (redrawn from Czihak 1960); lecithotrophic larva of *Heliocidaris erythrogramma* (redrawn from Williams and Anderson 1975). Holothuroidea: planktotrophic auricularia larva of *Parastichopus californicus* (Stimpson) (Friday Harbor Laboratories, WA, USA, July 1992); metamorphosing doliolaria larva of a holothuroid (plankton off Phuket Marine Biological Center, Thailand, March 1982); lecithotrophic larva of *Cucumaria elongata* (redrawn from Chia and Buchanan 1969).

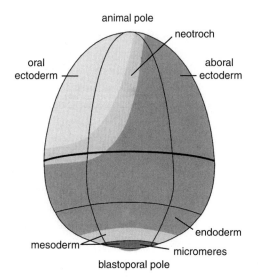

animal pole

neotroch

oral
ectoderm

aboral
ectoderm

endoderm

mesoderm

micromeres

blastoporal pole

Fig. 58.2. Fate map of an embryo of *Strongylocentrotus purpuratus* at a stage of about 400 cells. The position of the cleavage furrow between the apical (animal) and blastoporal (vegetal) cells of the 8-cell stage is indicated by the thick horizontal line. Note that the neotroch does not follow that line (compare with Fig. 22.10). (Adapted with permission from Davidson *et al.* 1998.) See Colour Plate 9.

lateral quadrants of the apical cells; the band crosses the apical pole and the apical organ becomes incorporated into the band. The apical end of the archenteron bends towards the oral field and a new mouth breaks through. This larval type is called dipleurula. The larva now assumes a more prismatic shape, with the first stages of the arms developing, and the mouth situated in the middle of the slightly concave oral field surrounded by the neotroch. A group of serotonergic nerve cells is located in the transversely elongated apical part of the ciliary band (Byrne *et al.* 2007).

The coelomic mesoderm forms as a pair of small lateral pockets at the apical end of the archenteron (enterocoely). The right and left coelomic pockets have quite different fates, with the water-vascular system developing from the left coelomic compartment. Abnormalities with the water-vascular system developing from the right side, or with a system with subsequent development of a set of tube feet on each side, have been observed several times, both in asteroids and echinoids (Gemmill 1915; Newman 1925; Herrmann 1981).

The left coelomic pouch elongates and divides into an anterior part, which becomes axocoel and hydrocoel, and a posterior metacoel, that becomes the main body cavity. The anterior sac extends towards the dorsal side and meets an invagination from the dorsal ectoderm, forming the hydropore. The narrow canal from the coelomic pouch to the exterior is ciliated and these cilia slowly transport coelomic fluid towards the hydropore; this is interpreted as a primitive excretory organ, the primordium of the adult axial gland (observed in *Asterias*; Ruppert and Balser 1986) (Fig. 57.1). The anterior part of this complex becomes the axocoel, with the axial gland and an oral ring in close contact with the metacoelomic oral ring described below. The posterior part becomes the hydrocoel with the stone canal; its posterior end elongates and curves into a circle with five small buds; this is the first stage of the stone canal and the ring canal with the radial canals of the water-vascular system.

The right coelomic pouch likewise divides into an anterior axo-hydrocoel and a posterior metacoel, which becomes the main body cavity. The anterior part disappears except for a small pulsating pouch, that later becomes the small, pulsating pericardium (dorsal sac).

The left and right metacoelomic pouches develop differently. The left pouch curves around the oral invagination of the adult rudiment and becomes the larger, oral body cavity. Both the metacoelomic oral ring with radial canals and (in asteroids) the aboral ring with extensions surrounding the gonads (Fig. 57.1) develop as five small extensions from the main coelom; these extensions become Y-shaped and the branches fuse into a ring.

The neotroch becomes extended onto long larval arms each with a lattice-like, calcareous skeletal rod. This larva is called echinopluteus (see Fig. 58.1). The neotroch is an upstream-collecting system, which is both particle-collecting and locomotory in the planktotrophic larvae. In normal swimming, the direction of the ciliary beat is away from the oral field (Strathmann 1975), and parts of the neotroch may be specialized for locomotion, for example as the epaulettes of some echinoid larvae. The direction of the beat can be reversed, so that the larva can swim backwards for some time (Strathmann 1971). The mechanism of particle 'capture' has been much discussed, but recent high-speed observations (Strathmann 2007) have shown that the cilia respond to the parcel of water surrounding a particle by reversing the direction of the beat and thereby redirecting the particle towards the oral field. Thus, there is no direct contact with the cilia and the mechanism cannot be described as mechanical filtering. The larval nervous systems of all five classes are discussed below.

The larvae then develop an adult rudiment with a new mouth surrounded by the first five tube feet on the left side of the larva (see Fig. 58.1), and the larval mouth and oesophagus become discarded or resorbed together with other parts of the larval body at metamorphosis. The oral side of the rudiment is formed from an invagination of the larval body wall, the vestibule or amnion, in most echinoids and asteroids. The rudiment gives rise to the adult nervous system (Davidson *et al.* 1998), and first five tube feet around the mouth each contain the tip of one of the radial canals of the water-vascular system. The larva is now competent for settling. In most species the larval arms with the ciliary bands degenerate and the small juveniles sink to the bottom. However, the competent larvae of some species with rather short arms test the substratum and swim up into the plankton again if is not found suitable (Gosselin and Jangoux 1998). The larval arms with the skeleton become resorbed, with exception of the basal parts of the skeletal arms that become transformed into the four genital plates (Emlet 1985). This indicates that also the adult skeletal system is formed by the descendants of the large micromeres.

The development of asteroids with planktotrophic larvae follow much the same pattern (Gemmill 1914), but there are differences in the development of the coeloms. In *Asterias* and *Astropecten* larvae, the anterior coelomic pouches fuse and remain as a large sac during the whole larval period (Gemmill 1914; Hörstadius 1939b), but one large pouch develops from the apical part of the archenteron in *Asterina* (Morris *et al.* 2009). The posterior part of the left metacoel curves around the rectum, and the posterior part of the right meta-

coel occupies the posterior pole of the larva. The median part of the left coelomic pouch, the mesocoel (hydrocoel), differentiates in an adult rudiment, as in the echinoids. The coelomic compartments become separated when the rudiment disc develops.

The dipleurula develops soft, flexible larval arms with loops of the neotroch; this stage is called bipinnaria. Some species later develop three short, pre-oral arms without ciliary bands, but with small attachment organs and extensions of the anterior coelomic sac. This type of larva is called brachiolaria; it can attach temporarily with the brachiolar arms, and this has been interpreted as an indication of an ancestral type of attachment (Grobben 1923). Most of the larval body is resorbed at metamorphosis, except in *Luidia sarsi*, where the juvenile detaches from the larval body (Tattersall and Sheppard 1934).

Ophiuroid embryology generally resembles that of the echinoids, and the dipleurula develops arms with skeletal spicules and ciliary bands, so that the larva, called ophiopluteus (Fig. 58.1), resembles the echinopluteus. The hydrocoel develops as a loop around the oesophagus, and the larval mouth is retained in the juvenile; the anus disappears. Near metamorphosis, the ciliary bands break up and the arms degenerate (Fig. 58.1). The small juveniles sink to the bottom.

Some holothurians have a planktotrophic auricularia larva with looped ciliary bands, which eventually develops into a lecithotrophic stage with five circular ciliary bands (a doliolaria larva; Fig. 58.3). A deviating mode of forming coelom and hydropore is seen in holothurians, such as *Labidoplax* (Selenka 1883) and *Stichopus* (Rustad 1938), in which the apical end of the archenteron first curves towards the aboral side and forms a connection with the exterior, thus forming the hydropore; the distal part of the archenteron then pinches off as the coelomic pouch, and the archenteron curves towards the oral side.

The development of the stalked crinoids, the sea lilies, which supposedly represent a more ancestral life cycle, are poorly known, but the lecithotrophic auricularia larva of *Metacrinus* has a curved neotroch that becomes broken up and reorganized into rings (Nakano *et al.* 2003), just like in the development of

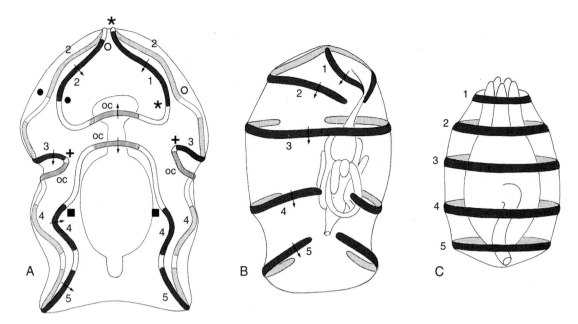

Figure 58.3. Transformation of the neotroch of the larva of *Synapta digitata* from an auricularia stage through an intermediate stage to an old doliolaria stage that has developed the first tube feet. The four parts of the neotroch marked move to the mouth and sink into the buccal cavity. The various signatures indicate points of the neotroch that fuse to form the circular bands (marked 1–5) of the doliolaria stage. (From Nielsen 1987.)

Synapta. The feather stars go through a gastrula stage, the coelomic pouches develop from the archenteron in a highly specialized pattern, and the embryo develops into a lecithotrophic doliolaria larva. The larva settles with an adhesive pit in front of the apical organ and develops into a small sea lily, but the body of the juvenile feather stars breaks away from the stalk and becomes free-living (Holland 1991). It is often claimed that the axocoel in crinoids develops an extension into the stalk, but this has been refuted by Grimmer *et al.* (1985).

The development of the larval nervous systems of the various types of larvae shows some variation, but a common pattern can be recognized (Byrne *et al.* 2007). Nerve cells develop either at the apical pole or more scattered over the gastrula, but soon become concentrated at the apical pole. In planktotrophic larvae, nerve cells are concentrated in two areas in the apical region just lateral to the apical loops of the neotroch; short axons span the narrow space. Axons from these ganglia extend along the ciliary bands of the arms.

Asteroid and holothurioid larvae have cell bodies on these nerves. The ultrastructure of the ciliary bands of *Pisaster* has been studied by Lacalli *et al.* (1990), who found three types of nerve cells in the bands, ciliated bipolar cells, ciliated sensory cells, and multipolar cells, with cell processes extending between the microvilli of the cuticle. Some of these cells resemble those in the developing neural tube of amphioxus (Lacalli and West 1993) (see also Chapter 63). The larval nervous systems degenerate at metamorphosis.

The many species with lecithotrophic larvae and especially those with direct development show interesting deviations from the general developmental pattern described above, but only some examples will be discussed here.

Some echinoids have planktonic larvae with facultative feeding (Hart 1996), and this is probably an intermediate stage in the evolution towards direct development (Wray 1996). Lecithotrophic larvae may be superficially similar to planktotrophic larvae of related species, but there are many examples of highly derived lecithotrophic types (Fig. 58.1). The ophiuroid genus *Macrophiothrix* contains species of all types, from planktotrophic larvae, through facultative feeders, to lecithotrophic larvae with longer or shorter arms (Allen and Podolsky 2007). The lecithotrophic doliolaria of some holothurians and crinoids is barrel-shaped with four or five ciliated rings. It is clearly derived from the planktotrophic type, as shown by the development from a planktotrophic auricularia, through a doliolaria, to the juvenile bottom stage (see above). This is further supported by the expression of *Otx* both in ciliated bands of planktotrophic larvae and in the ciliated bands of the holothurian doliolaria larva (Lowe *et al.* 2002).

As mentioned above, the planktotrophic type is considered ancestral, and this is further supported by observations on *Heliocidaris*, where the indirect developer *H. tuberculata* and the direct developer *H. erythrogramma* deposit similar yolk granules into the oocytes until it reaches the size of the ripe egg of *H. tuberculata*. The following, enormous growth of the egg of *H. erythrogramma* is characterized by a massive deposition of non-vitellogenetic material, such as maternal protein and lipid (Byrne *et al.* 1999). Also the presence of vestigial larval skeletons in lecithotrophic larvae of ophiuroids and echinoids supports this notion (Hendler 1982; Emlet 1995). Observations on fossil echinoids indicate that non-planktotrophy evolved independently in nine clades at the end of the Cretaceous (Jeffery 1997).

The origin of the gut, mouth, and coelomic compartments of the lecithotrophic larvae resembles that of the planktotrophic larvae to various degrees.

Direct development, often associated with brooding and in certain cases with ovovivipary (Komatsu *et al.* 1990), occurs in all eleuterozoan classes. The asteroid family Asterinidae comprises species with all types of development from planktotrophic larvae, through lecithotrophic larvae, non-planktonic larvae in an egg mass, to viviparity (Raff and Byrne 2006). An extremely derived developmental type is seen in the asteroid *Pteraster*, where the adult oral-aboral axis develops directly, parallel to the embryonic main axis; there is no sign of arms or ciliary bands and the rudiments of axocoel, somatocoels, and five radial elements of the hydrocoel develop as separate pouches from the archenteron (McEdward 1992; Janies and McEdward 1993).

The embryology of the direct-developing echinoid *Heliocidaris erythrogramma* has been compared with that of the planktotrophic *H. tuberculatus* and the general type described above (Wray and Raff 1990). The cleavage pattern is different, with *H. erythrogramma* having equal cleavage up to the 32-cell stage, and the fates of the corresponding blastomeres of the two types of embryos showing conspicuous differences. However, a comparison of the fate maps of the two types show complete identity of the spatial relationships between areas developing into identical organs in the larva.

The ophiuroid *Amphiodia* (Emlet 2006) spawns demersal eggs and the developing embryos show no signs of larval structures, but, as far as could be seen, the differentiation of the coelomic pouches proceeded in the usual way. The planktotrophic larval types described above from the four eleuterozoan classes, and the lecithotrophic *Metacrinus* larva, are clearly variations of a common theme: the dipleurula, with the ciliary bands extended on various types of arms with or without skeleton, or convoluted and broken up to form, for example, the circular bands of the doliolaria larvae (Fig. 58.3). Most of these larvae go through a complex metamorphosis during which the larval structures are abandoned. The numerous intermediate types, between the planktotrophic larvae and various types of lecithotrophic larvae, or direct development clearly shows that the change away from planktotrophy has taken place numerous times within all the classes (Strathmann 1974). It seems exceedingly improbable that the evolution has gone in the opposite direction and given rise to so many larval forms with identical ciliary feeding systems (Strathmann 1978, 1988; Wray 1996; Nielsen 1998; Villinski *et al.* 2002). The variations in cell lineage observed between different echinoids that have identical fate maps has a parallel in the annelids (Chapter 25) and it is clear that the cleavage patterns are much more labile than the spatial relationships between the different areas of the embryos.

Regeneration and autotomy occur in all classes and asexual reproduction by fission is known in asteroids, ophiuroids, and holothurians (Emson and Wilkie 1980). It has been presumed that the large larval body left after the liberation of the juvenile in *Luidia sarsi* should be able to survive and form a new sea star, but this has never been observed directly. However, ophiopluteus larvae of *Ophiopholis* have been observed to release the juvenile from the long posterolateral arms, which then regenerate the larval/juvenile organs and release a new juvenile (Balser 1998).

A completely different type of asexual reproduction has been observed in larvae of another species of *Luidia*, in which the posterior larval arms may form a new gut through a gastrulation process and then become pinched off so that a complete bipinnaria larva is formed (Bosch *et al.* 1989).

Larval types can sometimes give a clue to the understanding of phylogenetic relationships within a larger systematic category, but the larval types of the five classes of living echinoderms each show such wide variation that it is impossible to see a pattern (Strathmann 1988). The skeletons of echinoplutei and ophioplutei look very similar, but may not be homologous (Raff *et al.* 1987). The development of the adult from an adult rudiment on the left side of the larva is common to echinoids and asteroids and could indicate a sister-group relationship, but this is neither supported by morphological nor by molecular studies. In most echinoid and asteroid larvae, the rudiment disc is formed from the bottom of an epidermal invagination of the left side of the larva, the vestibule, or amniotic invagination. This has been interpreted as a synapomorphy of the two groups, but the elaborate, planktotrophic larva of the echinoid *Eucidaris* has a metamorphosis without trace of a vestibule (Emlet 1988), so the importance of this character is uncertain indeed. On the other hand, the early larval stage with apical organ, complete gut, slightly curved peri-oral ciliary band, and three pairs of coelomic pouches, i.e. the dipleurula, can be identified as common to all echinoderms. This is important for the understanding of the phylogenetic position of the phylum.

The highly derived character of the echinoderms is also revealed by the Hox genes. An almost complete 'ambulacrarian signature' is found (Fig. 21.3), but the *Hox1-3* genes are not expressed in early development, and at least in *Strongylocentrotus* the *Hox1-3* genes have been relocated to the 5´ end of the cluster and inverted (Cameron *et al.* 2006; Mooi and David 2008). *Hox3* was found to be expressed in the somatocoelar tooth sacs in the adult rudiment in good agreement with its posterior position in the Hox cluster (Arenas-Mena *et al.* 1998). The more posterior *Hox7-11/13b* are expressed in the somatocoels collinearly along the gut (Arenas-Mena *et al.* 2000).

The gene regulatory networks of the *Strongylocentrotus*, especially of the endoderm and mesoderm, are known in great detail (Davidson 2001), and this shows great possibilities for phylogenetic studies when other species become known in similar detail. An example is one of the 'kernels' of the gene regulatory network specifying the development of mesoderm in sea urchins and sea stars, where evolution from a common kernel can be deduced (Davidson and Erwin 2006; Peter and Davidson 2009).

The origin of the echinoderms has been discussed in numerous papers for more than a century. The bilaterally symmetrical larvae indicate that the ancestor of the echinoderms was bilateral, but the evolution of the asymmetric adults with the water-vascular system, and later on the pentamery, has been the subject of much speculation. Most of the theories compare the echinoderms with *Cephalodiscus* and propose that the ciliated tentacles with pinnules of the left side of the mesosome should have curved round the mouth and become specialized, as in the water-vascular system with podia (see for example Grobben 1923; Jefferies 1986; Holland 1988). However, it now appears that the echinoderms are not the sister group of the pterobranchs, and that the gill-slits of enteropneusts and chordates are homologous, which implies that the common ancestor of the deuterostomes had gill-slits, and therefore probably not tentacles (see Chapter 57). The pterobranch tentacles and the water-vascular system with podia of the echinoderms are both specializations of the mesosome, but the structures cannot be homologous. The Cambrian fossils are difficult to interpret and give no clue to the origin of the water vascular system, but it appears that the ancestral echinoderms had only one ray (arm), and that the pentamery possibly originated through duplications (Hotchkiss 1998).

The origin of the echinoderms is still enigmatic.

Interesting subjects for future research

1. Origin of the adult nervous system
2. Crinoid embryology

References

Allen, J.D. and Podolsky, R.D. 2007. Uncommon diversity in developmental mode and larval form in the genus *Macrophiothrix* (Echinodermata: Ophiuroidea). *Mar. Biol.* **151**: 85–97.

Arenas-Mena, C., Cameron, A.R. and Davidson, E.H. 2000. Spatial expression of *Hox* cluster genes in the ontogeny of a sea urchin. *Development* **127**: 4631–4643.

Arenas-Mena, C., Martinez, P., Cameron, R.A. and Davidson, E.H. 1998. Expression of the *Hox* gene complex in the indirect development of a sea urchin. *Proc. Natl. Acad. Sci. USA* **95**: 13062–13067.

Balser, E.J. 1998. Cloning by ophiuroid echinoderm larvae. *Biol. Bull.* **194**: 187–193.

Birenheide, R., Tamori, M., Motokawa, T., *et al.* 1998. Peptides controlling stiffness of connective tissue in sea cucumbers. *Biol. Rev.* **194**: 253–259.

Bosch, I., Rivkin, R.B. and Alexander, S.P. 1989. Asexual reproduction by oceanic echinoderm larvae. *Nature* **337**: 169–170.

Bottjer, D.J., Davidson, E.H., Peterson, K.J. and Cameron, A.R. 2006. Paleogenomics of echinoderms. *Science* **314**: 956–960.

Boveri, T. 1901. Über die Polarität von Ovocyte, Ei und Larve des *Strongylocentrotus lividus*. *Zool. Jahrb., Anat.* **14**: 630–653.

Burke, R.D., Angerer, L.M., Elphick, M.R., *et al.* 2006. A genomic view of the sea urchin nervous system. *Dev. Biol.* **300**: 434–460.

Byrne, M., Nakajima, Y., Chee, F.C. and Burke, R.D. 2007. Apical organs in echinoderm larvae: insights into larval evolution in the Ambulacraria. *Evol. Dev.* **9**: 432–445.

Byrne, M., Villinski, J.T., Cisternas, P., *et al.* 1999. Maternal factors and the evolution of developmental mode: evolution of oogenesis in *Heliocidaris erythrogramma*. *Dev. Genes Evol.* **209**: 275–283.

Cameron, A.R., Bowen, L., Nesbitt, R., *et al.* 2006. Unusual gene order and organization of the sea urchin Hox cluster. *J. Exp. Zool. (Mol. Dev. Evol.)* **306B**: 45–58.

Cameron, R.A., Fraser, S.E., Britten, R.J. and Davidson, E.H. 1989. The oral-aboral axis of a sea urchin embryo is specified by first cleavage. *Development* **106**: 641–647.

Cameron, R.A., Leahy, P.S. and Davidson, E.H. 1996. Twins raised from separated blastomeres develop into sexually mature *Strongylocentrotus purpuratus*. *Dev. Biol.* **178**: 514–519.

Chia, F.S. and Buchanan, J.B. 1969. Larval development of *Cucumaria elongata* (Echinodermata: Holothuroidea). *J. Mar. Biol. Assoc. U.K.* **49**: 151–159.

Cobb, J.L.S. 1988. A preliminary hypothesis to account for the neural basis of behaviour in echinoderms. In R.D. Burke (ed.): *Echinoderm Biology*, pp. 565–573. A.A. Balkema, Rotterdam.

Cobb, J.L.S. 1995. The nervous systems of Echinodermata: Recent results and new approaches. In O. Breidbach (ed.): *The Nervous Systems of Invertebrates: An Evolutionary and Comparative Approach*, pp. 407–424. Birkhäuser Verlag, Basel.

Czihak, G. 1960. Untersuchungen über die Coelomanlagen und die Metamorphose des Pluteus von *Psammechinus miliaris* (Gmelin). *Zool. Jahrb., Anat.* **78**: 235–256.

Davidson, E.H. 1999. A view from the genome: spatial control of transcription in sea urchin development. *Curr. Opin. Genet. Dev.* **9**: 530–541.

Davidson, E.H. 2001. *Genomic Regulatory Systems. Development and Evolution*. Academic Press, San Diego.

Davidson, E.H. and Erwin, D.H. 2006. Gene regulatory networks and the evolution of animal body plans. *Science* **311**: 796–800.

Davidson, E.H., Cameron, R.A. and Ransick, A. 1998. Specification of cell fate in the sea urchin embryo: summary and some proposed mechanisms. *Development* **125**: 3269–3290.

Delage, Y. and Hérouard, E. 1903. *Traité de Zoologie Concrète, vol. 3. Les Échinodermes*. Schleicher, Paris.

Emlet, R.B. 1985. Crystal axes in recent and fossil adult echinoids indicate trophic mode in larval development. *Science* **230**: 937–940.

Emlet, R.B. 1988. Larval form and metamorphosis of a 'primitive' sea urchin, *Eucidaris thouarsi* (Echinodermata: Echinoidea: Cidaroida), with implications for developmental and phylogenetic studies. *Biol. Bull.* **174**: 4–49.

Emlet, R.B. 1995. Larval spicules, cilia, and symmetry as remnants of indirect development in the direct developing sea urchin *Heliocidaris erythrogramma*. *Dev. Biol.* **167**: 405–415.

Emlet, R.B. 2006. Direct development of the brittle star *Amphiodia occidentalis* (Ophiuroidea, Amphiuridae) from the northeastern Pacific Ocean. *Invert. Biol.* **125**: 154–171.

Emson, R.H. and Wilkie, I.C. 1980. Fission and autotomy in echinoderms. *Oceanogr. Mar. Biol. Ann. Rev.* **18**: 155–250.

Frick, J.E. and Ruppert, E.E. 1996. Primordial germ cells of *Synaptula hydriformis* (Holothuroidea: Echinodermata) are epithelial flagellated-collar cells: their apical-basal polarity becomes primary egg polarity. *Biol. Bull.* **191**: 168–177.

Garré, J.M. and Bennett, M.V.L. 2009. Gap junctions as electrical synapses. In M. Hortsch and H. Umemori (eds): *The Sticky Synapse*, pp. 423–439. Springer, Dordrecht.

Gemmill, J.F. 1914. The development and certain points in the adult structure of the starfish *Asterias rubens*, L. *Phil. Trans. R. Soc. Lond. B* **205**: 213–294.

Gemmill, J.F. 1915. Double hydrocoele in the development and metamorphosis of the larva of *Asterias rubens*, L. *Q. J. Microsc. Sci., N. S.* **61**: 51–60.

Giese, A.C., Pearse, J.S. and Pearse, V.B. (eds.) 1991. *Reproduction of Marine Invertebrates*, vol. 6. The Boxwood Press, Pacific Grove, CA.

Gosselin, P. and Jangoux, M. 1998. From competent larva to exotrophic juvenile: a morphofunctional study of the perimetamorphic period of *Paracentrotus lividus* (Echinodermata, Echinoida). *Zoomorphology* **118**: 31–43.

Grimmer, J.C., Holland, N.D. and Hayami, I. 1985. Fine structure of the stalk of an isocrinoid sea lily (*Metacrinus rotundus*) (Echinodermata, Crinoidea). *Zoomorphology* **105**: 39–50.

Grobben, K. 1923. Theoretische Erörterungen betreffend die phylogenetische Ableitung der Echinodermen. *Sitzungsber. Akad. Wiss. Wien, 1. Abt.* **132**: 263–290.

Harrison, F.W. (ed.) 1994. *Microscopic Anatomy of Invertebrates, vol. 14 (Echinodermata)*. Wiley-Liss, New York.

Hart, M.W. 1996. Evolutionary loss of larval feeding: development, form and function in a facultatively feeding larva, *Brisaster latifrons*. *Evolution* **50**: 174–187.

Heinzeller, T. and Welsch, U. 2001. The echinoderm nervous system and its phylogenetic interpretation. In G. Roth and M.F. Wullimann (eds): *Brain Evolution and Cognition*, pp. 41–75. Wiley, New York.

Hendler, G. 1982. An echinoderm vitellaria with a bilateral larval skeleton: evidence for the evolution of ophiuroid vitellariae from ophioplutei. *Biol. Bull.* **163**: 431–437.

Henry, J.J., Klueg, K.M. and Raff, R.A. 1992. Evolutionary dissociation between cleavage, cell lineage and embryonic axes in sea urchin embryos. *Development* **114**: 931–938.

Herrmann, K. 1981. Metamorphose aberranter Formen biem Seeigel (*Psammechinus miliaris*). *Publ. Wiss. Film., Biol.*, 18. *Ser.* **4**: 1–12.

Holland, N.D. 1988. The meaning of developmental asymmetry for echinoderm evolution: a new interpretation. In C.R.C. Paul (ed.): *Echinoderm Phylogeny and Evolutionary Biology*, pp. 13–25. Oxford University Press, Oxford.

Holland, N.D. 1991. Echinodermata: Crinoidea. In A.C. Giese, J.S. Pearse and V.B. Pearse (eds): *Reproduction of Marine Invertebrates*, vol. 6, pp. 247–299. Boxwood Press, Pacific Grove, CA.

Hotchkiss, F.H.C. 1998. A 'rays-as-appendages' model for the origin of pentamerism in echinoderms. *Paleobiology* **24**: 200–214.

Hörstadius, S. 1939a. The mechanisms of sea urchin development studied by operative methods. *Biol. Rev.* **14**: 132–179.

Hörstadius, S. 1939b. Über die Entwicklung von *Astropecten aranciacus* L. *Pubbl. Stn. Zool. Napoli* **17**: 222–312.

Jamieson, B.G.M. 2000. Echinodermata. In K.G. Adiyodi, R.G. Adiyodi and B.G.M. Jamieson (eds): *Reproductive Biology of Invertebrates*, vol. 9C, pp. 163–246. John Wiley, Chichester.

Janies, D., Janet, V. and Daly, M. 2011. Echinoderm phylogeny including *Xyloplax*, a progenetic asteroid. *Syst. Biol.* **60**: 420–438.

Janies, D. and Mooi, R. 1998. *Xyloplax* is an asteroid. In M.D. Candia Carnevali (ed.): *Echinoderm Research 1998*, pp. 311–316. Balkema, Rotterdam.

Janies, D.A. and McEdward, L.R. 1993. Highly derived coelomic and water-vascular morphogenesis in a starfish with pelagic direct development. *Biol. Bull.* **185**: 56–75.

Jefferies, R.P.S. 1986. *The Ancestry of the Vertebrates*. British Museum (Natural History), London.

Jeffery, C.H. 1997. Dawn of echinoid nonplanktotrophy: coordinated shifts in development indicate environmental instability prior to the K-T boundary. *Geology* **25**: 991–994.

Komatsu, M., Kano, Y.T. and Oguro, C. 1990. Development of a true ovoviviparous sea star, *Asterina pseudoexigua pacifica* Hayashi. *Biol. Bull.* **179**: 254–263.

Komatsu, M., Murase, M. and Oguro, C. 1988. Morphology of the barrel-shaped larva of the sea-star, *Astropecten latespinosus*. In R.D. Burke (ed.): *Echinoderm Biology*, pp. 267–272. Balkema, Rotterdam.

Kominami, T. 1983. Establishment of embryonic axes in larvae of the starfish, *Asterina pectinifera*. *J. Embryol. Exp. Morphol.* **75**: 87–100.

Lacalli, T.C. and West, J.E. 1986. Ciliary band formation in the doliolaria larva of *Florometra*. I. The development of normal epithelial pattern. *J. Embryol. Exp. Morphol.* **96**: 303–323.

Lacalli, T.C. and West, J.E. 1993. A distinctive nerve cell type common to diverse deuterostome larvae: comparative data from echinoderms, hemichordates and amphioxus. *Acta Zool. (Stockh.)* **74**: 1–8.

Lacalli, T.C., Gilmour, T.H.J. and West, J.E. 1990. Ciliary band innervation in the bipinnaria larva of *Pisaster ochraceus*. *Phil. Trans. R. Soc. Lond. B* **330**: 371–390.

Lowe, C.J., Issel-Tarver, L. and Wray, G.A. 2002. Gene expression and larval evolution: changing roles of *distal-less* and *orthodenticle* in echinoderm larvae. *Evol. Dev.* **4**: 111–123.

MacBride, E.W. 1896. The development of *Asterina gibbosa*. *Q. J. Microsc. Sci., N. S.* **38**: 339–411.

MacBride, E.W. 1903. The development of *Echinus esculentus*, together with some points in the develoment of

E. miliaris and *E. acutus*. *Phil. Trans. R. Soc. Lond. B* **195**: 285–327.

Maruyama, Y.K. and Shinoda, M. 1990. Archenteron-forming capacity in blastomeres isolated from eight-cell stage embryos of the starfish, *Asterina pectinifera*. *Dev. Growth Differ.* **32**: 73–84.

Mashanov, V.S., Zueva, O.R., Heinzeller, T. and Dolmatov, I.Y. 2006. Ultrastructure of the circumoral nerve ring and the radial nerve cords in holothurians (Echinodermata). *Zoomorphology* **125**: 27–38.

McEdward, L.R. 1992. Morphology and development of a unique type of pelagic larva in the starfish *Pteraster tesselatus* (Echinodermata: Asteroidea). *Biol. Bull.* **182**: 177–187.

McEdward, L.R. and Miner, B.G. 2001. Larval and life-cycle patterns in echinoderms. *Can. J. Zool.* **79**: 1125–1170.

Mooi, R. 2001. Not all written in stone: interdisciplinary syntheses in echinoderm paleontology. *Can. J. Zool.* **79**: 1209–1231.

Mooi, R. and David, B. 1997. Skeletal homologies of echinoderms. *Paleontol. Soc. Pap.* **3**: 305–335.

Mooi, R. and David, B. 1998. Evolution within a bizarre phylum: homologies of the first echinoderm. *Am. Zool.* **38**: 965–974.

Mooi, R. and David, B. 2008. Radial symmetry, the anterior/posterior axis, and echinoderm Hox genes. *Annu. Rev. Ecol. Syst.* **39**: 43–63.

Morris, V.B. 1999. Bilateral homologues in echinoderms and a predictive model of the bilaterian echinoderm ancestor. *Biol. J. Linn. Soc.* **66**: 293–303.

Morris, V.B., Selvakumaraswamy, P., Whan, R. and Byrne, M. 2009. Development of the five primary podia from the coeloms of a sea star larva: homology with the echinoid echinoderms and other deuterostomes. *Proc. R. Soc. Lond. B* **276**: 1277–1284.

Mortensen, T. 1921. *Studies on the Development and Larval Forms of Echinoderms*. G.E.C. Gad, Copenhagen.

Mortensen, T. 1931. Contributions to the study of the development and larval forms of echinoderms. I–II. *K. Dan. Vidensk. Selsk. Skr., Naturvidensk. Math. Afd.*, 9. Rk. **4(1)**: 1–19.

Nakano, H., Hibino, T., Oji, T., Hara, Y. and Amemiya, S. 2003. Larval stages of a living sea lily (stalked crinoid echinoderm). *Nature* **421**: 158–160.

Newman, H.H. 1925. On the occurrence of paired madreporic pores and pore-canals in the advanced bipinnaria larvae of *Asterina (Patiria) miniata* together with a discussion of the significance of similar structures in other echinoderm larvae. *Biol. Bull.* **40**: 118–125.

Nielsen, C. 1987. Structure and function of metazoan ciliary bands and their phylogenetic significance. *Acta Zool. (Stockh.)* **68**: 205–262.

Nielsen, C. 1998. Origin and evolution of animal life cycles. *Biol. Rev.* **73**: 125–155.

Peter, I.S. and Davidson, E.H. 2009. Genomic control of patterning. *Int. J. Dev. Biol.* **53**: 707–716.

Primus, A.E. 2005. Regional specification in the early embryo of the brittle star *Ophiopholis aculeata*. *Dev. Biol.* **283**: 294–309.

Raff, R.A., Anstrom, J.A., Chin, J.E., *et al.* 1987. Molecular and developmental correlates of macroevolution. In R.A. Raff (ed.): *Development as an Evolutionary Process*, pp. 109–138. Alan R. Liss, New York.

Raff, R.A. and Byrne, M. 2006. The active evolutionary lives of echinoderm larvae. *Heredity* **97**: 244–252.

Rasmussen, H.W. 1977. Function and attachment of the stem in Isocrinidae and Pentacrinidae: review and interpretation. *Lethaia* **10**: 51–57.

Rowe, F.W.E., Baker, A.N. and Clark, H.E.S. 1988. The morphology, development and taxonomic status of *Xyloplax* Baker, Rowe and Clark (1986) (Echinodermata: Concentricycloidea), with description of a new species. *Proc. R. Soc. Lond. B* **233**: 431–459.

Rowe, F.W.E., Healy, J.M. and Anderson, D.T. 1994. Concentricycloidea. In F.W. Harrison (ed.): *Microscopic Anatomy of Invertebrates*, vol 14, pp. 149–167. Wiley-Liss, New York.

Ruppert, E.E. and Balser, E.J. 1986. Nephridia in the larvae of hemichordates and echinoderms. *Biol. Bull.* **171**: 188–196.

Rustad, D. 1938. The early development of *Stichopus tremulus* (Gunn.) (Holothuroidea). *Bergens Mus. Arbok, Naturvitensk. Rekke* **8**: 1–23.

Santos, R., Haesaerts, D., Jangoux, M. and Flammang, P. 2005. The tube feet of sea urchins and sea stars contain functionally different mutable collagenous tissues. *J. Exp. Biol.* **208**: 2277–2288.

Selenka, E. 1883. *Studien über Entwicklungsgeschichte der Thiere. II. Die Keimblätter der Echinodermen*. C.W. Kreidel, Wiesbaden.

Smith, A.B. 2004. Echinoderms (other than Echinoids). In R. Seely and R. Cocks (eds): *Encyclopaedia of Geology*, pp. 334–341. Elsevier, Oxford.

Smith, A.B. 2005. The pre-radial history of echinoderms. *Geol. J.* **40**: 255–280.

Smith, A.B. 2008. Deuterostomes with a twist: the origins of a radical new body plan. *Evol. Dev.* **10**: 493–503.

Strathmann, R. 1974. Introduction to function and adaptation in echinoderm larvae. *Thalassia Jugosl.* **10**: 321–339.

Strathmann, R.R. 1971. The feeding behavior of planktotrophic echinoderm larvae: mechanisms, regulation, and rates of suspension feeding. *J. Exp. Mar. Biol. Ecol.* **6**: 109–160.

Strathmann, R.R. 1975. Larval feeding in echinoderms. *Am. Zool.* **15**: 717–730.

Strathmann, R.R. 1978. The evolution and loss of feeding larval stages of marine invertebrates. *Evolution* **32**: 894–906.

Strathmann, R.R. 1988. Larvae, phylogeny, and von Baer's law. In C.R.C. Paul (ed.): *Echinoderm Phylogeny and Evolutionary Biology*, pp. 53–68. Oxford University Press, Oxford.

Strathmann, R.R. 2007. Time and extent of ciliary response to particles in a non-filtering feeding mechanism. *Biol. Bull.* **212**: 93–103.

Sumrall, C.D. and Wray, G.A. 2007. Ontogeny in the fossil record: diversification of body plans and the evolution of 'aberrant' symmetry in Paleozoic echinoderms. *Paleobiology* **33**: 149–163.

Swalla, B.J. and Smith, A.B. 2008. Deciphering deuterostome phylogeny: molecular, morphological and palaeontological perspectives. *Phil. Trans. R. Soc. Lond. B* **363**: 1557–1568.

Tattersall, W.M. and Sheppard, E.M. 1934. Observations on the bipinnaria of the asteroid genus *Luidia. James Johnstone Memorial Volume*, University Press of Liverpool, Liverpool.

Ubisch, L. 1913. Die Entwicklung von *Strongylocentrotus lividus* (*Echinus microtuberculatus, Arbacia pustulosa*). *Z. Wiss. Zool.* **106**: 409–448.

Vickery, M.S. and McClintock, J.B. 1998. Regeneration in metazoan larvae. *Nature* **394**: 140.

Villinski, J.V., Villinski, J.C., Byrne, M. and Raff, R.A. 2002. Convergent maternal provisioning and life-history evolution in echinoderms. *Evolution* **56**: 1764–1775.

Williams, D.H.C. and Anderson, D.T. 1975. The reproductive system, embryonic development, larval development and metamorphosis of the sea urchin *Heliocidaris erythrogramma* (Val.) (Echinoidea: Echinometridae). *Aust. J. Zool.* **23**: 371–403.

Wray, G.A. 1996. Parallel evolution of nonfeeding larvae in echinoids. *Syst. Biol.* **45**: 308–322.

Wray, G.A. 1997. Echinoderms. In S.F. Gilbert and A.M. Raunio (eds): *Embryology. Constructing the Organism*, pp. 309–329. Sinauer Associates, Sunderland, MA.

Wray, G.A. and Raff, R.A. 1990. Novel origins of lineage founder cells in the direct-developing sea urchin *Heliocidaris erythrogramma. Dev. Biol.* **141**: 41–54.

59

HEMICHORDATA

Pterobranchs and enteropneusts are usually treated together as the group Hemichordata, based mainly on the presence of the stomochord. This structure has been discussed in great detail in the literature because it has been interpreted as a homologue of the notochord of the chordates. However, three characters indicate that the structures are not homologous: the stomochord develops from the ectodermal pharynx in enteropneusts and pterobranchs, and the notochord from the archenteron roof in the chordates (for example Mayer and Bartolomaeus 2003); the function of the stomochord is uncertain, but that of the notochord is intimately connected with the undulatory movements caused by the lateral musculature, a function not found in the hemichordates; and gene expression (Takacs *et al.* 2002). Analyses of morphological characters usually show pterobranchs and enteropneusts as sister groups (for example Cameron 2005), and reviews treating the two groups together often become somewhat confused (for example Benito and Pardos 1997; Brusca and Brusca 2003). However, molecular phylogeny shows the pterobranchs as an ingroup of the enteropneusts, as the sister group of the Harrimaniidae (Cameron *et al.* 2000; Cannon *et al.* 2009).

If, as supposed here, the deuterostome ancestor was enteropneust-like (see for example Cameron 2005; Chapters 56 and 62), the tentacles of the pterobranchs must be an apomorphy, a specialization of structures of the mesosome. Their ciliary bands show the same structure and function as the neotroch of the

dipleurula larvae, and the tentacles with the ciliary band can perhaps best be interpreted as derived from the mesosome with an extended neotroch, not homologous with the tentacles on the prosome of the full-grown Krohn stage of an enteropneust (Fig. 60.1).

For practical reasons, the two groups are treated in separate chapters.

References

Benito, J. and Pardos, F. 1997. Hemichordata. In F.W. Harrison (ed.): *Microscopic Anatomy of Invertebrates*, Vol. 15, pp. 15–101. Wiley-Liss, New York.

Brusca, R.C. and Brusca, G.J. 2003. *Invertebrates*, 2nd ed. Sinauer, Sunderland, MA.

Cameron, C.B. 2005. A phylogeny of the hemichordates based on morphological characters. *Can. J. Zool.* **83**: 196–215.

Cameron, C.B., Garey, J.R. and Swalla, B.J. 2000. Evolution of the chordate body plan: new insights from phylogenetic analyses of deuterostome phyla. *Proc. Natl. Acad. Sci. USA* **97**: 4469–4474.

Cannon, J.T., Rychel, A.L., Eccleston, H., Halanych, K.M. and Swalla, B.J. 2009. Molecular phylogeny of Hemichordata, with updated status of deep-sea enteropneusts. *Mol. Phylogenet. Evol.* **52**: 17–24.

Mayer, G. and Bartolomaeus, T. 2003. Ultrastructure of the stomochord and the heart-glomerulus complex in *Rhabdopleura compacta* (Pterobranchia): phylogenetic implications. *Zoomorphology* **122**: 125–133.

Takacs, C.M., Moy, V.N. and Peterson, K.J. 2002. Testing putative hemichordate homologues of the chordate dorsal nervous system and endostyle: expression of *NK2.1* (*TTF-1*) in the acorn worm *Ptychodera flava* (Hemichordata, Ptychoderidae) *Evol. Dev.* **4**: 405–417.

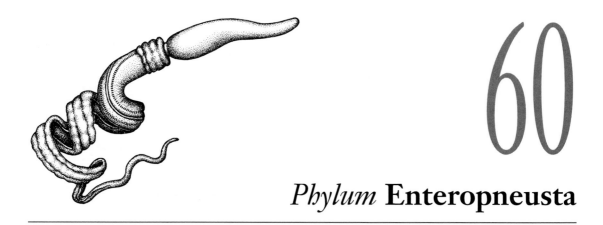

Phylum **Enteropneusta**

Enteropneusta, or acorn worms, is a well-defined phylum of about 90 burrowing or creeping, marine, worm-like species. Analyses of morphological characters indicate that Pterobranchia is the sister group of the three or four enteropneust families (Cameron 2005). However, molecular studies indicate the existence of two main lineages, Harrimaniidae and Ptychoderidae, with some of the specialized genera as ingroups. Quite surprisingly, the pterobranchs are placed as the sister group of the Harrimaniidae (Cannon *et al.* 2009). There is no reliable fossil record. Many species have a characteristic planktotrophic larval stage called tornaria. The spherical, centimetre-large, planktotrophic organism, with very complicated ciliary bands known as *Planctosphaera* (Hart *et al.* 1994), is possibly an enteropneust larva. The microscopic anatomy was reviewed by Benito and Pardos (1997), but the monograph of van der Horst (1927–1939) is still an indispensable source of information. For practical reasons, the pterobranchs are treated as a separate phylum (Chapter 61).

The body is archimeric (Chapter 56) with three clearly defined regions: an almost spherical to elongate conical proboscis (prosome), a short-collar region (mesosome), and a long-main body (metasome) with a lateral row of U-shaped gill slits on each side in the anterior part; some species have a pair of longitudinal, dorsolateral ridges, containing the gonads, in the branchial region.

The epithelia of the adult enteropneusts consist mainly of multiciliate cells, but monociliate cells are found, for example, in the mesothelial myocytes. In the multiciliate ectodermal cells of *Saccoglossus* each cilium has an accessory centriole, but such centrioles are lacking in *Glossobalanus*. In the pharynx of *Glossobalanus*, only the lateral cells of the gill bars have cilia with accessory centrioles. In the larvae, the neotroch consists of monociliate cells, whereas the perianal ciliary ring, the neurotroch, and the prominent opisthotroch (Fig. 60.1) consist of multiciliate cells (Nielsen and Hay-Schmidt 2007).

The proboscis has a coelomic cavity derived from the protocoel (see below), but the cavity is rather narrow because the mesoderm forms a thick layer of muscles. There is a small ciliated duct connecting the left side of the protocoel (or both sides for example in *Harrimania*) with the surface at the posterodorsal side of the proboscis. The dorsal side of the pharynx and the underside of the proximal part of the proboscis are supported by a thickened basal membrane, the so-called proboscis skeleton. An anterior extension of the pharynx, the stomochord, is another structure that apparently supports the proximal part of the proboscis. It is hollow with ciliated epithelium and surrounded by a strengthened basal membrane.

Protocoel, stomochord, and associated blood vessels form an axial complex, which is an ultrafiltration kidney (Balser and Ruppert 1990; Fig. 57.1). A

Chapter vignette: *Saccoglossus kowalevskii*. (Redrawn from Sherman and Sherman 1976.)

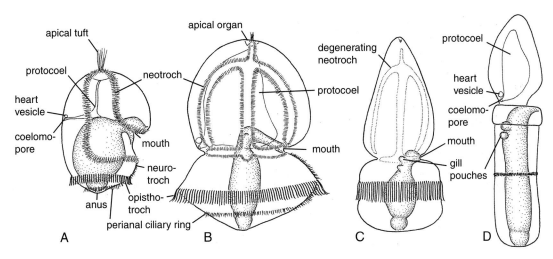

Figure 60.1. Developmental stages of enteropneusts with tornaria larvae seen from the left side. (A) A young larva (Heider stage). (B) A full-grown larva (Krohn stage). (C) A larva that has just commenced metamorphosis (Agassiz stage); the neotroch is degenerating and the gut is being pulled posteriorly. (D) A newly settled specimen; the collar with the collar cord has been formed, and the gill pouches are now located in the anterior end of the metasome but have not yet broken through to the exterior. (A modified from Stiasny 1914a; B–D modified from Morgan 1894.)

median blood vessel carries the blood anteriorly on the dorsal side of the stomochord and posteriorly on the ventral side. The anterior and ventral parts of the vessel have strongly folded walls consisting of peritoneal cells in the shape of podocytes resting on a basal membrane, and the structure, called the glomerulus, is the site of formation of primary urine. A modification of the urine before it leaves the proboscis pore has been suggested but not proven. A pericardial sac of myoepithelial cells and paired muscular extensions from the mesocoels surround the dorsal side of the blood vessel.

The collar region has a pair of mesocoelomic cavities, each with an anterodorsal ciliated coelomoduct, but the median septum is incomplete in many species. The coelomopores are situated in an ectodermal invagination in connection with the sac of the first gill slit. The function of the coelomoducts is unknown. The dorsal side shows a longitudinal strip of thickened, invaginated epithelium in the shape of the collar cord. The cord is hollow, with anterior and posterior pores, for example in many ptychoderids, but more or less compact in most other species.

The main body is long and comprises an anterior region, with numerous dorsal gill slits and gonads, a ventral 'digestive pharynx', and a long tail with a terminal anus. There is an undivided metacoelomic sac on each side, without nephridia or coelomopores.

The gill slits are U-shaped with a so-called tongue bar, an outgrowth from the dorsal side of the opening; the bars between adjacent gill slits are called gill bars or septa. The system is supported by a skeleton, which is a strengthened basal membrane with collagen-like fibrils in an amorphous matrix. The skeleton develops as curved bars or sheets along the anterior and posterior sides of the original gill slits and remains in this undifferentiated shape in *Protoglossus* (Burdon-Jones 1956). In most other species the skeletal parts differentiate further and send extensions into the tongue bars; the elements from the two sides finally fuse so that structures of very characteristic shapes are formed. The tongue bars are free and can be moved in some species, but, for example, the ptychoderids have transverse bars, synapticles, containing extensions of the skeletal system across the gill slits, so that a more rigid gill basket is formed. The main skeletal element of the tongue bars contains a coelomic extension, whereas the gill bars are solid.

The gill slits open into ectodermal invaginations called gill pouches; each U-shaped gill slit has its own gill pouch in some species, whereas other species have several gill slits opening into a common pouch.

The function of the gill slits is not well understood and their structure does not indicate a function in gas exchange. Filter feeding has been demonstrated in species of *Harrimania* and *Protoglossus*, where the filtering structure is the dense ciliation of the gill bars (Cameron 2002; Gonzalez and Cameron 2009). But filter feeding is not the only feeding method. Knight-Jones (1953) studied *Saccoglossus* and observed that particles are captured in mucus secreted from various gland cells on the proboscis and transported to the mouth and through the gut by a dense ciliation. Other particles were retained on the frontal side of the gill bars and transported ventrally along the grooves between the bars to the ventral part of the gut, where they were taken over by the general ciliation and transported towards the intestine. Burdon-Jones (1962) studied *Balanoglossus* and concluded that deposit feeding, by engulfing of substratum and by ciliary mechanisms of the proboscis, is the most important type of feeding, but that filter feeding may be of some importance too.

The dorsal wall of the gut of *Schizocardium* has a longitudinal epibranchial ridge, with 11 longitudinal rows of characteristic ciliated cells, resembling the rows of cells in the chordate endostyle (Fig. 56.2). However, secretion of mucus and concentration of iodine are not restricted to this ridge, but are found scattered over the whole pharyngeal epithelium (Ruppert 2005). Some species have a midventral strip of vacuolated cells (pygochord) extending into the mesentery (Willey 1899; van der Horst 1927–1939). It has been compared with the notochord (Nübler-Jung and Arendt 1999), but the homology seems unlikely (Takacs *et al.* 2002) (see Chapter 57).

The classical description of the nervous system as a well-developed nerve net with concentrations (Bullock 1946; Knight-Jones 1952) has been questioned by new investigations on *Ptychodera* using gene expression (Nomaksteinsky *et al.* 2009). This study showed the presence of scattered epidermal neurons, and dorsal and ventral concentrations. A strong bilateral concentration of presumptive sensory and motor neurons has been seen in the floor plate of the collar cord. A subepidermal nerve cord extends anteriorly into the proboscis stalk; posteriorly, a cord extends along the dorsal midline and a pair of cords follows the anterior end of the trunk and unites in a midventral cord. Some ptychoderids have giant unipolar nerve cells in the collar cord; their axons cross over to the opposite side and run posteriorly (Brown *et al.* 2008). There are apparently no special sense organs in the adults, but sensory cells are found scattered in the epithelia.

The haemal system is situated between the basal membranes and comprises the above-mentioned anterior heart complex that pumps the blood from the mediodorsal vessel, with the blood running in the anterior direction, to the ventral vessels surrounding the pharynx and uniting into a ventral longitudinal vessel, with the blood running posteriorly. The blood leaks through a subintestinal plexus in the ventral part of the metasome to a pair of lateral longitudinal vessels. From these, smaller vessels run dorsally along the gill bars to the dorsal end of the gill slits, where they divide and send one branch ventrally along the anterior and one along the posterior side of the neighbouring tongue bars. Numerous anastomosing narrow blood spaces connect these blind-ending vessels with a median vessel leading the blood to the median dorsal vessel. Many of these blood spaces are located inside the skeletal rods of the gills. A more enigmatic feature of the haemal system is the presence of podocytes in the coelomic lining of some of the blood spaces in the branchial sacs of *Glossobalanus* (Pardos and Benito 1988). The function of such structures is usually believed to be ultrafiltration from the blood and the formation of primary urine, but the metacoels have no ducts in enteropneusts, so the function of this structure is uncertain.

The gonads occur in rows along the sides of the metasome. Each gonad is sac-shaped with a short gonoduct; they are apparently formed from the peritoneum, but the exact nature is unknown. Fertilization is external in most species. The first cleavage is median and the second cleavage divides the embryo into a dorsal and a ventral half; the third cleavage separates the future prosomal and mesosomal ectoderm from the future metasomal ectoderm and mesoderm + endoderm (Colwin and Colwin 1951) (Fig. 62.2). A coeloblastula is soon formed and gastrulation is by invagination (Tagawa *et al.* 1998). In the planktotrophic forms, such as *Ptychodera* (Peterson *et al.* 1999; Nielsen and Hay-Schmidt 2007), the ciliated embryo hatches at this stage and the whole epithelium consists of monociliate cells. The blastopore closes and a perianal ring of multiciliate cells, with a ventral extension towards the area of the future stomodaeum, soon appears. Concomitantly, the hydropore breaks through from the bottom of the archenteron, and a convoluted ring of monociliate cells, the neotroch, becomes organized around a developing stomodaeal invagination. This band becomes highly complex in many larvae forming loops on small tentacles (Fig. 60.1). The neotroch is both the feeding and the locomotory organ of the youngest larvae, but a large, perianal ring of compound cilia, the opisthotroch, soon develops and takes over as the swimming organ (Fig. 60.1). The compound cilia are formed on a band of multiciliate cells that is several cells wide with the cells arranged in an regular rhomboid pattern in some species (Spengel 1893). The compound cilia are at first not much longer than the single cilia of the neotroch (about 20–25 μm), but with the increasing size of the larva the compound cilia increase in length, in some species up to about 200 μm. The neotroch is an upstream-collecting ciliary band (Strathmann and Bonar 1976; Nielsen 1987). The lecithotrophic larvae remain uniformly ciliated except for the apical tuft and the opisthotroch of compound cilia (Burdon-Jones 1952).

An apical organ with a ciliary tuft is found in all larvae; some species have a pair of eyes at the sides of the apical organ, and some species have both a pigment cup and a lens. Each photosensitive cell has both a modified cilium and an array of microvilli, which makes this type of cell unique among the photoreceptor cells (Brandenburger *et al.* 1973). The development of the nervous system in the indirect developing

Balanoglossus was studied by Miyamoto *et al.* (2010). They found an apical organ consisting mostly of serotonergic cells and a larval nervous system along the neotroch and later on along the opisthotroch of the tornaria. This system degenerates at metamorphosis. In a fully differentiated tornaria (the late Krohn stage) a mid-dorsal nerve extending from the apical organ to the opisthotroch was observed, and in a later stage with degenerating neotroch (the Agassiz stage) the whole adult nervous system, including the nerves around the anterior part of the mesosome and the ventral nerve, could be distinguished.

The development of the coelomic sacs (Fig. 57.2) has attracted much attention for over a century. The protocoel is apparently always pinched off as a spacious pouch from the anterior part of the archenteron, and the hydropore develops at an early stage (Ruppert and Balser 1986). The meso- and metacoels have different origins in different species, but some of the reports seem to be based on observations of only a few stages and should be checked.

The development in indirect-developing species has been studied in *Balanoglossus* and *Ptychodera*. In *Balanoglossus*, meso- and metacoel develop together from a pair of lateral pockets from the posterior part of the gut, with the two elongate pockets dividing afterwards into meso- and metacoel (Stiasny 1914a,b; Urata and Yamaguchi 2004). *Ptychdera* shows formation of meso- and metacoels from isolated cells from the gut (Peterson *et al.* 1999). Tornaria larvae from plankton of several oceans have revealed that the feeding tornaria larvae generally form meso- and metacoel when the larvae have already begun to feed, and that the early coeloms are therefore compact instead of hollow (Morgan 1891, 1894; Dawydoff 1944; 1948). The protocoel soon develops a connection with the dorsal side, with a ciliated canal leading to the hydropore and a compact string extending to the apical organ. The main part of the protocoelomic wall consists of podocytes, and the coelomic sac functions as a kidney with filtration of primary urine from the blastocoel. At metamorphosis the blastocoel shrinks and becomes transformed to the haemal system (Ruppert and Balser 1986). The heart vesicle is variously reported to originate from ectoderm (Stiasny 1914b) or mesoderm (Morgan 1894; Ruppert and Balser 1986). It soon develops muscle cells, and can then be seen making small pulsating movements.

Species of *Saccoglossus* have direct development, and two species have been reared from eggs. Davis (1908) studied *S. pusillus* and found that the lateral parts of the anterior coelomic sac extend posteriorly along the sides of the gut, and that meso- and metacoel became established when these long extensions became isolated from the protocoel and broke up into meso- and metacoels. Bateson (1884) studied *S. kowalevskii* and observed that both the meso- and the metacoelomic sacs develop as separate pockets from the lateral walls of the gut.

The primary body cavity of older larvae is filled by a gelatinous material that is of importance both for the stability of the shape and for several of the developmental processes (Strathmann 1989).

At metamorphosis, the neotroch degenerates rapidly, while the opisthotroch can be recognized for a while (Peterson *et al.* 1999; Nielsen and Hay-Schmidt 2007) (Fig. 60.1). The metamorphosis of the lecithotrophic larvae is more gradual, and the perianal ring becomes pulled out onto a small, temporary 'tail' ventral to the anus (Burdon-Jones 1952).

Large tornaria larvae show two to three pairs of gill pockets on the pharynx near the mouth. At metamorphosis the gut becomes pulled backwards so that the pockets become situated in the anterior part of the metasome where the gill slits break through (Fig. 60.1). Additional gill slits develop in a series behind the first few openings. The newly formed gill openings are oval, but a tongue bar soon develops from the dorsal side of the pore (Burdon-Jones 1956).

The Hox genes show the characteristic 'ambulacrarian signature' (Aronowicz and Lowe 2006; Ikuta *et al.* 2009; Urata *et al.* 2009) (Fig. 21.3).

The hollow nerve cord in the mesosome resembles the neural tube of the chordates also at the ultrastructural level (Kaul and Stach 2010), but as it has none of the functions of a brain it could perhaps better be compared to the infolded nerves of some of the echinoderms (Chapter 58). There is no specific indication of homology with the neural tube complex of the chordates. Gene expression supports the 'inversion' hypothesis, i.e. that the chordate neural tube develops from a region homologous to the ventral nerve cord of the enteropneusts (Holland 2003; Lowe *et al.* 2003, 2006) (see Chapter 62).

The possession of both an axial complex and the series of gill slits place the enteropneusts in a key position within the deuterostomes. The homology of the axial complexes of pterobranchs, echinoderms, and enteropneusts can hardly be doubted. The gill slits are regarded as a synapomorphy of the deuterostomes, and the hypobranchial ridge is possibly homologous with the chordate endostyle (see Chapters 56 and 62).

Interesting subjects for future research

1. Functions of the gill slits
2. Iodine in various tissues, especially in the cells of the epibranchial ridge

References

Aronowicz, J. and Lowe, C.J. 2006. *Hox* gene expression in the hemichordate *Saccoglossus kowalevskii* and the evolution of deuterostome nervous system. *Integr. Comp. Biol.* **46**: 890–901.

Balser, E.J. and Ruppert, E.E. 1990. Structure, ultrastructure, and function of the preoral heart-kidney in *Saccoglossus kowalevskii* (Hemichordata, Enteropneusta) including new data on the stomochord. *Acta Zool. (Stockh.)* **71**: 235–249.

Bateson, W. 1884. The early stages of the development of *Balanoglossus* (sp. incert.). *Q. J. Microsc. Sci., N. S.* **24**: 208–236.

Benito, J. and Pardos, F. 1997. Hemichordata. In F.W. Harrison (ed.): *Microscopic Anatomy of Invertebrates*, vol. 15, pp. 15–101. Wiley-Liss, New York.

Brandenburger, J.L., Woollacott, R.M. and Eakin, R.E. 1973. Fine structure of eyespots in tornarian larvae. *Z. Zellforsch.* **142**: 89–102.

Brown, F.D., Prendergast, A. and Swalla, B.J. 2008. Man is but a worm: Chordate origins. *Genesis* **46**: 605–613.

Bullock, T.H. 1946. The anatomical organization of the nervous system of enteropneusts. *Q. J. Microsc. Sci., N. S.* **86**: 55–111.

Burdon-Jones, C. 1952. Development and biology of the larva of *Saccoglossus horsti* (Enteropneusta). *Phil. Trans. R. Soc. Lond. B* **236**: 553–590.

Burdon-Jones, C. 1956. Observations on the enteropneust, *Protoglossus koehleri* (Caullery & Mesnil). *Proc. Zool. Soc. Lond.* **127**: 35–58.

Burdon-Jones, C. 1962. The feeding mechanism of *Balanoglossus gigas. Bol. Fac. Filos. Cienc. Let. Univ S. Paulo, Zool.* **24**: 255–280.

Cameron, C.B. 2002. The anatomy, life habits, and later development of a new species of enteropneust, *Harrimania planktophilus* (Hemichordata: Harrimaniidae) from Barkley Sound. *Biol. Bull.* **202**: 182–191.

Cameron, C.B. 2005. A phylogeny of the hemichordates based on morphological characters. *Can. J. Zool.* **83**: 196–215.

Cannon, J.T., Rychel, A.L., Eccleston, H., Halanych, K.M. and Swalla, B.J. 2009. Molecular phylogeny of Hemichordata, with updated status of deep-sea enteropneusts. *Mol. Phylogenet. Evol.* **52**: 17–24.

Colwin, A.L. and Colwin, L.H. 1951. Relationships between the egg and larva of *Saccoglossus kowalevskii* (Enteropneusta): axes and planes; general prospective significance of the early blastomeres. *J. Exp. Zool.* **117**: 111–137.

Davis, B.M. 1908. The early life-history of *Dolichoglossus pusillus* Ritter. *University Calif. Publ. Zool.* **4**: 187–226.

Dawydoff, C. 1944. Formation des cavités coelomiques chez les tornaria du plancton indochinois. *C. R. Hebd. Seanc. Acad. Sci. Paris* **218**: 427–429.

Dawydoff, C. 1948. Classe des Entéropneustes + Classe des Ptérobranches. *Traité de Zoologie*, vol. 11, pp. 369–489. Masson, Paris.

Gonzalez, P. and Cameron, C.B. 2009. The gill slits and pre-oral ciliary organ of *Protoglossus* (Hemichordata: Enteropneusta) are filter-feeding structures. *Biol. J. Linn. Soc.* **98**: 898–906.

Hart, M.W., Miller, R.L. and Madin, L.P. 1994. Form and feeding mechanism of a living *Planctosphaera pelagica* (phylum Hemichordata). *Mar. Biol.* **120**: 521–533.

Holland, N.D. 2003. Early central nervous system evolution: an area of skin brains? *Nature Rev. Neurosci.* **4**: 1–11.

Ikuta, T., Miyamoto, N., Saito, Y., *et al.* 2009. Ambulacrarian prototypical Hox and ParaHox gene complements of the indirect-developing hemichordate *Balanoglossus simodensis. Dev. Genes Evol.* **219**: 383–389.

Kaul, S. and Stach, T. 2010. Ontogeny of the collar cord: neurulation in the hemichordate *Saccoglossus kowalevskii. J. Morphol.* **271**: 1240–1259.

Knight-Jones, E.W. 1952. On the nervous system of *Saccoglossus cambrensis* (Enteropneusta). *Phil. Trans. R. Soc. Lond. B* **236**: 315–354.

Knight-Jones, E.W. 1953. Feeding in *Saccoglossus* (Enteropneusta). *Proc. Zool. Soc. Lond.* **123**: 637–654.

Lowe, C.J., Terasaki, M., Wu, M., *et al.* 2006. Dorsoventral patterning in hemichordates: insights into early chordate evolution. *PLoS Biol.* **4**: e291.

Lowe, C.J., Wu, M., Salic, A., *et al.* 2003. Anteroposterior patterning in hemichordates and the origins of the chordate nervous system. *Cell* **113**: 853–865.

Miyamoto, N., Nakajima, Y., Wada, H. and Saitoh, Y. 2010. Development of the nervous system in the acorn worm *Balanoglossus simodensis*: insights into nervous system evolution. *Evol. Dev.* **12**: 416–424.

Morgan, T.H. 1891. The growth and metamorphosis of tornaria. *J. Morphol.* **5**: 407–458.

Morgan, T.H. 1894. The development of *Balanoglossus. J. Morphol.* **9**: 1–86.

Nielsen, C. 1987. Structure and function of metazoan ciliary bands and their phylogenetic significance. *Acta Zool. (Stockh.)* **68**: 205–262.

Nielsen, C. and Hay-Schmidt, A. 2007. Development of the enteropneust *Ptychodera flava*: ciliary bands and nervous system. *J. Morphol.* **268**: 551–570.

Nomaksteinsky, M., Röttinger, E., Dufour, H.D., *et al.* 2009. Centralization of the deuterostome nervous system predates chordates. *Curr. Biol.* **19**: 1264–1269.

Nübler-Jung, K. and Arendt, D. 1999. Dorsoventral axis inversion: enteropneust anatomy links invertebrates to chordates turned upside down. *J. Zool. Syst. Evol. Res.* **37**: 93–100.

Pardos, F. and Benito, J. 1988. Ultrastructure of the branchial sacs of *Glossobalanus minutus* (Enteropneusta) with special reference to podocytes. *Arch. Biol.* **99**: 351–363.

Peterson, K.J., Cameron, R.A., Tagawa, K., Satoh, N. and Davidson, E.H. 1999. A comparative molecular approach to mesodermal patterning in basal deuterostomes: the expression pattern of *Brachyury* in the enteropneust hemichordate *Ptychodera flava. Development* **126**: 85–95.

Ruppert, E.E. 2005. Key characters uniting hemichordates and chordates: homologies or homoplasies? *Can. J. Zool.* **83**: 8–23.

Ruppert, E.E. and Balser, E.J. 1986. Nephridia in the larvae of hemichordates and echinoderms. *Biol. Bull.* **171**: 188–196.

Sherman, I.W. and Sherman, V.G. 1976. *The Invertebrates: Function and Form. A Laboratory Guide.* Macmillan, New York.

Spengel, J.W. 1893. Die Enteropneusten des Golfes von Neapel. *Fauna Flora Golf. Neapel* **18**: 1–758.

Stiasny, G. 1914a. Studien über die Entwicklung des *Balanoglossus clavigerus* Delle Chiaje. I. Die Entwicklung der Tornaria. *Z. Wiss. Zool.* **110**: 36–75.

Stiasny, G. 1914b. Studien über die Entwicklung des *Balanoglossus clavigerus* Delle Chiaje. II. Darstellung der weiteren Entwicklung bis zur Metamorphose. *Mitt. Zool. Stn. Neapel* **22**: 255–290.

Strathmann, R. and Bonar, D. 1976. Ciliary feeding of tornaria larvae of *Ptychodera flava* (Hemichordata: Enteropneusta). *Mar. Biol.* **34**: 317–324.

Strathmann, R.R. 1989. Existence and functions of a gel filled primary body cavity in development of echinoderms and hemichordates. *Biol. Bull.* **176**: 25–31.

Tagawa, K., Nishino, A., Humphreys, T. and Satoh, N. 1998. The spawning and early development of the Hawaiian acorn worm (Hemichordate), *Ptychodera flava. Zool. Sci. (Tokyo)* **15**: 85–91.

Takacs, C.M., Moy, V.N. and Peterson, K.J. 2002. Testing putative hemichordate homologues of the chordate dorsal nervous system and endostyle: expression of *NK2.1 (TFF-1)* in the acorn worm *Ptychodera flava* (Hemichordata, Ptychoderidae) *Evol. Dev.* **4**: 405–417.

Urata, M. and Yamaguchi, M. 2004. The development of the enteropneust hemichordate *Balanoglossus misakiensis* Kuwano. *Zool. Sci. (Tokyo)* **21**: 533–540.

Urata, M., Tsuchimoto, J., Yasui, K. and Yamaguchi, M. 2009. The *Hox8* of the hemichordate *Balanoglossus misakiensis. Dev. Genes Evol.* **219**: 377–382.

van der Horst, C.J. 1927–1939. Hemichordata. *Bronn's Klassen und Ordnunges des Tierreichs*, 4. Band, 4. Abt., 2. Buch, 2. Teil, pp. 1–737. Akademische Verlagsgesellschaft, Leipzig.

Willey, A. 1899. Enteropneusta from the South Pacific, with notes on the West Indian species. In A. Willey (ed.): *Zoological Results based on Material from New Britain, New Guinea, Loyalty Islands and elsewhere*, pp. 223–334. Cambridge University Press, Cambridge.

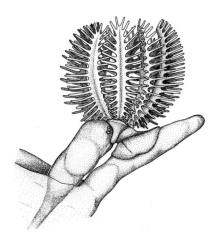

$Phylum$ **Pterobranchia**

61

Pterobranchia is one of the smallest animal phyla, comprising only three genera of marine, benthic organisms: *Rhabdopleura*, with four species forming small adnate colonies; *Cephalodiscus*, with 15–20 species forming sometimes quite extensive aggregations of tubes, called coenecia, housing solitary individuals with lively budding; and the apparently solitary *Atubaria*, which has only been recorded once. The fossil record of *Rhabdopleura*-like forms goes back to the Middle Cambrian (Bengtson and Urbanek 1986), and also the highly diverse fossil group Graptolithina (Cambrian-Carboniferous) is now believed to be closely related to the rhabdopleurids (Sato *et al.* 2008). The group is clearly monophyletic and has traditionally been considered as the sister group of the Enteropneusta, but some molecular studies indicate that the pterobranchs are an ingroup of the enteropneusts (Cameron *et al.* 2000; Bourlat *et al.* 2003; Cannon *et al.* 2009). However, the body plans for the two groups are so different that is seems most practical to discuss them separately.

The individual zooids have an archimeric body (Chapter 56) consisting of a preoral shield (prosome), used in creeping and in secreting the tubes/coenoecia, a short perioral collar (mesosome), carrying ciliated tentacles, and an elongated globular body (metasome), posteriorly extended into a narrow tail or stolon from which budding takes place. The gut is U-shaped with a short oesophagus, a globular stomach, and a narrow rectum passing dorsally from the posterior side of the stomach to the anus, which is situated a short distance behind the mesosome. Many ectodermal, endodermal, and peritoneal cells are ciliated, and there is always only one cilium per cell; this is also the case with the cells of the intraepithelial ganglion. The microscopic anatomy was reviewed by Benito and Pardos (1997).

The prosome is a flat shield with a rather narrow neck. It is used as a creeping sole and its thick, ventral epithelium is ciliated with many mucus cells. There is a pigmented transverse stripe without cilia on the ventral side, and the prosome can be folded along this zone when material secreted for tube building is being added to the edge of the tube (Dilly 1988). The zooids can move around in the tubes, and *Cephalodiscus* may even leave the coenoecium and start to build a new one if conditions become too adverse (Lester 1985). The unpaired protocoel is lined by a monolayered peritoneum and opens to the exterior through a pair of dorsal, ciliated ducts. Mid-dorsally, the protocoel is filled by a heart surrounded by a pericardial sac. The heart is an anterior extension of a median, U-shaped vessel, in the usual position between basal membranes; the dorsal vessel probably carries blood to the heart and the ventral vessel leads the blood posteriorly. The pericardial sac consists of a monolayered myoepithelium that develops from a cluster of mesenchymal cells

---

Chapter vignette: *Cephalodiscus gracilis* in feeding position. (Redrawn from Lester 1985.)

just behind the protocoel in *Rhabdopleura* (Lester 1988b), but that may be interpreted as an isolated pocket of protocoelomic peritoneum. The ventral part of the ventral vessel, called glomerulus, has convoluted ventrolateral walls with podocytes and is supposed to be a site for ultrafiltration of primary urine (Mayer and Bartolomaeus 2003). The protocoel thus functions like a Bowman's capsule in the vertebrate kidney. The primary urine in the protocoel could be modified during the passage through the coelomoducts, but this has not been investigated.

The mesosome is quite short, forming a collar around the mouth and foregut and carrying one to nine pairs of dorsal tentacles. The tentacles are feather-shaped with a row of pinnules on each side. In feeding specimens of *Cephalodiscus*, the several tentacles are held in a curved position so that an almost spherical shape is formed (Lester 1985; see the chapter vignette). The pinnules and tentacles are ciliated, with a double row of narrow, ciliated cells on each side of the pinnules (Dilly *et al.* 1986; Nielsen 1987). Gilmour (1979) and Halanych (1993) reported a row of laterofrontal, probably sensory cells along the frontal side of the double row of cilia, but these are not documented in their photos, and special laterofrontal cells are not indicated in the transmission electron microscopy sections pictures of Dilly *et al.* (1986) and Benito and Pardos (1997). The lateral cilia form an upstream-collecting system that pumps water into the sphere and out through a distal opening between the tentacle tips (Lester 1985). Particles strained from the water are passed to the mouth along the frontal side of the pinnules/tentacles, i.e. the side of the tentacles at the outside of the sphere. A similar mechanism was reported from *Rhabdopleura* by Halanych (1993), who described the two tentacles as having a row of holes on each side along the bases of the pinnules; however, the existence of these openings is contradicted by one of his illustrations that shows cells with pigment granules extending across the 'hole' (Halanych 1993: fig. 3C). The particle-collecting mechanism has not been studied in detail.

The paired mesocoelic cavities extend into the tentacles and are surrounded by a monolayered peritoneum. Dilly (1972) observed a blood sinus in the basement membrane along the frontal side of the tentacles in *Cephalodiscus*, while Gilmour (1979) described a small vessel in each tentacle in the shape of a longitudinal fold of the frontal part of the peritoneum in *Rhabdopleura*. There is a pair of dorsal, ciliated coelomoducts that open posterolaterally (just in front of the gill pores of *Cephalodiscus*). The walls of the coelomoducts contain groups of cells with cross-striated muscular filaments, and the function of these muscles appear to be an opening of the duct, perhaps in connection with rapid retractions of the zooids (Dilly *et al.* 1986). The main nervous concentration is situated at the dorsal side of the mesosome (see below). *Cephalodiscus* and *Atubaria* have a pair of gill pores from the oesophagus to the lateral sides of the mesosome, where the ciliated canals continue posteriorly in shallow furrows. Gilmour (1979) was of the opinion that the gill pores developed to allow the escape of excess water from the filter-feeding process entering the oesophagus, but direct observations are lacking.

A stomochord runs anteriorly between the pharynx and the peritoneum of the protocoel. It is compact or has a central cavity in connection with the pharynx, consists of monociliated, vacuolated cells, and is surrounded by a thickened extracellular sheath (Balser and Ruppert 1990; Mayer and Bartolomaeus 2003). Both during metamorphosis and budding it develops as a pocket from of the anterior part of the pharynx (see below). The function of this structure is uncertain, but it appears to support the neck region between prosome and metasome, and perhaps also to support the muscular heart (Balser and Ruppert 1990); a glandular function was suggested by Mayer and Bartolomaeus (2003).

The large metasome contains the major part of the gut, which is suspended in mesenteria formed by the median walls of the paired metacoelomic sacs. One or two gonads are situated at the dorsal side of the metacoel, covered by the peritoneum. The two metacoelomic sacs have no connection to the exterior, i.e. no metanephridia, and the gametes are shed directly through short canals separate from the coelom. The sexes of the individual zooids are usually separate, but some species of *Cephalodiscus* have one gonad of each sex. The metacoelomic sacs extend into the stalk region, where the septum between the two cavities are lacking in *Cephalodiscus*; the blood vessels and the nerves are the same in the stalk of the two genera.

The posterior end of the stalk has a somewhat different structure in the two genera. In *Cephalodiscus*, the tip has a small attachment organ, and the budding takes place from this area (see below). In *Rhabdopleura*, the stalk ends in a genuine branched stolon from which new buds arise. The stolon is a narrow string of vacuolated tissue surrounded by a black tube. The zooids bud off from the growing tips of the stolon and form a stalk and a main body that become contained in larger chambers that are partially erect. The stalk region appears to be homologous with the stalk of *Cephalodiscus*. The stalks are highly contractile and may retract the zooids to the bottom of their chambers (Stebbing and Dilly 1972; Lester 1985).

There is a nervous concentration at the dorsal side of the mesosome between the tentacle crown and the anus. It is situated basally in an oval area of thickened epithelium. The peripheral nervous system comprises nerves to the tentacles, a median nerve to the prosome, and a pair of connectives around the oesophagus to a midventral nerve that continues along the ventral side to the stalk in *Rhabdopleura*, and along the ventral side and back along the dorsal side of the whole stalk in *Cephalodiscus* (Schepotieff 1907a,b).

The haemal system comprises the above-mentioned vessels associated with the heart and the tentacle vessels, but important parts of the system are more lacunar than vessel-like, especially around the gut. A sinus extends from the median glomerulus along the dorsal wall of the pre-oral shield to the pharynx, which is surrounded by a pair of vessels uniting again behind the pharynx into a mid-ventral vessel; this vessel extends all the way to the tip of the stalk where it curves around and follows the dorsal side of the body almost to the tentacle area.

The coenoecium of *Cephalodiscus* consists of individual tubes held together in a mass of interwoven fibres so that characteristic, species-specific structures are formed. *Rhabdopleura* has naked branching tubes extending from a hemispherical ancestral chamber. The tubes are made up of a double row of alternating U-shaped pieces that are secreted by special glandular areas of the cephalic shield. The tube material contains collagen, but keratin and chitin have not been found (Armstrong *et al.* 1984; Dilly 1986).

In *Cephalodiscus*, the buds develop from the small attachment plate at the tip of the stalk where the first stage is an outgrowth from the stalk, consisting of ectoderm and peritoneum of the paired coelomic cavities of the stalk; its tip becomes the very large oral shield, and the gut develops from an ectodermal invagination starting orally (Masterman 1898; Harmer 1905; Schepotieff 1908). At a later stage, the anus breaks through at the dorsal side, and the bud attains the proportions of the adult. The metacoels of the stalk pinch off the protocoel, and possibly the pericardium, in the oral shield, and the mesocoels in the collar with the tentacles. The stomochord is formed as a small, anteriorly directed diverticulum from the epithelium of the pharynx. The buds remain attached to the parent until an advanced stage, and large clusters of buds of varying ages are often found.

The buds of *Rhabdopleura* develop from the tips or along the sides of the branching stolons (Schepotieff 1907c) (see above). The early buds are simple evaginations of the stolon, consisting of ectoderm and extensions of the paired coelomic compartments of the stolon. The origin of the gut is in need of further investigation, but the coelomic compartments develop much like those of *Cephalodiscus*. A special type of buds are dormant with a thick cuticle (Stebbing 1970; Dilly 1975).

The gonads are mesodermally derived and have a pair of separate ducts formed from ectodermal invaginations. Sperm and eggs are shed through the narrow gonoducts. The sperm of *Rhabdopleura* is of the specialized type with a spindle-shaped head and a most unusual mitochondrial filament (Lester 1988b). Fertilization has not been observed, but Andersson (1907) observed sperm within the ovaries of a *Cephalodiscus*, which indicates internal fertilization, but it is not known if this is the case for other species.

All species appear to deposit the fertilized, yolk-rich eggs in the tubes, usually behind the zooids, where the early part of the development takes place (Harmer 1905; John 1932; Lester 1988b).

Early studies of the development of *Cephalodiscus* (Andersson 1907; Schepotieff 1909; John 1932) were rather incomplete, but showed gastrulation by invagination and formation of mesoderm and coelomic pouches that can be interpreted in accordance with the following description of *Rhabdopleura*. Schepotieff (1909) also observed a number of quite different larvae that, as pointed out by Hyman (1959), are so similar to bryozoan larvae that it must be taken for granted that a confusion had taken place.

Rhabdopleura (Dilly 1973; Lester 1988b) shows total, equal cleavage that leads to the formation of a spherical, completely ciliated larva; later stages are elongate and show a shallow, anteroventral concavity (Fig. 61.1). The spherical stage consists of a layer of ectodermal cells around a mass of endodermal cells with much yolk; a narrow cavity at one side has a thin layer of cells, interpreted as mesoderm, covering the outer, ectodermal side. The elongate larvae have a thin layer of mesoderm covering the inside of the whole ectodermal sheet, and the cavity is now enlarged and situated at the anterodorsal side of the endoderm. The mesoderm is monolayered except in two pairs of lateral areas, where flat coelomic sacs are found; these sacs are interpreted as mesocoel and metacoel.

Settlement and metamorphosis of *Rhabdopleura* have been studied by Stebbing (1970), Dilly (1973), and Lester (1988a). The larvae test the substratum,

creeping on the ventral side after a short period of swimming; the larva then settles with the ventral depression in close contact with the substratum, and secretes a thin-surrounding cocoon (Fig. 61.1). The metamorphosis involves the development of the adult body regions, with the tentacles and the early stages of the pinnules; the endoderm begins as a compact mass of cells, but a lumen develops first in the region that becomes the intestine and later in the stomach, while the pharynx originates as an invagination from the ectoderm. The pericardial vesicle forms from a small mass of mesodermal cells between the dorsal epidermis and the pharynx. The cocoon breaks open after a few days and the first tube is secreted; this first stage of the colony, the prociculum, harbours the first feeding zooid of the new colony.

It now appears that the ambulacrarian ancestor was enteropneust-like with gill slits (Chapter 60), which implies that the ciliated tentacles of the pterobranchs must be an apomorphy. The ciliary bands on the tentacles have the same structure as those of the neotrochs of tornaria larvae and the various echinoderm larvae, and the function in particle collection is probably the same. This could indicate that the lateral ciliary bands on the pterobranch tentacles are specializations of the neotroch of a dipleurula, but this can probably only be

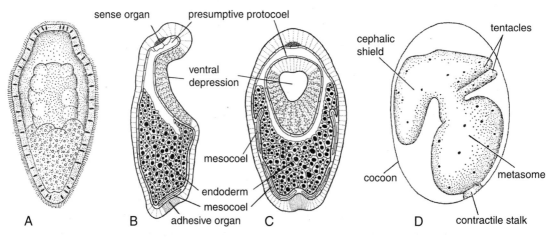

Figure 61.1. Development of *Rhabdopleura normani*. (A) Free-swimming larva in optical section. (B–C) Median and horizontal sections of swimming larvae; the mesocoelomic and metacoelomic cavities are well defined, but the extension of the protocoel is uncertain, because parts of the anterior cavity between mesoderm and endoderm may represent fixation artifacts. (D) Early settled larva in the cocoon. (A and D from Stebbing 1970; B and C modified from Lester 1988b.)

shown if some of the *Cephalodiscus* species turn out to have planktotrophic larvae.

The phylogenetic position within the Ambulacraria and Hemichordata is discussed further in Chapter 56.

Interesting subjects for future research

1. Embryology and larval development of both genera
2. Structure and biology of *Atubaria*

References

Andersson, K.A. 1907. Die Pterobranchier der schwedischen Südpolarexpedition. *Wissenschaftliche Ergebnisse der Schwedischen Südpolarexpedition* **5**: 1–122.

Armstrong, W.G., Dilly, P.N. and Urbanek, A. 1984. Collagen in the pterobranch coenecium and the problem of graptolite affinities. *Lethaia* **17**: 145–152.

Balser, E.J. and Ruppert, E.E. 1990. Structure, ultrastructure, and function of the preoral heart-kidney in *Saccoglossus kowalevskii* (Hemichordata, Enteropneusta) including new data on the stomochord. *Acta Zool. (Stockh.)* **71**: 235–249.

Bengtson, S. and Urbanek, A. 1986. *Rhabdotubus*, a Middle Cambrian rhabdopleurid hemichordate. *Lethaia* **19**: 293–308.

Benito, J. and Pardos, F. 1997. Hemichordata. In F.W. Harrison (ed.): *Microscopic Anatomy of Invertebrates*, vol. 15, pp. 15–101. Wiley-Liss, New York.

Bourlat, S.J., Nielsen, C., Lockyer, A.E., Littlewood, D.T.J. and Telford, M.J. 2003. *Xenoturbella* is a deuterostome that eats molluscs. *Nature* **424**: 925–928.

Cameron, C.B., Garey, J.R. and Swalla, B.J. 2000. Evolution of the chordate body plan: new insights from phylogenetic analyses of deuterostome phyla. *Proc. Natl. Acad. Sci. USA* **97**: 4469–4474.

Cannon, J.T., Rychel, A.L., Eccleston, H., Halanych, K.M. and Swalla, B.J. 2009. Molecular phylogeny of Hemichordata, with updated status of deep-sea enteropneusts. *Mol. Phylogenet. Evol.* **52**: 17–24.

Dilly, P.N. 1972. The structures of the tentacles of *Rhabdopleura compacta* (Hemichordata) with special reference to neurociliary control. *Z. Zellforsch.* **129**: 20–39.

Dilly, P.N. 1973. The larva of *Rhabdopleura compacta* (Hemichordata). *Mar. Biol.* **18**: 69–86.

Dilly, P.N. 1975. The dormant buds of *Rhabdopleura compacta* (Hemichordata). *Cell Tissue Res.* **159**: 387–397.

Dilly, P.N. 1986. Modern pterobranchs: observations on their behaviour and tube building. *Geol. Soc. Lond. Spec. Publ.* **20**: 261–269.

Dilly, P.N. 1988. Tube building by *Cephalodiscus gracilis*. *J. Zool.* **216**: 465–468.

Dilly, P.N., Welsch, U. and Rehkämper, G. 1986. Fine structure of tentacles, arms and associated coelomic structures of *Cephalodiscus gracilis* (Pterobranchia, Hemichordata). *Acta Zool. (Stockh.)* **67**: 181–191.

Gilmour, T.H.J. 1979. Feeding in pterobranch hemichordates and the evolution of gill slits. *Can. J. Zool.* **57**: 1136–1142.

Halanych, K.M. 1993. Suspension feeding by the lophophorate-like apparatus of the pterobranch hemichordate *Rhabdopleura normani Biol. Bull.* **185**: 417–427.

Harmer, S.F. 1905. The Pterobranchia of the Siboga-Expedition. *Siboga Expeditie* **26 bis**: 1–132.

Hyman, L.H. 1959. *The Invertebrates, vol. 5. Smaller Coelomate Groups*. McGraw-Hill, New York.

John, C.C. 1932. On the development of *Cephalodiscus*. *'Discovery' Reports* **6**: 191–204.

Lester, S.M. 1985. *Cephalodiscus* sp. (Hemichordata: Pterobranchia): observations of functional morphology, behavior and occurrence in shallow water around Bermuda. *Mar. Biol.* **85**: 263–268.

Lester, S.M. 1988a. Settlement and metamorphosis of *Rhabdopleura normani* (Hemichordata: Pterobranchia). *Acta Zool. (Stockh.)* **69**: 111–120.

Lester, S.M. 1988b. Ultrastructure of adult gonads and development and structure of the larva of *Rhabdopleura normani* (Hemichordata: Pterobranchia). *Acta Zool. (Stockh.)* **69**: 95–109.

Masterman, A.T. 1898. On the further anatomy and the budding process of *Cephalodiscus dodecalophus* (M'Intosh). *Trans. R. Soc. Edinb.* **39**: 507–527.

Mayer, G. and Bartolomaeus, T. 2003. Ultrastructure of the stomochord and the heart-glomerulus complex in *Rhabdopleura compacta* (Pterobranchia): phylogenetic implications. *Zoomorphology* **122**: 125–133.

Nielsen, C. 1987. Structure and function of metazoan ciliary bands and their phylogenetic significance. *Acta Zool. (Stockh.)* **68**: 205–262.

Sato, A., Rickards, B. and Holland, P.W.H. 2008. The origins of graptolites and other pterobranchs: a journey from 'Polyzoa'. *Lethaia* **41**: 303–316.

Schepotieff, A. 1907a. Die Pterobranchier. Die Anatomie von *Cephalodiscus*. *Zool. Jahrb., Anat.* **24**: 553–600.

Schepotieff, A. 1907b. Die Pterobranchier. Die Anatomie von *Rhabdopleura normanii* Allmann. *Zool. Jahrb., Anat.* **23**: 463–534.

Schepotieff, A. 1907c. Die Pterobranchier. Knospungsprozesse und Gehäuse von *Rhabdopleura*. *Zool. Jahrb., Anat.* **24**: 193–238.

Schepotieff, A. 1908. Die Pterobranchier. Knospungsprozess von *Cephalodiscus*. *Zool. Jahrb., Anat.* **25**: 405–486.

Schepotieff, A. 1909. Die Pterobranchier des Indischen Ozeans. *Zool. Jahrb., Syst.* **28**: 429–448.

Stebbing, A.R.D. 1970. Aspects of the reproduction and life cycle of *Rhabdopleura compacta* (Hemichordata). *Mar. Biol.* **5**: 205–212.

Stebbing, A.R.D. and Dilly, P.N. 1972. Some observations on living *Rhabdopleura compacta* (Hemichordata). *J. Mar. Biol. Assoc. U.K.* **52**: 443–448.

62

CHORDATA

The three living chordate phyla, Cephalochordata, Urochordata, and Vertebrata, have been interpreted as a natural, i.e. monophyletic group ever since Kowalevsky (1866, 1871) described the notochord (chorda dorsalis) and the neural tube in the larva of the ascidian *Ciona*, and followed their fate at settling and metamorphosis of the larva. There is a veritable avalanche of literature on all aspects of chordate biology and it is clearly not possible to cover this new information in any depth, so I have selected a few sets of characters that I consider important for our understanding of their phylogeny. Morphology-based studies on chordate origin and evolution proposed that the unsegmented urochordates were the sister group of the segmented groups. However, most of the newer studies on molecular phylogeny agree that urochordates and vertebrates are sister groups within the segmented Chordata (Fig. 62.1) (see for example Dunn *et al.* 2008; Holland *et al.* 2008; Putnam *et al.* 2008; Hejnol *et al.* 2009; Philippe *et al.* 2009; Pick *et al.* 2010), which implies that the urochordates have lost the segmentation. There is apparently no morphological sign of the reduced segmentation, but the genome studies show a massive loss of genes in the urochordates (Garcia-Fernàndez and Benito-Gutiérrez 2009) (Chapter 64). It could be interesting to identify genes involved in chordate segmentation and see if they are missing in the urochordates. The Hox genes show a characteristic 'chordate signature' (Fig. 21.3).

The recent molecular analyses agree that Urochordata and Vertebrata together form the clade Olfactores (a name introduced en passant by Jefferies 1991 in a discussion of his 'calcichordate' theory), but only a few morphological synapomorphies have been identified. The name refers to the presence of special olfactory areas in the buccal cavity (Jefferies *et al.* 1996), referring to the placodes originating from migrating neural crest cells (see also Holland *et al.* 2008). Additional evidence for the sister-group relationship comes from presence of special cadherins and collagens (Oda *et al.* 2002; Wada *et al.* 2006).

Several early fossils have been described during the last decade, but differential decay of organs before fossilization may result in 'stem-ward slippage', where the more advanced characters are lost first, and this makes it difficult to place the fossils on the phylogeny (Sansom *et al.* 2010). A number of the famous Lower Cambrian fossils from Chengjiang and Burgess Shale have probably been overinterpreted: *Pikaia* (Briggs *et al.* 1994) can apparently only be characterized as a chordate; *Nectocaris* can be a stem chordate, a stem cephalochordate, or a stem vertebrate (Sansom *et al.* 2010), and it has even been interpreted as a cephalopod (Smith and Caron 2010); *Myllokunmingia* (Shu *et al.* 1999) can be a stem vertebrate, a stem agnathan, or a stem gnathostome; the yunnanozoans (Chen *et al.* 1999) could be stem deuterostomes, stem enteropneusts, or stem chordates (Briggs 2010; Sansom *et al.* 2010); *Metaspriggina* could be a chordate (Conway Morris 2008); the pipiscids and vetulicolians seem entirely enigmatic (Shu *et al.* 1999; Aldridge *et al.* 2007). At this point, one can probably only conclude

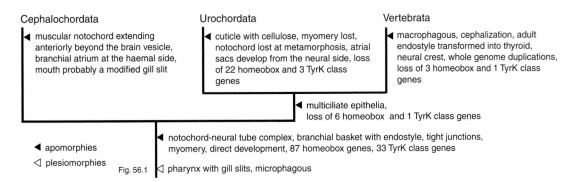

Cephalochordata

◄ muscular notochord extending anteriorly beyond the brain vesicle, branchial atrium at the haemal side, mouth probably a modified gill slit

Urochordata

◄ cuticle with cellulose, myomery lost, notochord lost at metamorphosis, atrial sacs develop from the neural side, loss of 22 homeobox and 3 TyrK class genes

Vertebrata

◄ macrophagous, cephalization, adult endostyle transformed into thyroid, neural crest, whole genome duplications, loss of 3 homeobox and 1 TyrK class genes

◄ multiciliate epithelia, loss of 6 homeobox and 1 TyrK class genes

◄ apomorphies
◁ plesiomorphies   Fig. 56.1

◄ notochord-neural tube complex, branchial basket with endostyle, tight junctions, myomery, direct development, 87 homeobox genes, 33 TyrK class genes

◁ pharynx with gill slits, microphagous

Figure 62.1. Phylogeny of the Chordata. Information on genes from Garcia-Fernàndez and Benito-Gutiérrez (2009).

that the Cambrian fossils demonstrate the presence of various early deuterostomes/chordates, but that their structure is so open to interpretation that they are of little help in reconstructing chordate evolution (see also Shu *et al.* 2010).

The dorsal-ventral orientation of the chordates has been much discussed. The traditional orientation, with neural tube and notochord on the dorsal side and the mouth on the ventral side, is clearly based on vertebrate morphology. The cephalochordates have a mouth that develops on the left side and moves to the 'ventral' side (Chapter 63). The urochordate larvae have the stomodaeum on the 'dorsal' side just in front of the neural fold (Chapter 64), but their orientation becomes obscured by the metamorphosis. The vertebrate stomodaeum develops from the anterior side of the neural fold (Chapter 65). The alternative orientation, with the neural tube on the ventral side, resembling the position of the nerve cord in arthropods and annelids, was proposed already by Geoffroy-Saint-Hilaire (1822) and later by Dohrn (1875) on more speculative grounds. More recently, the idea has been argued on the basis of new morphological and molecular characters, such as position of the main blood vessel (Malakhov 1977), gene expression (Arendt and Nübler-Jung 1994; Nübler-Jung and Arendt 1994), pharynx structure (Ruppert *et al.* 1999), and embryology (Nielsen 1999). Numerous studies on gene expression now support the interpretation of the 'dorsal' side of the chordates as homologous to the ventral side of protostomes and enteropneusts (Lowe *et al.* 2006).

The evolution of the new orientation is discussed below. I have earlier (Nielsen 1999) proposed to use the terms c-dorsal and c-ventral for the orientations in the chordates, but I will here use 'dorsal' for the neural side and 'ventral' for the haemal side.

Several apomorphies characterize the Chordata: (1) the notochord, a stiffening rod of tissue formed from the roof of the archenteron; (2) the neural tube, formed from a median, longitudinal fold of embryonic ectoderm mostly in contact with the notochord; (3) the longitudinal muscles along the notochord used in locomotion by creating wagging or undulatory movements of a finned body or tail; (4) the ciliated pharyngeal gill slits, which support a mucus filter secreted by the endostyle; in the vertebrates, this character is only found in the somewhat modified filtration system of the ammocoetes-larva of the lampreys, but the various cell types of the endostyle can be recognized in the thyroid of all the adult vertebrates; and (5) the very similar fate maps. Characters 1–3 are not independent; they form a functional unit, and intricate signalling between cells of the archenteron and the overlying ectoderm specify the differentiation of the various cell types in the neural tube-notochord-mesoderm complex (see for example Harland and Gerhart 1997; Stemple 2005).

The notochord is a cylinder of cells surrounded by a strengthened basal membrane. It differentiates from the neural side of the archenteron. However, the morphology of the notochord differs between the three phyla. In amphioxus, the notochord consists of a stack of flat muscle cells (Chapter 63). In the urochordates, the early larval notochord is a row of cylindrical cells, but in most of the older larvae it becomes transformed to a hollow tube of cells surrounding a core of elastic material (Chapter 64). In the vertebrates, the notochord is a rod of irregularly arranged cells with large vacuoles in the cyclostomes, but this structure is modified by the development of the vertebrae in the gnathostomes (Chapter 65).

The neural tube develops from the neural plate, an area of the ectoderm in contact with the notochord; its edges are thickened forming the neural fold. The lateral parts of the neural fold fuse medially so that the neural tube is formed, covered by the ectoderm. Posteriorly, the fold continues around the blastopore, which becomes enclosed by the fusion of the fold so that the archenteron and the neural canal become connected by the narrow neurenteric canal. Anteriorly, the infolding follows the notochord in cephalochordates and vertebrates, but in the urochordates, the short notochord in the tail apparently also induces the infolding of the anterior part of the nerve tube along the body (Chapter 64). The anterior opening from the neural tube, the neuropore, is retained in the cephalochordates, where it connects with Kölliker's pit (Chapter 63). In the urochordates, it is for some time in communication with the stomodaeal invagination, which forms the oral siphon (Chapter 64). In the vertebrates, it seems that the neuropore just disappears (Ruppert 1990). Reissner's fibre is an example of a homology of neural tube structures in the three phyla. It is a thin fibre situated in the neural tube, formed by the few cells of the infundibular organ in amphioxus, by one cell in *Oikopleura*, and by the subcommisural cells in vertebrates (Holmberg and Olsson 1984; Lemaire 2009).

Comparisons of the differentiation of the neural tube into the central nervous system (CNS) of the three chordate phyla are now based on both morphological, ontogenetic, and molecular data (its phylogenetic origin will be discussed below). All data indicate that the sensory vesicle of the ascidian larva and of the vertebrate forebrain + midbrain is related to the anterior somite of amphioxus (Wicht and Lacalli 2005).

Gene expression supports this interpretation. *Otx* is expressed in the cerebral vesicle of amphioxus, in the sensory vesicle of ascidian larvae, and in forebrain + midbrain in vertebrates. Hox genes are expressed behind the neck in amphioxus and ascidian larvae and a little further back in vertebrates (Wada and Satoh 2001; Cañestro *et al.* 2005; Lacalli 2006, 2008). The ciliary frontal eye of amphioxus has been proposed to be a homologue of the lateral eyes of the vertebrates (Wicht and Lacalli 2005). These observations indicate the amphioxus CNS has an ancestral chordate layout, and that vertebrate (and urochordate) evolution has proceeded through a whole series of specializations of the anterior part of the CNS, especially the enormous enlargement of the forebrain in the vertebrates, whereas amphioxus has retained the general layout but added a number of unique sensory organs (Chapter 63).

The longitudinal muscles originate from segmental mesodermal pockets from the archenteron in cephalochordates and vertebrates, but the segmentation has been lost in the urochordates (see below). Their muscles develop from longitudinal zones of the archenteron lateral to the notochord; they are restricted to the larval tail and lost at metamorphosis. However, these muscles are all primarily involved in swimming, and their functional connection with the notochord strongly indicates that the combined structure is a homologous unit in the three phyla. The more posterior coelomic pockets of cephalochordates are added alternately on the two sides, organized by genes such as Notch-Delta and Wnt/β-catenin, but concomitantly in the vertebrates, organized by an oscillating wave-front expression of a number of the same genes. The wave-front mechanism is apparently a vertebrate apomorphy (Beaster-Jones *et al.* 2008).

In cephalochordates and urochordates the branchial basket, with longitudinal rows of ciliated gill slits, pumps water from the mouth through the slits, and particles are filtered from the water by a mucus net secreted by the endostyle. The branchial basket is highly modified in the vertebrates, with the endostyle of the adults modified as an endocrine gland, which shows expression of some of the same thyroid-related genes as those found in cephalochordates and ascidian larvae (Ogasawara *et al.* 1999b; Holland *et al.* 2008; Holland and Short 2010). The endostyles of cephalochordates and urochordates comprise characteristic bands of cells with similar functions. The different bands of glandular and ciliated cells in the endostyles of urochordates, cephalochordates, and the ammocoetes-larva of the lampreys cannot be homologized directly, because the functions are not distributed uniformly in the various forms, but all forms apparently have a midventral zone of monociliate cells with very long cilia. The

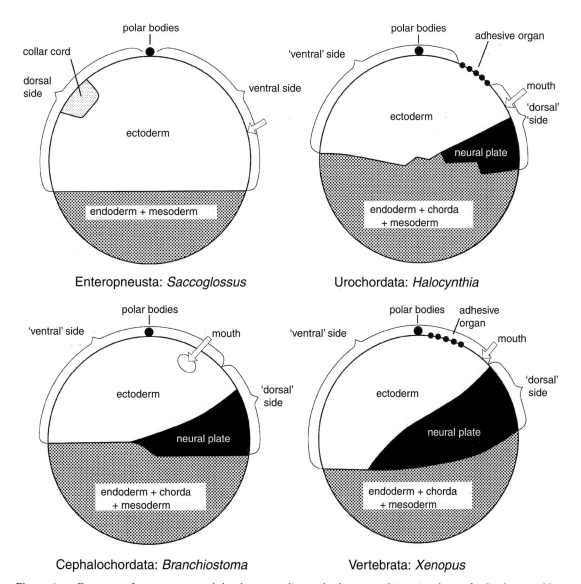

Figure 62.2. Fate maps of enteropneusts and chordates according to the dorsoventral inversion theory; fertilized eggs or blastula stages seen from the left side. Enteropneusta: *Saccoglossus kowalevskii.* (Based on Colwin and Colwin 1951.) Urochordata: *Halocynthia roretzi.* (Based on Nishida 1987.) Cephalochordata: *Branchiostoma lanceolatum.* (Based on Tung *et al.* 1962; see Holland and Holland 2007 for a more detailed map.) Vertebrata: *Bombinator.* (Based on Vogt 1929.)

ascidians appear to have the most complicated endostyle with eight bands of cells on each side (Ogasawara *et al.* 1999a), while the appendicularian *Oikopleura* has only three bands, with the major band performing both major secretory functions (Olsson 1963; Cañestro *et al.* 2008). The cell rows two and four generally secrete the mucus net with proteins.

The fate maps of representatives of all three chordate phyla show that the mouth develops from an area between the apical pole and the neural plate (Fig. 62.2). In some ascidians, a stomodaeal invagination originates from the anterior rim of the neural plate (Chapter 64). The fate maps support the interpretation of the 'dorsal' or neural side of the vertebrates as homologous with the ventral side of the non-chordates (see above).

The homologies of a number of anterior structures have been much discussed in the past, and a number of homologies have now been ascertained. Only the anterior pair of coelomic sacs, the protocoel, characteristic of the ambulacrarians, can be recognized with certainty in the chordates. The mesocoel has to my knowledge not been identified with certainty in urochordates and vertebrates, but is possibly found as the mouth coelom in the larva and the velar coelom in the adult amphioxus (Ruppert 1997). The axial complex so characteristic of the ambulacrarians (Fig. 57.3), is not found in any of the chordates, but various small coelomic pouches and ciliated ectodermal pits or pockets have been homologized with different parts of this complex. Ruppert (1990, 1997) contributed to our knowledge about structure and function of some of these organs and made a very informative summary and interpretation of the literature about the various organs (see

Table 62.1). He pointed out that two independent organ complexes should be distinguished: the anterior coelomopore complex and the anterior neuropore. The anterior coelomopore complex consists of the left protocoel and the ectodermal invagination forming part of its coelomoduct, while the anterior neuropore is derived from the anterior opening of the neural tube. The anterior coelomopore complex obviously represents the axial complex, while the anterior neuropore is a chordate apomorphy related to the formation of the neural tube. Further details about the cephalochordates can be found in Chapter 63.

Gastrulation and neurulation are two processes that are important in the comparisons of the groups under discussion. Gastrulation, the internalization of endodermal cells forming an archenteron, generally by invagination or epiboly, is a process that can easily be recognized in many eumetazoans. Neurulation is the differentiation of the neural tube on one side of the chordate gastrula, encompassing the blastopore. Amphioxus represents the supposedly ancestral type of gastrulation, with a thin layer of cells surrounding a spacious blastocoel that becomes obliterated by invagination; neurulation is initiated after completion of gastrulation. A more complicated type is represented by the amphibians with yolky eggs that go through a very oblique gastrulation, with a circular infolding surrounding a plug of endodermal cells; the infolding is deep on the dorsal side and shallow on the ventral side, but the endodermal plug finally disappears into the circular blastopore (Fig. 65.1). The first stages of the neural fold are formed at this stage and neurulation proceeds through the following phase of

Table 62.1. Structure/function of the anterior coelomopore complex (= left protocoel + ectodermal invagination) and of the right protocoel in the deuterostome phyla; based on Ruppert (1990, 1997). * Stach (2002) interprets Hatschek's nephridium as the homologue of the left protocoel.

	Anterior coelomopore complex (left)	Right protocoel
Echinodermata	axocoel + hydropore	dorsal sac
Enteropneusta	protocoel + proboscis pore	? pericard
Cephalochordata*	larva: left anterior head coelom + preoral pit	right head coelom
	adult: absent + Hatschek's pit	ventral rostral coelom
Urochordata	?	?
Vertebrata	embryo: premandibular somite + Rathke's pouch	
	adult: eye muscles + adenohypophysis	eye muscles

ontogeny. In both cases, gastrulation is the constriction of an equatorial zone of the blastula that becomes the blastopore. There is no sign of lateral blastopore closure in chordates, as suggested by Arendt and Nübler-Jung (1997) and Bergström (1997).

The complex of notochord, neural tube, and longitudinal muscles is obviously an adaptation for swimming, which indicates that the chordate ancestor was free living, as also concluded by Holland *et al.* (2008). The swimming was probably by undulatory movements of the tail, with the muscles on the two sides functioning as antagonists and the notochord as a length stabilizing rod (see below).

The well-founded homology of the gill slits of enteropneusts and chordates (Chapter 56; Fig. 56.2) indicates that the chordate ancestor was enteropneust-like, probably with a planktotrophic dipleurula larva. The larva of this ancestor could have evolved a wiggling tail based on longitudinal muscles of the metacoel. An analogous, anterior extension seen in the

larva of some species of *Luidia* (Fig. 62.3) shows that new locomotory structures can evolve in dipleurula larvae. The extension of its pre-oral lobe contains a pair of longitudinal muscles that are used in swimming, by creating more or less oblique dorsoventral bendings using the protocoel as the antagonist to the muscles (Tattersall and Sheppard 1934). The notochord could have evolved in a posterior extension in the larva, functioning as a stiffening rod and antagonist to contractions of lateral muscles in the metacoel. A tail with a metacoel on each side would have been able to make lateral or oblique bendings, but a longitudinal compartmentalization of the metacoels would enable directed swimming by undulatory movements of the tail. An ancestral larva of this type would resemble the 'somatic-visceral' ancestral chordate proposed by Romer (1972), but with the difference that his hypothetical organism was the larva of a sessile chordate ancestor. The evolution of the swimming adult ancestor with notochord and

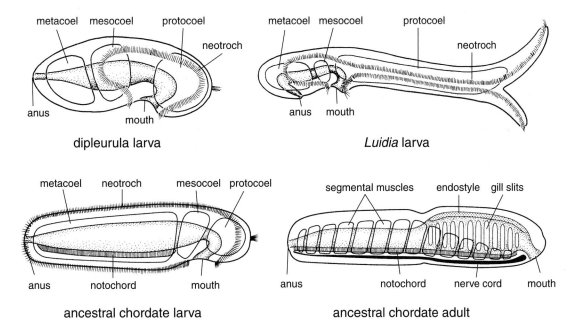

Figure 62.3. Evolution of chorda, tail, and segmentation. The dipleurula is a generalized drawing. The *Luidia* larva is based on Tattersall and Sheppard (1934) and own observations. The 'ancestral chordate larva' is feeding with the neotroch but is able to swim by lateral movements of the tail. The 'ancestral chordate adult' has developed gill slits and endostyle, and the movement of the tail with segmented muscles is coordinated by cells of the neural tube; the mouth is still in the ventral position (as in the urochordate larva).

lateral muscles should then have been through paedomorphosis. The proposed evolution and specialization of the posterior part of a dipleurula larva would explain the origin of the notochord and the segmented musculature, but not of the neural tube/spinal cord (CNS).

The classical auricularia/dipleurula theory for the evolution of the chordate CNS was proposed and refined by Garstang (1894, 1928) and Garstang and Garstang (1926), and the evolution of this idea and its modification through studies, for example, by Berrill (1955) have been reviewed by Gee (1996) and Lacalli (2004). It proposed that the chordate CNS evolved through modification of the circumoral ciliary band (neotroch) of a dipleurula, in which the two lateral sides of the ciliary band moved dorsally and fused, enclosing both apical organ and anus and forming the neural tube, the brain, and the neurenteric canal. This implies that the neural side of the chordates corresponds to the dorsal side of the non-chordates, and this is not in accordance with the present understanding of the orientation of the chordates (see above and for example Holland et al. 2008).

However, the general similarity between the ultrastructure of the ciliary bands of the echinoderm and enteropneust larvae and the neural folds and early neural tube of amphioxus indicates that the neural tube has evolved from a part of the neotroch (Lacalli et al. 1990; Crowther and Whittaker 1992; Lacalli and West 1993; Lacalli 1996). This finds support from the expression of genes involved in the development of the chordate CNS in the neotroch of echinoderm and enteropneust larvae (Shoguchi et al. 2000; Nieuwenhuis 2002; Miyamoto et al. 2010).

In an effort to combine most of the available data, I have proposed (Nielsen 1999) that the chordate neural tube is derived from a ventral loop of the neotroch of a dipleurula-type larva of the latest common ancestor of ambulacrarians and chordates (Fig. 62.4). This theory accommodates the observations of similarity between the ciliary bands of echinoderm and enteropneust larvae and the neural tube of amphioxus and the several morphological data that indicate that

the chordates are turned upside-down relative to the enteropneusts. It is further in accordance with the molecular data, provided that the side interpreted as ventral in the non-chordate deuterostomes corresponds to the ventral side of the protostomes. However, it implies that the brains and nerve cords of protostomes and chordates have evolved independently. The origin of the innervation of the lateral muscles by cells in the nerve cord remains unexplained.

This theory has not been much discussed. Satoh (2008) rejected it in proposing that the dorsal-ventral inversion is a misinterpretation. The theory implies that the apical organ of the dipleurula larvae is not homologous with structures in the brain vesicle in amphioxus larvae, as proposed for example by Lacalli et al. (1994), and this seems to be contradicted by the similar expressions of T-box genes in the apical plate of a tornaria larva (Tagawa et al. 2000) and in the telencephalon of various vertebrates (Nieuwenhuis 2002). However, amphioxus shows T-box gene expression in the early brain vesicle and a little later on in eyespots and endo- and mesoderm, but there is no specific similarity with the expression in the dipleurula larva that could indicate homology (Horton et al. 2008). Nieuwenhuis (2002) and Lacalli (2006, 2010) both found the theory to be in good accordance with data both from morphology and gene expression.

There are many detailed similarities in gene expression and neuron arrangement in the ventral neuronal epithelium of protostomian embryos and the neural plate of chordates (see for example Denes et al. 2007), and there are conspicuous similarities in the pattern of gene expression in the anterior protostomian CNS and in the brain of vertebrate embryos (Arendt et al. 2008). As mentioned above there is no morphological trace of amphistomy, i.e. the division of the blastopore into mouth and anus in the deuterostomes, but the first ciliary band to develop in the tornaria larva is the perianal ciliary ring, which has a midventral extension, the gastrotroch, reaching to the oral region behind the neotroch (Nielsen and Hay-Schmidt 2007). This would correspond to the midline cells in the vertebrate neural tube. Could this be a clue to the similarities between the ventral nerve cords originating

Development of an enteropneust (morphological = biological orientation)

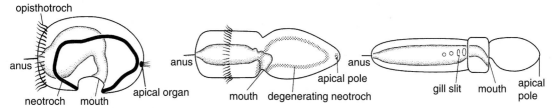

Hypothetical development of chordates (morphological orientation)

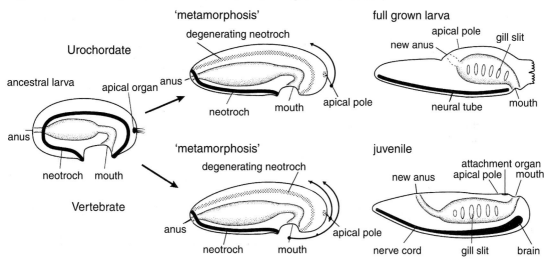

Figure 62.4. Evolution of the chordate CNS from a ventral loop of the neotroch of a dipleurula larva. The urochordates are represented by a tadpole larva. The arrows indicate the observed ontogenetic movements of the apical pole and the mouth (see Fig. 65.1). (Modified from Nielsen 1999.)

from the lateral blastopore lips in the protostomes and the neural tube of the chordates?

The gene expression data leaves ample room for discussions about the morphological evolution of the chordates.

References

Aldridge, R.J., Hou, X.-G., Siveter, D.J., Siveter, D.J. and Gabbott, S.E. 2007. The systematics and phylogenetic relationships of vetulicolians. *Palaeontology* **50**: 131–168.

Arendt, D., Denes, A.S., Jékely, G. and Tessmar-Raible, K. 2008. The evolution of nervous system centralization. *Phil. Trans. R. Soc. Lond. B* **363**: 1523–1528.

Arendt, D. and Nübler-Jung, K. 1994. Inversion of dorsoventral axis. *Nature* **371**: 26.

Arendt, D. and Nübler-Jung, K. 1997. Dorsal or ventral: similarities in fate maps and gastrulation patterns in annelids, arthropods and chordates. *Mech. Dev.* **61**: 7–21.

Beaster-Jones, L., Kaltenbach, S.L., Koop, D., *et al.* 2008. Expression of somite segmentation genes in amphioxus: a clock without a wavefront? *Dev. Genes Evol.* **218**: 599–611.

Bergström, J. 1997. Origin of high-rank groups of organisms. *Paleontol. Res.* **1**: 1–14.

Berrill, N.J. 1955. *The Origin of the Vertebrates.* Oxford University Press, Oxford.

Briggs, D.E.G. 2010. Decay distorts ancestry. *Nature* **463**: 741–743.

Briggs, D.E.G., Erwin, D.H. and Collier, F.J. 1994. *The Fossils of the Burgess Shale.* Smithsonian Institution Press, Washington.

Cañestro, C., Bassham, S. and Postlethwait, J. 2005. Development of the central nervous system in the larvacean *Oikopleura dioica* and the evolution of the chordate brain. *Dev. Biol.* **285**: 298–315.

Cañestro, C., Bassham, S. and Postlethwait, J.H. 2008. Evolution of the thyroid: anterior–posterior regionalization of the Oikopleura endostyle revealed by *Otx*, *Pax2/5/8*, and *Hox1* expression. *Dev. Dyn.* **237**: 1490–1499.

Chen, J.Y., Huang, D.Y. and Li, C.W. 1999. An early Cambrian craniate-like chordate. *Nature* **402**: 518–522.

Colwin, A.L. and Colwin, L.H. 1951. Relationships between the egg and larva of *Saccoglossus kowalevskii* (Enteropneusta): axes and planes; general prospective significance of the early blastomeres. *J. Exp. Zool.* **117**: 111–137.

Conway Morris, S. 2008. A redescription of a rare chordate, *Metaspriggina walcotti* Simonetta and Insom, from the Burgess Shale (Middle Cambrian), British Columbia, Canada. *Palaeontology* **82**: 424–430.

Crowther, R.J. and Whittaker, J.R. 1992. Structure of the caudal neural tube in an ascidian larva: vestiges of its possible evolutionary origin from a ciliated band. *J. Neurobiol.* **23**: 280–292.

Denes, A.S., Jékely, G., Steinmetz, P.R.H., *et al.* 2007. Molecular architecture of annelid nerve cord supports common origin of nervous system centralization in Bilateria. *Cell* **129**: 277–288.

Dohrn, A. 1875. *Der Ursprung der Wirbelthiere und das Princip des Funktionswechsels*. Wilhelm Engelmann, Leipzig.

Dunn, C.W., Hejnol, A., Matus, D.Q., *et al.* 2008. Broad phylogenomic sampling improves resolution of the animal tree of life. *Nature* **452**: 745–749.

Garcia-Fernàndez, J. and Benito-Gutiérrez, È. 2009. It's a long way from Amphioxus: descendants of the earliest chordate. *BioEssays* **31**: 665–675.

Garstang, S. and Garstang, W. 1926. On the development of *Botrylloides* and the ancestry of the vertebrates (preliminary note). *Proc. Leeds Philos. Soc.* **1926**: 81–86.

Garstang, W. 1894. Preliminary note on a new theory of the phylogeny of the Chordata. *Zool. Anz.* **17**: 122–125.

Garstang, W. 1928. The morphology of the Tunicata, and its bearings on the phylogeny of the Chordata. *Q. J. Microsc. Sci., N. S.* **72**: 51–187.

Gee, H. 1996. *Before the Backbone*. Chapman & Hall, London.

Geoffroy-Saint-Hilaire, E. 1822. Considérations générales sur la vertèbre. *Mem. Mus. Hist. Nat.* **9**: 89–119.

Harland, R. and Gerhart, J. 1997. Formation and function of Spemann's organizer. *Annu. Rev. Cell Dev. Biol.* **13**: 611–667.

Hejnol, A., Obst, M., Stamatakis, A., *et al.* 2009. Assessing the root of bilaterian animals with scalable phylogenomic methods. *Proc. R. Soc. Lond. B* **276**: 4261–4270.

Holland, L.Z. and Holland, N.D. 2007. A revised fate map for amphioxus and the evolution of axial patterning in chordates. *Integr. Comp. Biol.* **47**: 360–372.

Holland, L.Z. and Short, S. 2010. Alternative splicing in development and function of chordate endocrine systems: a focus on Pax genes. *Integr. Comp. Biol.* **50**: 22–34.

Holland, L.Z., Albalat, R., Azumi, K., *et al.* 2008. The amphioxus genome illuminates vertebrate origins and cephalochordate biology. *Genome Res.* **18**: 1100–1111.

Holmberg, K. and Olsson, R. 1984. The origin of Reissner's fibre in an appendicularian, *Oikopleura dioica*. *Vidensk. Medd. Dan. Naturhist. Foren.* **145**: 43–52.

Horton, A.C., Mahadevan, N.R., Minguillon, C., *et al.* 2008. Conservation of linkage and evolution of developmental function within the Tbx2/3/4/5 subfamily of T-box genes: implications for the origin of vertebrate limbs. *Dev. Genes Evol.* **218**: 613–628.

Jefferies, R.P.S. 1991. Two types of bilateral symmetry in the Metazoa: chordate and bilaterian. *Ciba Found. Symp.* **162**: 94–127.

Jefferies, R.P.S., Brown, N.A. and Daley, P.E. 1996. The early phylogeny of chordates and echinoderms and the origin of chordate left-right asymmetry and bilateral symmetry. *Acta Zool. (Stockh.)* **77**: 101–122.

Kowalevsky, A. 1866. Entwickelungsgeschichte der einfachen Ascidien. *Mem. Acad. Sci. St-Petersb.*, 7. Ser. **10**: 1–19.

Kowalevsky, A. 1871. Weitere studien über die Entwicklung der einfachen Ascidien. *Arch. Mikrosk. Anat.* **7**: 101–130.

Lacalli, T.C. 1996. Dorsoventral axis inversion: a phylogenetic perspective. *BioEssays* **18**: 251–254.

Lacalli, T.C. 2004. Sensory systems in amphioxus: a window on the ancestral chordate condition. *Brain Behav. Evol.* **64**: 148–164.

Lacalli, T.C. 2006. Prospective protochordate homologs of vertebrate midbrain and MHB, with some thoughts on MHB origins. *Int. J. Dev. Biol.* **2**: 104–109.

Lacalli, T.C. 2008. Basic features of the ancestral chordate brain: a protochordate perspective. *Brain Res. Bull.* **75**: 319–323.

Lacalli, T.C. 2010. The emergence of the chordate body plan: some puzzles and problems. *Acta Zool. (Stockh.)* **91**: 4–10.

Lacalli, T.C. and West, J.E. 1993. A distinctive nerve cell type common to diverse deuterostome larvae: comparative data from echinoderms, hemichordates and amphioxus. *Acta Zool. (Stockh.)* **74**: 1–8.

Lacalli, T.C., Gilmour, T.H.J. and West, J.E. 1990. Ciliary band innervation in the bipinnaria larva of *Pisaster ochraceus*. *Phil. Trans. R. Soc. Lond. B* **330**: 371–390.

Lacalli, T.C., Holland, N.D. and West, J.E. 1994. Landmarks in the anterior central nervous system of amphioxus larvae. *Phil. Trans. R. Soc. Lond. B* **344**: 165–185.

Lemaire, P. 2009. Unfolding a chordate developmental program, one cell at a time: Invariant cell lineages, short-range inductions and evolutionary plasticity in ascidians. *Dev. Biol.* **332**: 48–60.

Lowe, C.J., Terasaki, M., Wu, M., *et al.* 2006. Dorsoventral patterning in hemichordates: insights into early chordate evolution. *PLoS Biol.* **4**: e291.

Malakhov, V.V. 1977. The problem of the basic structural plan in various groups of Deuterostomia. *Zh. Obshch. Biol.* **38**: 485–499.

Miyamoto, N., Nakajima, Y., Wada, H. and Saitoh, Y. 2010. Development of the nervous system in the acorn worm *Balanoglossus simodensis*: insights into nervous system evolution. *Evol. Dev.* **12**: 416–424.

Nielsen, C. 1999. Origin of the chordate central nervous system - and the origin of the chordates. *Dev. Genes Evol.* **209**: 198–205.

Nielsen, C. and Hay-Schmidt, A. 2007. Development of the enteropneust *Ptychodera flava*: ciliary bands and nervous system. *J. Morphol.* **268**: 551–570.

Nieuwenhuis, R. 2002. Deuterostome brains: synopsis and commentary. *Brain Res. Bull.* **57**: 257–270.

Nishida, H. 1987. Cell lineage analysis in ascidian embryos by intracellular injection of a tracer enzyme III. Up to the tissue restricted stage. *Dev. Biol.* **121**: 526–541.

Nübler-Jung, K. and Arendt, D. 1994. Is ventral in insects dorsal in vertebrates? *Roux's Arch. Dev. Biol.* **203**: 357–366.

Oda, H., Wada, H., Tagawa, K., *et al.* 2002. A novel amphioxus cadherin that localizes to epithelial adherens junctions has an unusual domain organization with implications for chordate phylogeny. *Evol. Dev.* **4**: 428–434.

Ogasawara, M., Di Lauro, R. and Satoh, N. 1999a. Ascidian homologs of mammalian thyroid peroxidase genes are expressed in the thyroid-equivalent region of the endostyle. *J. Exp. Zool.* **285**: 158–169.

Ogasawara, M., Di Lauro, R. and Satoh, N. 1999b. Ascidian homologs of mammalian thyroid transcription factor-1 gene are expressed in the endostyle. *Zool. Sci. (Tokyo)* **16**: 559–565.

Olsson, R. 1963. Endostyles and endostylar secretions: a comparative histochemical study. *Acta Zool. (Stockh.)* **44**: 299–328.

Philippe, H., Derelle, R., Lopez, P., *et al.* 2009. Phylogenomics revives traditional views on deep animal relationships. *Curr. Biol.* **19**: 706–712.

Pick, K.S., Philippe, H., Schreiber, F., *et al.* 2010. Improved phylogenomic taxon sampling noticeably affects nonbilaterian relationships. *Mol. Biol. Evol.* **27**: 1983–1987.

Putnam, N.H., Butts, T., Ferrier, D.E.K., *et al.* 2008. The amphioxus genome and the evolution of the chordate karyotype. *Nature* **453**: 1064–1071.

Romer, A.S. 1972. The vertebrate as a dual animal—somatic and visceral. *Evol. Biol.* **6**: 121–156.

Ruppert, E.E. 1990. Structure, ultrastructure and function of the neural gland complex of *Ascidia interrupta* (Chordata, Ascidiacea): clarification of hypotheses regarding the evolution of the vertebrate anterior pituitary. *Acta Zool. (Stockh.)* **71**: 135–149.

Ruppert, E.E. 1997. Cephalochordata (Acrania). In F.W. Harrison (ed.): *Microscopic Anatomy of Invertebrates*, vol. 15, pp. 349–504. Wiley-Liss, New York.

Ruppert, E.E., Cameron, C.B. and Frick, J.E. 1999. Endostyle-like features of the dorsal epibranchial ridge of an enteropneust and the hypothesis of dorsal-ventral axis inversion in chordates. *Invertebr. Biol.* **118**: 202–212.

Sansom, R.S., Gabbott, S.E. and Purnell, M.A. 2010. Non-random decay of chordate characters causes bias in fossil interpretation. *Nature* **463**: 797–800.

Satoh, N. 2008. An aboral-dorsalization hypothesis for chordate origin. *Genesis* **46**: 614–622.

Shoguchi, E., Harada, Y., Numakunai, T. and Satoh, N. 2000. Expression of the *Otx* gene in the ciliary bands during sea cucumber embryogenesis. *Genesis* **27**: 58–63.

Shu, D., Conway Morris, S., Zhang, X.L., *et al.* 1999. A pipiscid-like fossil from the Lower Cambrian of south China. *Nature* **400**: 746–749.

Shu, D.G., Conway Morris, S., Zhang, Z.-F. and Han, J. 2010. The earliest history of the deuterostomes: the importance of the Chengjiang Fossil-Lagerstätte. *Proc. R. Soc. Lond. B* **277**: 165–174.

Smith, M.R. and Caron, J.-B. 2010. Primitive soft-bodied cephalopods from the Cambrian. *Nature* **465**: 469–472

Stach, T. 2002. Minireview: On the homology of the protocoel in Cephalochordata and 'lower' Deuterostomia. *Acta Zool. (Stockh.)* **83**: 25–31.

Stemple, D.L. 2005. Structure and function of the notochord: an essential organ for chordate development. *Development* **132**: 2503–2512.

Tagawa, K., Humphries, T. and Satoh, N. 2000. T-Brain expression in the apical organ of hemichordate tornaria larvae suggests its evolutionary link to the vertebrate forebrain. *J. Exp. Zool.* **288**: 23–31.

Tattersall, W.M. and Sheppard, E.M. 1934. Observations on the bipinnaria of the asteroid genus *Luidia*. *James Johnstone Memorial Volume*, University Press of Liverpool, Liverpool.

Tung, T.C., Wu, S.C. and Tung, Y.Y.F. 1962. The presumptive areas of the egg of amphioxus. *Sci. Sin.* **11**: 629–644.

Vogt, W. 1929. Gestaltungsanalyse am Amphibienkeim mit Örtlicher Vitalfärbung. II. Teil. Gastrulation und Mesodermbildung bei Urodelen und Anuren. *Arch Entwicklungsmech. Org.* **120**: 384–706.

Wada, H. and Satoh, G. 2001. Patterning the protochordate neural tube. *Curr. Opin. Neurobiol.* **11**: 16–21.

Wada, H., Okuyama, M., Satoh, N. and Zhang, S. 2006. Molecular evolution of fibrillar collagen in chordates, with implications for the evolution of vertebrate skeletons and chordate phylogeny. *Evol. Dev.* **8**: 370–377.

Wicht, H. and Lacalli, T.C. 2005. The nervous system of amphioxus: structure, development, and evolutionary significance. *Can. J. Zool.* **83**: 122–150.

63

Phylum **Cephalochordata**

Cephalochordata, or lancelets, is a very small phylum, comprising only about 30 species representing two families (the excellent treatise by Ruppert 1997 is the general reference for this chapter). All species are marine, mostly living more or less buried in coarse sand. The pelagic larvae are almost like juveniles. The fossil record contains no unquestionable cephalochordates (Sansom *et al.* 2010).

The most common genus, *Branchiostoma*, more commonly known as amphioxus, has been in the focus of evolutionary studies for more than a century, and only selected topics from the literature will be touched upon here.

The body is lanceolate, and an archimeric regionation can neither be distinguished from the outer shape nor from the embryology or morphology of the mesoderm.

All epithelia are monolayered, and all the ectoderm of the juvenile stages, as well as many areas of the gut of the adult stages, are ciliated; occasional cilia have been observed also on the peritoneum; only monociliate cells have been observed.

The mouth is situated at the bottom of an antero 'ventral' invagination, the vestibule. The opening of the vestibule is surrounded by a horseshoe of cirri, and there is a system of ciliated ridges, the wheel organ, at the bottom of the vestibule in front of the mouth opening. A ciliated structure at the 'dorsal' side of the vestibule, called Hatschek's pit, secretes mucus that flows over the wheel organ; the mucus with trapped particles becomes ingested. The mouth is surrounded by a ring of ciliated velar tentacles. The gut consists of a spacious pharynx with numerous parallel gill slits (the branchial basket), an oesophagus, an intestine with an anterior, digestive diverticulum, and a short rectum. The gill slits open into a latero-'ventral' atrial chamber that has a posterior, mid-'ventral' opening some distance in front of the anus. The anus is situated 'ventrally' a short distance from the posterior end.

The gill slits are U-shaped with a median tongue bar originating from the 'dorsal' side of the gill opening; gill slits are subdivided into vertical rows of gill pores. The system is supported by a skeleton formed by a thickened basal membrane with stiffening rods. The basal membrane contains collagen, but this is absent in the rods that consist of structural proteins and acid mucopolysaccharides, probably including chondroitin sulphate, and thus possibly representing a cartilage-like composition; chitin is absent. There are several blood vessels both in the basal membrane and in the rods, but no special respiratory areas have been reported. The gill slits represent one type of segmentation, branchiomery, that is reflected in the excretory organs and in parts of the haemal system; it is not the same as the myomeric segmentation defined by the coelomic compartments giving rise to the longitudinal muscles.

The branchial basket is the feeding organ. The cilia of the gill slits create the water current that

Chapter vignette: *Branchiostoma lanceolatum* or amphioxus. (Based on Drach 1948 and Pearse *et al.* 1987.)

enters the mouth, passes through the slits to the atrial chamber, and leaves through the atriopore. A 'ventral' endostyle secretes a fine mucus filter that is transported along the gill bars to the 'dorsal' side, where the filters of the two sides with the captured particles are rolled together and transported posteriorly to the oesophagus. The endostyle is a longitudinal groove with parallel bands of different cell types (Fig. 56.2). One pair of bands secretes the muco-proteinaceous filter and another pair binds iodine and secretes iodinated tyrosine to the filter, but the construction of the net is in need of further study.

The almost cylindrical notochord extends anteriorly beyond the mouth and the 'dorsal' nerve tube and posteriorly to the tip of the tail. It consists of a stack of coin-shaped cells with transverse, striated myofibrils and is surrounded by a thick basal membrane. This chordal sheath has two rows of latero-'dorsal' pits where the membrane is very thin, and extensions from the notochordal cells extend to these thin areas, where synapses with cells in the apposing spinal chord are formed. The function of the muscle cells is debated, but it is believed that the contraction stiffens the notochord, so that it becomes more efficient as an antagonist to the lateral muscles when the animal swims and burrows.

The central nervous system (Wicht and Lacalli 2005) consists of a neural tube or spinal cord with a small anterior brain vesicle. Several cells have retained the epithelial character and are monociliate. A mid-'ventral' group of such cells at the posterior side of the brain vesicle form the infundibular organ, which secretes Reissner's fibre extending through the neural canal to its posterior end. The neural tube contains four types of photoreceptors, the numerous Hesse ocelli and the Joseph cells being of the rhabdomeric (microvillar) type, and the frontal pigment-cup ocellus and the lamellar body of the ciliary type; all the cells are monociliate. None of them bear special resemblance to other chordate photoreceptors, but the frontal ocellus has been interpreted as a homologue of the paired vertebrate eyes. Several giant (Rohde) cells are found in the anterior and the posterior 'dorsal' parts of the cord; the axons of the anterior cells cross over

'ventrally' to the opposite side and extend posteriorly, while the posterior cells have anterior axons. The two sides have alternating 'dorsal' nerve roots, which pass between the myomeres and contain both sensory and viscero-motory elements.

The coelomic compartments are quite complicated, and their morphology is best explained in connection with the ontogeny (see below).

The principal muscles are the segmented lateral muscles. Each muscle originates from one coelomic sac, whose median side becomes transformed into a large longitudinal muscle segment; the muscle segments of the two sides alternate. The septa between the segments are conical so that the characteristic interlocking muscular lamellae, also known for example from teleosts, are formed although the shape is more simple. Each muscle cell has one nucleus and contains a stack of flat, cross-striated muscle fibres. There are two types of muscle cells, deep muscle cells that are possibly engaged in fast movements such as swimming, and superficial muscle cells that could be 'slow' cells (Flood 1968). The muscle cells are connected with the neural tube through a 'dorsal tail', which forms a synapse with the lateral parts of the spinal cord.

Neural crest cells are not obvious during organogenesis, but some of the genes expressed in vertebrate neural-crest cells are found in migratory cells at the border of the neural plate. These cells could represent early stages in the evolution of the neural crest (Baker 2008).

The haemal system resembles that of vertebrates in general layout, and a number of vessels have been homologized with vertebrate blood vessels. The larger vessels may have a lining of haemocytes, but they are not connected by junctions and are therefore not an endothelium. Some of the vessels are contractile, but the muscle cells are part of the surrounding peritoneum and there is no heart. The contractile vessels create a slow circulation of the blood in constant directions, generally with anterior flow 'ventrally' and posterior flow 'dorsally'.

A row of nephridia is found on each side of the branchial region; each organ has a nephridiopore in the atrial chamber at the base of a tongue bar. The nephridia are made up of cells of a unique type, the cyrtopodocyte, which consists of one part forming a usual podocytic lining of a blood vessel and another part that resembles a protonephridial solenocyte with a ring of long microvilli surrounding a cilium. The solenocyte part of the cells traverses the subchordal coelom, and the tips of the microvilli and the cilium penetrate between the cells of the nephridial canals. The cyrtopodocytes appear to be of coelomic origin (see below), and the ring of microvilli surrounding a cilium could perhaps be interpreted as a specialization of the corresponding structures seen, for example, in peritoneal cells of echinoderms (Walker 1979). The organ called Hatcheck's nephridium is a ciliated, tubular structure with groups of solenocyte-like cells that opens 'dorsally' into the left side of the pharynx.

The gonads are situated along the atrium. They develop from the 'ventral' parts of the myocoels, where the gametes originate from the epithelium. The early oocytes retain the apical cilium, and the apical pole of the oocyte becomes the apical pole of the fertilized egg. The ripe gametes break through the thin body wall to the atrial cavity; fertilization is external.

The development of amphioxus has been the subject of several classical studies (Hatschek 1881; Cerfontaine 1906; Conklin 1933b), and these papers are the sources of the following description when nothing else is stated.

The first cleavage is median and the two first blastomeres are able to develop into small, but apparently completely normal larvae if isolated. The second cleavage separates the antero-'dorsal' and postero-'ventral' halves of the embryo, and isolated blastomeres of this stage are not capable of forming complete embryos (Tung *et al.* 1962b). Cell-lineage studies (Tung *et al.* 1962a,b) have made it possible to construct fate maps (Fig. 62.2), which show that the ectoderm is formed from the apical quartet, the endoderm from the blastoporal quartet; mesoderm and notochord-precursor cells form a ring-shaped area at the apical half of the blastoporal quartet, and the neural-tube cells form a 'dorsal' crescent at the apical quartet. The cell-lineage studies have shown that the apical pole moves to

become situated mid-'ventrally' behind the mouth (Fig. 62.2). The following development goes through a coeloblastula and an embolic gastrula, in which the blastocoel soon becomes obliterated. The ectodermal cells develop one cilium each at the stage when the blastopore narrows, and the endodermal cells become monociliate at a later stage. The second polar body is situated inside the fertilization membrane and can be used as a marker of the apical pole until a late gastrula stage when the embryo hatches. The gastrula becomes bilaterally symmetrical with a flattened 'dorsal' side (Fig. 63.1). The dorsomedian zone of the endoderm becomes thick and induces the overlying ectoderm to form the neural plate, which becomes overgrown by a pair of lateral ectodermal folds (Tung *et al.* 1962a). The lateral parts of the neural plate soon fold up and fuse so that the neural tube is formed. The neural folds leave an anterior opening, the neuropore, but at the posterior end they continue around the blastopore, which thus becomes enclosed and the neurenteric canal forms. The medio-'dorsal' zone of endoderm folds up forming the notochord, and the latero-'dorsal' zones of endoderm form longitudinal folds, which soon break up into rows of coelomic sacs.

The anterior part of the embryo shows many asymmetries (Holland *et al.* 1997; Yasui and Kaji 2008). The anterior part of the archenteron forms a pair of small lateral pockets that become pinched off as the anterior pair of coelomic sacs (in the first metamere). This is very similar to the formation of the protocoelomic sacs in ambulacrarians (Fig. 57.1), and the two sacs are generally believed to represent the protocoel (Table 62.1). The left sac (the left anterior head coelom) becomes connected with the ectoderm through the pre-oral pit; the coelom is lost in the adult, but the pre-oral pit becomes differentiated into Hatschek's pit. The right head coelom becomes the 'ventral' rostral coelom in the adult. The more posterior coelomic sacs become added sequentially, though not from teloblasts as supposed by Hatschek (1881). Only the first eight sacs develop through enterocoely, the more posterior sacs are compact at first. The coelomic sacs of the two sides are symmetrically aligned until a stage of about seven to eight pairs of sacs, but at this stage the left anterior

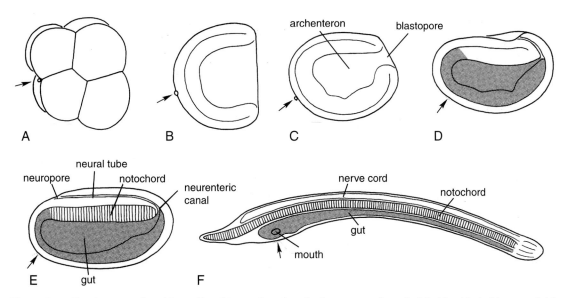

Figure 63.1. Development of amphioxus *(Branchiostoma lanceolatum)*; all stages seen from the left side with the blastoporal side oriented towards the right; the position of the polar bodies/apical pole is indicated by arrows. (A) 8-Cell stage. (B) Early gastrula. (C) Late gastrula. (D) Blastopore closure with early stage of the neurenteric canal. (E) Neurula with mesodermal sacs (lateral to the notochord and therefore not seen in the drawing). (F) Juvenile with lateral mouth. (A redrawn from Conklin 1933b; B–E redrawn from Conklin 1933a; F redrawn from Tung *et al.* 1962b.)

sac remains small while the right becomes a larger, thin-walled sac, and the somites of the right side thus become situated a little more posteriorly than those of the left side so that the somites of the two sides alternate. New somites differentiate alternately from the two sides of the posterior growth zone, organized by a number of genes (Beaster-Jones *et al.* 2008).

A small mouth opening breaks through from the archenteron on the left side between metameres 2 and 3. It expands largely to become an elongate opening stretching from the head to metamere 8. It develops small ciliary 'oral spines' along the edge. Subsequently, it moves to the 'ventral' midline and becomes more circular, with tentacles developing at its edge forming the velum. Hatschek's nephridium is a large organ with cyrtopodocytes, as those of the more posterior nephridia; it opens into the upper posterior side of the early mouth opening and can be followed along the movements of the mouth (Ruppert 1996; Yasui and Kaji 2008; Lacalli 2010). All evidence points to the mouth opening being the modified first gill opening and consequently not homologous with the mouth of the other chordates (Ruppert 1997; Yasui and Kaji 2008).

The club-shaped gland (Olsson 1983; Holland *et al.* 2009) develops as an evagination from the upper-right side of the pharynx and develops into a long 'ventral' tube, with an external opening at the left side below the mouth. It transports water to the pharynx and secretes material into the pharynx. Its function and homology are much discussed; it degenerates at metamorphosis.

A row of gill pores break through at the right side of the archenteron from the level of the early mouth. They gradually move 'ventrally' to end up at the left side, and a new row of gill pores develop on the right side. This whole complex of movements, including a rotation of the early, V-shaped endostyle, establishes a quasi-bilateral symmetry (Lacalli 2008).

The wall of the more posterior coelomic sacs in contact with the notochord differentiates into muscle cells while the remaining parts remain as a thin peritoneum. Each sac then divides into a 'dorsal' and a 'ventral' sac. The 'dorsal' sacs, somites, comprise the muscles and a thin peritoneal part that covers the body wall and the 'dorsal' part of the gut (Hatschek 1888).

The 'ventral' sacs are situated lateral to the 'ventral' parts of the gut and soon fuse to a pair of longitudinal sacs, and finally to one sac surrounding the gut except mid-'dorsally'. The muscular part of the 'dorsal' sacs extend 'dorsally' around the neural tube and 'ventrally' around the gut, but the 'ventral' part of the thin peritoneum has developed an extension, the sclerocoel, separating the median side of the muscles from the notochord; the coelomic compartments lateral to the longitudinal muscles are called myocoels. This arrangement is seen in the region between the atrial pore and the anus; the anterior part of the body becomes more complicated owing to the development of the gill pores and the atrium. The gill slits (see below) divide the 'ventral' coelom into a number of more narrow spaces: paired longitudinal epibranchial (or subchordal) coeloms at the neural side of the gill slits, narrow channels through the gill bars, and a mid-'ventral' hypobranchial (or subendostylar) coelom. The cyrtopodocytes apparently develop from the peritoneum of the subchordal coelom (Goodrich 1934), but the differentiation of the cells and the origin of the nephridial ducts are in need of further investigation. The notochord elongates anteriorly in front of both mouth and neural tube (Fig. 63.1).

The endostyle develops from a ciliated field at the anterior end of the archenteron anterior to the club-shaped gland; it elongates on the left side and differentiates into two strips of thickened glandular cells, separated by a narrow band of ciliated cells. This conspicuous structure becomes indented anteriorly and bends into a V-shape (Willey 1891; Olsson 1983). Finally, the endostyle moves to a mid-'ventral' position behind the mouth and its two branches move close together so that the anterior indentation becomes the midline along the bottom of the endostylar groove, where the two rows of glandular cells can be recognized as cell rows two and four. The peripharyngeal ciliated bands extend from the anterior tips of the endostyle branches to the 'dorsal' side (van Wijhe 1914).

The gill pores soon become heart-shaped and then U-shaped by the development of an extension from the 'dorsal' side that becomes the tongue bar; each vertical slit becomes divided into a row of gill pores by

the development of transverse bars, synapticles, between the gill bars and tongue bars. The branchial skeleton develops first along the anterior and posterior sides of the gill pores; in later stages the two curved rods fuse 'dorsally' and extend into the tongue bar, and finally transverse bars are formed in the synapticles (Lönnberg 1901–1905).

Larvae about 10 days old have a large mouth on the left side and three gill slits on the right side. The club-shaped gland secretes mucus threads that become stretched out as a vertical filter by the 'dorsal' and 'ventral' ciliary pharyngeal bands and transported posteriorly to the intestine. The cilia of the gill pores propel a water current through the mouth, and particles get caught by the filter and carried to the intestine (Gilmour 1996). Ciliary 'spines' consisting of closely apposed, immobile cilia extend from the 'dorsal' edge of the mouth across the opening; contact with larger particles elicit a contraction of the gill slit and pharynx, which blows the particle away from the mouth (Lacalli et al. 1999). The innervation of these cells resembles that of cells in vertebrate taste buds and even perhaps the oral plexus in echinoderm larvae.

The atrium develops by the formation of a pair of lateral metapleural folds that grow towards the 'ventral' side covering the branchial basket and finally fuse in the midline leaving only a posterior atrial pore; these folds contain coelomic compartments that were originally parts of the 'ventral' coelom (Lankester and Willey 1890).

The nervous system of the 10–14 day old juveniles with three to four gill openings has been documented in elegant details based on serial transmission electron microscopy sectioning (Lacalli et al. 1994; Lacalli 1996a,b; Lacalli and Kelly 1999). These studies throw much light on the understanding of the adult nervous system and on its phylogenetic importance.

Three small papillae have been described from the mouth region of larvae in the one-gill-pore stage (van Wijhe 1926; Berrill 1987): an oral papilla is situated on the left side in front of the mouth just below the preoral organ; a right papilla on the right side opposite from the mouth; and a median papilla mid-'ventrally' just 'ventral' to the gill opening. The right and median papillae are in need of further studies. Some of the earlier authors supposed that the papillae are adhesive, and proposed a homology to the adhesive papillae of ascidian and fish larvae. However, the ultrastructure of the oral papilla shows no signs of secretory activity, cilia, or innervation (Andersson and Olsson 1989), which makes it unlikely that they are homologues of the adhesive papillae of the ascidian larvae (Chapter 64).

The series of asymmetries in the juvenile stages of amphioxus have been the subject of many phylogenetic speculations. A central question is of course whether the mouth is homologous with the mouth of the other chordates, or whether it is the anterior left gill slit of the embryo that has taken over the function of a mouth and become the bilaterally symmetrical mouth of the adult. The ontogenetic origin of the mouth as an opening on the left side, its innervation from two 'dorsal' nerves of the left side, and the origin of the velar coelom from left second coelomic sac together point to the mouth actually being a specialized gill slit (see above), and this could explain most of the enigmatic asymmetries; but it does not give any hint of how an ancestral mouth should have been lost. Willey (1894) suggested that when the notochord of the cephalochordate ancestor extended forward, the mouth 'came in the way' and became displaced to the left side. The asymmetries of several other anterior organs should have evolved as a consequence of this change. Bone (1958) and Gilmour (1996) observed the feeding of early juveniles and found that the ectodermal cilia create a posteriorly directed current along the body, and that the cilia of the gill slit create a current that enters the mouth and leaves through the single gill slit. Bone (1958) interpreted the lateral position of the juvenile mouth as an adaptation that made an enlargement of the mouth possible, which should enhance the flow of water through the filtering structure of the juvenile. A similar expansion of a 'ventral' mouth on a laterally compressed, fish-like organism was considered mechanically unsound. The interpretation of the asymmetries as 'larval' specializations has been followed for example by Presley *et al.* (1996), but the question is still open for speculation.

The whole genome of amphioxus (Holland *et al.* 2008; Putnam *et al.* 2008) shows a very complete set of genes, several of which have been lost in the vertebrates, and especially in the urochordates. It contains one complete set of the chordate Hox genes (Fig. 21.3). This is discussed further in Chapter 62. Cephalochordates show a long series of apomorphies, such as the notochord consisting of muscle cells innervated directly from the neural tube, the 'ventral' atrium, and the asymmetries of the juveniles. They do indeed show many features that are supposed to be characteristic also of the ancestral vertebrate, but these characters must be interpreted as chordate apomorphies. Further discussion can be found in Chapter 62.

Interesting subjects for future research

1. Coelomogenesis
2. Development of nephridia and cyrtopodocytes
3. Secretion of the mucus net by the endostyle

References

Andersson, A. and Olsson, R. 1989. The oral papilla of the lancelet larva (*Branchiostoma lanceolatum*) (Cephalochordata). *Acta Zool. (Stockh.)* **70**: 53–56.

Baker, C.B.H. 2008. The evolution and elaboration of vertebrate neural crest cells. *Curr. Opin. Genet. Dev.* **18**: 536–543.

Beaster-Jones, L., Kaltenbach, S.L., Koop, D., *et al.* 2008. Expression of somite segmentation genes in amphioxus: a clock without a wavefront? *Dev. Genes Evol.* **218**: 599–611.

Berrill, N.J. 1987. Early chordate evolution Part 2. Amphioxus and ascidians. To settle or not to settle. *Int. J. Invertebr. Reprod. Dev.* **11**: 15–28.

Bone, Q. 1958. The asymmetry of the larval amphioxus. *Proc. Zool. Soc. Lond.* **130**: 289–293.

Cerfontaine, P. 1906. Recherches sur le développement de l'Amphioxus. *Arch. Biol.* **22**: 229–418.

Conklin, E.G. 1933a. The development of isolated and partially separated blastomeres of amphioxus. *J. Exp. Zool.* **64**: 303–375.

Conklin, E.G. 1933b. The embryology of amphioxus. *J. Morphol.* **54**: 69–151.

Drach, P. 1948. Embranchement des Céphalochordés. *Traité de Zoologie*, vol. 11, pp. 931–1037. Masson, Paris.

Flood, P.R. 1968. Structure of the segmental trunk muscles in amphioxus. *Z. Zellforsch.* **84**: 389–416.

Gilmour, T.H.J. 1996. Feeding methods of cephalochordate larvae. *Isr. J. Zool.* **42** (**Suppl.**): S87–S95.

Goodrich, E.S. 1934. The early development of the nephridia in Amphioxus: Part II, The paired nephridia. *Q. J. Microsc. Sci., N. S.* **76**: 655–674.

Hatschek, B. 1881. Studien über Entwicklung des Amphioxus. *Arb. Zool. Inst. University Wien.* **4**: 1–88.

Hatschek, B. 1888. Über den Schichtenbau von Amphioxus. *Anat. Anz.* **3**: 662–667.

Holland, L.Z., Albalat, R., Azumi, K., *et al.* 2008. The amphioxus genome illuminates vertebrate origins and cephalochordate biology. *Genome Res.* **18**: 1100–1111.

Holland, L.Z., Kene, M., Williams, N.A. and Holland, N.D. 1997. Sequence and embryonic expression of the amphioxus *engralied* gene (*AmphiEn*): the metameric pattern of transcription resembles that of its segment-polarity homolog in *Drosophila. Development* **124**: 1723–1732.

Holland, N.D., Paris, M. and Koop, D. 2009. The club-shaped gland of amphioxus: export of secretion to the pharynx in pre-metamorphic larvae and apoptosis during metamorphosis. *Acta Zool. (Stockh.)* **90**: 372–379.

Lacalli, T.C. 1996a. Frontal eye circuitry, rostral sensory pathways and brain organization in amphioxus larvae: evidence from 3D reconstructions. *Phil. Trans. R. Soc. Lond. B* **351**: 243–263.

Lacalli, T.C. 1996b. Landmarks and subdomains in the larval brain of *Branchiostoma* vertebrate homologs and invertebrate antecedents. *Isr. J. Zool.* **42** (**Suppl.**): S131–S146.

Lacalli, T.C. 2008. Basic features of the ancestral chordate brain: a protochordate perspective. *Brain Res. Bull.* **75**: 319–323.

Lacalli, T.C. 2010. The emergence of the chordate body plan: some puzzles and problems. *Acta Zool. (Stockh.)* **91**: 4–10.

Lacalli, T.C. and Kelly, S.J. 1999. Somatic motoneurons in amphioxus larvae: cell types, cell positions and innervation patterns. *Acta Zool. (Stockh.)* **80**: 113–124.

Lacalli, T.C., Gilmour, T.H.J. and Kelly, S.J. 1999. The oral nerve plexus in amphioxus larvae: Function, cell types and phylogenetic significance. *Phil. Trans. R. Soc. Lond. B* **266**: 1461–1470.

Lacalli, T.C., Holland, N.D. and West, J.E. 1994. Landmarks in the anterior central nervous system of amphioxus larvae. *Phil. Trans. R. Soc. Lond. B* **344**: 165–185.

Lankester, E.R. and Willey, A. 1890. The development of the atrial chamber of Amphioxus. *Q. J. Microsc. Sci., N. S.* **31**: 445–466.

Lönnberg, E. 1901–1905. Leptocardii. *Bronn's Klassen und Ordnungen des Tierreichs*, 6. Band, 1. Abt., 1. Buch, pp. 99–249. Akademsiche Verlagsgesellschaft, Leipzig.

Olsson, R. 1983. Club-shaped gland and endostyle in larval *Branchiostoma lanceolatum* (Cephalochordata). *Zoomorphology* **103**: 1–13.

Pearse, V., Pearse, J., Buchsbaum, M. and Buchsbaum, R. 1987. *Living Invertebrates.* Blackwell Scientific Publ., Palo Alto, CA.

Presley, R., Horder, T.J. and Slípka, J. 1996. Lancelet development as evidence of ancestral chordate structure. *Isr. J. Zool.* **42** (**Suppl.**): S97–S116.

Putnam, N.H., Butts, T., Ferrier, D.E.K., *et al.* 2008. The amphioxus genome and the evolution of the chordate karyotype. *Nature* **453**: 1064–1071.

Ruppert, E.E. 1996. Morphology of Hatschek's nephridium in larval and juvenile stages of *Branchiostoma virginiae* (Cephalochordata). *Isr. J. Zool.* **42** (**Suppl.**): S161–S182.

Ruppert, E.E. 1997. Cephalochordata (Acrania). In F.W. Harrison (ed.): *Microscopic Anatomy of Invertebrates*, vol. 15, pp. 349–504. Wiley-Liss, New York.

Sansom, R.S., Gabbott, S.E. and Purnell, M.A. 2010. Non-random decay of chordate characters causes bias in fossil interpretation. *Nature* **463**: 797–800.

Tung, T.C., Wu, S.C. and Tung, Y.Y.F. 1962a. Experimental studies on the neural induction in amphioxus. *Sci. Sin.* **11**: 805–820.

Tung, T.C., Wu, S.C. and Tung, Y.Y.F. 1962b. The presumptive areas of the egg of amphioxus. *Sci. Sin.* **11**: 629–644.

van Wijhe, J.W. 1914. Studien über Amphioxus. I. Mund und Darmkanal während der Metamorphose. *Verh. K. Akad. Wet. Amsterd.*, 2. Sect. **18(1)**: 1–84.

van Wijhe, J.W. 1926. On the temporary presence of the primary mouth-opening in the larva of amphioxus, and the occurrence of three postoral papillae, which are probably homologous with those of the larva of ascidians. *Proc. Sect. Sci. K. Ned. Akad. Wet.* **29**: 286–295.

Walker, C.W. 1979. Ultrastructure of the somatic portion of the gonads in asteroids, with emphasis on flagellated-collar cells and nutrient transport. *J. Morphol.* **162**: 127–162.

Wicht, H. and Lacalli, T.C. 2005. The nervous system of amphioxus: structure, development, and evolutionary significance. *Can. J. Zool.* **83**: 122–150.

Willey, A. 1891. The later larval development of Amphioxus. *Q. J. Microsc. Sci., N. S.* **32**: 183–234.

Willey, A. 1894. *Amphioxus and the Ancestry of the Vertebrates.* Macmillan, New York.

Yasui, K. and Kaji, T. 2008. The lancelet and ammocoete mouths. *Zool. Sci. (Tokyo)* **25**: 1012–1019.

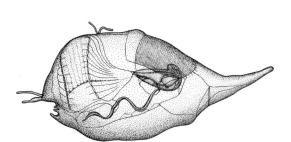

64

Phylum **Urochordata**

Urochordates or tunicates are a very distinct, marine phylum of about 3000 described species. They have traditionally been arranged in three classes, Ascidiacea, Thaliacea and Appendicularia (also called Larvacea); with the sorberaceans and *Octacnemus*, which lack the branchial basket, as specialized ascidians. However, studies on molecular phylogeny and combined morphological and molecular data have shown a different picture, with the planktonic thaliaceans as an ingroup of various of the ascidians (Stach and Turbeville 2002; Zeng *et al.* 2006; Tsagkogeorga *et al.* 2009). The appendicularians, where *Oikopleura* with a life cycle of less than one week has now become a model organism (Bouquet *et al.* 2009), are in some studies seen as the sister group of the ascidians, but in other studies in an ingroup position (Zeng *et al.* 2006), and their embryology shows traces of sessility (Stach *et al.* 2008). So the pelago-benthic life cycle with a tadpole larva is probably a urochordate apomorphy.

Other urochordate apomorphies include the highly determined embryology, with a bilateral cleavage pattern (see below) and the cuticle with cellulose Iβ (Hirose *et al.* 1999). This compound is found in the tunic of the ascidians and in the secreted house of the appendicularians (Kimura *et al.* 2001). It has been suggested that the genes involved in the cellulose synthesis have been acquired through lateral gene transfer from protists (Matthysse *et al.* 2004), but it has also been suggested that the genes are ancient and have been lost in most metazoan groups (Davison and Blaxter 2005).

There is a meagre fossil record, but appendicularians have been reported from the Early Cambrian of China (Zhang 1987), and a fossil resembling a solitary ascidian has been reported too (Chen *et al.* 2003).

The shape of the adult urochordates is highly variable, and many species form colonies or other aggregates of highly characteristic shapes. The ascidian colonies show various degrees of integration between the zooids, from forms with independent zooids budded from stolons to compound forms with a common tunic and zooids arranged around common excurrent openings. The various types of coloniality have evolved independently in a number of lineages (Zeng *et al.* 2006).

The appendicularians are treated separately below, whereas the remaining urochordates are discussed as ascidians, with some remarks on special features of salps and doliolids. Burighel and Cloney (1997) is the general reference for ascidian morphology, excluding the salps and doliolids that are discussed briefly in Godeaux *et al.* (1998). The appendicularians are briefly described in Fenaux (1998). An archimeric regionation cannot be recognized, and the mesoderm does not form the coelomic pouches characteristic of the other deuterostomes. The body is in most cases globular or elongate, but the neotenic appendicularians have a long, laterally compressed tail, which is

---

Chapter vignette: The appendicularian *Oikopleura dioica* in its house. (Redrawn from Lohmann 1903.)

twisted 90° at the base, so that its movements are in the vertical plane. All urochordate epithelia are monolayered. The ectoderm is unciliated, but many areas of the endoderm consist of multiciliate cells. Monociliate epithelial cells are found in certain zones of the endostyle and of the stomach. Myoepithelial cells have been found in the adhesive papillae of ascidian tadpole larvae.

Adult ascidians have a thick tunic, primarily secreted by the epithelial cells. Ectodermal extensions from a zone near the heart form a more or less extensive system of blood vessels in the tunic, and mesodermal cells from the blood wander through the ectoderm or through the walls of the vessels into the tunic where they take part in the formation of additional tunic material. The tunic is only adhering to the epithelium in the areas around the siphons and at the blood vessels. The tunic contains cellulose, mucopolysaccharides, and protein, and may accumulate unusual substances such as vanadium. The cellulose fibrils are synthesized in special glomerulocytes in the epithelium (Kimura and Itoh 1995). Colonial forms have a common tunic, in some cases with interzooidal blood vessels. The tunic of didemnids contains a network of myocytes that can cause contractions of the tunic for example around the common excurrent openings. Doliolids have a thin ectoderm that secretes a thin cuticle without cells. Large areas of the cuticle are moulted regularly, so that the surface of the animal is kept free from detritus (Uljanin 1884).

The appendicularian ectoderm comprises areas with a number of very specialized cell groups, the oikoplast epithelium, that secretes the various parts of a more or less complicated house (Fenaux 1998; Flood and Deibel 1998) (see the chapter vignette); this structure is used as a filtering device that concentrates the plankton particles in the water before it becomes pumped through the small filter secreted by the endostyle (see below). The water current is created by undulating movements of the tail (Bone *et al.* 2003). The house is discarded periodically and a new one secreted.

The ascidian gut consists of a spacious pharynx or branchial basket with gill slits (stigmata) and a 'ventral'

endostyle, a narrow oesophagus, a stomach with various digestive diverticula and glands, and an intestine that opens in the left atrial chamber. The finest gill bars consist of one layer of ectodermal cells with a thin basal membrane surrounding a blood vessel, but the longitudinal gill bars are thicker and have unciliated zones with muscle cells surrounded by a basal membrane. Each stigma is surrounded by seven (six to eight) rings of very narrow cells, each with a row of cilia. The cilia of the gill slits transport water out of the branchial basket and new water with particles is sucked in through the mouth, which forms the incurrent siphon. The particles are caught by a fine mucus net produced continuously by the endostyle and transported along the wall of the basket by the cilia of the peripharyngeal bands and of various structures on the gill bars to the 'dorsal' lamina, which rolls the net together and passes it posteriorly to the oesophagus. The endostyle generally consists of eight rows of characteristic cell types on each side, three of which are glandular; the doliolids have only two rows of gland cells (Compere and Godeaux 1997). Certain of the rows of cells secrete the mucus net, which is a rectangular meshwork consisting primarily of proteins and polysaccharides, and some also concentrate iodine and show expression of thyroid genes (Ogasawara *et al.* 1999a,b; Bone *et al.* 2003). The filtered water passes through the stigmata into the lateral atrial chambers and exits through the mid-'dorsal' excurrent siphon.

Salps have a highly specialized branchial basket with only one huge gill opening on each side. A normal mucus filter is formed by the endostyle and pulled 'dorsally' by the peripharyngeal bands and then posteriorly to the oesophagus. The current through the filter is created by the muscular swimming movements of the body; the transverse muscles constrict the body, and the flow of water is regulated by one-way valves consisting of a system of flaps at each siphon (Bone *et al.* 1991, 2003). Appendicularians lack the atrial chambers and have a 'ventral' anus. They have a tail with a notochord consisting of a central hyaline mass surrounded by flattened cells and a strengthened basal membrane. The pharynx has only one circular gill opening on each side, and a first concentration of the food particles takes place in the filter structure formed by the oikoplast epithelium (see above). The final capture of the particles takes place in a fine, normal mucus filter formed by the endostyle (Bone *et al.* 2003). Only *Kowalevskaia* lacks the endostyle completely, and the food particles are apparently captured by cilia on some elaborate folds surrounding the unusually large gill pores (Fenaux 1998). The short endostyle is usually curved, in *Fritillaria* so strongly that there is only a narrow opening from the endostyle to the pharynx. The endostyle of *Oikopleura* has only one row of gland cells, which produce both the mucus net and the iodo-tyrosines (Olsson 1963; Cañestro *et al.* 2008).

The central nervous system of the adult consists of a cerebral ganglion or brain at the 'dorsal' side of the pharynx just behind the peripharyngeal ciliary bands (Mackie and Burighel 2005). A number of paired and unpaired nerves innervate various organs. In front of the brain, a convoluted ciliated funnel leads from the pharynx via a ciliated duct to the neural gland, and water is pumped through this neural-gland complex to the haemal system; a glandular function is not indicated (Ruppert 1990). The gland is situated 'dorsal' or 'ventral' to the brain, or in a few cases at its right side. The connection of this complex with the nervous system is described below. A posterior tubular extension from the posterior end of the neural gland, called the 'dorsal' strand, is surrounded by a nerve net communicating with the visceral nerve. The salps have a large rhabdomeric eye at the 'dorsal' side of the compact brain (McReynolds and Gorman 1975; Lacalli and Holland 1998). Communication between zooids in colonies of ascidians is not through nerve cells, but epithelial cells along the blood vessels conducting electrical impulses (Mackie 1995). A special communication system is found in *Pyrosoma*, where light flashes from a stimulated zooid are detected by receptors in the neighbouring zooids, which then react. Many types of sensory organs have been described.

The appendicularians have a small but unexpectedly complicated central nervous system that resembles that of the ascidian tadpole larvae (Olsson *et al.* 1990; Bone 1998; Cañestro *et al.* 2005). *Oikopleura* has an elongate brain consisting of about 70 cells and with an extended brain vesicle with a statocyte, a neck region, a small caudal ganglion, and a hollow nerve cord consisting of four rows of thin epithelial cells. The central canal of the nerve cord contains a Reissner's fibre that is formed by a 'dorsal' cell in the caudal ganglion (Holmberg and Olsson 1984). A ciliated brain duct connects the haemocoel near the brain with the buccal cavity. The 'ventral' ganglion consists of about 36 cells, and there are small groups of nerve cells along the neural tube; these cells are, in some species, arranged in small groups corresponding roughly with the muscle cells, whereas other species have higher numbers of cell groups or a more irregular arrangement. The ganglionic cells innervate the lateral muscle cells; the innervation is dual, with cholinergic and GABA-ergic nerve endings, which enables the simple tail to make the complicated movements connected with the formation of the house (Bone 1992). The mesoderm does not form the coelomic cavities usually recognized in deuterostomes. Some of the ascidians have a pair of more or less extensive cavities, called epicardial sacs, which are extensions from a postero-'ventral' part of the pharynx. These two pockets could be interpreted as mesodermal sacs formed by enterocoely, and they surround parts of the gut as the metacoels. However, they are not formed until a late ontogenetic stage, and their homology with the coelomic sacs of other deuterostomes is questionable. A heart situated 'ventral' to the posterior part of the pharynx pumps the blood, alternating in two directions. It consists of a mesodermal sac folded around a blood lacuna; its inner wall forms the circular musculature of the heart and the lumen forms the pericardial sac. The haemal system is a complex of channels or lacunae surrounded by a basement membrane and without endothelium; the ectodermal extensions in the tunic form well-defined vessels. The blood contains various cell types, such as lymphocytes, macrophages, vanadocytes, and nephrocytes.

Appendicularian musculature is restricted to the heart and the two longitudinal tail muscles, which each consist of one row of seven to ten flat cells. The number of muscle cells is species specific and is apparently the number laid down during embryonic development (Fenaux 1998). There is no excretory system, like metanephridia or an axial complex, but waste products accumulate in special blood cells that may become deposited in the tunic or in special regions of the body as renal organs. One or more mesodermal gonads are situated near the stomach and their common or separate ducts open into the atrial chamber. In appendicularians the ripe eggs disrupt the maternal body wall and the animal dies (Berrill 1950).

There are numerous studies of ascidian embryology (reviews in Cloney 1990; Satoh 1994; Jeffery and Swalla 1997; Lemaire *et al.* 2008), from classical studies using direct observations of the embryos (as those of Van Beneden and Julin 1887: *Clavelina*, and Conklin 1905: *Cynthia* and *Ciona*) to the recent, elegant cell-lineage studies using various marking techniques, especially on the Japanese *Halocynthia roretzi* (see for example Kumano and Nishida 2007) and the North Atlantic *Ciona intestinalis* (see for example Lemaire 2009); there seems to be very little variation between cell lineages of the solitary ascidian species. The cell lineage shows a very strict, invariant pattern. Isolated blastomeres of the 2-cell stage can develop into complete, competent larvae (Nakauchi and Takeshita 1983), but blastomeres of the 4- and 8-cell stages cannot (Cloney 1990). There is considerable interaction between blastomeres, and the fate of the blastomeres is not restricted until the 7th or 8th cleavage in several cell types (Nishida 1994, 1997).

The unfertilized egg has a fixed apical-blastoporal axis, and fertilization triggers a characteristic reorganization of the cortex of the egg, called ooplasmic segregation, followed by the formation of the polar bodies at the apical pole (Kumano and Nishida 2007; Lemaire 2009). A bilateral organization, often indicated by variously coloured areas and with the polar bodies in the median plane, becomes established after a few minutes (Nishida 1994).

The first cleavage is median, the second transverse, and the third obliquely equatorial (Fig. 64.1); the polar bodies indicating the apical pole can usually be followed through the early development. The following development goes through a coeloblastula and an embolic gastrula in which the blastocoel becomes obliterated. The fate maps (Figs. 62.2, 64.2) show that the whole ectodermal epithelium originates from the apical half of the embryo and the whole endoderm from the blastoporal half. The nervous system develops from a crescentic area at the anterior part of the equatorial zone of both halves, while the mesodermal organs develop from an equatorial zone consisting almost exclusively of posterior areas of the blastoporal half. The cell lineage, with the blastomeres named in a rather awkward system introduced by Conklin (1905), has been traced all the way to the organs of the tadpole larva (see for example Lemaire 2009). The 110-cell stage marks the beginning of gastrulation (Fig. 64.2) with ten presumptive endodermal cells forming a shallow depression and two additional cells (possibly representing a proctodaeum) situated a little further posteriorly. Ten presumptive notochord cells form a crescent in front of the endodermal cells, and the blastomeres of the neural plate form a crescent anterior and lateral to the notochord cells. The brain and the adhesive palps develop from the two descendants of the a6.5 cell (Lemaire 2009). The blastomeres from which the tail muscles of the larva differentiate form a posterior horseshoe, behind and lateral to the endodermal cells.

During gastrulation, the crescentic neural plate and the notochord cells change shape dramatically, becoming the elongate neural plate and presumptive notochord (Nicol and Meinertzhagen 1988a,b). The cell lineage of the larval nervous system has been studied in a number of species (Meinertzhagen *et al.* 2004; Meinertzhagen 2005; Imai and Meinertzhagen 2007; Kumano and Nishida 2007). The approximate 80 cells of the neural plate are destined to become the neural tube, but an induction from the notochord cells is needed to trigger the differentiation (Satoh 1994). The differentiation of the notochord is in turn induced by cells of the endoderm (Nakatani and Nishida 1994). The lateral neural folds move together 'dorsally', forming a narrow neural tube. Posteriorly, the folds encompass the blastopore so that a neurenteric canal is formed when the folds fuse. Anteriorly, where the neural plate extends anterior to the notochord, the neuropore becomes situated at the posterior side of

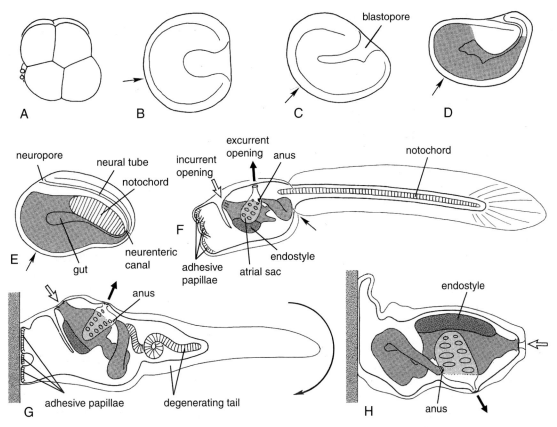

Figure 64.1. Ontogeny of ascidians with the primitive larval type (A, *Styela partita*; B–H, *Clavelina lepadiformis*), all stages seen from the left; the position of the polar bodies are seen in A and their approximate position is indicated by a thin arrow in the later stages (based on the observations on *Halocynthia roretzi* by Nishida 1987). (A) 8-cell stage. (B) Gastrula. (C) Late gastrula, the blastopore has turned slightly to the dorsal side). (D) Beginning of neural tube formation. (E) Early tailbud stage with chorda, mesoderm and neural tube. (F) Newly hatched larva with attachment papillae, endostyle, atrium with two rows of gill openings, and the rectum opening into the left atrium. (G) Newly settled larva, the tail is almost completely retracted and the rotation of the gut (indicated by the long arrow) has begun. (H) Juvenile ascidian with completed rotation of the gut. (A redrawn from Conklin 1905; B–E redrawn from Van Beneden and Julin 1887; F redrawn from Julin 1904; G–H redrawn from Seeliger 1893–1907.)

the stomodaeal invagination, which forms the oral siphon. Its opening into the oral siphon becomes differentiated into the ciliated funnel (Manni *et al.* 2005; Veeman *et al.* 2010). The neural tube elongates and differentiates into an anterior brain vesicle, a visceral ganglion, and a narrow, posterior nerve tube along the notochord, with the cells arranged in four longitudinal rows. The large horseshoe of mesodermal cells develops into the tail muscles along the notochord; other anterior mesodermal cells become the lateral trunk

mesenchyme. The endoderm differentiates into an anterior almost globular mass of cells and a posterior endodermal strand below the notochord.

Differential growth displaces the apical pole 'ventrally' to a position at the base of the tail, so that the whole epithelium of the body is derived from the anterior pair of cells of the apical quartet, while the tail has ectoderm derived from the posterior part (Fig. 64.1). The embryo curves 'ventrally', while the tail elongates, and this becomes very pronounced during the late

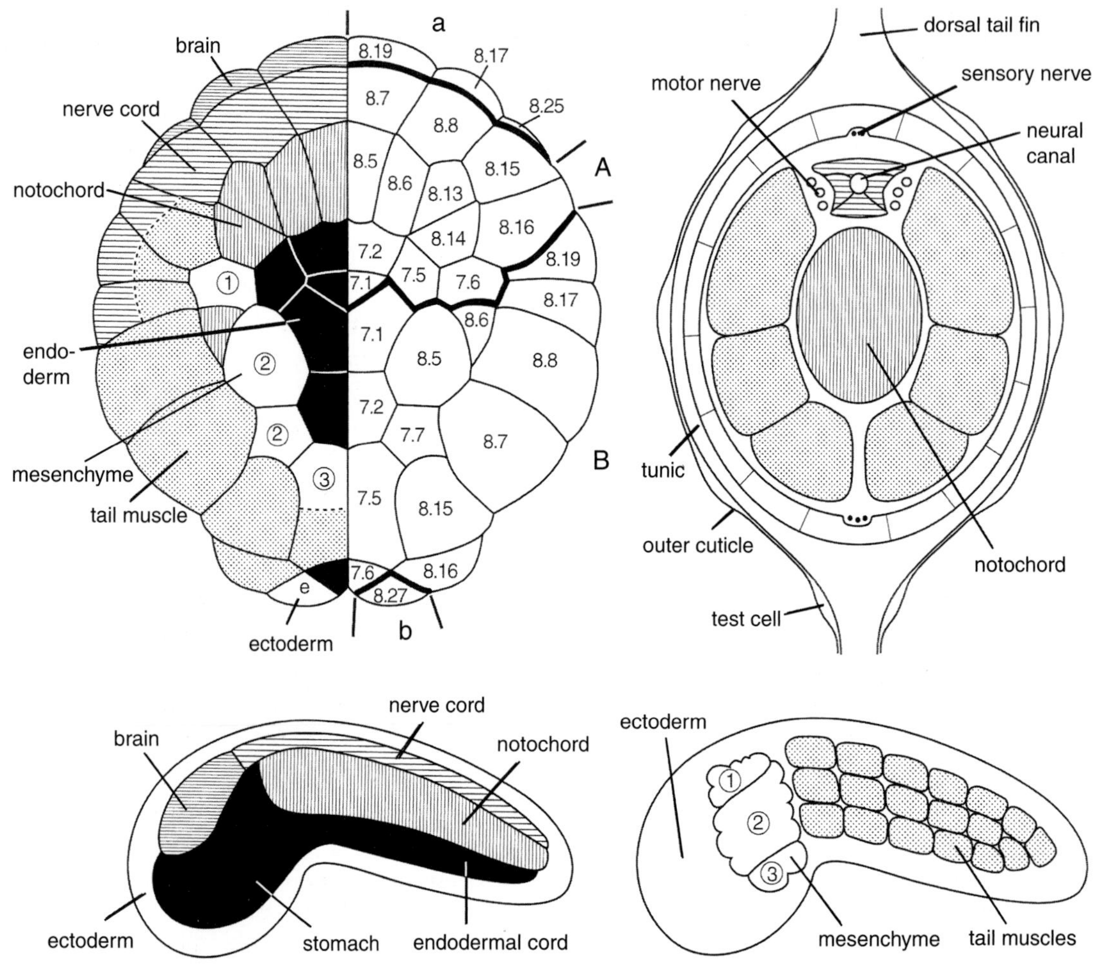

Figure 64.2. Cell lineage of ascidians. Upper left, blastoporal side of a 110-cell embryo of *Halocynthia roretzi*, thick lines denote boundaries between the quadrants A, B, a, and b; the names of the blastomeres are shown on the right side and their fates at the left side. Upper right, cross-section of the larval tail of *Dendrodoa grossularia*. Below, organization of the median (left) and lateral (right) groups of endodermal and mesodermal cells in the tailbud stage of *H. roretzi*. (Based on Bone 1992; Hirano and Nishida 1997.)

embryonic stages where the tail encircles the body. Neural-crest-like cells in the 'dorsal' midline in the region of the brain move into the embryo and differentiate into pigment cells (Jeffery 2007).

The ascidian larva (general reference Burighel and Cloney 1997) consists of a globular body and a slim tail with a cuticular dorsal-ventral fin. The cuticle comprises a thin, outer larval layer, a narrow space with scattered cells, and an inner cuticle (Fig. 64.2). There is considerable variation in the internal organization between species, with tail-less larvae (see below)

or larvae with juvenile/adult characters, including budding, developing precociously in many species (Jeffery 2007). The following description of the large embryos is mainly based on a type like *Clavelina* or *Ciona*, with a rather complete gut and without the precocious development of juvenile/adult structures.

The body contains the rather voluminous gut with mouth and anus. The gut shows the general regions of the adult gut. In the pharynx, the endostyle has differentiated from a 'ventral' longitudinal groove, and the epicardial sacs of uncertain function should develop from the oral side of the pharynx behind the endostyle (Berrill 1950). The pericardium develops from paired 'ventral' parts of the lateral trunk mesoderm that migrate anteriorly, develop a cavity at each side, and fuse around a haemal sinus (Berrill 1950). Other parts of this mesoderm have migrated laterally to the anterior part of the body in the region of the adhesive papillae; this region swells up at metamorphosis and forms a rather large cavity with scattered mesodermal cells, which do not form a mesodermal sac (Willey 1893). The trunk lateral cells also give rise to blood cells and to the cells in the tunic (Nishide et al. 1989).

The atrium develops from a pair of 'dorso'-lateral ectodermal invaginations that later on fuse 'dorsally' to form one median anal siphon (Seeliger 1893; Mazet and Shimeld 2005). The gill openings break through between the pharynx endoderm and the ectoderm of the atrial invaginations. In many species there are two or three primary gill openings, protostigmata, on each side, and these openings then divide and form the sometimes very complicated patterns of round, elongate or curved stigmata (Julin 1904; Brien 1948); the whole branchial basket may form complicated folds. The first stage of the protostigmata is often circular and they then become elongate, and in some species curve into a J-shape and become divided into series of stigmata.

The larval brain shows some variations between species (Burighel and Cloney 1997). In *Ciona*, an early stage in the differentiation shows an almost symmetrical brain vesicle, which subsequently becomes displaced to the right when the adult brain rudiment and the neural-gland complex develops from the left side. The cell lineage of each part of the differentiating brain vesicle has been traced in *Halocynthia* (Taniguchi and Nishida 2004).

The right brain vesicle may contain three types of sense organs, an ocellus, a statocyte, and a group of modified cilia that may be pressure receptors, but only few species have all three types of sensory cells. The ocellus and the statocyte develop from a pair of cells (left and right a8.25 in the cell lineage), but there is no fixed right/left origin of the two organs (Nishida 1987). The ocellus has ciliary photoreceptor cells that show some resemblance to those of vertebrates, but the membrane folds are arranged parallel to the cilium instead of perpendicular to it. *Botryllus* lacks both ocellus and statocyte but has a photolith, which appears to be an independent specialization (Sorrentino et al. 2000).

The left brain vesicle elongates anteriorly as a ciliated duct, and an opening through the antero-'dorsal' wall of the pharynx is retained or becomes established; the ciliated funnel originates from the anterior cells of the duct (see above). A group of cells of the 'dorsal' or 'ventral' side of the vesicle differentiates as the nerve cells of the cerebral ganglion, which becomes separated from the vesicle; the remaining part of the vesicle becomes the neural gland.

Three frontal adhesive papillae are the attachment organs used in settling. In addition to the adhesive cells they contain two types of primary sensory cells, both monociliate with the cilium entering the tunic. The axons from these cells unite to form one nerve from each papilla, and the three nerves merge into a nerve that winds through the anterior haemocoel to the visceral ganglion.

The notochord is the most conspicuous structure in the centre of the tail. In a few genera, such as *Dendrodoa*, it consists of a stack of coin-shaped cells, with many yolk globules and surrounded by a strengthened basal membrane as in the older embryos; in most other genera, extracellular matrix is secreted between the cells, and these matrix lenses finally fuse into a central rod surrounded by a continuous layer of flat notochordal cells. The two lateral bands of muscle cells are arranged in two to several rows according to the species (Jeffery and Swalla 1992), but the cells in the rows are not aligned, so there is no indication of segmentation. The nerve tube consists of four longitudinal rows of cells that retain their epithelial structure with one cilium each, with axons from nerve cells in the visceral ganglion extending along the lateral sides of the tube (Fig. 64.2), innervating the muscle cells across the basal membrane. In *Dendrodoa*, which has three rows of muscle cells, the 'dorsal' row of cells and the anteriormost cells of the 'ventral' row are innervated directly, and all the cells are coupled through gap junctions (Bone 1989). A 'dorsal' and sometimes also a 'ventral' sensory nerve is found in the basal part of the ectoderm below the cuticular tail fins; these nerves are bundles of axons from primary sense organs each with a cilium extending into the fin; these cells can be recognized in the tail at an early stage and their axons pass to the visceral ganglion without contact with the motor axons of the cord (Torrence and Cloney 1982). The neural canal contains a Reissner's fibre, but its origin is unknown (Olsson 1972). The only trace of the original gut is a row of endodermal cells along the 'ventral' side of the nerve cord. A layer of ectodermal cells surround the tail and secrete a thin, double tunic that extends into a thin 'dorsal', posterior, and 'ventral' fin. The tail is twisted 90° at the base in many species.

The type described above is considered ancestral and has been found in a number of families of mainly solitary forms, but many modifications involving delayed development of the gut or precocious development of buds have been described (Millar 1971). Several molgulids and stylelids have tail-less (anural) larvae, but cell lineages of both notochord and neural tube are present, as in the normal tadpoles (Berrill 1931). The closure of the neural tube is especially clearly shown in *Molgula*, where the neural folds fuse along the midline leaving both the neural pore and the blastopore open for a short time; the blastopore finally becomes constricted and the neurenteric canal formed. Swalla and Jeffery (1996) showed that the shift from the tailed larva to the anuran type is caused by a change in the gene responsible for the elongation of the tissues of the tail. This change has apparently evolved several times independently within the two families (Hadfield *et al.* 1995), which once again demonstrates that larval characters can easily be lost, whereas a complicated genetic apparatus must be involved in organizing all the tissues present in the organs.

Most species have a short larval phase and settle after reaching competency. At settling, the larva attaches by the adhesive papillae, the larval cuticle is shed, and the tail becomes retracted by methods that vary between species (Cloney 1978; 1990). The ectodermal material of the tail, the notochord, and of the whole neural tube including the sensory vesicle with its sense organs become resorbed; the endodermal strand becomes incorporated into the gut. Just after metamorphosis, the zone between the attachment and the mouth expands strongly so that the gut rotates 90–180° and the oral siphon finally points away from the substratum (Fig. 64.1).

Sexual reproduction and development of the oozooid of salps is complicated and not well known (Brien 1948). The eggs become fertilized in the gonoduct where the development takes place, each embryo being nourished by a placenta, which is very different from the mammalian placenta in that its two layers are both of maternal origin (Bone *et al.* 1985). The embryology is complicated through the invasion of follicular cells, called calymmocytes, which conceal some of the developing organs (Brien 1948; Sutton 1960). The usual gastrulation and neurulation is not observed, and the various tissues and organs, such as ganglion, gut, and atrium, become organized directly; the muscles differentiate from a continuous mesodermal sheet. The first stages of the developing ganglion are compact, but a neural canal soon develops. The anterior end of the ganglion rudiment contacts the pharyngeal epithelium and a ciliated neural duct is formed, which soon opens into the neural canal; the connection breaks at a later stage. The ganglion becomes compact again and the neurons innervating the various organs differentiate; a horseshoe-shaped eye develops in the 'dorsal' side of the ganglion (Lacalli and Holland 1998). It is unclear whether a notochord is actually present during ontogeny, and the morphology of the newborn oozooid is undescribed.

Also doliolid development is known only fragmentarily (Uljanin 1884; Neumann 1906; Godeaux 1958; Braconnot 1970). *Doliolum* is a free spawner. The embryo goes through a blastula and a gastrula stage, and the next stage that has been observed already has a body and a tail, which is bent 'dorsally' in the middle like a hairpin; the notochord, consisting of one row of cells, is recognized but the other structures are not well described. A later stage has the tail stretched out and the notochord separated from the developing gut by a mass of mesodermal cells; a short anterior neural tube is formed by the infolding of a neural plate. The tail has three rows of muscle cells on each side, and the embryo is able to make weak swimming movements, but the movements seem mainly to be sharp bendings of the base of the tail, while the tail itself is kept stiff (Braconnot 1970). There are no nerve tube or nerves in the tail. The large atrium develops from a pair of 'dorsal' ectodermal invaginations that fuse medially. The endoderm has become a hollow tube that breaks through anteriorly and postero-'dorsally' forming mouth and anus. The anterior part of the gut becomes the spacious pharynx with the 'ventral' endostyle; four pairs of gill openings break through at each side. The tail is resorbed and the young oozooid hatches from the egg membrane. The larva of *Dolioletta* lacks the tail (Godeaux *et al.* 1998).

Knowledge of the development of appendicularians has exploded during recent years where *Oikopleura*, with a generation time of only about five days, has become a model genetic organism (Fujii *et al.* 2008). Its cell lineage up to the hatching stage, when the cells are tissue restricted, is well known (Nishida 2008; Stach *et al.* 2008). The pattern is very similar to that of the 'ascidians' tadpole larvae described above.

The colonial ascidians show a whole series of different types of budding, which may take place both in the adult stage and precociously in the larval stage (Nakauchi 1982). The buds may develop through a strobilation-like process involving parts of the gut, but buds may also originate from epicardial or peribranchial areas; stolonial budding involving only vascular elements in the tunic is found in several types. The budding gives rise to characteristic colonies with many identical zooids. A special type of budding is associated with dormant buds, which may similarly develop from various types of tissues.

The salps have more complicated life cycles: The zygotes develop into solitary oozooids, which lack gonads, but have a 'ventral' budding zone that gives rise to a stolon; this stolon gives rise to a chain of blastozooids, which have gonads but lack the stolon (Ihle 1935; Brien 1948). The doliolids have still more intricate cycles with several types of individuals formed through budding (Neumann 1906; Brien 1948; Braconnot 1971). The buds of salps and doliolids are

formed through processes that resemble various types known in the ascidians, and it seems difficult to find any phylogenetic pattern in the distribution of the types.

The ascidian genomes are small in general and very small in the highly modified *Oikopleura* (Cañestro *et al.* 2007; Denoeud *et al.* 2010), which corresponds well with the massive loss of genes demonstrated by whole genome analyses (see Holland *et al.* 2008; Garcia-Fernàndez and Benito-Gutiérrez 2009). *Ciona* has *Hox1–6*, *10*, *12*, and *13* on two chromosomes (Ikuta and Saiga 2005). In the larva, there is no Hox expression in the brain vesicle or in the neck region; *Hox1* is expressed in the 'ventral' ganglion and the anterior part of the nerve tube, *Hox3* in the anterior part of the visceral ganglion, *Hox5* and *10* in the anterior parts of the nerve tube, and *Hox12* in the in the tail tip. Surprisingly, knockdown of *Hox1–5* did not affect the development of the larvae. In the juveniles, *Hox10*, *12*, and *13* are expressed along the intestine (Ikuta and Saiga 2005; Ikuta *et al.* 2010). *Oikopleura* has *Hox1*, *2*, *4*, *9A*, *9B*, and *10–13* completely dispersed (Seo *et al.* 2004). In *Ciona*, *Otx* is expressed in the brain vesicle and *pax2/5/8* in the neck region (Lacalli 2006).

Urochordate monophyly is generally undisputed, although the number of synapomorphies may be small, because many characters turn out to be plesiomorphies. The presence of cellulose in the cuticle (see above) seems a strong apomorphy, and the atrium developing from a pair of 'dorsal' invaginations is unique. Also the haemal system that pumps the blood alternating in two opposite directions is an unusual character. The lack of mesodermal segmentation is probably an apomorphy, as indicated by the phylogeny (see Chapter 62).

The recent phylogenetic studies indicate that the ancestral urochordate was a sessile, ascidian-like form with a tadpole larva, and that the holopelagic thaliaceans are specialized 'ascidians'. The old idea that the appendicularians are neotenic (Garstang 1928; Berrill 1955) has found new support both through morphological and molecular studies (see above), but there seems to be no support for an evolution of the other chordates from an appendicularian-like ancestor.

The 'new' chordate phylogeny (Fig. 62.1), with urochordates and vertebrates as sister groups, strongly indicates that the urochordates are derived from free-swimming chordates, with the complex of notochord, neural tube, and segmented lateral muscles. The ancestor may have been 'amphioxus-like' (but without an atrium); it probably settled with an attachment area in front of the mouth, which was on the neural side, and retained the free-living stage as a larva. Lecithotrophy of the larva developed at a later stage.

The common ancestor of the Olfactores (Urochordata + Vertebrata) is very difficult to visualize (see Chapter 62).

Interesting subjects for future research

1. Loss of segmentation genes

References

Berrill, N.J. 1931. Studies in tunicate development. Part II. Abbreviation of development in the Molgulidae. *Phil. Trans. R. Soc. Lond.* B **219**: 225–346.

Berrill, N.J. 1950. *The Tunicata with an Account of the British Species*. Ray Society, London.

Berrill, N.J. 1955. *The Origin of the Vertebrates*. Oxford University Press, Oxford.

Bone, Q. 1989. Evolutionary patterns of axial muscle systems in some invertebrates and fish. *Am. Zool.* **29**: 5–18.

Bone, Q. 1992. On the locomotion of ascidian larvae. *J. Mar. Biol. Assoc. U.K.* **72**: 161–186.

Bone, Q. 1998. Nervous system, sense organs, and excitable epithelia. In Q. Bone (ed.): *The Biology of Pelagic Tunicates*, pp. 55–80. Oxford University Press, Oxford.

Bone, Q., Braconnot, J.C. and Ryan, K.P. 1991. On the pharyngeal feeding filter of the salp *Pegea confoederata* (Tunicata: Thaliacea). *Acta Zool. (Stockh.)* **72**: 55–60.

Bone, Q., Carré, C. and Chang, P. 2003. Tunicate feeding filters. *J. Mar. Biol. Assoc. U.K.* **83**: 907–919.

Bone, Q., Pulsford, A.L. and Amoroso, E.C. 1985. The placenta of the salp (Tunicata: Thaliacea). *Placenta* **6**: 53–64.

Bouquet, J.-M., Spriet, E., Troedsson, C., *et al.* 2009. Culture optimization for the emergent zooplanktonic model organism *Oikopleura dioica*. *J. Plankton Res.* **31**: 359–370.

Braconnot, J.C. 1970. Contribution a l'étude des stades successifs dans le cycle des Tuniciers pélagiques Doliolides I. Les stades larvaire, oozooide, nourrice et gastrozoide. *Arch. Zool. Exp. Gen.* **111**: 629–668.

Braconnot, J.C. 1971. Contribution a l'étude des stades successifs dans le cycle des Tuniciers pélagiques Doliolides II. Les stades phorozoide et gonozoide des doliolides. *Arch. Zool. Exp. Gen.* **112**: 5–31.

Brien, P. 1948. Embranchement des Tuniciers. Morphologie et reproduction. *Traité de Zoologie*, vol. 11, pp. 545–930. Masson, Paris.

Burighel, P. and Cloney, R.A. 1997. Urochordata: Ascidiacea. In F.W. Harrison (ed.): *Microscopic Anatomy of Invertebrates*, vol. 15, pp. 221–347. Wiley-Liss, New York.

Cañestro, C., Bassham, S. and Postlethwait, J. 2005. Development of the central nervous system in the larvacean *Oikopleura dioica* and the evolution of the chordate brain. *Dev. Biol.* **285**: 298–315.

Cañestro, C., Bassham, S. and Postlethwait, J.H. 2008. Evolution of the thyroid: anterior–posterior regionalization of the *Oikopleura* endostyle revealed by *Otx*, *Pax2/5/8*, and *Hox1* expression. *Dev. Dyn.* **237**: 1490–1499.

Cañestro, C., Yokoi, H. and Postlethwait, J.H. 2007. Evolutionary developmental biology and genomics. *Nature Rev. Genet.* **8**: 932–942.

Chen, J.-Y., Huang, D.-Y., Peng, Q.-Q., *et al.* 2003. The first tunicate from the Early Cambrian of South China. *Proc. Natl. Acad. Sci. USA* **100**: 8314–8318.

Cloney, R.A. 1978. Ascidian metamorphosis: review and analysis. In F.S. Chia and M.E. Rice (eds): *Settlement and Metamorphosis of Marine Invertebrate Larvae*, pp. 255–282. Elsevier, New York.

Cloney, R.A. 1990. Urochordata - Ascidiacea. In K.G. Adiyodi and R.G. Adiyodi (eds): *Reproductive Biology of Invertebrates*, vol. 4B, pp. 391–451. Oxford & IBH Publishing, New Delhi.

Compere, P. and Godeaux, J.E.A. 1997. On endostyle ultrastructure in two new species of doliolid-like tunicates. *Mar. Biol.* **128**: 447–453.

Conklin, E.G. 1905. The organization and cell-lineage of the ascidian egg. *Journal of the Academy of Natural Sciences of Philadelphia*, 2. ser. **13**: 1–119.

Davison, A. and Blaxter, M. 2005. Ancient origin of glycosyl hydrolase family 9 cellulase genes. *Mol. Biol. Evol.* **22**: 1273–1284.

Denoeud, F., Henriet, S., Mungpakdee, S., *et al.* 2010. Plasticity of animal genome architecture unmasked by rapid evolution of a pelagic tunicate. *Science* **330**: 1381–1385.

Fenaux, R. 1998. Anatomy and functional morphology of the Appendicularia. In Q. Bone (ed.): *The Biology of Pelagic Tunicates*, pp. 25–34. Oxford University Press, Oxford.

Flood, P.R. and Deibel, D. 1998. The appendicularian house. In Q. Bone (ed.): *The Biology of Pelagic Tunicates*, pp. 105–124. Oxford University Press, Oxford.

Fujii, S., Nishio, T. and Nishida, H. 2008. Cleavage pattern, gastrulation and neurulation in the appendicularian, *Oikopleura dioica Dev. Genes Evol.* **218**: 69–79.

Garcia-Fernàndez, J. and Benito-Gutiérrez, È. 2009. It's a long way from Amphioxus: descendants of the earliest chordate. *BioEssays* **31**: 665–675.

Garstang, W. 1928. The morphology of the Tunicata, and its bearings on the phylogeny of the Chordata. *Q. J. Microsc. Sci., N. S.* **72**: 51–187.

Godeaux, J. 1958. Contribution á la connaissance des Thaliacés (*Pyrosome* et *Doliolum*). *Ann. Soc. R. Zool. Belg.* **88**: 5–285.

Godeaux, J., Bone, Q. and Braconnot, J.C. 1998. Anatomy of Thaliacea. In Q. Bone (ed.): *The Biology of Pelagic Tunicates*, pp. 1–24. Oxford University Press, Oxford.

Hadfield, K.A., Swalla, B.J. and Jeffery, W.R. 1995. Multiple origins of anural development in ascidians inferred from rDNA sequences. *J. Mol. Evol.* **40**: 413–427.

Hirano, T. and Nishida, H. 1997. Developmental fates of larval tissues after metamorphosis in ascidian *Halocynthia roretzi*. I. Origin of mesodermal tissues of the juvenile. *Dev. Biol.* **192**: 199–210.

Hirose, E., Kimura, S., Itoh, T. and Nishikawa, J. 1999. Tunic morphology and cellulosic components of pyrosomas, doliolods, and salps (Thaliacea, Urochordata). *Biol. Bull.* **196**: 113–120.

Holland, L.Z., Albalat, R., Azumi, K., *et al.* 2008. The amphioxus genome illuminates vertebrate origins and cephalochordate biology. *Genome Res.* **18**: 1100–1111.

Holmberg, K. and Olsson, R. 1984. The origin of Reissner's fibre in an appendicularian, *Oikopleura dioica*. *Vidensk. Medd. Dan. Naturhist. Foren.* **145**: 43–52.

Ihle, J.E.W. 1935. Desmomyaria. *Handbuch der Zoologie*, 5. Band, 2. Hälfte, pp. 401–532. Walter de Gruyter, Berlin.

Ikuta, T. and Saiga, H. 2005. Organization of Hox genes in ascidians: present, past, and future. *Dev. Dyn.* **233**: 382–389.

Ikuta, T., Satoh, N. and Saiga, H. 2010. Limited functions of Hox genes in the larval development of the ascidian *Ciona intestinalis*. *Development* **137**: 1505–1513.

Imai, J.H. and Meinertzhagen, I.A. 2007. Neurons of the ascidian larval nervous system in *Ciona intestinalis*: I. Central nervous system. *J. Comp. Neurol.* **501**: 316–334.

Jeffery, W.R. 2007. Chordate ancestry of the neural crest: New insights from ascidians. *Semin. Cell Dev. Biol.* **18**: 481–491.

Jeffery, W.R. and Swalla, B.J. 1992. Evolution of alternate modes of development in ascidians. *BioEssays* **14**: 219–226.

Jeffery, W.R. and Swalla, B.J. 1997. Tunicates. In S.F. Gilbert and A.M. Raunio (eds): *Embryology. Constructing the Organism*, pp. 331–364. Sinauer Associates, Sunderland, MA.

Julin, C. 1904. Recherches sur la phylogenèse des Tuniciers. *Z. Wiss. Zool.* **76**: 544–611.

Kimura, S. and Itoh, T. 1995. Evidence for the role of the glomerulocyte in cellulose synthesis in the tunicate, *Metandrocarpa uedai*. *Protoplasma* **186**: 24–33.

Kimura, S., Ohshima, C., Hirose, E., Nishikawa, J. and Itoh, T. 2001. Cellulose in the house of the appendicularian *Oikopleura rufescens*. *Protoplasma* **216**: 71–74.

Kumano, G. and Nishida, H. 2007. Ascidian embryonic development: An emerging model system for the study of cell fate specification in chordates. *Dev. Dyn.* **236**: 1732–1747.

Lacalli, T.C. 2006. Prospective protochordate homologs of vertebrate midbrain and MHB, with some thoughts on MHB origins. *Int. J. Dev. Biol.* **2**: 104–109.

Lacalli, T.C. and Holland, L.Z. 1998. The developing dorsal ganglion of the salp *Thalia democratica*, and the nature of the ancestral chordate brain. *Phil. Trans. R. Soc. Lond. B* **353**: 1943–1967.

Lemaire, P. 2009. Unfolding a chordate developmental program, one cell at a time: Invariant cell lineages, short-range inductions and evolutionary plasticity in ascidians. *Dev. Biol.* **332**: 48–60.

Lemaire, P., Smith, W.C. and Nishida, H. 2008. Ascidians and the plasticity of the chordate developmental program. *Curr. Biol.* **18**: R620–R631.

Lohmann, H. 1903. Neue Untersuchungen über den Reichtum des Meeres an Plankton und über die Brauchbarkeit der verschiedenen Fangmethoden. *Wiss. Meeresunters., Kiel., N. F.* **7**: 1–86.

Mackie, G.O. 1995. Unconventional signalling in tunicates. *Mar. Freshw. Behav. Physiol.* **26**: 197–205.

Mackie, G.O. and Burighel, P. 2005. The nervous system in adult tunicates: current research directions. *Can. J. Zool.* **83**: 151–183.

Manni, L., Agnoletto, A., Zaniolo, G. and Burighel, P. 2005. Stomodeal and neurohypophysial placodes in *Ciona intestinalis*. insights into the origin of the pituitary gland. *J. Exp. Zool.* **304B**: 324–339.

Matthysse, A.G., Deschet, K., Williams, M., *et al.* 2004. A functional cellulose synthase from ascidian epidermis. *Proc. Natl. Acad. Sci. USA* **101**: 986–991.

Mazet, F. and Shimeld, S.M. 2005. Molecular evidence from ascidians for the evolutionary origin of vertebrate cranial sensory placodes. *J. Exp. Zool.* **305B**: 340–346.

McReynolds, J.S. and Gorman, A.L.F. 1975. Hyperpolarizing photoreceptors in the eye of a primitive chordate, *Salpa democratica* Vision Res. **15**: 1181–1186.

Meinertzhagen, I.A. 2005. Eutely, cell lineage, and fate within the ascidian larval nervous system: determinacy or to be determined? *Can. J. Zool.* **83**: 184–195.

Meinertzhagen, I.A., Lemaire, P. and Okamura, Y. 2004. The neurobiology of the ascidian tadpole larva: Recent developments in an ancient chordate. *Annu. Rev. Neurosci.* **27**: 453–485.

Millar, R.H. 1971. The biology of ascidians. *Adv. Mar. Biol* **9**: 1–100.

Nakatani, Y. and Nishida, H. 1994. Induction of notochord during ascidian embryogenesis. *Dev. Biol.* **166**: 289–299.

Nakauchi, M. 1982. Asexual development of ascidians: its biological significance, diversity, and morphogenesis. *Am. Zool.* **22**: 753–763.

Nakauchi, M. and Takeshita, T. 1983. Ascidian one-half embryos can develop into functional adult. *J. Exp. Zool.* **227**: 155–158.

Neumann, G. 1906. *Doliolum. Wissenschaftliche Ergebnisse der Deutschen Tiefsee-Expedition auf dem Dampfer 'Valdivia' 1898–1899* **12**: 93–243.

Nicol, D. and Meinertzhagen, I.A. 1988a. Development of the central nervous system of the larva of the ascidian, *Ciona intestinalis* L. I. The early lineages of the neural plate. *Dev. Biol.* **130**: 721–736.

Nicol, D. and Meinertzhagen, I.A. 1988b. Development of the central nervous system of the larva of the ascidian, *Ciona intestinalis* L. II. Neural plate morphogenesis and cell lineages during neurulation. *Dev. Biol.* **130**: 737–766.

Nishida, H. 1987. Cell lineage analysis in ascidian embryos by intracellular injection of a tracer enzyme III. Up to the tissue restricted stage. *Dev. Biol.* **121**: 526–541.

Nishida, H. 1994. Localization of determinants for formation of the anterior-posterior axis in eggs of the ascidian *Halocynthia roretzi. Development* **120**: 3093–3104.

Nishida, H. 1997. Cell fate specification by localized cytoplasmic determinants and cell interactions in ascidian embryos. *Int. Rev. Cytol.* **176**: 245–306.

Nishida, H. 2008. Development of the appendicularian *Oikopleura dioica*: Culture, genome, and cell lineages. *Dev. Growth Differ.* **50**: S239–S256

Nishide, K., Nishikata, T. and Satoh, N. 1989. A monoclonal antibody specific to embryonic trunk-lateral cells of the ascidian *Halocynthia roretzi* stains coelomic cells of juvenile and basophilic blood cells. *Dev. Growth Differ.* **31**: 595–600.

Ogasawara, M., Di Lauro, R. and Satoh, N. 1999a. Ascidian homologs of mammalian thyroid peroxidase genes are expressed in the thyroid-equivalent region of the endostyle. *J. Exp. Zool.* **285**: 158–169.

Ogasawara, M., Di Lauro, R. and Satoh, N. 1999b. Ascidian homologs of mammalian thyroid transcription factor-1 gene are expressed in the endostyle. *Zool. Sci. (Tokyo)* **16**: 559–565.

Olsson, R. 1963. Endostyles and endostylar secretions: a comparative histochemical study. *Acta Zool. (Stockh.)* **44**: 299–328.

Olsson, R. 1972. Reissner's fiber in ascidian tadpole larvae. *Acta Zool. (Stockh.)* **53**: 17–21.

Olsson, R., Holmberg, K. and Lilliemarck, Y. 1990. Fine structure of the brain and brain nerves of Oikopleura dioica (Urochordata, Appendicularia). *Zoomorphology* **110**: 1–7.

Ruppert, E.E. 1990. Structure, ultrastructure and function of the neural gland complex of *Ascidia interrupta* (Chordata, Ascidiacea): clarification of hypotheses regarding the evolution of the vertebrate anterior pituitary. *Acta Zool. (Stockh.)* **71**: 135–149.

Satoh, N. 1994. *Developmental Biology of Ascidians*. Cambridge University Press, Cambridge.

Seeliger, O. 1893. Über die Entstehung des Peribranchialraumes in den Embryonen der Ascidien. *Z. Wiss. Zool.* **56**: 365–401.

Seeliger, O. 1893–1907. Die Appendicularien und Ascidien. *Bronn's Klassen und Ordnungen des Tierreichs*, 3. Band (Suppl.), 1. Abt., pp. 1–1280. Akademische Verlagsgesellschaft, Leipzig.

Seo, H.C., Edvardsen, R.B., Maeland, A.D., *et al.* 2004. *Hox* cluster disintegration with persistent anterioposterior order of expression in *Oikopleura dioica. Nature* **431**: 67–71.

Sorrentino, M., Manni, L., Lane, N.J. and Burighel, P. 2000. Evolution of cerebral vesicles and their sensory organs in an ascidian larva. *Acta Zool. (Stockh.)* **81**: 243–258.

Stach, T. and Turbeville, J.M. 2002. Phylogeny of Tunicata inferred from molecular and morphological characters. *Mol. Phylogenet. Evol.* **25**: 408–428.

Stach, T., Winter, J., Bouquet, J.M., Chourrout, D. and Schnabel, R. 2008. Embryology of a planktonic tunicate reveals traces of sessility. *Proc. Natl. Acad. Sci. USA* **105**: 7229–7234.

Sutton, M.F. 1960. The sexual development of *Salpa fusiformis* (Cuvier). *J. Embryol. Exp. Morphol.* **8**: 268–290.

Swalla, B.J. and Jeffery, W.R. 1996. Requirement of the *Manx* gene for expression of chordate features in a tailless ascidian larva. *Science* **274**: 1205–1208.

Taniguchi, K. and Nishida, H. 2004. Tracing cell fate in brain formation during embryogenesis of the ascidian *Halocynthia roretzi. Dev. Growth Differ.* **46**: 163–180.

Torrence, S.A. and Cloney, R.A. 1982. Nervous system of ascidian larvae: caudal primary sensory neurons. *Zoomorphology* **99**: 103–115.

Tsagkogeorga, G., Turon, X., Hopcroft, R.R., *et al.* 2009. An updated 18S rRNA phylogeny of tunicates based on mixture and secondary structure models. *BMC Evol. Biol.* **9**: 187.

Uljanin, B. 1884. Die Arten der Gattung *Doliolum* im Golfe von Neapel. *Fauna Flora Golf. Neapel* **10**: 1–140.

Van Beneden, E. and Julin, C. 1887. Recherches sur la morphologie des Tuniciers. *Arch. Biol.* **6**: 237–476.

Veeman, M.T., Newman-Smith, E., El-Nachef, D. and Smith, W.C. 2010. The ascidian mouth opening is derived from the anterior neuropore: Reassessing the mouth/neural tube relationship in chordate evolution. *Dev. Biol.* **344**: 138–149.

Willey, A. 1893. Studies in the Protochordata. *Q. J. Microsc. Sci., N. S.* **34**: 317–369.

Zeng, L., Jacobs, M.W. and Swalla, B.J. 2006. Coloniality has evolved once in Stolidobranch Ascidians. *Integr. Comp. Biol.* **46**: 255–268.

Zhang, A. 1987. Fossil appendicularians in the Early Cambrian. *Sci. Sin., Ser. B* **30**: 888–896

65

Phylum **Vertebrata (Craniata)**

The literature about the morphology and embryology of our own phylum is absolutely overwhelming, and studies on molecular biology is exploding, so the list of subjects that could be taken up in a phylogenetic discussion seems endless. I have therefore chosen not to follow the outline used in the preceding chapters, but to arrange selected characters in two main groups: general chordate characters and autapomorphies of the vertebrates. Most of the information can be checked in the common textbooks, so references have been kept to a minimum.

It is now universally agreed that the cyclostomes (hagfish and lampreys) show a whole suite of characters that must be regarded as ancestral to the vertebrates, although the few living representatives are at the same time specialized for unusual feeding modes. These groups are therefore of special interest for the phylogenetic discussion, but it should be remembered that many of the characters that show similarities between vertebrates and the other chordates are found in the ammocoetes larva of the lampreys, and in embryonic stages of the various vertebrate groups. The conodont animals are now known to have been vertebrates, possibly the sister group of the gnathostomes (Aldridge and Briggs 2009).

In Chapter 62, the following character complexes were listed as important apomorphies of the chordates: notochord, neural tube, longitudinal muscles along the chorda, ciliated pharyngeal gill slits functioning as a mucociliary filtering structure with the mucous net secreted by the endostyle. The expressions of these characters in the vertebrates should be commented briefly.

The notochord develops from the 'dorsal' side of the archenteron or from the primitive streak, also when the development is complicated through large amounts of yolk or a placenta. In the lampreys, the notochord is formed as a median fold from the roof of the archenteron (see below); the first stages show an irregular arrangement of cells, but the later larval stages have a notochord consisting of one row of flat cells. The adults again show an irregular arrangement of cells with large vacuoles. The notochord is surrounded by a thickened basal membrane. The vertebral column replaces the notochord more or less completely in the gnathostomes.

The neural tube is formed from the ectoderm in contact with the notochord, and the induction of the ectoderm from the notochord cells is documented through numerous studies from the beginning of the last century. The adult central nervous system consists of a neural tube, the spinal cord, and a highly complex, anterior brain. Comparisons with amphioxus show that the spinal cord is in principle very

---

Chapter vignette: *Petromyzon marinus.* (Redrawn from Muus and Dahlstrøm 1964.)

similar in the two groups, and that the brain is an enormously enlarged and specialized brain vesicle (see also Chapter 62).

The mesoderm develops from the archenteron (see below) or from corresponding masses of cells from the 'dorsal' blastopore lip.

The pharynx of the ammocoetes larva has a row of gill slits on each side and a 'ventral' endostyle, but although the general morphology is rather similar to that of amphioxus (except that the slits are not U-shaped), a number of differences have been pointed out (Mallatt 1981). An important difference is that the water flow through the pharynx is set up by pumping movements of the pharynx rather than by the beat of the cilia of the gill slits. Another important difference is that although the endostyle has bands of cells secreting proteinaceous mucus and iodinated compounds to the pharynx, it does not organize a mucous filtering net; the mucus appears to be secreted mainly from goblet cells on the gill bars. At metamorphosis, the endostyle becomes transformed into the thyroid, which is found in all other vertebrates (see also Chapter 62).

The main axis of the egg is determined already in the ovary, and the sperm entry point determines the plane of the first cleavage. Detailed fate maps of the various regions of the egg have been constructed for many species. The fate maps of the three chordate phyla show identical spatial relations between the various areas of the egg (Fig. 62.2).

For general reviews of embryology see Chapters 19–22 in Gilbert and Raunio (1997). The first cleavage is median in anurans with total cleavage, but other groups, such as bony fish and mammals, that have partial cleavage or placentally nourished embryos, show no specific relationship of the first cleavages and the body symmetry. In *Lampetra* (Damas 1944) cleavage leads to a coeloblastula with a narrow blastocoel at the apical side; gastrulation is through invagination, and the narrow, tubular archenteron lies close to the 'dorsal' ectoderm. The 'dorsal' cells of the archenteron become the notochord and the two plates of somewhat smaller cells lateral to the notochord become the mesoderm. This general type of development can with more or less modification be recognized in all vertebrates.

The mesoderm usually develops through ingression of cells from the 'dorsal' blastopore lip or corresponding areas, but the supposedly primitive formation of mesodermal sacs as pockets from the roof of the archenteron lateral to the notochord, has only been observed in a few groups, for example lampreys. Their mesoderm is at first a pair of compact longitudinal cell masses lateral to the notochord and neural tube, but the anterior parts of the mesodermal plates become divided into about 20 segments, which subsequently divide into 'dorsal' somites and 'ventral' sacs (lateral plates); the posterior part of the mesodermal bands form segmental somites, but the 'ventral' part remains undivided. The primitive vertebrate nephridia develop from the narrow stalk (nephrotome) between the somites and the 'ventral' sacs. In other vertebrates, the somites can always be recognized, but are formed as compact cell masses without connection with the archenteron, and the 'ventral' mesoderm is undivided. So although the origin of the mesoderm is in most cases quite different from that found in amphioxus, the early morphology of the coeloms is quite similar, and the segmentation of the mesoderm in the two phyla must be regarded as homologous.

The anuran embryo (just as the ascidian embryo) shows the primordial neural plate and notochord cells in crescentic zones in front of the blastopore just before the gastrulation, and both areas become elongate through rearrangement of the cells (Keller *et al.* 1992). Mouth and nostrils develop from an area at the frontal side of the neural fold just after closure of the neural tube and move to the 'ventral' side through differential growth (Vogt 1929; Drysdale and Elinson 1991) (Fig. 65.1).

There are important differences between the anterior parts of the neural tubes of amphioxus and the vertebrates (see Chapter 62), but the posterior parts are more similar, and Bone (1960) found 'rather striking' resemblances between the arrangements of neurons in amphioxus and the young ammocoetes-larvae, with amphioxus representing a more primitive type.

Numerous morphological studies discuss vertebrate apomorphies, but only some of the more conspicuous characters will be mentioned.

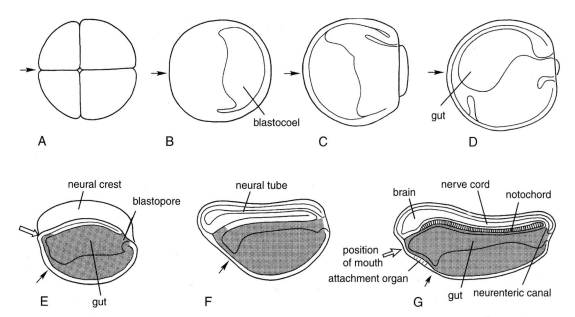

Figure 65.1. Development of *Bombinator platypus*; all stages are seen from the left side with the blastoporal pole to the right; the position of the polar bodies/apical pole is indicated by black arrows. (A) 8-cell stage. (B) Blastula. (C) Early gastrula. (D) Late gastrula. (E) Neural crest formation. (F) Early neural tube stage. (G) Fully developed neural tube stage. (Redrawn from Vogt 1929; see also Dickinson and Sivea 2007; Veeman *et al.* 2010.)

One of the most important complexes appears to be related to the evolution of the neural crest and the epidermal placodes (Baker and Sclhosser 2005; Baker 2008). These structures develop from cells at the edges of the neural plate and give rise to a number of structures that are unknown in the non-vertebrates. Many of the structures are related to the macrophagous habits of the vertebrates, and the related evolution of a more active life style with complicated sense organs and a complex brain. Cells of the neural crest give rise to sensory nerves with ganglia, peripheral motor ganglia, and higher-order motor neurons, and cells of the placodes form sense organs like eyes (not the sensory tissues that are parts of the brain), ears, lateral line organs, and gustatory organs. Also important parts of the skeleton, for example the cartilage in the gill bars, are derived from the neural crest.

The vertebrate haemal system has an inner layer of cells, the endothelium, a character that is only found in a few scattered 'invertebrate' groups, where it is regarded as apomorphic within the phyla (Ruppert and Carle 1983).

Another vertebrate apomorphy is the multilayered ectoderm, where cell divisions in the basal layer give rise to new cells that replace old cells worn off at the surface. Similar epithelia are only known from certain body regions of chaetognaths (Chapter 55), where a homology is not indicated. Monociliate epithelia are found in an area of the neural plate in front of the blastopore lip (Nonaka *et al.* 1998; Song *et al.* 2010) and such cells are also found in amphibian kidney tubules (Møbjerg *et al.* 1998). Most other ciliated epithelia are multiciliate.

The anterior end of the neural tube is greatly enlarged and specialized as a brain in all vertebrates. It is difficult to make comparisons with the very simple 'brain' of amphioxus, but the position of the cells secreting Reissner's fibre, the infundibular organ at the posteroventral side of the brain vesicle in amphioxus, and the flexural organ in larval salmon (Olsson 1956) indicate that it is the areas in front of this region that have become the vertebrate brain with several new multicellular sense organs and centres for processing of information from the new sense organs and the coordination of more complex movements, and this is now supported by several morphological and molecular studies.

The fully developed vertebrate nephridia are of the glomerular type, with podocyte-lined blood vessels surrounded by a small coelomic compartment, the Bowmann capsule; there is a common nephridial duct on each side. They develop from approximately the same position as those of amphioxus, but the vertebrate nephridia are situated behind the branchial region and are aligned with the somites, i.e. myomeric, while those of amphioxus are aligned with the gill slits, i.e. branchiomeric. This makes it unlikely that the two types of nephridia are homologous.

The name vertebrates points to another apomorphy of the phylum, the presence of a calcified skeleton at least in connection with specializations of the notochord, but in all gnathostome groups with additional bones both in the head region and in the paired extremities.

The concept of a highly conserved 'phylotypic' stage, which should be passed during ontogeny of all members of a phylum, was proposed by Haeckel (1874), with the well-known example of the vertebrate 'pharyngula'. However, as emphasized by several contemporary and recent authors (see Richardson *et al.* 1997), Haeckel's illustrations were inaccurate, overemphasizing the similarities between the embryos. Embryos of various vertebrates do of course resemble each other, but there is no stage where all organ systems are at the same ontogenetic stage in all species.

A wealth of molecular studies demonstrates the monophyly of the vertebrates. Whole genome studies indicate that the vertebrate genome is derived from an amphioxus-like genome through some gene losses and two rounds of genome duplications, which apparently took place before the split between cyclostomes and gnathostomes (Kuraku *et al.* 2009). Further gene losses and duplications have occurred for example in the teleosts.

References

Aldridge, R.J. and Briggs, D.E.G. 2009. The discovery of conodont anatomy and its importance for understanding the early history of vertebrates. In D. Sepkoski and M. Ruse (eds): *The Paleobiological Revolution: Essays on the Growth of Modern Paleontology*, University of Chicago Press, Chicago.

Baker, C.B.H. 2008. The evolution and elaboration of vertebrate neural crest cells. *Curr. Opin. Genet. Dev.* **18**: 536–543.

Baker, C.V.H. and Schlosser, G. 2005. Editorial: The evolutionary origin of neural crest and placodes. *J. Exp. Zool.* **304B**: 269–273.

Bone, Q. 1960. The central nervous system in amphioxus. *J. Comp. Neurol.* **115**: 27–64.

Damas, H. 1944. Recherches sur le développement de *Lampetra fluviatilis* L. *Arch. Biol.* **55**: 1–284.

Dickinson, A. and Sivea, H. 2007. Positioning the extreme anterior in *Xenopus*: Cement gland, primary mouth and anterior pituitary. *Semin. Cell Dev. Biol.* **18**: 525–533.

Drysdale, T.A. and Elinson, R.P. 1991. Development of the *Xenopus laevis* hatching gland and its relationships to surface ectoderm patterning. *Development* **111**: 469–478.

Gilbert, S.F. and Raunio, A.M. (eds) 1997. *Embryology. Constructing the Organism.* Sinauer, Sunderland, MA.

Haeckel, E. 1874. *Anthropogenie oder Entwicklungsgeschichte des Menschen.* Engelmann, Leipzig.

Keller, R., Shih, J. and Sater, A. 1992. The cellular basis of the convergence and extension of the *Xenopus* neural plate. *Dev. Dyn.* **193**: 199–217.

Kuraku, S., Meyer, A. and Kuratani, S. 2009. Timing of genome duplications relative to the origin of the vertebrates: did cyclostomes diverge before or after? *Mol. Biol. Evol.* **26**: 47–59.

Mallatt, J. 1981. The suspension feeding mechanism of the larval lamprey *Petromyzon marinus. J. Zool.* **194**: 103–142.

Møbjerg, N., Larsen, E.H. and Jespersen, Å. 1998. Morphlogy of the nephron in the mesonephros of *Bufo bufo* (Amphibia, Anura, Bufonidae). *Acta Zool. (Stockh.)* **79**: 31–50.

Muus, B.J. and Dahlstrøm, P. 1964. *Havfisk og Fiskeri.* Gad, Copenhagen.

Nonaka, S., Tanaka, Y., Okada, Y., *et al.* 1998. Randomization of left-right asymmetry due to loss of nodal cilia generating leftward flow of extraembryonic fluid in mice lacking KIF3B motor protein. *Cell* **95**: 829–837.

Olsson, R. 1956. The development of the Reissner's fibre in the brain of the salmon. *Acta Zool. (Stockh.)* **37**: 235–250.

Richardson, M.K., Hanken, J., Gooneratne, M.L., *et al.* 1997. There is no highly conserved embryonic stage in the vertebrates: implications for current theories of evolution and development. *Anat. Embryol.* **196**: 91–106.

Ruppert, E.E. and Carle, K.J. 1983. Morphology of metazoan circulatory systems. *Zoomorphology* **103**: 193–208.

Song, H., Hu, J., Chen, W.C., *et al.* 2010. Planar cell polarity breaks bilateral symmetry by controlling ciliary positioning. *Nature* **466**: 378–382.

Veeman, M.T., Newman-Smith, E., El-Nachef, D. and Smith, W.C. 2010. The ascidian mouth opening is derived from the anterior neuropore: Reassessing the mouth/neural tube relationship in chordate evolution. *Dev. Biol.* **344**: 138–149.

Vogt, W. 1929. Gestaltungsanalyse am Amphibienkeim mit Örtlicher Vitalfärbung. II. Teil. Gastrulation und Mesodermbildung bei Urodelen und Anuren. *Arch Entwicklungsmech. Org.* **120**: 384–706.

66

Problematica

After the publication of the second edition of this book, new information, especially from molecular studies, has made it possible to assign well-founded phylogenetic positions to two of the problematic groups, viz. *Buddenbrockia* (and the Myxozoa as a whole), which have been included in the Cnidaria (Chapter 13), and *Symbion*, which has been placed in its own phylum (Chapter 37), closely related to Entoprocta and Ectoprocta. However, *Salvinella* and the 'Mesozoa', i.e. the Orthonectida and Dicyemida (Rhombozoa), remain enigmatic, and a new taxon must be added, viz. *Diurodrilus*.

The completely perplexing *Salinella salve*, which was obtained by Frenzel (1892) from a saline culture of material from Córdoba, Argentina, has never been found again. The description shows a tube of cells with cilia both on the inner and the outer side and with special cilia around both openings. Various developmental stages and an encystation after 'conjugation' were also described. The whole description could well be a complete misunderstanding, and it seems futile to discuss it further.

Dicyemida and Orthonectida are well-established 'Problematica', which are discussed in all major textbooks and encyclopedia. Their position is sometimes regarded as close to the metazoan stem (as indicated by the name Mesozoa) or even as multicellular organisms evolved separately from the metazoans, whereas other authors regard them as specialized parasitic flatworms. The latter view is supported by some molecular studies. The two groups may not be sister groups. They do not contribute to our understanding of animal evolution.

Dicyemids are parasites in the kidneys of cephalopods. A review of their ultrastructure and life cycle is given by Horvath (1997). The polarized epithelia with cell junctions show that they are indeed metazoans, but they lack nerves, and their structure appears highly modified, probably in connection with the parasitic life style. Molecular data support the view that they are bilaterians and perhaps more specifically spiralians, but not related to the platyhelminths (Kobayashi *et al.* 1999; Aruga *et al.* 2007; Kobayashi *et al.* 2009; Suzuki *et al.* 2010).

Orthonectids are parasites of various marine invertebrates (Kozloff 1990). The free-swimming females consist of an outer layer of ciliated and unciliated cells in species-specific patterns, an inner mass of cells, and longitudinal muscle cells (Slyusarev 2003); nervous cells have not been observed. The much smaller males have a similar structure. Small ciliated larvae enter the host, but this part of the life cycle needs more study. A study of 18S rDNA (Hanelt *et al.* 1996) indicates that the orthonectids are bilaterians, not related to platyhelminths, but without further details.

Diurodrilus, with six described species, is a genus of small interstitial worms found in tidal and subti-

Chapter vignette: *Archirrhinos haeckelii*. (Redrawn from Stümpke 1961.)

dal sands around the world (Worsaae and Kristensen 2005). They were originally placed in the 'Archiannelida', but new investigations (Worsaae and Rouse 2008) demonstrate that they have a unique morphology, lacking several key annelid characteristics and showing a few similarities with micrognathozoans. Molecular studies were restricted by the material available, but showed no support for an annelid relationship.

The Rhinogradentia (see the chapter vignette) were obviously vertebrates (Stümpke 1961), but the dramatic annihilation of the whole group makes further studies impossible.

References

Aruga, J., Odaka, Y.S., Kamiya, A. and Furuya, H. 2007. *Dicyema* Pax6 and Zic: tool-kit genes in a highly simplified bolaterian. *BMC Evol. Biol.* **7**: 201.

Frenzel, J. 1892. Untersuchungen über die mikroskopische fauna Argentiniens. *Salvinella salve* nov. gen. nov. spec. *Arch. Naturgesch.* **58**: 66–96.

Hanelt, B., Van Schyndel, D., Adema, C.M., Lewis, L.A. and Loker, E.S. 1996. The phylogenetic position of *Rhopalura*

ophiocomae (Orthonectida) based on 18S ribosomal DNA sequence analysis. *Mol. Biol. Evol.* **13**: 1187–1191.

Horvath, P. 1997. Dicyemid mesozoans. In S.F. Gilbert and A.M. Raunio (eds): *Embryology. Constructing the Organism*, pp. 31–38. Sinauer Associates, Sunderland. MA.

Kobayashi, M., Furuya, H. and Holland, P.W.H. 1999. Dicyemids are higher animals. *Nature* **401**: 762.

Kobayashi, M., Furuya, H. and Wada, H. 2009. Molecular markers comparing the extremely simple body plan of dicyemids to that of lophotrochozoans: insight from the expression patterns of *Hox, Otx*, and *brachyury*. *Evol. Dev.* **11**: 582–589.

Kozloff, E.N. 1990. *Invertebrates*. Saunders College Publishing, Philadelphia.

Slyusarev, G.S. 2003. The fine structure of the muscle system in the female of the orthonectid *Intoshia variabili* (Orthonectida). *Acta Zool. (Stockh.)* **84**: 107–111.

Stümpke, H. 1961. *Bau und Leben der Rhinogradentia*. Gustav Fischer, Stuttgart.

Suzuki, T.G., Ogino, K., Tsuneki, K. and Furuya, H. 2010. Phylogenetic analysis of dicyemid mesozoans (Phylum Dicyemida) from innexin amino acid sequences: Dicyemids are not related to Platyhelminthes. *J. Parasitol.* **96**: 614–625.

Worsaae, K. and Kristensen, R.M. 2005. Evolution of interstitial Polychaeta (Annelida). *Hydrobiologia* **235/236**: 319–340.

Worsaae, K. and Rouse, G.W. 2008. Is *Diurodrilus* an annelid? *J. Morphol.* **269**: 1426–1455.

Systematic index

Page numbers in *italics* refer to illustrations or tables

Subject index

Page numbers in *italics* refer to illustrations or tables